Natural Computing Series

Founding Editor
Grzegorz Rozenberg

The Natural Computing book series covers theory, experiment, and implementations at the intersection of computation and natural systems. This includes:

- **Computation inspired by Nature**: Paradigms, algorithms, and theories inspired by natural phenomena. Examples include cellular automata, simulated annealing, neural computation, evolutionary computation, swarm intelligence, and membrane computing.
- **Computing using Nature-inspired novel substrates**: Examples include biomolecular (DNA) computing, quantum computing, chemical computing, synthetic biology, soft robotics, and artificial life.
- **Computational analysis of Nature**: Understanding nature through a computational lens. Examples include systems biology, computational neuroscience, quantum information processing.

Xin Yao • Siang Yew Chong

Coevolutionary Computation and Its Applications

Springer

Xin Yao
School of Data Science
Lingnan University
Tuen Mun, Hong Kong

Siang Yew Chong
Department of Computer Science and Engineering
Southern University of Science and Technology
Shenzhen, Guangdong, China

ISSN 1619-7127 ISSN 2627-6461 (electronic)
Natural Computing Series
ISBN 978-981-96-2840-7 ISBN 978-981-96-2841-4 (eBook)
https://doi.org/10.1007/978-981-96-2841-4

This Springer imprint is published by the registered company Springer Nature Singapore Pte Ltd.
The registered company address is: 152 Beach Road, #21-01/04 Gateway East, Singapore 189721, Singapore

Preface

Coevolution shares many design and construction aspects with its more established and well-studied evolution counterpart. Some historical examples in Evolutionary Computation literature would point to Coevolutionary Algorithms (CEAs) repurposing the two main operators of variation and selection developed for Evolutionary Algorithms (EAs). Despite this, the key difference in the working principle of their respective approaches on evaluating a candidate solution's quality or fitness is significant. A coevolutionary process is guided by means of *relative* comparison on candidates. There is no *objective* function that assigns any candidate a fitness value to be used as an absolute quality measure.

This crucial difference we highlight here is not just for theoretical clarity but to illustrate a practical issue one encounters in applications. Consider problems for which the search for their solutions are formulated as optimization tasks. The main operating assumption in most EA applications is that there is a closed-form equation that acts as a surrogate for the actual, possibly unknown fitness function. What is computationally expensive then is the large number of calls to this equation over an EA's lifetime. This applies to cooperative CEAs that are used for to solve optimization problems. However, in typical competitive CEA applications, test cases are used to determine the preference of one candidate over the other. A single test case evaluation is often more computationally expensive than what is expended to obtain the output of a typical, closed-form fitness function in EA. Furthermore, the set of test cases can be very large and a priori no representative subset of test cases can be reasonably obtained. However, practitioners of CEAs will inform you the main point is not to obtain such representative subset in advance for use but instead to have coevolution find such test cases and when necessary, update them in response to improving candidate solutions so as to differentiate the latter.

Aims There are two main aims of this book on Coevolutionary Computation. The first is to present advancements made to understand the underlying principles of coevolutionary processes that allow one to better *characterize* the *arms-race* dynamics of coevolving solutions and test cases. This is important as the key to develop effective coevolutionary systems and algorithms are those that exhibit such dynamics for search of increasingly higher performing solutions over a reasonable time. The second is to showcase how principles of coevolution can be exploited and integrated with specific domain knowledge to develop new problem-solving methodologies. This is perhaps of greater interest to those who at first want to explore alternative approaches but with added familiarity then seek to incorporate coevolution into their main problem-solving arsenals.

There are two obvious choices in presenting this book, its contents drawn from a collection of works most of which we have contributed, to cover both aims. They differ over which of the presentation perspectives we prioritize more, that is, between the algorithmic development of coevolution and the application of coevolution principles. We have chosen the latter, for reasons beyond simply appealing to a broader audience. In particular, this book could serve as a modern guide to these readers who want to understand and apply coevolution as a problem-solving methodology in their own problem domains. Notwithstanding the first part of the book being an introduction and overview, we have structured the book in the manner such that individual chapters in the latter two parts are mostly self-contained. As such, some repetitions are unavoidable when this book is read in its entirety.

Structure and Organization This book is structured in three main parts. Part I will introduce the fundamentals in Coevolutionary Computation with no assumption made on familiarity with Evolutionary Computation literature. These fundamentals include key concepts and operational principles of the (co)evolutionary processes that are modelled as iterative algorithms and systems implementable in computing machines as well as specific terminologies introduced and commonly used in the field. Parts II and III contain applications of coevolution on problems that are framed in the context of optimization and learning, respectively. This structure is designed to support the anticipated general and frequent usage of this book. In particular, the reader is able to obtain a quick and general introduction on the principles of coevolution in Part I, and then go over in detail the specifics how coevolutionary principles are exploited and applied to solve problems in the relevant chapters of Parts II and III.

Part I is organized into five chapters. Chapter 1 starts with a broad overview of Evolutionary Computation, giving attention to how EAs can be framed as a generate-and-test approach to problem-solving. The chapter will also highlight the main difference between classical numerical optimization and evolutionary optimization. Chapter 2 introduces CEA by giving a thorough presentation of main concepts that differentiate its underlying search process against EA. This includes the notion of relative fitness evaluations as well as an overview of the various implementations introduced in the literature. It will be made clear why for some problems the use of two or more populations in a CEA is required. Chapter 3 will further develop the two principal approaches of coevolution (competitive and cooperative) from an algorithmic perspective that highlights the main difference of their working principles. Each approach is further illustrated through an example-driven presentation how it can be applied in problem settings involving optimization and learning. Chapter 4 consists of a literature review of applications of coevolution in different problem domains that are organized along sectors in real-world settings (e.g. business, games, engineering design, and others). Chapter 5 will define and describe known pathologies that affect coevolutionary processes through an informal, example-driven presentation first. The second half will present current understanding of these challenges more formally with the aid of the latest theoretical tools.

Part II consists of five chapters on various applications of the principles of coevolution to complex real-world problem-solving that are formulated as optimization problems. Chapter 6 introduces Parallel Algorithm Portfolio construction approaches for combinatorial optimization problems that exploits the adversarial nature of relations between solvers and problem instances. This can be explicitly implemented as a procedure that considers both portfolio construction and instance generation. Further theoretical analysis on the generalization performance of such portfolio solvers leads to the development of a two-population competitive coevolution framework for automatic construction of portfolios. The next two chapters focus on coevolutionary approaches to solve large-scale or high-dimensional optimization problems. Chapter 7 introduces a cooperative coevolutionary framework with grouping mechanism that enables automatic identification of variable interactions underlying additively separable, large-scale continuous optimization problems for problem decomposition amenable to Divide-and-Conquer search. All the associated procedures implemented for operations in the differential grouping mechanisms are informed by theoretical studies. They include identification of pairwise interactions of two decision variables, reduction of computational requirements for complete pairwise interactions of variables identification via sample point reuse, and error analysis of floating-point arithmetic in differential grouping operations. Chapter 8 introduces a framework that employs both cooperative and competitive coevolutionary principles to solve high-dimensional optimization problems as computationally expensive optimization problems. This is informed by an analysis that identifies one crucial scope to improve the coevolutionary Divide-and-Conquer problem-solving approach that involves the use of meta-models to approximately rank candidate solutions for the decomposed subcomponents. Chapter 9 introduces a cooperative coevolution framework for effective and efficient Design Space Exploration method that is used to search for optimal multiprocessor mappings in modern signal processing and multimedia applications operating in platforms with heterogeneous processors. The cooperative coevolution framework takes advantage of a new problem-specific decomposition strategy in which solutions of node mapping of each mode are assigned to an individual population. Chapter 10 presents a study on cooperative coevolutionary principle to develop two parallel cooperative CEAs to solve two types of Concurrent Engineering problems, which include quasi-separable Multidisciplinary Design Optimization. In particular, the issue of communication frequency between the populations of candidate solutions for decomposed subproblems is investigated through an extensive computational study, which then informs the development of a self-adaptive method that can set the optimal frequency.

Part III consists of three chapters on studies in competitive coevolution for complex strategic decision-making problems. In particular, the search for solutions of these problems can be formulated as a learning process. Chapter 11 presents a case study in which multi-agent coevolutionary learning systems can be used as simulations to better understand complex real-world interactions that can be formulated as noncooperative games. These coevolutionary simulations of Iterated Prisoner's Dilemma (IPD) interactions have uncovered how cooperative strategies could arise in populations of adaptive agents despite challenges to the behavioural learning process due to increasing number of alternative choices for the interactions. The coevolved agents learn to implement cooperative strategies when they leverage on the additional mechanism of indirect reciprocity that allows them to ascertain the reputation of the other player during interactions. Chapter 12 presents a series of studies that focus on the issue of rigorous quantitative performance analysis of competitive coevolutionary learning systems. In particular, a theoretical framework of performance analysis based on the notion of generalization from Machine Learning will be presented. Subsequently, various generalization performance estimators are constructed for which computational studies are conducted to demonstrate how these statistical estimators can be applied to competitive coevolutionary learning. Chapter 13 presents another series of studies that focus on qualitative information in competitive coevolution that are informative towards subsequent development of quantitative analysis tools. The two main formulations that are key to such transfer are: (1) Directed graph representation to capture complete and relevant structures in coevolutionary problems modelled as two-player competitive games. (2) Finite-state Markov chain models for the search processes of abstract $(1 + 1)$ coevolution on these directed graphs. The second half showcases how a suitable Markov chain formulation (PageRank) establishes an intimate link between coevolutionary problem structures and processes whereby the visitation probability of a vertex (corresponding to a specific pure strategy or solution) can also be interpreted as its importance or performance with respect to the set of strategies of the directed graph.

Tuen Mun, Hong Kong SAR — Xin Yao
Shenzhen, PR China — Siang Yew Chong
November 22, 2024

Acknowledgements

The materials that have contributed to this book are developed over more than a decade. They are realized through years of scientific discussions, feedbacks, and suggestions with colleagues and students at the University of Birmingham (UK), SUSTech (PR China), and elsewhere. The following list of persons is not exhaustive and any omission is much regretted. We fully acknowledge and thank the following persons who being our collaborators have led to joint-works that underlie the materials with which this book is based on. They are (not in the list of importance but following source materials with which the chapters are organized): Shengcai Liu, Ke Tang, Peng Yang, Zhengyu Yang, Mohammad Nabi Omidvar, Xiaodong Li, Yi Mei, Ming Yang, Bo Yuan, Xiaofen Lu, Stefan Menzel, Peter Tiňo, and Day Chyi Ku. Further acknowledgement is accorded to Day Chyi Ku for suggestions involving the procedural presentations of the various algorithms in this book to make them more readable to modern practitioners who are used to high-level programming languages, and for various new plots and illustrations. The preparation of this book is made possible through research grant support under the National Natural Science Foundation of China (Grant No. 62250710682) and the Guangdong Provincial Key Laboratory (Grant No. 2020B121201001) to the first author, and to the second author in particular the Shenzhen Pengcheng Peacock Special Post.

Contents

Part I Coevolutionary Computation—Fundamentals, Applications, and Challenges

1 Evolutionary Computation: An Overview ... 3
1.1 Evolutionary Computation as a Modern Search Heuristic ... 3
1.2 The Framework of Evolutionary Algorithms ... 7
1.2.1 On the Use of Populations ... 11
1.2.2 Representations for Solutions ... 14
1.2.3 Variation Operators ... 19
1.2.4 Selection Operators ... 24
1.3 Parameter Control in Evolutionary Algorithms ... 28
1.3.1 Parameter Control on Population Size ... 29
1.3.2 Parameter Control on Variation Operators ... 30
1.3.3 Parameter Control on Selection Operators ... 33
1.4 Concluding Remarks and Notes for Further Reading ... 34
References ... 38

2 Introduction to Coevolutionary Computation ... 45
2.1 Coevolutionary Systems: Overview ... 45
2.2 Development of Coevolutionary Computation ... 46
2.2.1 A Categorical View of Key Developments ... 46
2.2.2 Evolutionary Biology and Its Relationship to Coevolutionary Computation ... 52
2.3 The Main Principle of Coevolution ... 54
2.4 Concepts in Coevolution ... 57
2.4.1 Population Structures in Coevolution ... 57
2.4.2 Interactions in Coevolution ... 59
2.5 Concluding Remarks and Notes for Further Reading ... 61
References ... 65

3 Principal Approaches of Coevolution: Competitive and Cooperative 69
3.1 Coevolutionary Algorithms: An Introduction 69
3.2 The Frameworks of Coevolutionary Algorithms 70
3.2.1 Competitive Coevolution 71
3.2.2 Cooperative Coevolution 74
3.3 Coevolutionary Problem Solving 77
3.3.1 Competitive Coevolution for Learning 78
3.3.2 Cooperative Coevolution for Optimization 81
3.4 Concluding Remarks and Notes for Further Reading 86
References 89

4 Applications of Coevolution in Real-World Problem Solving 93
4.1 General Introduction 93
4.2 Business, Economics, and Public Policy 95
4.3 Entertainment Games 103
4.4 Engineering, Logistics, and Transportation 107
4.5 Software, Systems, and Services 113
4.6 Concluding Remarks and Notes for Further Reading 118
References 122

5 Pathologies and Analysis in Coevolutionary Search 127
5.1 General Challenges in Coevolution 127
5.2 Pathologies in Coevolutionary Search 129
5.2.1 Overspecialization, Focusing, and Intransitivity 129
5.2.2 Cycling and Forgetting 132
5.2.3 Disengagement and Mediocre Stable States 134
5.3 Analysis of Coevolutionary Search 136
5.3.1 Cycle Structures in Coevolutionary Problems 136
5.3.2 Population Dynamics of Coevolutionary Processes 139
5.4 Concluding Remarks and Notes for Further Reading 143
References 147

Part II Coevolutionary Computation on Optimization Problems

6 Coevolutionary Construction of Parallel Algorithm Portfolio Optimization 153
6.1 Introduction to Parallel Portfolio Optimization 153
6.2 Construction of Parallel Algorithm Portfolio Through a Generative Adversarial Approach 157
6.2.1 Parallel Algorithm Portfolios 158
6.2.2 Generative Adversarial Solver Trainer (GAST) 160
6.2.3 GAST on TSP and SAT 164
6.2.4 Computational Study of GAST on TSP and SAT 167
6.3 Coevolutionary Framework of Parallel Algorithm Portfolio Construction 172

6.3.1 Parameterized Solvers Made Generalizable........ 173
6.3.2 The Framework of Coevolution of Parameterized Search........ 177
6.3.3 Instantiations for TSP and VRPSPDTW........ 179
6.3.4 Computational Study on CEPS........ 184
6.4 Discussion and Conclusion........ 191
References........ 193

7 Cooperative Coevolution for Large-Scale Optimization........ 199
7.1 Main Issues in Solving Large-Scale Optimization via Cooperative Coevolution........ 199
7.2 Random Grouping in Cooperative Coevolution........ 203
7.2.1 Theoretical Motivation for Random Grouping........ 203
7.2.2 Cooperation Coevolution Framework with Random Grouping........ 207
7.2.3 Differential Evolution Cooperative Coevolution with Random Grouping (DECC-G)........ 210
7.2.4 Computational Study of DECC-G........ 212
7.3 Differential Grouping in Cooperative Coevolution........ 221
7.3.1 Theoretical Studies and Procedural Construction of Differential Grouping........ 222
7.3.2 Computational Studies on Cooperative Coevolution with Differential Grouping........ 226
7.3.3 Improved Differential Grouping (DG2)........ 242
7.3.4 Computational Studies on Cooperative Coevolution with Improved Differential Grouping........ 253
7.4 Discussion and Conclusion........ 263
References........ 266

8 High-Dimensional Optimization as Computationally Expensive Optimization........ 271
8.1 Solving High-Dimensional Optimization Problems via Cooperative Coevolution........ 271
8.2 The Main Challenge in Coevolutionary Divide-and-Conquer........ 273
8.3 Self-Evaluation Evolution........ 275
8.3.1 The Framework of SEE........ 276
8.3.2 Meta-Modelling in SEE........ 277
8.3.3 A Specific Implementation of SEE........ 280
8.4 Computational Study of SEE........ 283
8.5 Discussion and Conclusion........ 295
References........ 297

9 Design Space Exploration for Multimode Dataflow Mapping........ 301
9.1 Introduction to Heterogeneous Multiprocessor Mapping........ 301
9.2 Problem Model........ 304
9.3 Cooperative Coevolutionary Framework........ 309

9.4 Computational Study ... 317
9.5 Discussion and Conclusion ... 328
References ... 330

10 Cooperative Coevolutionary Design Optimization in Concurrent Engineering ... 333
10.1 An Introduction to Concurrent Engineering in Relation to Cooperative Coevolution ... 333
10.2 Problem Formulation and Evolutionary-Based Problem-Solving Approaches ... 336
10.2.1 Complex CE Design Problem Formulation ... 336
10.2.2 Design Problems for the Computational Studies ... 338
10.2.3 Use of Evolutionary Algorithms in MDO ... 344
10.3 Novelty-Driven Cooperative Coevolution ... 347
10.3.1 Novelty-Driven Cooperative CEA ... 347
10.3.2 Novelty-Driven Cooperative Coevolution with Stochastic Selection ... 350
10.3.3 Coevolutionary Concurrent Design Method for Quasi-Separable MDO ... 352
10.3.4 Communication Frequency Self-Adaptation ... 353
10.4 Computational Studies ... 356
10.4.1 Results on Performance Comparisons ... 357
10.4.2 The Effect of Communication Frequency ... 360
10.5 Discussion and Conclusion ... 365
References ... 366

Part III Coevolutionary Computation on Learning Problems

11 Coevolutionary Learning in Noncooperative Games ... 373
11.1 Introduction to Coevolutionary Learning in Games ... 373
11.2 Complex Iterated Prisoner's Dilemma for Real-World Interactions ... 376
11.3 Challenges to Cooperation via Direct Reciprocity with More Choices ... 382
11.3.1 Multi-Agent Coevolutionary Learning System for IPD ... 383
11.3.2 Revisiting the IPD with More Choices: Evolution of Defection ... 386
11.3.3 More Choices Create More Opportunities to Exploit Partners ... 388
11.3.4 Is Cooperative Play Still Possible? ... 391
11.4 Incorporating the Mechanism of Indirect Reciprocity in Coevolution ... 393
11.4.1 Implementation of Reputation ... 394

11.4.2 Reputation Promotes Cooperative Behaviours.............. 396
11.4.3 Why and How Reputation Helps with Cooperation........ 397
11.4.4 Does It Pay to Maintain Good Reputation?.................. 400
11.5 Discussion and Conclusion.. 401
References.. 404

12 Generalization in Coevolutionary Learning.................................. 407
12.1 Introduction to Performance Analysis in Competitive Coevolution.. 407
12.2 Generalization Performance in Coevolutionary Learning........... 409
12.2.1 Theoretical Framework of Generalization Performance... 410
12.2.2 Constructions of Generalization Performance Measurements.. 413
12.2.3 Computational Studies of Chebyshev's Bound for Competitive Coevolution.................................. 417
12.3 Improving Generalization Performance Estimates in Coevolutionary Learning.. 422
12.3.1 Obtaining Tighter Chebyshev's Bound by Estimating the Variance σ^2...................................... 422
12.3.2 Error Estimations for Gaussian-Distributed Generalization Estimates....................................... 425
12.3.3 Statistical Testing for Comparison of Strategies............ 428
12.3.4 Computational Studies of Improved Generalization Estimates in Coevolutionary Learning..... 430
12.4 Discussion and Conclusion.. 438
References.. 441

13 Impact of Problem Structures in Coevolution.............................. 445
13.1 Introduction of Coevolutionary Problem Structures and Their Effects on Search... 445
13.2 Random Walks on Digraphs as Models of Coevolutionary Search Processes.. 447
13.2.1 Coevolutionary Search Processes as Random Walks on Digraphs... 448
13.2.2 Characterization of Coevolutionary Markov Chains........ 453
13.2.3 Computational Studies on Quantitative Measures of Coevolutionary Markov Chains............................. 459
13.3 Duality of Solution Performance and Visitation Probability in Competitive Coevolution................................. 463
13.3.1 Invariant Measures and PageRank Authority for Abstract Coevolution.. 464
13.3.2 Connection Between PageRank and Coevolutionary Markov Chains................................. 466

13.3.3 Computational Studies in PageRanking Competitive Coevolution 470
13.4 Discussion and Conclusion 479
References 483

Index 487

Acronyms and Abbreviations

ANN	Artificial Neural Network
CBCC	Contribution-Based Cooperative Coevolution
CCDM	Coevolutionary Concurrent Design Method
CCGA	Cooperative Coevolutionary Genetic Algorithm
CCL	Classical Coevolutionary Learning
CE	Concurrent Engineering
CEA	Coevolutionary Algorithm
CEPS	Co-Evolution of Parameterized Search
CFS	Communication Frequency Self-Adaptation
CMC	Coevolutionary Markov Chain
DECC	Differential Evolution Cooperative Coevolution
DSE	Design Space Exploration
EA	Evolutionary Algorithm
GA	Genetic Algorithm
GAST	Generative Adversarial Solver Trainer
ICL	Improved Coevolutionary Learning
IPD	Iterated Prisoner's Dilemma
MDO	Multidisciplinary Design Optimization
MLP	Multilayer Perceptron
MMDF	Multi-Mode Dataflow
MOP	Multi-Objective Optimization Problem
NDCC-SS	Novelty-Driven Cooperative Coevolution with Stochastic Selection
PAP	Parallel Algorithm Portfolio
SAT	Boolean Satisfiability Problem
SDF	Synchronous Dataflow
SEE	Self-Evaluated Evolution
TSP	Traveling Salesman Problem
UEM	Universal Electric Motor
VRPSPDTW	Vehicle Routing Problem with Simultaneous Pickup-Delivery and Time Windows
WPC	Weighted Piece Counter

Part I
Coevolutionary Computation—Fundamentals, Applications, and Challenges

Chapter 1
Evolutionary Computation: An Overview

1.1 Evolutionary Computation as a Modern Search Heuristic

As a modern branch of ***Artificial Intelligence*** (***AI***) method, ***Evolutionary Computation*** encapsulates computational systems and procedures that are inspired from natural evolutionary processes. They are used to solve complex, real-world problems and provide solutions of sufficient quality in a reasonable time. They are applied across various domains that range from physical, biological to social sciences for problems framed as optimization and learning tasks to those involving modelling and simulations. In these situations, they offer working alternatives to classical exact methods for which the latter are unable to provide guaranteed optimal solutions within the specified time-frame, requiring substantial augmentations to deal with specific problem structures, or relaxation of strong assumptions that underpin much of the formal constructions of those methods.

Despite principled working differences between typical approaches used in Evolutionary Computation with classical methods, there are shared similarities as well. Indeed, the notion of problem solving by search is general and covers many families of methods [114]. These ***Evolutionary Algorithms*** (***EAs***) are formulated within a generate-and-test framework that employs an iterative procedure implementable on a computing machine [131]. As such, problem solving simply involves an iterative search on the space of solutions. In each step, new candidates are generated and then tested against a set of criteria that determines whether the goal of the search has been met, thereby stopping it.

A direct approach would have been to use the problem definition so as to formulate the test criteria. One could come out with different, systematic ways to generate successive candidates. However, there is a potential limitation to such ***uninformed*** search. By way of a simple illustration, consider a problem involving discrete structures with its solution representation parameterized in the manner

X. Yao, S. Y. Chong, *Coevolutionary Computation and Its Applications*,
Natural Computing Series, https://doi.org/10.1007/978-981-96-2841-4_1

whereby the entire finite solution space is laid out as a list of unique candidates. Suspending the stopping of search and keeping the problem fixed, the behaviour of uninformed search with a particular generating approach is akin to operating on a specific permutation of this list. The search behaviour can be understood as going left-to-right in sequence and testing the candidate at each step. The worst-case search scenario is when the candidate that best satisfies the test criteria is generated last, that is, located at the rightmost end of the list.

The same argument can be made when the strategy to generate successors is fixed and the problem is changed. What matters is that if one hopes to avoid the worst-case scenario in return for obtaining a possibly suboptimal solution of acceptable quality in reduced time, then additional albeit partial knowledge about the nature of the problem is required. In practice, this involves a form of ***heuristic function*** that indicates how far the current candidate is to the optimum to guide the selection of successive candidates. Furthermore, one could limit the generation of successive candidates to a smaller neighbourhood defined by ***transformation*** operators. Modern heuristics methods combine these two methodological features to produce iterative search algorithms that in effect restricts search to a subset of the solution space. Continuing on with our simple illustration, the search space is now a significantly shorter list.

Of course, modern heuristic search methods employ sophisticated procedures for which even their theoretical models when applied to abstractions of real-world problems require well-developed mathematical machineries for analysis. Indeed, this applies to EAs as well. One example is the probabilistic treatments used for theoretical analysis of EAs as a class of ***randomized algorithms*** on discrete optimization problems [47]. The challenges faced arise from the nature of an EA's search behaviour. The analysis needs to (i) formulate ***state*** descriptions of an EA at each time step that reflect the entire population of candidates as a whole and (ii) track their changes as the population evolves. The state space in which a population-based EA operates and where analysis has to be made is different and more complex than the underlying search space.

Although the development of these tools has led to the rigorous analysis of EAs as well as new insights in the design and construction of more effective EAs, the field of Evolutionary Computation has some of its roots in the empirical. Observations of natural evolutionary processes account for some of the motivating principles underlying these nascent approaches in the 50s to 60s [13, 14, 20, 22, 64, 65]. The interested readers would note these studies are not restricted to investigating applications of evolutionary approaches on some abstract optimization and decision-making problems. The study [20] has proposed an EA-like procedure applied to industrial manufacturing in the context of ***Operations Research*** at the time of machines with limited computing power. Yet others have explored evolutionary processes implemented on finite computing machines for their emergent properties in the manner similar to ***Complex Adaptive Systems*** and ***Artificial Life*** [11, 12, 61].

Other independent studies then follow on, forming precursors that later in the nineties when the field is established and unified under a single umbrella term would historically be referred then as the three main independent branches of

Evolutionary Computation [9, 58]. They are ***Genetic Algorithm*** (***GA***), ***Evolutionary Programming*** (***EP***), and ***Evolution Strategies*** (***ES***). Other new approaches would be introduced, first having similarities with the three main identified branches, but are then significantly developed in their own right to form other classes upon conceptual maturity. Some of these classes adopt nomenclatures such as ***Particle Swarm Optimization*** (***PSO***) [19, 34, 87] and ***Estimation of Distribution Algorithm*** (***EDA***) [77, 96, 110], among others.

At this point and perhaps of interest to those familiar and new to Evolutionary and Neural Computation is that one of the founders of AI and the inventor of the ***Turing Machine*** and ***Turing Test*** [124] has proposed some models and methodologies of these two computational approaches in 1948 (the report is published 20 years later in 1968) [36]. In this report on *Intelligent Machinery* [125], three computational models that are collectively called ***unorganized machines*** are conceptualized at the time when one of the first computers called Automatic Computing Engine is designed by the very same author and its pilot version is being physically constructed. In particular, the proposed A-type machine is reminiscent of a ***Boolean Network*** [86] consisting of a fixed network of basic computing units, each of which is capable of performing a ***Boolean function*** with two inputs and a single output and has functional completeness (e.g. a NAND gate) [25]. Unlike the A-type machines with fixed interconnections that allow signals to pass through unaltered, the B-type machines have modifiable switch-type interconnections whereby signals can be altered. The main conceptual purpose of endowing B-type machines with such interconnections is so that the initial unorganized machine can be transformed through changes in the functionality of those interconnections to an *organized machine* capable of performing a computational task via learning procedures that include evolutionary and cultural search [124, 125].

In the modern setting, GA is thought of and applied as an evolutionary approach that uses binary-encoded solution representation with associated genetic operators [40, 66]. As a simple illustration how GA search is prepared, consider problems with non-hierarchical solution representations. Candidates are points in the Cartesian product of elementary sets that form the search space. Each set can be discrete or continuous of a type of floating point number, supported by an encoding scheme with a binary string of a specified length so as to range over all distinct elements. This coordinate addressing of points ensures that the encoding is reversible (since it is a one-to-one mapping). For the user, what practically matters is that the parameterization of solutions now appears as one long contiguous binary string, which is analogous to the biological chromosomes. Genetic operators can now be defined such as crossovers that swap adjacent blocks of same lengths delineated at randomly selected positions of the binary string from a pair of candidates to form a new pair. However, the theory of GA that was introduced and further developed by Holland [80, 81] calls for a theory of computation via genetic adaptation. The modern usage as we know it is only one applied interpretation of this theory. For example, the solutions or programs are not restricted to being function or parametric optimizers [42].

In the same manner, modern success of EP might point to the approach being one developed for parametric optimization. As an example how such a perspective could have come about, one could focus on the end results in [60] about an AI agent trained to play a two-player zero-sum board game without preprogrammed knowledge to near human-expert-player level. What matters for practical applications are the system design to train the game-playing agent. In this case, the parameterized solution representation takes the form of a large feedforward neural network that functions as board evaluations to a shallow minimax for game-play. The training success might appear to hinge on the effectiveness of EP to optimize the neural network via specific mutation operators that make small perturbation to its connection weights. However, the approach introduced by Fogel [56] and investigated by others [5, 23] is originally conceived as a broader approach to develop Artificial Intelligence. The early experiments that predominantly feature ***Finite State Machines*** for solution representation (e.g. even for game-playing strategies [23]) would serve: (i) To investigate the idea that intelligence, particularly behavioural intelligence as abstracted in the form of symbolic prediction can be evolved. (ii) To showcase that intelligent behaviours can be adapted, i.e. evolutionary change can directly act at the level of individuals.

Of the three, the motivation in developing ES is driven more towards the practical problem-solving applications. The early studies include both the iterative procedures and their accompanying analysis [17, 112, 115]. In the original setting of experimental parameter optimization, candidate solutions undergo physical adjustments and then performance evaluations through actual measurements. These random heuristics are designed to introduce small and variable adjustments on design parameters that are then in turn evaluated for their performance with the possibility of noise corruption. The values of these adjustments are drawn from some probability distributions, resulting in stochastic procedures that outperform gradient-based, numerical optimization methods in those actual, physical engineering optimization settings that are multimodal and obviously noisy [17]. The advent and greater accessibility of higher performance computing machines made it possible subsequently to increase the development rate. Combining with the appropriate abstractions whereby solution design parameterizations and their subsequent fitness function formulations are made, the evolutionary procedure can be computationally simulated.

Despite real-world success of Evolutionary Computation and this subsequent temptation to view EAs as a collection of specific prescriptions targeted to remedy challenging problems having distinct structures, our motivation here as with others much earlier [40, 58, 66] is to highlight what the field can offer to the practitioners in other aspects of problem solving beyond simply being a search methodology directly applied on tangibly formulated problems. One particularly powerful principle that forms the main thematic presentation of this book is one that involves ***coadaptation*** between interacting agents as the means to guide their search.

This principle idea underlying ***Coevolutionary Computation*** is not new [61]. However, its development over the last 20 years has brought about better understanding and opened up avenues for new applications. In the case of the former whereby coevolution has been naturally applied to drive the learning of decision-making agents, we will showcase new theoretical insights and analysis tools for ***Coevolutionary Algorithms*** (***CEAs***). They can inform practitioners when coevolutionary search is affected with pathological issues and how to go about mitigating those effects. In the case of the latter, we will demonstrate how one can think about the principle of coadaptation as a framework for problem solving at a higher level of search.

For the practitioners, this is not the simple case of a ***metaheuristic***-like extension to a niche class of EAs. Instead, this is on the systemization of some higher-level processes human experts employ in algorithm tuning and design in problem solving that can be represented as coevolutionary procedures. Starting with the construction of the space of solvers through appropriate parameterization allowing for distance computation and associated transformation operator design of primitive solvers, coevolution can then act on this higher-level search space. The challenge of formulating an appropriate fitness evaluation scheme is also addressed through a similar construction of the space of problem test cases. This formulation of coevolution between a population of EAs and another population of test cases forms the basis towards addressing known challenges relating to the generalization capabilities of ***Portfolio Optimization*** approaches.

In the following sections, we will present in more detail the specifics of EAs. Section 1.2 will focus on the key aspects that form the basis of the design of a generic EA: the use of a population, the solution representation, and the evolutionary operator of variation and selection. Specific terminologies that are commonly used in Evolutionary Computation will be described and emphasized. Some examples of EAs and, in particular, those with similar characteristics to the CEAs that are presented in later chapters will be introduced. After that, Sect. 1.3 will introduce the notion of parameter control of EAs and describe some of the advanced methodologies that enable the optimization of parameters associated with the operations of EAs to improve performance.

1.2 The Framework of Evolutionary Algorithms

EAs belong to a broad class of population-based, stochastic search algorithms that are inspired by natural evolution. The common abstraction that is adopted in the field to relate both the algorithmic and natural search is to cast them both within a ***generate-and-test*** framework. Be it computational or natural, the procedure operates on a state space whereby its constituents represent a finite population of candidates. We provide three perspectives with which one can understand EAs and how their design aspects mirror and are influenced by the important facets within natural evolution.

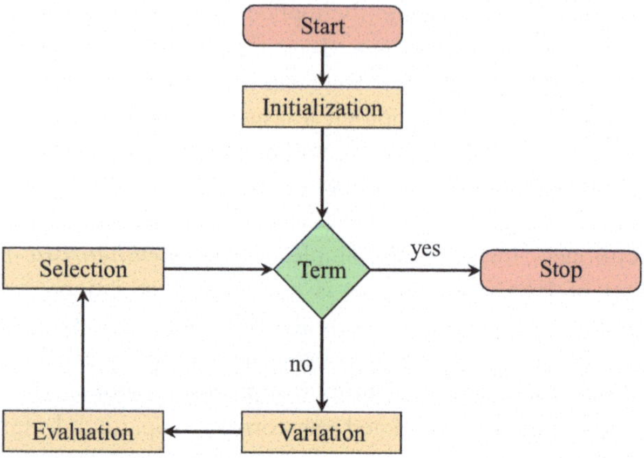

Fig. 1.1 Flowchart of a generic EA

The flowchart in Fig. 1.1 illustrates an informal, system view whereby both the generic EA and the simplified neo-Darwinian model of natural evolution comprise of three main, interdependent components that interact in a loop. Starting with the initialization of parent candidate solutions in the population, they are then subjected to repeated applications of evolutionary operators of ***variation*** and ***selection*** in turn until the loop is stopped. The variation operator is responsible for the generation of new candidates from existing ones. For example, the offspring candidate that is generated from a parent candidate but has some of its inherited solution parameters subjected to small perturbations that change their values would be analogous to genetic mutation in a biological asexual reproduction process.

The selection operator chooses which of the candidates in the current population to be carried forward and form part of the population in the next generation. Selection depends on the fitness of candidates obtained through an evaluation process. These two components are separated here from a system's view for clarity but are more intertwined in models with complex selection processes. They may reflect natural models of selection that acts on the genotypic or phenotypic level [101]. For example, the frequency proportional selection involves specific candidates chosen for reproduction at rates dependent on the proportion of their fitness values with respect to the cumulative fitness values of the entire population. This is adopted in the canonical GA and is similar to the biological analog of selection pressure acting at the genotypic level of a population of the same specie (e.g. the frequencies with which alternative forms of genes at specific locations in the chromosome persist) [81]. Other EA approaches such as those in the EP family use models of localized competitions for reproduction at the phenotypic level whereby selection acts on subsets of the population [57, 62].

Algorithm 1.1 A general EA

```
Input:
  Y        Population of candidate solutions
  s_term   Termination criterion selection
Output:
  X        Population of evolved candidate solutions

procedure EA(Y, s_term)
    t := 0                                   ▷ Initialize time-stamp
    b_term := true                           ▷ Initialize loop
    X := initialize(Y)                       ▷ Initialize population of parents in X
    while b_term do
        X := variation(X)                    ▷ Apply variation operator on X
        evaluation(X)                        ▷ Fitness Evaluation of X
        X := selection(X)                    ▷ Apply selection operator on X
        t := t + 1
        b_term := termination(X, t, s_term)  ▷ Check for termination
    end while
    return X
end procedure
```

The shared common framework with intrinsic design flexibility in its implementation is emphasized from an algorithmic perspective. This can be seen in the manner with which the data structure is formulated within Algorithm 1.1 for the general procedure **EA**. The main structured data $\boldsymbol{X} = \big((\boldsymbol{x}, f)_i : i = 1, 2, 3, \ldots, n\big)$ contains information of each candidate's parameter values $\boldsymbol{x}$ that collectively describe the constituents of the population of size n. In other words, $\big((\boldsymbol{x}, f)_i : i = 1, 2, 3, \ldots, n\big)$ constitutes the configuration of the population $\boldsymbol{X}$. There are two main data transformations represented by the two procedures that return data—**variation** and **selection**—that are written to make explicit they act on $\boldsymbol{X}$ as a whole, its instance taking on a specific population configuration from $\{X_1, X_2, X_3, \ldots\} = \mathcal{X}$. The one minor exception is the procedure **initialization**. It is called only once and could be called outside of **EA** and pass in via the input parameter $\boldsymbol{Y}$ and takes a value from $\mathcal{X}$. Despite the restriction to having as input and output data of the type $\mathcal{X}$, the internal operations within **variation** and **selection** can be as involved as necessary to implement their specific design requirements.

The other procedures can be grouped into two categories. The first changes only specific data contained in $\boldsymbol{X}$. For procedure **evaluation**, this data will be the fitness of the individual candidate, which is addressable (e.g. $X[i].f$). Indeed, **evaluation** can be called within **selection** itself. The second refers to those that operate on local variables that control the loop. The primitive arithmetic operation on the local variable time-stamp t keeps track of the iteration (generation) count. This binary operator $t + 1$ is written using the infix notation. The procedure **termination** checks for the termination condition of the loop. It returns false when the condition holds, and true otherwise. It causes the loop to terminate by a change in value of the local boolean flag $\mathrm{b_{term}}$, which is used in place of the loop condition. Note that

termination accepts three inputs. The first two $\boldsymbol{X}$ and t represent information. The third input $\mathrm{s}_{\mathrm{term}}$ represents the choice that relates to how information is used to assess termination of the loop.

Finally, one can describe an abstract evolutionary process that acts on the space of population configurations $\mathcal{F} : \mathcal{X} \rightarrow \mathcal{X}$ as a ***discrete-time dynamical system***. The study [127] has mathematically formulated GA as a dynamical system with explicit geometric modelling of the evolutionary operators. Another study [85] has modelled the search behaviour of a population-two EA applied to one-dimensional problem as a discrete-time dynamical system generating trajectories of expected population states values. Here, we detail only the basic formalism to focus on the main constructions to describe the search behaviour as trajectories in $\mathcal{X}$. The abstract evolutionary process can be expressed by the following difference equation:

$$\begin{aligned} \boldsymbol{X}^{t+1} &= \mathcal{F}(\boldsymbol{X}^t), \ t = 0, 1, 2, \ldots \\ \mathcal{F} &= \mathcal{S} \circ \mathcal{V}. \end{aligned} \tag{1.1}$$

The functions $\mathcal{S} : \mathcal{X} \rightarrow \mathcal{X}$ and $\mathcal{V} : \mathcal{X} \rightarrow \mathcal{X}$ represent the selection and variation operators, respectively. Essentially, the processes involve changing the constituents of the population $\boldsymbol{X}$. In this manner, an evolutionary process can be observed from its trajectory in $\mathcal{X}$ as the iterates $\boldsymbol{X}^0, \boldsymbol{X}^1, \boldsymbol{X}^2, \ldots$. Its search behaviour can be ascertained from the regimes describing the trajectories. For example, let the fixed point $\boldsymbol{X}^*$ be defined as $\boldsymbol{X}^* = \mathcal{F}(\boldsymbol{X}^*)$. Then, trajectories that fall into one of the $\boldsymbol{X}^*$s would describe a search towards a local optimum.

Note that implicit in this model is the incorporation of fitness evaluation within the selection operator. For the individual candidate $i = 1, 2, 3, \ldots, n$ described by its solution parameters $\boldsymbol{x}_i = \boldsymbol{X}[i].\boldsymbol{x} \in \mathcal{R}$, with $\mathcal{R}$ being the appropriate subset of the solution space (e.g. $\mathcal{R} \in \mathbb{R}^D$ in a constrained continuous optimization setting), this involves computing $f(\boldsymbol{x}_i)$ with $f : \mathcal{R} \rightarrow \mathbb{R}_{\geq 0}$. Since selection acts on the entire population $\boldsymbol{X}$, then within the operator itself there are two processes. The first process $\mathcal{S}_{\mathrm{fe}} : \mathcal{X} \rightarrow \mathbb{R}^n$ is on fitness evaluations of all n individuals. The second process $\mathcal{S}_{\mathrm{cs}} : \mathbb{R}^n \rightarrow \mathcal{X}$ is on using these fitness values to guide the selection of candidates for the population in the next generation. The composition of these two processes $\mathcal{S}_{\mathrm{cs}} \circ \mathcal{S}_{\mathrm{fe}} : \mathcal{X} \rightarrow \mathbb{R}^n \rightarrow \mathcal{X}$ will result in the selection operator $\mathcal{S} = \mathcal{S}_{\mathrm{cs}} \circ \mathcal{S}_{\mathrm{fe}}$ having the right type $\mathcal{X} \rightarrow \mathcal{X}$.

However, our main presentation focus here is to show how even such an abstract model can reveal an important aspect in the design of EAs. Note earlier in our presentation we have that the evolutionary process begins with the variation operator after the initialization of the parent candidates in the population. This is an arbitrary choice, and instead one can implement the selection operator first, then the variation operator next, in turn. There is no qualitative difference between the two. The main distinguishing aspect is with regard to how the population is initialized. Indeed, we can make this argument more formal and exploit the properties of only working with functions of type $\mathcal{X} \rightarrow \mathcal{X}$ and that function composition is associative to demonstrate this. The population at some arbitrary iterate is the result of repeated applications of $\mathcal{G}$. An argument by induction will show that $X^{m+1} = \mathcal{F}(\mathcal{F}^m(X^0)) = \mathcal{F} \circ \mathcal{F}^m(X^0)$. Since $\mathcal{F} = \mathcal{S} \circ \mathcal{V}$, we have

$$\begin{aligned}
\mathcal{F} \circ \cdots \circ \mathcal{F}(X^0) &= \mathcal{S} \circ \mathcal{V} \circ \cdots \circ \mathcal{S} \circ \mathcal{V}(X^0) \\
&= \mathcal{S} \circ \mathcal{V} \circ \cdots \circ \mathcal{S}(\mathcal{V}(X^0)) \\
&= \mathcal{S} \circ \mathcal{V} \circ \mathcal{S} \circ \cdots \circ \mathcal{V} \circ \mathcal{S}(\mathcal{V}(X^0)) \\
&= \mathcal{S} \circ \mathcal{G} \circ \cdots \circ \mathcal{G}(\mathcal{V}(X^0)),
\end{aligned}$$

where $\mathcal{G} = \mathcal{V} \circ \mathcal{S}$. If one chooses the convention that $\mathcal{F}$ models the evolutionary process, then the population must be completely initialized—parent and offspring candidates—for selection to operate on. The convention we use would have X^0 partially initialized with parent candidates since $\mathcal{V}$ acts to generate offsprings from parents, thereby completing the population initialization.

Lastly, one can note as an example formalized in [127] for the abstract GA, even the behaviour of a simple population search towards the global optimum in function optimization can be descriptively subtle. This is due to the construction of the GA map $\mathcal{F} = \mathcal{S} \circ \mathcal{V}$. On the one hand, there are many other unstable fixed points of $\mathcal{S}$, except the one associated with the global optimum that is located within the interior of its basin of attraction. On the other hand, the unique fixed point X^v of $\mathcal{V}$ corresponds to the population configuration with which all candidates are represented equally. This fixed point has as its basin of attraction $\mathcal{X} \setminus X^v$. When one combines both operators, $\mathcal{S}$ with its focusing effect and $\mathcal{V}$ with its diffusion-like effect in $\mathcal{X}$, the resulting GA map $\mathcal{S} \circ \mathcal{V}$ demonstrates a punctuated-equilibria-like search. Here, there are periods of search that the population moves away from one equilibrium configuration to another (these being unstable fixed points of $\mathcal{S}$) due to the action of variation $\mathcal{V}$. The trajectory will eventually fall into the fixed point corresponding to the global optimum and cycle nearby it.

1.2.1 On the Use of Populations

One characteristic that differentiates typical EAs in comparison to other families of heuristics (when one considers their classical formulations and not includes specific augmentations or hybridizations with other heuristic families) is in the use of ***populations*** of candidates. A main defining feature of GA is the use of crossover operators, which by definition and their constructions require a population of candidates. The use of a single ***panmictic*** population whereby candidates compete for reproduction is typical in EAs though by no means the only form that evolutionary process acts on. Parallel EAs have been proposed as well, which consist of multiple populations undergoing a more involved, loosely coupled process analogous to the natural evolution observed on neighbouring ***islands***. In a simple parallel EA construction, each population is placed on its own island and mostly independently evolves, except for specific and typically over a short duration of exchanges of candidates between the population [15, 121]. This ***migration*** process can be scheduled periodically or in other more involved manner.

These ***Island-based EAs*** are implemented in both parallel and sequential computing setups. In a parallel setup whereby operations on an individual island (population) are placed under a specific computing core, the additional and direct argument could be made for the practical speed-up in the evolutionary process to obtain the solutions through increased computational resources. However, in a sequential setup whereby the evolutionary process within each island has to be run in turn on a single computing machine, the argument for a speed-up in successful search will be more subtle and involved. There is the additional overhead in computational cost from having to process more candidates. When these parallel EAs implemented over a sequential setup clearly demonstrate improved search performance over the conventional EAs, there would be some intrinsic quality in having to use a larger pool of candidates.

This line of argument could be applied to the classic single panmictic population setup of EAs. In particular, one might ask whether there is any significant difference between an EA and an augmented hill climber with stochastic and parallel search (e.g. ***stochastic beam search*** as described in [114]). Indeed, the answer as to why the use of a population is intrinsic to the success of EAs has to do with having specific diversity in the population during evolutionary search.

One informal approach to illustrate this issue is to start with the simple opposite case whereby the population has specialized to just one specific candidate. The rest of the candidates in the population are just copies of that candidate. The worst-case scenario is that the candidate corresponds a local optimum where there is no escape because the current design of the EA via the variation and selection operators act in a concerted manner such that only copies of those candidate are carried forward to the population in subsequent generations. The EA is said to have undergone ***premature convergence*** in this situation.

One might note other population configurations in conjunction with specific EA designs and problem structures giving rise to this search behaviour. For example, the candidates in the population are distinct but all correspond to points within a wide attracting basin of the local optimum. The effect of crossover operators is nullified because of the problem structure, that is, any crossover between a pair of mostly similar candidates will generate a new pair within this basin. Mutation is also insufficient in that it will only be a very rare event that a point outside of this basin is generated and subsequently selected because it has a better fitness value.

Indeed, increasingly large populations can be detrimental to an evolutionary optimizer especially where specific problem structures are concerned. Despite the specific EA configuration and problem setup that necessitates the technical proofs in [31], there is the wider implication that the theoretical results would suggest. Informally, increasing population size can act to increase selection pressure acting on the population to remain trapped inside the basins of some local optima. This is because of the higher fitness values that are associated with those basins. One can mitigate this with a change in the variation operator although care needs to be taken that upon escape it would not impair the search when it finds the basin of the global optimum.

As such, the inherent utility for using of populations in EAs to solve complex, real-world problems [134] is tied to having the appropriate diversity at the right time or phase of evolutionary search. The benefits of population usage with diversity have been established in the field [16, 27, 38, 41, 68, 105, 111, 117, 126] and need to be considered together in any discussion of EAs. We list four of them here, following the same introductory discussion made [47] that summarizes current state-of-the-art in theoretical understanding on population diversity in EAs.

First, EAs are designed with ***global exploration*** in mind whereby the population allows for different regions of the search space to be explored. Global search is important in cases where problem structures in conjunction with evolutionary operators introduce additional challenges. For example, those with a highly ***multimodal*** fitness landscape (see Sect. 1.4 for details) require some intrinsic mechanism in the evolutionary search via a diverse population to escape from those local optima.

Second, some EAs are designed to primarily use the crossover type of variation operators, with the mutation type operators taking a secondary role in generating new offspring candidates. In particular, the effect of crossover of largely similar parent candidates would be similar to that of mutation with low strength and rate. Here, almost similar offsprings would be generated. As such, a diverse population might ***facilitate use of crossover*** to generate offspring candidates for effective global search.

Third, there are some family of real-world problems such as those formulated as ***Multi-objective Optimization*** that require solutions to address several and possibly also conflicting objectives simultaneously [134]. Implicit in these problems is the aspect of ***decision-making*** whereby an agent (human or artificial) would select from a set of solutions obtained from a solver. In this situation, generating a population of solutions is a requirement. As such, EAs must be designed to maintain a diverse population of candidate solutions that reflect trade-offs associated with meeting those objectives.

Fourth, the issue of ***robustness*** is crucial in some real-world problem solving in which certain aspects of the problem can change (with time), that is, they are dynamic rather than static. This could be a result of noisy measurements in experimental evaluation of solution fitness or changes in the parameters of the problem formulation. In these situations, the EAs need to produce robust solutions rather than those whose optimality is tied to current experimental measurement setup or problem formulation. This is more easily or quickly achievable if a diverse population is produced that has discovered multiple optima or able to track any movement in the optima as a result of change in fitness evaluations.

1.2.2 Representations for Solutions

One result of the generic design framework of EAs and subsequently their wide-ranging applications in different problem domains is that there are various forms solution ***representations*** available for use. Any discussion and elaboration of these representations are tied to their associated variation operators, which we detail more in the next section. Here, we present two of the common solution representations that have been in use in conjunction with EAs and provide brief discussions how a particular representation form is applied over various problem settings. However, to have a better appreciation how solution representations and the additional use of encodings can influence the search behaviour of EAs, we first introduce the biological correspondence and concept on ***phenotype*** and ***genotype*** to start the discussion.

Unlike the experimental parameter optimization setting in early ES studies whereby one is able to work on a model that is close to the actual object to be optimized [10], most of the EA studies in these problem settings work with simulated evolutionary optimization. An abstraction process will have been carried out to model these objects so that their behaviour or input-output responses can be validated, thus enabling their use. There are parameters that are associated with the models and their responses. The parameters that are being considered for optimizing responses of the models are the phenotypes. In this case, the ***parameter space*** under consideration constitutes the ***phenotype space***.

The nature of genotypes arises because of these simulated evolutionary processes within computing machines. In this situation, the models need to be represented in some form of computational (representational) models. Typically, specific mathematical objects for which transformation operators can be constructed and applied are used. The genotypes refer to instances of these computational models, each associated to a candidate solution under consideration for optimization. As such, the ***genotype space*** corresponds to the ***search space***, with the variation operators being the transformation operators that change a member of the population from one genotype to another during search [8, 9].

The need for a mapping or ***encoding function*** that relates genotypes and phenotypes is obvious and usually induced from the construction of the representational models. However, this ***genotype-phenotype mapping*** gives rise to various issues in the use of EA for problem solving. In most applications, the EA has no direct access to the phenotype space. Instead, the EA operates on the genotype space underlying the phenotype space. One can refer to the illustration in [59] for a graphical view. The complexity arising from this mapping not only affects evolutionary search performance but also makes analysis challenging. The challenges associated with these issues are perhaps as much as the nature of the problem and the EA that are used to solve it, as well as the differences in the philosophy that motivates their constructions in the first place.

In particular, the emphasis and the call to use natural representation are not as straightforward as one might realize at first. Informally, the essence for using a natural representation for effective search might be construed as having the EA to operate directly on the phenotype (parameter) space. One example is the case of using real-valued evolutionary optimizers such as ES and EP on continuous optimization problems of known mathematically constructed benchmarks [75, 76]. Here, one effectively has a one-to-one (bijection) mapping of the genotype and phenotype spaces. If one considers for simplicity only mutation operators that transform a parent candidate to an offspring candidate, then neighbourhood structures induced from such operators (the reachable neighbourhood of points) in the genotype space are consistently translated to the phenotype space. Observed trajectories arising from the evolutionary process in the genotype space are identical to those evolving in the phenotype space.

In other situations where natural representation is not used, and instead involve some encoding schemes (e.g. in the case of GAs), then the more complex situation arises whereby in effect the genotype-to-phenotype mapping is many to one [59]. Despite this, the important thing to note here is about the notion of predictability of the evolutionary process, whether in terms of its analysis and control of its search behaviour. Inherent problem structures such as convexity (e.g. unimodal continuous optimization problems) can present a case for which EAs with encoding can be used [113].

For fully discrete optimization problems, the issues of having a natural representation as well as using an encoding scheme that allows for effective search are obviously challenging. In addition to the combinatorial nature of these problems, one has to deal with representational issue. For example, consider the classic ***Travelling Salesman Problem*** (***TSP***) that involves searching for the Hamiltonian cycle of a given weighted graph with the least cost (e.g. total distances). Informally, a solution constitutes the tour or path that travels through every vertex of the graph once, except the starting and ending vertex (which is the same). Specifying a string encoding scheme for the TSP problem is straightforward, but an additional step for tour verification is required. It is very difficult if at all possible to come out with some schemes that inherently encode valid tours for a large family of graphs. Repair operators are typically designed in conjunction with variation operators that together act to define the neighbourhood structure of valid tours [26].

This is unlike the case of continuous optimization problems for which the parameter and search space could just involve the standard Euclidean vector space. Nevertheless, one can still specify the requirements necessary for a natural-like representation and encoding scheme. The incorporation of the relevant problem-specific knowledge can be viewed as an exercise in formulating the appropriate metric for the space by way of specifying distance measures, allowing for consistent translation of neighbourhood structures in phenotype and genotype spaces [26, 49, 50].

On the issue within string encoding, it has been theoretically argued [63] that there is no intrinsic advantages of using scheme specified by an alphabet A of any cardinality $k = |A|$ (e.g. using base value of $k \geq 2$ in the case of the k-ary string representation). The search space is formed as a result of using a particular k-ary string representation of length l given by the tuple $(a_i \in A : 1 \leq i \leq l)$ with the variation operator that defines the neighbourhood structures of each point (unique instance of the k-ary string) in the space A^l. The probability mass function (the "dice") that defines how likely a candidate is generated can be formulated on this search space. Crucially, a variation operator can always be specifically designed when any other k value is used in the k-ary string representation so as to preserve those probabilities.

Armed with a more comprehensive perspective how solution representation and encoding act as an interface between what one can observe about an EA's search behaviour and its underlying operational processes, we now turn our attention to presenting those that are used in real-world problem solving. We focus on representations that will be discussed in later chapters so that the section here would also act to provide a brief introduction. The first involves a particular type of real-valued representation that constitutes a broad family of connectionist systems. As one example, ***Artificial Neural Networks*** (***ANNs***) themselves form a major branch of AI methodologies. However, it soon became apparent evolutionary processes constitute not just an alternative training approach but one that can address some weaknesses in classical approaches [132].

ANNs consist of a large variety of networks of nodes, each of which is a neural processing element (neuron). A typical neuron $y(\boldsymbol{x}, \boldsymbol{w})$ takes as inputs a finite set of reals $\{x_i\}$ stored in $\boldsymbol{x}$, which are first linearly combined. The sum is controlled by real-valued coefficients or parameters stored in the set of weights $\boldsymbol{w}$. This weighted sum a is forwarded to a non-linear transfer (activation) function h to output a real value $\{y\}$. One has $y(\boldsymbol{x}, \boldsymbol{w}) = h\big(a(\boldsymbol{x})\big)$, $a(\boldsymbol{x}) = \sum_i w_i x_i + w_0$, with the term w_0 representing an independent bias term. A standard ANN that is typically applied to solve a sufficiently challenging learning problem involves a ***feedforward*** architecture, whereby neurons are placed in multiple layers that are connected in a specific manner. In particular, the network is a specific ***directed acyclic graph***. The lack of cycles is due to the constraints of not having connections between neurons in the same layer and not having feedback connections between neurons in subsequent layers back to neurons in earlier layers. However, recurrent connections can be incorporated that then give rise to recurrent neural networks.

We describe a typical feedforward neural network with a single hidden layer and an output layer following the usual expositions in the literature such as in [18]. The ANN model takes as input the set $\{x_i : i = 1, \ldots, I\}$ and output the set $\{y_k : k = 1, \ldots, K\}$, with a single hidden layer of J neurons. The construction is straightforward and one only needs to be careful with indexing. We need to construct twice, calculations for outputs of neurons in the hidden layer, and then of those in the output layer. Some diagrams might illustrate those representing the inputs $\{x_i : i = 1, \ldots, I\}$ as nodes. However, they do not do any processing other than just passing in real values to the ANN (they act as identity functions). Let z_j be the output for the jth neuron in the single hidden layer. The outputs of J neurons indexed as $j = 1, \ldots, J$ can be expressed as follows:

$$z_j = h\big(a(\boldsymbol{x})\big) = h\left(\sum_{i=1}^{I} w_{ji}^{(1)} x_i + w_{j0}^{(1)}\right).$$

Then for the neurons $k = 1, \ldots, K$ in the output layer:

$$y_k = h\big(a(z)\big) = h\left(\sum_{j=1}^{J} w_{kj}^{(2)} z_j + w_{k0}^{(2)}\right).$$

The superscript index (l) is used to explicitly indicate that the weights belong to neurons associated with the first and only hidden (1) and output (2) layers. Crucially, note the compositional nature of this particular ANN because the outputs of each layer become the inputs in the subsequent layer. In this case, we have a composition of two functions mapping $\{x_i\}$ to $\{z_j\}$ to $\{y_k\}$.

However, modern ANNs such as ***Deep Neural Networks*** (***DNNs***) [1] employ deep, many-layered architectures with diverse processing capabilities such as those that involve convolutional operations that capture spatial structures (e.g. images) and others that can handle sequential dependencies of the attributes in the data (e.g. time series and textual). These DNNs are used for problem solving in a variety of domains, having applications that are grouped and properly formulated in the settings of computer vision, speech recognition, natural language processing, and many others [70]. Two simple illustrations of neural network capabilities are those in which the problem-solving settings are formulated as ***regression*** and ***classification*** tasks. A simple regression task involves approximating the relationship between continuous dependent and target variables with a vector of input variables. The straightforward binary classification task involves assigning an observation given by a vector of input variables to one of the two discrete categories or labels by way of establishing a decision boundary.

The power of these feedforward neural networks in problem solving can then be viewed as their abilities as ***universal approximators*** to any continuous function on a compact domain (such as the unit hypercube $[0, 1]^n$) to any arbitrary precision with mild constraints over its architecture have been well established (e.g. see [37, 82, 104, 106]). Establishing such theoretical guarantees when neural networks are allowed to grow in width (a fixed number of hidden layers with an arbitrary number of neurons in them) or depth (a fixed number of neurons in each hidden layer with an arbitrary number of layers) would provide the research impetus to address the challenge of establishing effective training approaches.

This forms the second aspect of the framework of neural network for problem solving, namely their ***Machine Learning*** capability. A simple illustration is to consider the setting of ***supervised learning***, whereby iterative procedures can be formulated that changes the internal structures of a neural network that are guided with the use of a series of labelled training examples. In this manner, the neural network is said to acquire knowledge necessary to solve the problem by way of pattern extraction from raw data [70]. Crucially, this learning ability by way of training can also be achieved through adaptation. EAs are developed to optimize both the parameters and architectures of ANNs [132]. Parameter optimization in ANNs can be done directly using EAs that employ real-valued variation operators. However, the architectures of ANNs have discrete structures and in this situation some form of encoding might be used. Given the complexity in the design of these DNNs, EAs have been developed for the automatic construction of these DNNs that comes with evolutionary optimization of architecture and parameters [136, 137].

The next solution representation we introduce has the form of a ***tree structure***. A tree is a connected acyclic graph for which any pair of nodes is connected by only a single edge. It is typically illustrated with a single node at the top called the root, which then connects to other subsets of nodes in a single path. In Evolutionary Computation, ***Genetic Programming*** (***GP***) forms a family of algorithms that evolves computer programs taking the form of a tree structure representation. The conception of evolving computer programs goes back to the seminal work on GAs [80] but is then formalized and applied as GP [91]. The ***syntax tree*** representation that GP uses provides the means to directly express computer programs that are written as texts in formal languages with syntactic structures incorporated in them.

A simple ***GP tree*** consists of connected nodes that are used to represent a primitive set. This primitive set is a simple program, consisting of the leaf nodes called terminals that represent variables and constants and internal nodes (including the root) that represent executable instructions of simple functions (e.g. binary relations). The structure is such that the arguments for the instruction indicated by the current node are represented by the immediate connections that link it to its child nodes. One simple illustration of a GP tree is the program that computes the minimum of two numbers given by the mathematical expression $\min(x*y, 2*x+y)$. In this example, instruction min is represented by the root node. Since the arguments to min are not simple numbers but constitute the mathematical expressions $x * y$ and $2 * x + y$, the latter themselves form subtrees. This is one of the main advantages of using a tree representation as it is recursive by nature.

Note the original expression $\min(x * y, 2 * x + y)$ is typical in standard mathematical writing that uses a combination of infix (for addition and multiplication operators) and prefix (for the minimum operator) notations. It is common to use a prefix-only notation such as the symbolic expression (s-expression) used in LISP that makes the tree structure clear from the syntax in GP [95]. Here, one has (min $(* \; x \; x)$ $(+ \; (* \; 2 \; x) \; y)$). Given a tree representation with shallow depth, it is more apparent to human readers the branch structures in the example when s-expression is used, e.g. the right branch $(+ \; (* \; 2 \; x) \; y)$ contains a subtree as well.

This tree structure can be extended to form more advanced GP trees to represent sophisticated programs that are typical of those produced by human programmers. It is common in standard programming for specific sequences of instructions to be reused. These instructions can be created as templates that then form components or subroutines. They are subsequently used in the larger program in the form of different instantiations taking on different input values in their arguments. One of the more popular extension is the ***Automatically Defined Functions*** [92]. Other types of subroutine representations are developed as well such as Automatically Defined Iterations, Automatically Defined Loops, Automatically Defined Recursions, and others [93]. They form the branches of the advanced GP trees when in use, which are grouped under a special root node. These defined subroutines can operate to either perform work or produce results.

Note that there are other GPs that use different forms of directed acyclic graphs such as those in Cartesian GP (CGP) [128] and Metric-Based GP (MBGP) [49, 50]. These solution representations with specific encodings are originally motivated and applied for the evolution of digital circuits in CGP and boolean functions in MBGP. Given the broad scope of what computer programs can be represented by GP trees, it is not surprising to see them being used in problem solving across different domains in various types of applications. However, the main differentiating aspect between GP trees and graphs with ANNs is that of the difference between symbolic and connectionist models. As an example, GP performs ***symbolic regression*** as opposed to ***parametric regression*** in evolutionary ANNs [91]. The symbolic approach in GP calls for constructing a general function with its coefficients through an evolutionary process instead of finding the coefficients of predefined functions as in the case of evolving ANNs with fixed architectures. Similarly, the tree structure representation used in GPs lends itself to the task of ***symbolic classification*** [89] and other formulations such as those, e.g. decision-making in the form of a policy for routing problems in Operations Research [129].

1.2.3 Variation Operators

In the preceding section, we discuss how upon taking simplifying assumptions, the parametric representation of the solutions is analogous to the biological chromosome. They are the basis of the coordinate system underlying the solution space so that information describing a complex, distinct individual can be structured and positioned from each other in a specific manner as unique points in the solution space. After that, a ***variation operator*** can be defined that describes the neighbourhood structure of those points that leads to the construction of the search space. In this section, we discuss the more practical aspects of the variation operator, that is, in the context of the mechanism to transform one distinct candidate solution to another. Given that the design choice of any variation operator is tied to the solution representation that one has considered for use, we will focus on common examples that have been used for real-valued and discrete representations, specifically for ANNs and GP trees we have introduced earlier.

For a conceptual introduction, we first describe the two forms of variation operators that are used for the ***binary string*** representation used in the canonical GA. They are mutation and recombination (or crossover) operators [40, 66]. We start the simple ***mutation*** operator by way of an example. Consider a fixed-length binary string of six bits given by the sequence $(b_5, \ldots, b_0)$, $b_i \in \{0, 1\}$. Note that we reverse the addressing index i so as to have b_5 being the most significant bit and b_0 being the least significant bit. Here, a simple uniform mutation can be implemented as a ***bit-flip mutation*** operator, with each bit b_i having the same probability that its current value is changed. In a canonical GA, mutation is considered as a secondary variation operator and that there should only be a small chance that a bit-flip occurs. In this example, a value is drawn from a uniform distribution such as $r \sim U(0, 1)$

for every bit b_i that then has its current value changed whenever $r < p_m$. Here, $p_m \in [0, 1]$ is the parameter specifying the probability of mutation since bit-flip occurs with probability $\mathbb{P}(r < p_m)$.

Crossover operators are the main variation operators in a canonical GA [40, 66]. As with the mutation operator, there is a parameter $p_c \in [0, 1]$ that specifies the probability that crossover occurs. The value for p_c is usually taken to be significantly higher than that of p_m. It is not unusual in actual applications to have p_c set at values at least an order of a magnitude higher than p_m (e.g. $p_c = 0.5, \ p_m = 0.01$) as an indication that the task for generating new candidate solutions is taken up by crossovers. Furthermore, in a crossover operation typically two parent candidates are randomly selected from the population to undergo a specific process to produce two child candidates. This involves having values stored in specific subsequences (or alleles in GA terminology) of each of the parent candidates being exchanged during the process of generating the pair of child candidates. In the standard k-point crossover, k crossover points are randomly selected to indicate which subsequences are exchanged.

The simplest version is when $k = 1$ to obtain the ***one-point crossover*** operator, with a crossover point being selected with uniform probability. We will illustrate this operation as follows. Consider two parents (x_i^1), (x_i^2) and two children (y_i^1), (y_i^2) with the index $i = 5, \ldots, 0$ and their bits taking values from $\{0, 1\}$. The one-point crossover with a crossover point selected at bit $\{j\}$ is represented by the symbol $\otimes_{\{j\}}^{(1)}$. In the following, we use a prefix notation to express the operation. We drop the index term i for the bits and rely on natural ordering in a tuple notation (reversed in our case) to remove indexing clutter and increase clarity how the exchange occurs. Then, we have as an example where $j = 3$ that $\otimes_{\{3\}}^{(1)}$ will generate two children as follows:

$$
\begin{aligned}
\{(y_i^1), (y_i^2)\} &= \otimes_{\{3\}}^{(1)} \{(x^1, x^1, x^1, x^1, x^1, x^1), (x^2, x^2, x^2, x^2, x^2, x^2)\} \\
&= \{(x^1, x^1, x^1, x^2, x^2, x^2), (x^2, x^2, x^2, x^1, x^1, x^1)\}.
\end{aligned}
$$

A specific instance of a two-point crossover $\otimes_{\{4,2\}}^{(2)}$ having crossover points at bit locations $\{4, 2\}$ will generate

$$
\begin{aligned}
\{(y_i^1), (y_i^2)\} &= \otimes_{\{4,2\}}^{(2)} \{(x^1, x^1, x^1, x^1, x^1, x^1), (x^2, x^2, x^2, x^2, x^2, x^2)\} \\
&= \{(x^1, x^1, x^2, x^2, x^1, x^1), (x^2, x^2, x^1, x^1, x^2, x^2)\}.
\end{aligned}
$$

There are various other forms of crossovers [40, 66]. In particular, some of these variants are designed to be used with binary encoding representation of solutions for specific discrete structure problems such as TSP incorporating repair mechanisms to generate valid candidates. However, even for problems such as those involving continuous function optimizations of higher dimensionality when optimal solutions are known to take on integer values, standard crossover and mutation operators applied to binary string representation of solutions might not suffice. Encodings together with the associated variation operators may introduce local optima in the new search space due to changes in the neighbourhood structure where none is present in the original solution space formed by real-valued representations [43, 113].

In situations where one needs to work on continuous problem domains such as those that are involved in parameter optimizations of ANNs, there are EAs that use real-valued mutation and recombination operators. Historically, ES and EP are the two main branches of Evolutionary Computation that emphasize adaptation at the individual (phenotypic) level. They provide mutation operators that typically take a single parent candidate to generate a single child candidate. These operators are modelled as a perturbation process. Let $\boldsymbol{x}, \boldsymbol{x}' \in \mathbb{R}^n$ be the solution vectors for the parent and child candidates having n real-valued design parameters. Consider a perturbation with the source described by a unimodal multivariate distribution. A simple mutation operator using an ***isotropic perturbation*** $\boldsymbol{z}$ is given as follows:

$$\boldsymbol{x}' = \boldsymbol{x} + \boldsymbol{z},$$

$$\boldsymbol{z} \sim \sigma \boldsymbol{N}(\boldsymbol{0}, \mathbf{I}).$$

Here, $\boldsymbol{z}$ is a multivariate random variable having value drawn from the spherical Gaussian distribution $\sigma \boldsymbol{N}(\boldsymbol{0}, \mathbf{I})$ that is scaled to the standard deviation σ. It has the form of a standard normal random vector. Each component is an independent and identically distributed (iid) random variable, with value drawn from the same Gaussian distribution $z_i \sim N(0, \sigma)$ for all $i = 1, \ldots, n$ [17].

The ***Gaussian mutation*** operator above can be expressed as follows, taking note for design parameters indexed by $i = 1, \ldots, n$

$$x_i' = x_i + \sigma N_i(0, 1).$$

Since the z_is are iid random variables, one can just sample anew from the standard normal distribution $N(0, 1)$ for each i. One can move away from using a spherical distribution, and consider those with their surfaces of constant density being elliptical. In this case, the z_is are independent but not identical random variables. Each z_i is still drawn from a Gaussian distribution but with different σ_i, resulting in

$$x_i' = x_i + \sigma_i N_i(0, 1).$$

For example, this allows one to increase σ_i so as to generate larger perturbations better suited for some design parameters x_i. A more general version of this mutation operator considers ***multivariate Gaussian distributions with correlations*** in the random variables $\boldsymbol{N}(\boldsymbol{0}, \boldsymbol{\Sigma})$, which is expressed by a non-diagonal, positive-definite covariance matrix $\boldsymbol{\Sigma}$. This has the effect of rotating the coordinate system describing the multivariate Gaussian, which allows the mutation operator to introduce more directed perturbation in the solution space [17].

We start with a more formal description above so as to clearly describe the mechanistic details and design approaches of some early mutation operators that are used for high-dimensional continuous optimization problems. This would be less of an issue if the problem is simply one dimensional. In particular, the mutation operators we have introduced so far require a user-imposed perturbation model (e.g. a multivariate Gaussian). A more involved mutation operators used in ES and EP implements ***self-adaptation*** on the parameter σ_i so that the shape of the distribution is evolved as well to some extent. For design parameters indexed as $i = 1, \ldots, n$, this self-adaptive mutation operator is described as follows:

$$\sigma_i' = \sigma_i \exp\left(\tau' N(0,1) + \tau N_i(0,1)\right),$$
$$x_i' = x_i + \sigma_i' N_i(0,1),$$

where the fixed parameters are given as $\tau' = 1/\sqrt{2\sqrt{n}}$ and $\tau = 1/\sqrt{2n}$. This self-adaptive mutation operator has been applied in evolving ANNs [28, 32, 33].

In addition, the perturbation-based mutation operator can be extended in other aspects. As one example, there are those designed to provide more frequent, large step size mutations via the use of heavy-tailed distributions such as the Cauchy distribution, as well as the more generalized family of Lévy distribution [97, 133]. Note that self-adaptation can be incorporated to these heavy-tailed perturbation-based mutation operators. There are yet other mutation operators that are not based on perturbations modelled around the use of multivariate random distributions. For example, in ***Differential Evolution (DE)*** [39, 120], mutation operators have as the source a scaled difference vector obtained from two randomly selected candidates from the parent population. As such, a typical implementation requires three different parent candidates $\boldsymbol{x}$, $\boldsymbol{y}$, $\boldsymbol{z}$, with $\boldsymbol{x}$ being the immediate parent to the generated child $\boldsymbol{x}'$, and two other parents $\boldsymbol{y}$ and $\boldsymbol{z}$ that provide the difference vector. The standard DE mutation operator can be described as follows:

$$x_i' = x_i + \upsilon(y_i - z_i),$$

with $\upsilon \in [0, 2]$.

Recombination operators in EAs for continuous optimization problems take a secondary role in comparison to mutation operators in generating new candidate solutions. The design principles of real-valued recombination operators can be similar to those underlying crossover operators in GA. Here, the recombination operators act on the real-valued vector representation of the solutions. This can be encoded in the form of a sequence of reals after one has fixed the coordinate system by labelling the axis and associating them with specific positions in the sequence. In this case, one has $\boldsymbol{x} \in \mathbb{R}^n$ encoded as $(x_i \in \mathbb{R} : i = 1, \ldots, n)$.

Two types of recombination operators can be designed. They generate a child candidate $\boldsymbol{x}$ from a set of randomly selected donor parent candidates $\{\boldsymbol{y}^1, \ldots, \boldsymbol{y}^m\}$. The first type is called ***discrete recombination***. For each of the design parameter $i = 1, \ldots, n$, the mechanism involves copying components associated with the parents of the donor set as follows:

$$x_i = y_i^j,$$
$$j \sim U\{1, \dots, m\},$$

where j is drawn from the discrete uniform distribution $U\{1, \dots, m\}$, with each value in $\{1, \dots, m\}$ having equal probability of being chosen. The second type is called ***intermediate recombination***. The mechanism simply involves taking the centroid of the solution vectors representing the parent donor set. For each of the design parameter $i = 1, \dots, n$, this can be expressed as

$$x_i = \frac{1}{m} \sum_{j=1}^{m} y_i^j.$$

We will close this section with a brief discussion on standard variation operators that are used in GP. In particular, we will focus on recombination and mutation operators for GP tree representations. The mechanism used in GP crossover operators is similar to those in GAs on binary string representations, but with constraints imposed on the design of the operators. These constraints are required to address two common issues found in evolving GP trees. The first being runaway growth in the structures leading to bloated trees that are behaviourally similar to more compact ones, e.g. having similar fitness values. The second and more critical issue has to do with the generation of non-legal trees. This occurs when the problem domain to be solved requires the use of a primitive set having functions with multiple types. As an example, this could involve functions of $\mathbb{R} \times \mathbb{R} \rightarrow \mathbb{R}$ (e.g. arithmetic operations such as addition $x + y = z$) and those of $\mathbb{R} \times \mathbb{R} \rightarrow \mathbb{B}$ (e.g. inequalities such as $x \leq y \equiv b,\ b \in \{\text{true, false}\}$). One approach to address this issue of generating non-legal trees is to incorporate data types as part of the descriptions of the elements in the primitive set. Such an extension has been introduced and studied as ***Strongly Typed GP*** (***STGP***) [109]. STGP has been applied successfully to evolve computer programs in [3, 4].

Once these constraints have been identified, then they can be incorporated as part of the design of the variation operators in GP to generate legal trees. One simple recombination operator is the standard subtree crossover [95]. Here, the operator takes as input two parent candidates to generate a single child candidate. A crossover point in each of the two parent GP trees is first randomly selected. The crossover point indicates the subtree that is rooted under it. The next step of the operator involves replacing the identified subtree in the first parent GP tree with the identified subtree taken from the second parent GP tree. Note that the process of random selection of crossover points is designed to be biased in applications. This is because most GP trees have at least an average branching factor of two (e.g. from the use of binary operators and relations). There are fewer crossover points associated with larger subtrees. For example, in a complete binary GP tree of depth two, four out of six crossover points have leaves (terminals) rooted under them. A uniform random selection will more likely choose these leaves rather than the subtree above them, leading to small exchange in genetic materials to generate the child candidate. Other more sophisticated crossover operators typically involve a matching process to ensure subtrees that are exchanged are structurally similar, e.g. as in size-fair crossover operator introduced in [94].

Mutation operators in GP also require a more thorough design with respect to making changes to a tree structure that is legal and can be interpreted. Depending on what is selected in the GP tree to undergo a mutation process, one can first start with the simplest structure (i.e. the leaves) and then proceed to more complex subtrees. If selected leaves only contain constant values in them, then one can use a simple perturbation process to change those values. If an internal node is selected, then a point mutation operator can be applied to that node. This is achieved by changing the element in the node with one from the primitive set that satisfies the arity constraint (i.e. the number of arguments). For example, a node that represents subtraction can be changed by point mutation to addition. Next, one can have the subtree mutation operator that can generate new subtree structures. A mutation point is first selected. After that, a randomly grown subtree is attached to it. Constraints can be imposed on this operator, for example, such as fixing the maximum depth the subtree can be grown [3, 4].

1.2.4 Selection Operators

The main task of the ***selection*** operator in an EA is to form the new population in the next generation. In general, this usually involves selecting from the current full population, a collection of candidate solutions that will become the parents in the next iteration of the evolutionary process. The selection operator uses assigned fitness values of candidates after their evaluations to perform this task. This differentiates it from the variation operator that uses the genetic information encoded within the candidates. This simple description belies the full extent of what the mechanisms in selection operators do to transform the configurations of the population. For example, the group of parent candidates selected for the next generation is typically a ***multiset*** (or ***bag***) with elements drawn from the current population. Individual candidates with higher fitness value are more likely to be selected and also more often in comparison to other candidates. As such, the new parent population might consist of copies of these individual candidates. However, the selection process does not have to be ***elitist***. In this case, a more involved mechanism can occasionally choose a candidate that is inferior to those with higher fitness values so as to increase and maintain some level of diversity in the evolving population.

Despite this, one could develop a more in-depth understanding of selection operators using a simple process that is based only on fitness values. Consider an EA that is applied to solve an optimization problem with a global optimum.

As a population-based search methodology, this suggests the first discovery and subsequent convergence of the population specializing to this optimal solution to describe the phases in a successful search. One would like to ascertain whether the progress towards a successful evolutionary search is related to the setting of the few parameters associated with the selection operators. This is studied in the context of ***selection pressure*** (or selective pressure) [7]. Selection pressure is a property of the operator that characterizes the relationship between fitness values of individual candidates and the likelihood they are sampled for replications to form the new population.

Theoretical studies have been made to investigate the impact of selection pressure on evolutionary search. Early approach for analysis uses notions such as takeover time that quantifies the time it takes for the population to converge to the optimal solution [67]. The modern, well-established runtime analysis of EAs as randomized algorithms provides a rigorous approach to quantify the speed with which the population in EAs ***drift*** towards the optimal region [47]. It has been used to study the impact of selection operators exhibiting different selection pressure on the runtime performance of EAs [30, 45]. In applied and empirical studies, a loose and informal usage is considered whereby selection pressure can be changed by adjusting the setting of parameters or as part of the design of the associated operator to affect the phase with which an evolutionary search undertakes. A strong selection pressure can bias evolutionary search in exploiting regions in the search space described by the high-performing candidates. A weak selection pressure could encourage evolutionary search in exploring other regions associated with lower performing candidates, which potentially might be several evolutionary steps away from new and more promising regions. It is with this context that one might consider among the various selection operators that have been designed for use in EAs. In the following, we will describe several common selection operators.

The first is the ***fitness proportional*** (or roulette wheel) selection that is typically used in GAs. Consider the solution space $\boldsymbol{S}$ (e.g. $\boldsymbol{S} \subseteq \mathbb{R}^n$). This specific operator uses a simple probabilistic approach to determine the replication rates of individuals $\boldsymbol{x}_i$, $i = 1, \ldots, s$, $\boldsymbol{x}_i \in \boldsymbol{S}$ in the current population of size s. The rates are based on the proportion of their individual assigned fitness values $f(\boldsymbol{x}_i)$ with respect to the cumulative fitnesses of the population $\sum_i^s f(\boldsymbol{x}_i)$. The probability that an individual is chosen for replication and forms the population in the next generation is simply

$$\mathbb{P}_i = \frac{f(\boldsymbol{x}_i)}{\sum_i^s f(\boldsymbol{x}_i)},$$

with $\sum_i^s \mathbb{P}_i = 1$. This normalization of fitness values makes them interpretable as probabilities. They define a discrete probability distribution over the set $\{\boldsymbol{x}_1, \ldots, \boldsymbol{x}_s\}$ that one can then draw a random sample to form the parent population in the next generation. Note that fitness values have to be positive reals. The specification for fitness proportional selection is straightforward for maximization problems with fitness functions $f : \boldsymbol{S} \to \mathbb{R}_{\geq 0}$. Other problems such as those in the context of minimization problems or those that have negative fitness values will require reformulation of the fitness function or to use an appropriate scaling function. The latter in general takes the gap between the candidate's original fitness value and that of the worst candidate to be the new assigned fitness value [7].

Next is the ***tournament*** selection operator. The mechanism is straightforward and requires two positive integer parameters: tournament size q and parent population size of the next generation p. The operator will randomly sample from the current population p times (with replacement). The best candidate in each sample will then be copied over to form the parent candidate in the next generation. The study [7] gives a formal combinatorial description of this operator and derives the probabilities $\mathbb{P}_i$, $i = 1, \ldots, s$ associated with each candidate i in the current population of size s being selected for replication. Here, we describe an informal implementation of the operator. One assumes a discrete uniform distribution over the current population and then draws iid random samples from this distribution p times. This produces p samples $(\boldsymbol{x}_k)^j$, $j = 1, \ldots, p$. Each sample $(\boldsymbol{x}_k)^j$ is a collection of candidate solutions, and we consider a permutation $(\boldsymbol{x}_i, \ldots, \boldsymbol{x}_q)$ satisfying $f(\boldsymbol{x}_i) \leq \cdots \leq f(\boldsymbol{x}_q)$, i.e. they are ordered according to increasing fitness values. Let $\boldsymbol{x}_{k,j}$ be the kth element in the jth ordered list taken from $(\boldsymbol{x}_k)^1, \ldots, (\boldsymbol{x}_k)^p$. For a maximization problem example, the parent population in the next generation is obtained by taking $\boldsymbol{y}_j = \boldsymbol{x}_{q,j}$. For a minimization problem, one can then just take $\boldsymbol{y}_j = \boldsymbol{x}_{1,j}$.

There are several ***rank-based*** selection operators that have been designed [71]. In general, these operators use assigned ranks instead of fitness values of the candidates in the population as the main criterion for selection. There are two advantages to using ranks. The first being that assigning ranks to candidates might be preferable in situations whereby it is difficult to obtain a useful objective function or any approximation to act as the surrogate function to determine the quality of the candidates. In this case, ranks can still be obtained via pairwise comparisons of the candidates in the population. The second advantage is that it removes the need to perform any scaling that might arise to construct the discrete probability distribution (e.g. negative fitness values) and other scaling issues that depend on the population configuration (e.g. when the majority of the candidates have similar fitness values, or when the population small groups of candidates with substantial and disproportionate differences in fitness values). One particular version is the ***linear ranking*** selection operator that assigns probabilities to candidates $i = 1, \ldots, s$ in the current population of size s, which are already ordered or ranked in increasing fitness values, in a linear fashion to their ranks. The probabilities are

$$\mathbb{P}_i = \frac{1}{s}\left(\alpha + (\beta - \alpha)\frac{i-1}{s-1}\right).$$

There are two parameters for the operator with $1 \leq \beta \leq 2$ and $\alpha = 2 - \beta$. β and α are associated with the expected number of replicates given to the best and worst candidates in the current population, respectively. Other rank-based selection operators use different functions having geometric and exponential forms to scale the probabilities [71]. Once this is done, the operator can then proceed to draw random samples from the discrete probability distribution that is obtained from the calculated probabilities.

We next describe a family of selection operators based on deterministic mechanisms. In particular, the $(\boldsymbol{\mu},\boldsymbol{\lambda})$, and $(\boldsymbol{\mu} + \boldsymbol{\lambda})$ selection operators are typically used in EP and ES. In both cases, the variation operator is first applied to generate $\lambda = k\mu,\ k \geq 1$ child candidates from μ parent candidates. That is, each parent candidate will generate k child candidates. In (μ, λ) selection, k is typically set a value larger than one so that the best μ out of λ generated child candidates can be chosen to form the parent population in the next generation. In $(\mu+\lambda)$ selection, the plus + symbol is to indicate that combined population of μ parents and λ children candidates is considered as the pool with which replication can operate. Here, the best μ candidates out of $\mu + \lambda$ in this pool are chosen to form the parent population in the next generation.

There are various other selection operators, but we would close this section with a particular operator used in EP that mechanism wise would be a hybrid between the tournament selection described earlier and $(\lambda + \lambda)$ selection (i.e. $k = 1$ so that $\mu = \lambda$). The reference to $(\lambda + \lambda)$ comes about from the variation operator using λ parent candidates to generate λ child candidates, which are both taken as a whole to provide the full population for selection. The study [7] has given a technical argument as to the nature of this selection operator being a probabilistic $(\lambda + \lambda)$. We give an informal and intuitive account here by taking some simplifying assumptions. Consider a maximization problem and that the combined population $\boldsymbol{x}_1, \ldots, \boldsymbol{x}_{2\lambda}$ is naturally ordered according to increasing and distinct fitness values of the candidates. However, the operator uses ranks obtained from an ordering based on assigned tournament scores of the candidates. These scores are calculated from pairwise comparisons of the candidate $\boldsymbol{x}_i$ against a sample of other candidates in the combined population. If the full population is used, then the ranks preserve the natural ordering. Otherwise, the use of a smaller sample from the population to calculate the tournament scores permutes this natural ordering to realize the tournament ranks.

In the following, we describe the mechanistic aspects of the operator. Associated to each candidate $\boldsymbol{x}_i,\ i = 1, \ldots, 2\lambda$ is a random sample of q candidates $T_q^i = \{\boldsymbol{y}_j : j = 1, \ldots, q\}$ drawn with a discrete uniform probability distribution over the set of the combined population. That is, $\boldsymbol{y}_j \sim U\{\boldsymbol{x}_1, \ldots, \boldsymbol{x}_{2\lambda}\}$ and that $\boldsymbol{y}_1, \ldots, \boldsymbol{y}_q$ are iid random variables each taking value from $\{\boldsymbol{x}_1, \ldots, \boldsymbol{x}_{2\lambda}\}$. Then, the tournament score of each $\boldsymbol{x}_i,\ i = 1, \ldots, 2\lambda$ can be calculated as follows:

$$g\left(\boldsymbol{x}_i, T_q^i\right) = \sum_{j \in T_q^i} \mathbf{1}_{\mathbb{R}_{\geq 0}}\left(f(\boldsymbol{y}_j) - f(\boldsymbol{x}_i)\right),$$

$$\mathbf{1}_{\mathbb{R}_{\geq 0}}\left(f(\boldsymbol{y}_j) - f(\boldsymbol{x}_i)\right) = \begin{cases} 1 & \text{if } f(\boldsymbol{x}_i) \geq f(\boldsymbol{y}_j), \\ 0 & \text{otherwise.} \end{cases}$$

Note that the tournament score for each $\boldsymbol{x}_i$ can be interpreted as the number of the wins the candidate has in a two-player game against each opponent in the tournament sample $\boldsymbol{y}_j \in T_q^i$. The indicator function $\mathbf{1}_{\mathbb{R}_{\geq 0}}$ essentially realizes the game and gives its output for the first player $\boldsymbol{x}_i$. The game involves a pairwise comparison of both players' fitness values, with $f(\boldsymbol{x}_i) \geq f(\boldsymbol{y}_j)$ indicating $\boldsymbol{x}_i$ wins, and obtains a score of one, otherwise $\boldsymbol{x}_i$ loses and obtains a score of zero. Tournament selection can then be performed by taking the best λ candidates with highest tournament scores $g(\boldsymbol{x}_i, T_q^i)$.

1.3 Parameter Control in Evolutionary Algorithms

Earlier at the start of Sect. 1.2, we have detailed how EAs can be described through a framework of population-based, stochastic search approaches to problem solving. Subsequently in individual sections from 1.2.1 to 1.2.4, we have described the main components required in implementing EAs: the population, the use of solution representation, and both the evolutionary operators of variation and selection. One could use the standard setup such as the canonical forms of GAs, EP, and ES, or even subsequent modifications that are designed to improve them in general or customized to address specific problem structures (e.g. GPs and their symbolic evolutionary approach to evolving programs). It should be noted that an EA has a number of parameters that must be initialized for its operations. They are associated with the main components in an EA (i.e. Algorithm 1.1) such as population size and those within the evolutionary operators that alter variation strength and selection pressure, thus influencing search performance. What this means is that there is a significantly larger degree of customizations in EAs beyond just deciding in advance for each components, which particular settings, design of representation, and operators to use.

We first distinguish two main aspects with which parameters associated with EA operations can be altered following the taxonomy put forth in [54] that is applied within the context of EAs. There are other taxonomies such as the one proposed in [83] that applies to the wider contexts of metaheuristic methods. First, ***parameter control*** refers to specific changes made on those parameters during the operation of an EA. The general argument put forth is that in different phases of an EA search process, different parameter values are required for more effective search. Second, ***parameter tuning*** refers to the setting of these parameters of EAs statically before their execution. In general, the performance of an EA with respect to a target problem depends on the parameter values the EA are initialized with. This correspondence necessitates preliminary attempts to set up an EA instead of arbitrary parameter initializations.

In [54], there are three broad categories of parameter control approaches in EAs. A ***deterministic*** approach typically uses some form of fixed schedules to change these parameters. The other two, ***adaptive*** and ***self-adaptive***, endow EAs with mechanisms for changing those parameters, typically via feedback based on measurements on the search process. The self-adaptive approach can be thought as a further refinement of the two. The parameters of an EA will be represented in a suitable form for which mechanisms that change them are incorporated into the evolutionary process. Note that this classification of parameter control have been further refined and extended in [46] for which there are five categories.

The main difference between these two categories has to do with when parameters are allowed to change. In parameter tuning this is done in an offline fashion, while in parameter control the changes to parameters occur online during executing of EAs. Despite this, commonalities in the general design of their implementations as well as challenges to be addressed can be found. For the remainder of this section, we further discuss in separate subsections how parameter tuning and control are applied on EAs through their population, variation, and selection operators. Our discussion summarizes and reflects on studies that have been surveyed earlier in [84], with developments that are detailed in new studies such as in [46].

1.3.1 Parameter Control on Population Size

The main issue of finding the appropriate population size for a given problem is motivated towards developing an understanding on the effects of population size in successful EA search that can be informative to developing procedures for population size control. A number of early theoretical studies have employed the ***Building Block*** model [69] with its simplifying assumptions. In the context of discrete optimizations, the nature of the problem structures is such that it is assumed to be additively decomposable to individual partitions. In each partition, there are various subsolutions or configurations, one of which is superior to others and crucially belongs to the global optimum of the full solution. The analysis will focus on how GA assembles all the building blocks to form the right solution. Some studies have formulated this process as a Markov chain [100], specifically a one-dimensional random walk [103]. One can study such a model and interpret the effects of population size towards either making a successful search more likely than a failed one.

As is common in the historical and modern development of Evolutionary Computation, theoretical studies are supported and complemented by extensive, systematic empirical studies. The strong assumptions on problem structures or EA designs required to make mathematical analysis tractable can be relaxed in these studies. There have been three main issues identified in the literature on the parameter tuning and control of population size [84]. One involves removing the need for setting population size with most studies investigated control mechanisms that simulate lifetime in individuals that then act to remove them from the population after initialization of sufficient size [2, 35]. Another issue involves approximating the required population size for the search, and most mechanisms focus on increasing population size in a controlled manner, e.g. doubling them through greedy [118] and stochastic [72, 73] approaches. Lastly, various studies have investigated the effectiveness of variable population size during the evolutionary process under deterministic [90], adaptive [116], and self-adaptive [122] control.

1.3.2 Parameter Control on Variation Operators

As noted in [84], the tuning and control of parameters associated with the variation operator have been investigated extensively in comparison to other components in EAs. This is not surprising given that the main role of the variation operator to generate new candidate solutions for evaluation, which are then passed on to the selection operator that decides those that will form the population in the next generation for replication. In this case, the various motivations for these studies have focused on addressing two main related issues. The first issue has to do with maintaining an effective balance between encouraging global exploration and local exploitation in search when it is required. The second related issue has to do with the effectiveness of the variation operator, for example, in having the appropriate step sizes for the EA to better search the space [29]. Our following discussion will focus on two main design mechanisms with which parameter control can be applied on variation operators in EAs.

Modern runtime analysis of EAs has made it possible to study in-depth both static parameter settings (e.g. mutation rates) [78] and those that involve dynamic parameter settings in variation operators of EAs [46]. Despite the constraint of having to work with simpler EA implementations that are correspondingly matched to well-designed benchmark problems to ensure that the mathematical analysis becomes tractable, theoretical studies have reported positive results on the use of parameter tuning and control in variation operators. Crucially, they have provided further insights why parameter tuning and control do work. For example, the early study in [51] has demonstrated a deterministic approach that uses a simple scheduling scheme to double the mutation rate in every iteration of the classic $(1+1)$EA in discrete optimization problem is superior compared to the same EA but with fixed mutation rate for problems exhibiting certain structures. Informally, such a problem possesses the global optimum that is surrounded by local optima acting as traps because they can be searched easily. Once a local optimum is reached, large jumps by direct mutation are required to reach the global optimum. In this situation, the $(1 + 1)$ EA has a better chance of reaching the global optimum if it is allowed to increase the mutation rates of its bit-flip operator. However, the choice of using a fixed mutation rate might be better for other problems with different characteristics.

These strict but explainable findings are reflected in empirical studies that have investigated other more involved mechanisms for parameter control of variations operators in EAs. One methodology involves mechanisms that change the parameter settings of the variation operators as a result of some measurable state changes in the EA process. They are called state-dependent parameter control mechanisms in [46]. One of these types is referred to as diversity maintenance techniques. An early study [119] has used such a mechanism that changes the crossover and mutation rates based on a particular measurement indicating the level of convergence in the population. This measure takes the difference in the fitness values between the best candidate and the population average $f_{\max} - f_{\mathrm{avg}}$. A simple scaling function that takes the inverse relationship between the parameters for the variations operators and the convergence level is used by the operator to change and set crossover and mutation rates. However, additional checks are included whereby high crossover and mutation rates are applied to low fitness candidates when the population has converged to configurations indicating low diversity (e.g. given by the low $f_{\max} - f_{\mathrm{avg}}$ values).

Mechanisms we have just described above require user design schedules, functions, and any other operation to change the parameters. Their effectiveness would depend on some prior knowledge about the problem structures. An alternative to this approach would be to use mechanisms that employ an adaptive control approach. This approach associates parameters in the variation operator to candidates in the population. Their changes are done typically through some perturbation process, but they are selected only when the associated candidates are selected for replication in the subsequent generation. As an example, the study [135] has introduced such adaptive control for DE. In this case, changes to the crossover rate and mutation factor in DE variation operators are done through perturbations. For example, a Cauchy-distribution-based perturbation is used to change the mutation factor. Note that because the mechanisms to change the parameters in the variation operator and the parameters in the solution representation differ, it is considered as an adaptive control approach. When the evolutionary process is used directly to make the change to both sets of parameters, then it is said to use a self-adaptive parameter control mechanism. We have introduced a particular self-adaptive parameter control in variation operators for real-valued mutation operators based on independent perturbations via symmetric univariate probability distributions independently for each component of the vector representation of the solution in Sect. 1.2.3.

The last approach we will discuss employs more sophisticated mechanisms whereby the changes in the variation operators occur not just at the level of parameters but also that of their symbolic operational descriptions. The main operational issue underlying the design of these mechanisms is the allocation of limited and fixed computational resources to specific variation operators during the evolutionary process that maximize gain of the EA by way of subsequently generating child candidates with higher fitness values. There are two main methodologies that are used in the design and implementation of such mechanisms. The first is based from those in Machine Learning that has formalized this allocation process in the ***multi-armed bandit*** framework. The second is based on those in metaheuristics that has introduced the notion of ***hyper-heuristics***, whereby one has in addition to selection from an existing set of operators, a generator that produces more complex operators from basic versions.

An early study in [107] has provided a detailed review and conducted empirical investigations on such mechanisms applied to variation operators in EA. The study has focused on discrete optimization problems, specifically the classic ***Boolean Satisfiability Problem*** (commonly abbreviated as ***SAT***) that involves finding the appropriate assignments into the variables that satisfy a propositional logic expression (i.e. the expression evaluates to true). The solution can be represented in the standard ***Conjunctive Normal Form*** (***CNF***). This consists of a conjunction of clauses, with clauses themselves being in the form of a disjunction of the literals or variables (i.e. they can be assigned a value from the Boolean set $\mathbb{B} = \{\text{true, false}\}$). For example, the 2SAT problem restricts the propositional expression in CNF to two variables per clause with the requirement that one of the two variables must be true. A candidate solution is then the combination of Boolean values to be assigned into variables that then solve or satisfy the given propositional expression. Unlike the case of exact methods, EAs constitute an approximate approach that uses a surrogate function (e.g. minimize the number of false clauses) to guide the evolutionary search for the solutions.

In [107], a sophisticated operator management system is introduced and incorporated into the EA, which is capable of both selection and generation of crossover operators. In principle, this requires the construction of the space of crossover operators. This is achieved after identifying four basic features, each consisting of several choices of actions that can be made to effect changes in the parent candidates representing instances of possible solutions in a CNF form. By combining an action for each of the four features, a crossover operator can be generated. In effect, a search in this space of crossover operators can be made in an online fashion, after which the mechanism will pass the operator that is deemed best suited for the particular phase of evolutionary search for use. This operator assessment involves three measurements indicative of the population diversity, solution quality, and execution time. Not all adaptive operator selection and hyper-heuristic approaches to parameter control in variation operators are as complex as what is developed in [107]. For example, due to the inherent technical challenges involved, one of the few theoretical studies for this type of parameter control [48] has investigated a simple bit-flip mutation operator that can vary its mutation strength via a single parameter.

In general, a portfolio or a set of k parameter values is constructed with which a mechanism can select a value for use in the mutation operator. The basic approach to selection of these parameter values involves ***probability matching*** [46, 107]. Here, each $i = 1, \ldots, k$ parameter value is associated with a confidence value c_i that indicates the suitability of the parameter value during the current phase of the EA search at time step t. The approach assigns the probability $\mathbb{P}_i$ of selecting the ith parameter proportionally to the associated confidence value c_i as follows:

$$\mathbb{P}_i(t) = \mathbb{P}_{\min} + (1 - k\mathbb{P}_{\min}) \frac{c_i(t)}{\sum_{j=1}^{k} c_j(t)},$$

$$c_i(t+1) = (1-\alpha)c_i(t) + \alpha r(t).$$

The confidence value is weight updated with the normalized reward or gain $r(t)$ to the EA as a result of the application of the mutation operator with ith parameter value setting. The hyper-parameter $\alpha \in (0, 1)$ is used to control the adaptation rate of the confidence value. Note that each ith parameter value has at least a minimum probability $\mathbb{P}_{\min}$ of being selected.

There are other more sophisticated approaches to probability matching that consider a more involved formulation to the selection of the parameter values. In adaptive pursuit, the update rule increases the confidence value of the selected parameter value while simultaneously decreases those associated with parameter values not chosen during an iteration of the EA search. In the upper confidence bound algorithm, selection is based on choosing the parameter value that maximizes a weighted sum of two terms. The first term involves the expected reward over a fixed distribution associated to the parameter value. It models the exploitative search of the mechanism for the parameter value that brings about the largest fitness gain. The second term involves a function whose output is inversely proportional to the number of times a parameter value has been applied in the past but proportional to the logarithm of the iteration in the evolutionary process. It models the exploratory search of the mechanism and acts to balance and select parameter values applied less often in the early stages of EAs [46].

1.3.3 Parameter Control on Selection Operators

The mechanisms of parameter control on selection operators seek to change their selection pressure by way, for example, of making specific changes to the operators' parameter settings. As before with other components in EAs, one basic approach involves deterministic parameter control that is user designed to make scheduled changes to selection pressure. One mechanism that has been studied and used is based on the approach employed in the stochastic optimizer known as ***Simulated Annealing*** (***SA***). The original study [88] has made a deep link between the approach that SA uses to solve of the TSP problem with that of the annealing process in physical systems. Informally, the latter employs a statistical description of a many-body thermodynamic physical system. There is a probability associated with a particular state configuration i of the ensemble of particles that is described by the ***Boltzmann distribution***. The probability $\mathbb{P}_i$ is proportionate to $\exp(-E_i/kT)$, whereby E_i is the energy of the state configuration i, k is the Boltzmann constant, and T is the temperature of the system. The physical annealing process to obtain the ground state of a thermodynamic system via a cooling schedule is exploited to design a stochastic search process for optimization. In addition, the generate-and-test search process involves probabilistic acceptance. In a minimization setting, the generated candidate is accepted if it has an improve fitness difference given by $\Delta f \leq 0$. Otherwise, a lower quality generated candidate has a probability proportional to $\exp(-\Delta f/kT)$ to be accepted.

In [44], the study has designed a Boltzmann selection operator that modifies the standard fitness proportional selection operator. This modification involves applying a scaled exponential transformation to the original, assigned fitness values of the candidates in the population. The function describing this transformation has the form of the Boltzmann distribution. The probability for selection the candidate $i = 1, \ldots, s$ in the population of s candidates is given by

$$\mathbb{P}_i = \frac{\exp\big(f(\boldsymbol{x}_i)/T\big)}{\sum_i^s \exp\big(f(\boldsymbol{x}_i)/T\big)},$$

for a maximization problem. For a minimization problem, $-\exp\big(f(\boldsymbol{x}_i)/T\big)$ transformation is used instead. Note that if the identity function is used instead, the Boltzmann selection operator reduces to that of the fitness proportional selection. As such, there is now a parameter T that can control the selection pressure of this operator. Selection pressure changes between two extremes of the parameter settings of T. At low values, the transformation amplifies the differences in the original fitness values, thus increasing the chance for replication to high fitness candidates. At the other end of high values, the transformation reduces any difference in the original fitness values, thus acting to more evenly allocate replication opportunities. As such, a linear cooling schedule is designed in [44] to promote a more exploratory search of the EA at the initial phase before then switching it over to more exploitative search as evolutionary search progresses.

Next, the study in [108] employs an adaptive, state-dependent parameter control approach on the tournament selection operator. In particular, a diversity maintenance mechanism is used to control the size of the tournament samples as the means to control selection pressure during different phases of the evolutionary search. A specific diversity measurement called Healthy Population Diversity (HPD) is introduced. The tournament size is then varied and made proportional to the HPD value. The HPD measure takes the form of a fitness-value-weighted Euclidean distance of each candidate against the population average. In [55], a self-adaptive approach is used that can change the selection pressure in the tournament selection operator. Each candidate is endowed with an additional parameter that is subjected to the usual adaptive change as a result of the evolutionary process. The parameter setting for the tournament size is changed based on the aggregation of these values contributed by the evolving candidates.

1.4 Concluding Remarks and Notes for Further Reading

In our presentation and discussion of the main aspects of EAs, we have emphasized this broad family of population-based, stochastic, iterative search approaches as a general problem solver. In particular, EAs can be understood through the generate-and-test framework as a computational system consisting of three main components. Exhaustive empirical and rigorous theoretical studies have been made to develop different mechanisms underlying the operations of these components as well as to investigate the impact of parameter settings on the performance of EAs. As such, an EA can be suitably designed with matching components that are targeted towards addressing the underlying problem structures and properties. One could refer to the existing literature of the family of EAs that have demonstrated effectiveness in solving a class of similar problems and then make the changes based on suggestions provided in the literature or one own's preliminary studies that might bring about increased performance.

The ease with which one can develop an EA to solve a new problem belies the underlying complexity of its search behaviour. It is perhaps understandable that for EAs to gain greater acceptance as one of the main heuristic approaches and primary alternative to exact methods to problem solving, that the behaviour of the evolutionary search process is well understood. However, the complexity in the design of the mechanisms used in components of real-world EAs makes this task challenging. As one example, consider the notion of ***fitness landscape***, which is a common tool that is used for both conceptual understanding and analysis of optimization problems in which EAs are used to solve. It is formally defined as the triple $\boldsymbol{L} = (\boldsymbol{S}, f, d)$ consisting of the search space $\boldsymbol{S}$, the fitness function $f : \boldsymbol{S} \to \mathbb{R}$, and the distance measure $d : \boldsymbol{S} \times \boldsymbol{S} \to \mathbb{R}_{\geq 0} \cup \{+\infty\}$. The fitness landscape for continuous optimization problems is straightforward, being mostly described by the fitness function (e.g. $f : \mathbb{R}^n \to \mathbb{R}$ for an unconstrained optimization problem). However, note that even for the more straightforward case of a single-objective optimization problem, one can develop functions having various problem characteristics [76], which is to be collectively used as a benchmark set to assess the performance of EAs [79].

The case for discrete optimization problems or even in situations involving encoding of solutions in the continuous domain can be significantly more challenging. One can refer to [43] for a more detailed discussion. We will only give a brief discussion here. Informally, the full $(\boldsymbol{S}, f, d)$ is required because the discrete search space $\boldsymbol{S}$ differs depending on the topological relationship of points $\boldsymbol{s}$ within it. One can gain a better understanding by representing the search space $\boldsymbol{S}$ as a graph $G = (V, E)$, with the set of vertices V representing all the candidate solutions, and the set of connectivities $E \subseteq V \times V$ representing transformations by the variation operator on the candidates. As such, the topological relationship of $\boldsymbol{s} \in \boldsymbol{S}$ is described by its neighbourhood structure, which depends on the action of the variation operator. For example, a binary string encoded representation together with a one-bit mutation will induce a hypercube structure on the search space. When a multi-bit mutation of the full length of the binary string is used, a complete graph structure is induced on the search space. In this context, the structures of the problem can change (e.g. introducing new local optima) when different solution representations and encodings are used because of the changes in the neighbourhood structures [43].

As such, developing a deep understanding through rigorous analysis of the search behaviours of real-world EAs can be very challenging. This also applies to theoretical studies where significant efforts have been made to construct more abstract versions of real-world EAs as well as benchmark functions with detailed problem structures. For example, one can refer to the recent text [47] that has provided a comprehensive overview and summary of 20 years of research on modern runtime analysis of EAs on discrete optimization problems to gain an appreciation of the technical challenges involved. Here, EAs are modelled as randomized algorithms and the task of their runtime analysis involves three steps [98]. One starts with identifying a suitable distance measure X_t on the progress of the EA after iteration or time step t. The next step involves modelling and developing an understanding of the nature of the one-step change of the measure as the random variable $X_t - X_{t+1}$ for any value X_t takes. This knowledge is then exploited to characterize the runtime T of the EA, which is the number of iterations until EA solves the problem. As an example of the challenges involved even for well-defined search behaviour of simple EAs on minimization problems, consider the case whereby the process can be constrained to the form of a Markov chain $(X_t \in \boldsymbol{S} : t \in \mathbb{N}_0)$ with X_t taking values from the set $\boldsymbol{S} = \{0\} \cup [1, \infty)$ whereby $\mathbb{N}_0 = \{0\} \cup \mathbb{N}$. Even in such a construction, it can be nontrivial to derive statements about the expected first hitting time $\mathbb{E}[T]$ (i.e. given the random variable T, the earliest time $t \geq 0$ it takes with which the EA solves the minimization problem having the fitness measure $X_t = 0$) from knowledge about the drift $\mathbb{E}[X_t - X_{t+1} \mid X_t = s]$ for all $s \in \boldsymbol{S}$ [99].

We will close this chapter with a brief remark on the conceptual differences and similarities between a modern evolutionary optimizer against the classical numerical optimizer. Going back to one might argue the simpler single-objective optimization problems, there is still significant sophistication in the evolutionary search process of modern real-world EAs. Consider the problem setting of an unconstrained minimization of $f : \mathbb{R}^n \rightarrow \mathbb{R}$ that is convex and twice differentiable. In particular, we focus on the family of convex quadratic minimization problem (e.g. see Chapter 9 in [21]). There are various of such quadratic minimization problems that are used in most standard benchmarks problems [76], such as the simple sphere problem formulated as $f(\boldsymbol{x}) = \|\boldsymbol{x}\|^2$ with the global solution at the origin $\boldsymbol{x}_{\text{opt}} = \boldsymbol{0}$.

In this case, one could use a ***gradient descent method*** [21] to solve this problem since the negative gradient can be derived. For the simple sphere problem and discounting the computational effort required for using an exact line search, one would then observe this method descending along the direction from the current point towards the origin. However, EAs as search methods that are ***derivative-free*** treat the problem as ***Blackbox Optimization*** (i.e. knowledge about the objective function is limited to only making call or query to evaluate the fitness of the candidate) [6] and make no effort to model the problem structure that would exhibit different search behaviours. Although convex structure in the sphere problem ensures that an elitist $(1+1)$EA descents the fitness landscape (i.e. since a generated child candidate $\boldsymbol{x}'$ is only accepted whenever $f(\boldsymbol{x}') < f(\boldsymbol{x})$), the search pattern of the EA could appear in a zig-zag-like manner instead of a direct line from the initial point to the origin.

However, the search behaviour of some modern, larger than population-one EAs could qualitatively appear more like those exhibited by descent-based methods. In general, this involves mechanisms that incorporate information about the problem structures to guide the evolutionary search process of the entire population as a whole. Recall that EAs such as ES and EP use real-valued, perturbation-based variation operators to generate a child candidate directly from a parent candidate. There are those such as the ***Covariance Matrix Adaptation-ES*** (***CMA-ES***) that directly generates child candidates by sampling from a multivariate Gaussian distribution defined over the search space $\mathbb{R}^n$ [74]. A sophisticated mechanism is designed to adapt this multivariate distribution by updating the covariance matrix $\mathbf{\Sigma}$ through information obtained from the evolutionary path of the search process. Each iterate in this evolutionary path is a point in $\mathbb{R}^n$ and is described by the weighted mean of the best μ out of λ candidates of the evolving population in the current generation. As such, the child candidates are simply samples drawn from the adapted multivariate Gaussian distribution centred at the population mean in each generation.

Despite the more mechanistic construction of the evolutionary search procedure in CMA-ES, it provides the conceptual link between the family of ES and methods that employ gradient estimations in retrospect. This link is made closer with the development of ***Natural ES*** (***NES***) [130]. The main mechanistic working principle of CMA-ES to extract information from the fitness landscape of the problem using the population as it changes under influence of the evolutionary process that favours those having higher fitness values is formalized in NES [102]. In principle, this requires reformulating the original minimization problem to one that is about minimizing the expectation $\min_\theta J(\theta) = \mathbb{E}_\theta[f(\boldsymbol{x})]$ over the space of the parametric search distributions, e.g. typically multivariate Gaussians with parameters θ. The search behaviour by the evolving population on the fitness landscape that one can observe is governed by an underlying process that operates as a search on the parameter space of the multivariate Gaussian distributions. The latter requires as a guide ***search gradients*** to update the multivariate Gaussian distribution. Instead of using sampled gradients of the expected fitness by the evolving population as mechanistically done in CMA-ES, it is possible to formally construct the parameter space of the multivariate Gaussian distributions and define an optimization process here. The optimization over the parameter space is achieved by way of using the ***natural gradient***, which can be derived to give the steepest search direction. This is conceptually illustrated as Figure 2 in [130] to highlight the differences between sampled fitness gradients and the natural gradient.

Before closing this chapter, we revisit our earlier introduction of Turing's works on Turing Machines and unorganized machines as general precursors of Evolutionary and Neural Computation. It would be many years later after their introductions that a series of theoretical studies (e.g. [24, 52, 53]) combine all these computational concepts and associated approaches to deeply investigate the paradigm of Evolutionary Computation. In particular, ***Evolutionary Turing Machine*** is introduced as a formal computational model to study EA. This allows as one example the means with which one can establish the expressiveness of EAs by answering the general question whether EAs can perform better than ***algorithms***, i.e. those that operate via digital machines and hence can be represented as Turing Machines. The ***halting problem*** serves to provide a formal setting for investigation. Here, the problem involves determining whether an arbitrary computer program (algorithm) will continue to run indefinitely or terminates from its description and a given input that it runs on. It has been proven that algorithms represented by Turing Machine cannot solve the halting problem of the ***Universal Turing Machine*** (one capable of computing any computable sequence, i.e. in the sense it can represent any Turing machine) [123]. In other words, no algorithm can be found that solves this halting problem and thus expresses or characterizes the limit of computability of Turing Machine as a class of algorithms.

In [52], Evolutionary Turing Machine is defined as a sequence (of possibly infinite) Turing Machines $\boldsymbol{T}(t)$ operating on populations of solutions $\boldsymbol{x}(t)$, $t = 0, 1, 2, \ldots$. Turing's framework is used for which claims can be made about how expressive EAs (represented as Evolutionary Turing Machine) are. Among the various theoretical results is one that indicates higher expressivity of EAs compared to classical algorithms (which by definition terminate and produce required solutions) by virtue of the former's ***nonalgorithmic*** search behaviour. EAs can be shown to evolve solutions that converge to the optimum (to within a finite distance) asymptotically. Other provable arguments can be made. In [53], EAs represented as ***Finite State Automata*** (***FSA***) could be used to evolve solutions with FSA representation such that those solutions expressively could be outside of its class (i.e. what FSA is computationally capable of). Moving forward, such theoretical insights are important for design of EAs for practical problem solving in the context that rather than develop more elaborate EAs that arguably could not exceed a certain expressivity limit, there are now increasing research towards applying the principles of EAs as ***metaalgorithms***, e.g. evolving EAs such as in Portfolio Optimization.

References

1. Aggarwal, C.C.: Neural Networks and Deep Learning: A Textbook. Springer, Switzerland (2018)
2. Arabas, J., Michalewicz, Z., Mulawkai, J.J.: Gavaps: a genetic algorithm with varying population size. In: Proceedings of the First IEEE Conference on Evolutionary Computation. Piscataway, pp. 73–78 (1994)
3. Arcuri, A., Yao, X.: A novel co-evolutionary approach to automatic software bug fixing. In: 2008 IEEE Congress on Evolutionary Computation (IEEE World Congress on Computational Intelligence), Piscataway, pp. 162–168 (2008)
4. Arcuri, A., Yao, X.: Co-evolutionary automatic programming for software development. Inf. Sci. **259**, 412–432 (2014)
5. Atmar, J.W.: Speculation on the evolution of intelligence and its possible realization in machine form. Ph.D. thesis, New Mexico State University, Las Cruces (1976)
6. Audet, C., Hare, W. (eds.): Derivative-Free and Blackbox Optimization. Springer, Switzerland (2017)
7. Bäck, T.: Selective pressure in evolutionary algorithms: a characterization of selection mechanisms. In: Proceedings of the First IEEE Conference on Evolutionary Computation. IEEE World Congress on Computational Intelligence, Piscataway, pp. 57–62 (1994)
8. Bäck, T.: Evolutionary Algorithms in Theory and Practice: Evolution Strategies, Evolutionary Programming, Genetic Algorithms. Oxford University Press, New York (1996)

9. Bäck, T., Hammel, U., Schwefel, H.-P.: Evolutionary computation: comments on the history and current state. IEEE Trans. Evol. Comput. **1**(1), 3–17 (1997)
10. Balci, O.: Guidelines for successful simulation studies. In: 1990 Winter Simulation Conference Proceedings, pp. 25–32 (1990)
11. Barricelli, N.A.: Esempi numerici di processi di evoluzione. In: Methodos, pp. 45–68 (1954)
12. Barricelli, N.A.: Symbiogenetic evolution processes realized by artificial method. Methodos **9**(35–36), 145–182 (1957)
13. Barricelli, N.A.: Numerical testing of evolution theories: part I: theoretical introduction and basic tests. Acta Biotheor. **16**(1–2), 69–98 (1962)
14. Barricelli, N.A.: Numerical testing of evolution theories: part II: preliminary tests of performance, symbiogenesis and terrestrial life. Acta Biotheor. **16**(3–4), 99–126 (1963)
15. Belding, T.C.: Parallel genetic algorithms for a hypercube. In: Eshelman, L.J. (ed.) Proceedings of the Sixth International Conference on Genetic Algorithms, pp. 114–121. Morgan Kaufmann, San Francisco (1995)
16. Bessaou, M., Pétrowski, A., Siarry, P.: Island model cooperating with speciation for multimodal optimization. In: The 6th International Conference on Parallel Problem Solving from Nature PPSN VI. Lecture Notes in Computer Science, vol. 1917, pp. 437–446. Springer, Paris (2000)
17. Beyer, H.-G., Schwefel, H.-P.: Evolution strategies: a comprehensive introduction. Nat. Comput. **1**, 3–52 (2002)
18. Bishop, C.M.: Pattern Recognition and Machine Learning. Springer, New York (2006)
19. Bonyadi, M.R.: A theoretical guideline for designing an effective adaptive particle swarm. IEEE Trans. Evol. Comput. **24**(1), 57–68 (2020)
20. Box, G.E.P.: Evolutionary operation: a method for increasing industrial productivity. J. R. Stat. Soc. Ser. C (Appl. Stat.) **6**(2), 81–101 (1957)
21. Boyd, S., Vandenberghe, L. (eds.): Convex Optimization. Cambridge University Press, Cambridge (2004)
22. Bremermann, H.J.: Optimization through evolution and recombination. In: Yovits, M.C., Jacobi, G.T., Goldstein, G.D. (eds.) Self-Organizing Systems. Spartan Books, Washington, DC (1962)
23. Burgin, G.H.: On playing two-person zero-sum games against nonminimax players. IEEE Trans. Syst. Sci. Cybernet. **5**(4), 369–370 (1969)
24. Burgin, M., Eberbach, E.: Cooperative combinatorial optimization: evolutionary computation case study. Biosystems **91**, 34–50 (2007)
25. Burgin, M., Eberbach, E.: Evolutionary Turing in the context of evolutionary machines (2013). https://arxiv.org/abs/1304.3762
26. Burke, E.K., Kendall, G. (eds.): Search Methodologies: Introductory Tutorials in Discrete Optimization and Decision Support Techniques, 2nd edn. Springer, New York (2014)
27. Burke, E.K., Gustafson, S., Kendall, G.: Diversity in genetic programming: an analysis of measures and correlation with fitness. IEEE Trans. Evol. Comput. **8**(1), 47–62 (2004)
28. Chellapilla, K., Fogel, D.B.: Evolution, neural networks, games, and intelligence. Proc. IEEE **87**(9), 1471–1496 (1999)
29. Chellapilla, K., Fogel, D.B.: Fitness distributions in evolutionary computation: motivation and examples in the continuous domain. Biosystems **54**(1–2), 15–29 (1999)
30. Chen, T., He, J., Chen, G., Yao, X.: Choosing selection pressure for wide-gap problems. Theor. Comput. Sci. **411**, 926–934 (2010)
31. Chen, T., Tang, K., Chen, G., Yao, X.: A large population size can be unhelpful in evolutionary algorithms. Theor. Comput. Sci. **436**, 54–70 (2012)
32. Chong, S.Y., Yao, X.: Multiple choices and reputation in multiagent interactions. IEEE Trans. Evol. Comput. **11**(6), 689–711 (2007)
33. Chong, S.Y., Tan, M.K., White, J.D.: Observing the evolution of neural networks learning to play the game of Othello. IEEE Trans. Evol. Comput. **9**(3), 240–251 (2005)
34. Clerc, M., Kennedy, J.: The particle swarm–explosion, stability, and convergence in a multidimensional complex space. IEEE Trans. Evol. Comput. **6**(1), 58–73 (2002)

35. Cook, J.E., Tauritz, D.R.: An exploration into dynamic population sizing. In: Proceedings of the 2010 Genetic and Evolutionary Computation Conference (GECCO'10), pp. 807–814. Portland (2010)
36. Copeland, J., Bowen, J., Sprevak, M., Wilson, R.: The Turing Guide. Oxford University Press, Oxford (2017)
37. Cybenko, G.: Approximation by superpositions of a sigmoidal function. Math. Control Signals Syst. **2**, 303–314 (1989)
38. Darwen, P.J., Yao, X.: Speciation as automatic categorical modularization. IEEE Trans. Evol. Comput. **1**(2), 101–108 (1997)
39. Das, S., Suganthan, P.N.: Differential evolution: a survey of the state-of-the-art. IEEE Trans. Evol. Comput. **15**(1), 4–31 (2011)
40. Davis, L.D. (ed.): Handbook of Genetic Algorithms. Van Nostrand Reinhold, New York (1991)
41. DeJong, K.A.: An analysis of the behavior of a class of genetic adaptive systems. Ph.D. thesis, University of Michigan, Ann Arbor, MI (1975)
42. De Jong, K.A.: Genetic algorithms are not function optimizers. In: Whitley, L.D. (ed.) Foundations of Genetic Algorithms 2, pp. 5–17. Morgan Kaufmann, San Mateo (1993)
43. De Jong, K.A.: Fitness landscapes. In: Burke, E.K., Kendall, G. (eds.) Search Methodologies: Introductory Tutorials in Discrete Optimization and Decision Support Techniques, pp. 681–705. Springer, New York (2014)
44. de la Maza, M., Tidor, B.: Boltzmann weighted selection improves performance of genetic algorithms. Technical Report. AI Memo No. 1345, Artificial Intelligence Laboratory, Massachusetts Institute of Technology, USA (1991)
45. Doerr, B.: Does comma selection help to cope with local optima? Algorithmica **84**, 1659–1693 (2022)
46. Doerr, B., Doerr, C.: Theory of parameter control for discrete black-box optimization: provable performance gains through dynamic parameter choices. In: Doerr, B., Neumann, F. (eds.) Theory of Evolutionary Computation: Recent Developments in Discrete Optimization, pp. 271–321. Springer, Switzerland (2020)
47. Doerr, B., Neumann, F. (eds.): Theory of Evolutionary Computation: Recent Developments in Discrete Optimization. Springer, Switzerland (2020)
48. Doerr, B., Doerr, C., Yang, J.: k-bit mutation with self-adjusting k outperforms standard bit mutation. In: Proceedings of Parallel Problem Solving from Nature PPSN 2016. Lecture Notes in Computer Science, vol. 9921, pp. 824–834. Springer, Amsterdam (2016)
49. Droste, S., Wiesmann, D.: On representation and genetic operators in evolutionary algorithms. Technical Report. Technical Report CI-4198, Collaborative Research Center 531, University of Dortmund (1998)
50. Droste, S., Wiesmann, D.: Metric based evolutionary algorithms. In: European Conference on Genetic Programming (EuroGP 2000). Lecture Notes in Computer Science, vol. 1802, pp. 29–43. Springer, Edinburgh (2000)
51. Droste, S., Jansen, T., Wegener, I.: Dynamic parameter control in simple evolutionary algorithms. In: Martin, W.N., Spears, W.M. (eds.) Foundations of Genetic Algorithms 6, pp. 275–294. Morgan Kaufmann, San Mateo (2001)
52. Eberbach, E.: Towards a theory of evolutionary computation. Biosystems **82**, 1–19 (2005)
53. Eberbach, E., Burgin, M.: Evolutionary automata as foundation of evolutionary computation: Larry Fogel was right. In: Proceedings of the IEEE Conference on Evolutionary Computation (CEC2009), Piscataway, pp. 2149–2156 (2009)
54. Eiben, A.E., Hinterding, R., Michalewicz, Z.: Parameter control in evolutionary algorithms. IEEE Trans. Evol. Comput. **3**(2), 124–141 (1999)
55. Eiben, A.E., Schut, M.C., de Wilde, A.R.: Is self-adaptation of selection pressure and population size possible? – a case study. In: Proceedings of Parallel Problem Solving from Nature PPSN IX. Lecture Notes in Computer Science, vol. 4193, pp. 900–909. Springer, Berlin (2006)
56. Fogel, L.J.: Autonomous automata. Ind. Res. **4**, 14–19 (1962)

57. Fogel, D.B.: Evolutionary programming: an introduction and some current directions. Stat. Comput. **4**, 113–129 (1994)
58. Fogel, D.B.: Evolutionary Computation: Toward a New Philosophy of Machine Intelligence. IEEE Press, Piscataway (1995)
59. Fogel, D.B.: Phenotypes, genotypes, and operators in evolutionary computation. In: Proceedings of the 1995 IEEE Conference on Evolutionary Computation, Piscataway, pp. 193–198 (1995)
60. Fogel, D.B.: Blondie24: Playing at the Edge of AI. Morgan Kaufmann, San Francisco (2002)
61. Fogel, D.B.: Nils Barricelli–artificial life, coevolution, self-adaptation. IEEE Comput. Intell. Mag. **1**(1), 41–45 (2006)
62. Fogel, D.B., Fogel, L.J.: An introduction to evolutionary programming. In: Alliot, J.M., Lutton, E., Ronald, E., Schoenauer, M., Snyers, D. (eds.) Artificial Evolution. Lecture Notes in Computer Science, vol. 1063, pp. 21–33. Springer, Berlin (1996)
63. Fogel, D.B., Ghozeil, A.: A note on representations and variation operators. IEEE Trans. Evol. Comput. **1**(2), 159–161 (1997)
64. Friedberg, R.M.: A learning machine: part I. IBM J. Res. Dev. **2**(1), 2–13 (1958)
65. Friedberg, R.M., Dunham, B., North, J.H.: A learning machine: part II. IBM J. Res. Dev. **3**(3), 282–287 (1959)
66. Goldberg, D.E.: Genetic Algorithms in Search, Optimization and Machine Learning. Addison-Wesley, Reading (1989)
67. Goldberg, D.E., Deb, K.: A comparative analysis of selection schemes used in genetic algorithms. In: Rawlins, G.J.E. (ed.) Foundations of Genetic Algorithms 1, pp. 69–93. Morgan Kaufmann, San Mateo (1994)
68. Goldberg, D.E., Richardson, J.: Genetic algorithms with sharing for multimodal function optimization. In: Proceedings of the Second International Conference on Genetic Algorithms and Their Applications, pp. 41–49 (1987)
69. Goldberg, D.E., Deb, K., Clark, J.H.: Genetic algorithms, noise, and the sizing of populations. Complex Syst. **6**(4), 333–362 (1992)
70. Goodfellow, I., Bengio, Y., Courville, A.: Deep Learning. The MIT Press, Cambridge (2016)
71. Grefenstette, J.: Rank-based selection. In: Bäck, T., Fogel, D.B., Michalewicz, Z. (eds.) Evolutionary Computation 1: Basic Algorithms and Operators, pp. 187–194. Taylor & Francis, New York (2000)
72. Hansen, N.: Benchmarking a bi-population CMA-ES on the BBOB-2009 function testbed. In: Proceedings of the 11th Annual Conference Companion on Genetic and Evolutionary Computation Conference: Late Breaking Papers, pp. 2389–2396 (2009)
73. Hansen, N.: Benchmarking a bi-population CMA-ES on the BBOB-2009 noisy testbed. In: Proceedings of the 11th Annual Conference Companion on Genetic and Evolutionary Computation Conference: Late Breaking Papers, pp. 2397–2402 (2009)
74. Hansen, N., Ostermeier, A.: Completely derandomized self-adaptation in evolution strategies. Evol. Comput. **9**(2), 159–195 (2001)
75. Hansen, N., Auger, A., Ros, R., Finck, S., Pošík, P.: Comparing results of 31 algorithms from the black-box optimization benchmarking BBOB-2009. In: Proceedings of the 2010 Genetic and Evolutionary Computation Conference (GECCO'10), Portland, pp. 1689–1696 (2010)
76. Hansen, N., Auger, A., Ros, R., Tušar, T., Brockhoff, D.: COCO: a platform for comparing continuous optimizers in a black-box setting. Optim. Methods Software **36**(1), 114–144 (2021)
77. Hauschild, M., Pelikan, M.: An introduction and survey of estimation of distribution algorithms. Swarm Evol. Comput. **1**, 111–128 (2020)
78. He, J., Yao, X.: Drift analysis and average time complexity of evolutionary algorithms. Artif. Intell. **127**, 57–85 (2003)
79. Hellwig, M., Beyer, H.-G.: Benchmarking evolutionary algorithms for single objective real-valued constrained optimization – a critical review. Swarm Evol. Comput. **44**, 927–944 (2019)
80. Holland, J.H.: Outline for a logical theory of adaptive systems. J. ACM **9**(3), 297–314 (1962)

81. Holland, J.H.: Adaptation in Natural and Artificial Systems. University of Michigan Press, Ann Arbor (1975)
82. Hornik, K., Stinchcombe, M., White, H.: Multilayer feedforward networks are universal approximators. Neural Networks **2**, 359–366 (1989)
83. Huang, C., Li, Y., Yao, X.: A survey of automatic parameter tuning methods for metaheuristics. IEEE Trans. Evol. Comput. **24**(2), 201–216 (2020)
84. Karafotias, G., Hoogendoorn, M., Eiben, A.E.: Parameter control in evolutionary algorithms: trends and challenges. IEEE Trans. Evol. Comput. **19**(2), 167–187 (2015)
85. Karcz-Duleba, I.: Dynamics of two-element populations in the space of population states. IEEE Trans. Evol. Comput. **10**(2), 199–209 (2006)
86. Kauffman, S.A., McCulloch, W.S.: Random nets of formal genes. Technical Report. Quarterly Progress Report No. 88, Research Laboratory of Electronics, Massachusetts Institute of Technology (1968)
87. Kennedy, J., Eberhart, R.: Particle swarm optimization. In: Proceedings of ICNN'95 – International Conference on Neural Networks, Perth, pp. 1942–1948 (1995)
88. Kirkpatrick, S., Gelatt, C.D., Vecchi, M.P.: Optimization by simulated annealing. Science **220**(4598), 671–680 (1983)
89. Korns, M.F.: Genetic programming symbolic classification: a study. In: Genetic Programming Theory and Practice XV, Genetic and Evolutionary Computation, pp. 39–54. Springer, Berlin (2018)
90. Koumousis, V.K., Katsaras, C.P.: A saw-tooth genetic algorithm combining the effects of variable population size and reinitialization to enhance performance. IEEE Trans. Evol. Comput. **10**(1), 19–28 (1999)
91. Koza, J.R.: Genetic Programming: On the Programming of Computers by Means of Natural Selection. The MIT Press, Cambridge (1992)
92. Koza, J.R.: Genetic Programming II: Automatic Discovery of Reusable Programs. The MIT Press, Cambridge (1994)
93. Koza, J.R., Bennett III, F.H., Andre, D., Keane, M.A.: Genetic Programming III: Darwinian Invention and Problem Solving. Morgan Kaufmann, San Francisco (1999)
94. Langdon, W.B.: Size fair and homologous tree crossovers for tree genetic programming. Genet. Program. Evolvable Mach. **1**, 95–119 (2000)
95. Langdon, W.B., Poli, R., McPhee, N.F., Koza, J.R.: Genetic programming: an introduction and tutorial, with a survey of techniques and applications. In: Computational Intelligence: A Compendium. Studies in Computational Intelligence, vol. 115, pp. 927–1028. Springer, Berlin (2008)
96. Larrañaga, P., Lozano, J.L. (eds.): Estimation of Distribution Algorithms: A New Tool for Evolutionary Computation. Springer, New York (2002)
97. Lee, C.-Y., Yao, X.: Evolutionary programming using mutations based on the Lévy probability distribution. IEEE Trans. Evol. Comput. **8**(1), 1–13 (2004)
98. Lengler, J.: Drift analysis. In: Doerr, B., Neumann, F. (eds.) Theory of Evolutionary Computation: Recent Developments in Discrete Optimization, pp. 89–132. Springer, Switzerland (2020)
99. Lengler, J., Steger, A.: Drift analysis and evolutionary algorithms revisited. Comb. Probab. Comput. **27**(4), 643–666 (2018)
100. Levin, D.A., Peres, Y., Wilmer, E.L.: Markov Chains and Mixing Times. American Mathematical Society, Providence (2009)
101. Lewontin, R.C.: A general method for investigating the equilibrium of gene frequency in a population. Genetics **43**(3), 419–434 (1958)
102. Li, Z., Li, X., Zhang, Q., Liu, H.: Evolution strategies for continuous optimization: a survey of the state-of-the-art. Swarm Evol. Comput. **56** (2020). https://doi.org/10.1016/j.swevo.2020.100694
103. Lobo, F.G.: Idealized dynamic population sizing for uniformly scaled problems. In: Proceedings of the 13th Annual Conference on Genetic and Evolutionary Computation (GECCO'11), Dublin, pp. 917–924 (2011)

104. Lu, Z., Pu, H., Wang, F., Hu, Z., Wang, L.: The expressive power of neural networks: a view from the width. In: Advances in Neural Information Processing System, vol. 30, pp. 6231–6239. Curran Associates, Inc., Red Hook (2017)
105. Mahfoud, S.W.: Niching methods for genetic algorithms. Ph.D. thesis, University of Illinois, Urbana-Champaign, IL (1995)
106. Maiorov, V., Pinkus, A.: Lower bounds for approximation by mlp neural networks. Neurocomputing **25**(1–3), 81–91 (1999)
107. Maturana, J., Lardeux, F., Saubion, F.: Autonomous operator management for evolutionary algorithms. J. Heuristics **16**(6), 881–909 (2016)
108. McGinley, B., Maher, J., O'Riordan, C., Morgan, F.: Maintaining healthy population diversity using adaptive crossover, mutation, and selection. IEEE Trans. Evol. Comput. **15**(5), 692–714 (2011)
109. Montana, D.J.: Strongly typed genetic programming. Evol. Comput. **3**(2), 199–230 (1995)
110. Mühlenbein, H., Paaβ, G.: From recombination of genes to the estimation of distributions I. Binary parameters. In: Parallel Problem Solving from Nature PPSN IV. Lecture Notes in Computer Science, vol. 1141, pp. 178–187. Springer, Berlin (1996)
111. Pétrowski, A.: A new selection operator dedicated to speciation. In: Proceedings of the Seventh International Conference on Genetic Algorithms (ICGA'97), pp. 144–151 (1997)
112. Rechenberg, I.: Evolutionsstrategie: Optimierung technischer systeme nach prinzipien der biologischen evolution. Dr.Ing. thesis, Technical University of Berlin (1971)
113. Rowe, J., Whitley, D., Barbulescu, L., Watson, J.-P.: Properties of gray and binary representations. Evol. Comput. **12**(1), 47–76 (2004)
114. Russell, S., Norvig, P.: Artificial Intelligence: A Modern Approach, 4th edn. Pearson, New Jersey (2021)
115. Schwefel, H.-P.: Kybernetische evolution als strategie der exprimentellen forschung in der strömungstechnik. Master's thesis, Technical University of Berlin (1971)
116. Smith, R.E., Smuda, E.: Adaptively resizing populations: algorithm, analysis, and first results. Complex Syst. **9**(1), 47–72 (1995)
117. Smith, R.E., Forrest, S., Perelson, A.S.: Searching for diverse, cooperative populations with genetic algorithms. Evol. Comput. **1**(2), 127–149 (1992)
118. Smorodkina, E., Tauritz, D.R.: Greedy population sizing for evolutionary algorithm. In: Proceedings of the IEEE Conference on Evolutionary Computation (CEC2007), Piscataway, pp. 2181–2187 (2007)
119. Srinivas, M., Patnaik, L.M.: Adaptive probabilities of crossover and mutation in genetic algorithms. IEEE Trans. Syst. Man Cybernet. **24**(4), 656–667 (1994)
120. Storn, R., Price, K.: Differential evolution – a simple and efficient heuristic for global optimization over continuous spaces. J. Global Optim. **11**, 341–359 (1997)
121. Tanese, R.: Parallel genetic algorithms for a hypercube. In: Grefenstette, J.J. (ed.) Proceedings of the Second International Conference on Genetic Algorithms on Genetic Algorithms and their Application, pp. 177–183. Lawrence Erlbaum, Hillsdale (1987)
122. Teo, J.: Exploring dynamic self-adaptive populations in differential evolution. Soft Comput. **10**(8), 673–686 (2006)
123. Turing, A.M.: On computable numbers, with an application to the Entscheidungsproblem. Proc. London Math. Soc. **42**(1), 230–265 (1937)
124. Turing, A.M.: Computing machinery and intelligence. Mind **59**(236), 433–460 (1950)
125. Turing, A.M.: Intelligent machinery. In: Ince, D.C. (ed.) Mechanical Intelligence: Collected Works of A. M. Turing. North Holland, Amsterdam (1992)
126. Ursem, R.K.: Diversity-guided evolutionary algorithms. In: The 7th International Conference on Parallel Problem Solving from Nature PPSN VII. Lecture Notes in Computer Science, vol. 2439, pp. 462–474. Springer, Granada (2002)
127. Vose, M.D., Liepins, G.E.: Punctuated equilibria in genetic search. Complex Syst. **5**(1), 31–44 (1991)
128. Walker, J.A., Miller, J.F.: The automatic acquisition, evolution and reuse of modules in cartesian genetic programming. IEEE Trans. Evol. Comput. **12**(4), 397–417 (2008)

129. Wang, S., Mei, Y., Zhang, M., Yao, X.: Genetic programming with niching for uncertain capacitated arc routing problem. IEEE Trans. Evol. Comput. **26**(1), 73–87 (2022)
130. Wierstra, D., Schaul, T., Glasmachers, T., Sun, Y., Peters, J., Schmidhuber, J.: Natural evolution strategies. J. Mach. Learn. Res. **15**, 949–980 (2014)
131. Yao, X.: Evolutionary Computation: Theory and Applications. World Scientific Publication, Singapore (1999)
132. Yao, X.: Evolving artificial neural networks. Proc. IEEE **87**(9), 1423–1447 (1999)
133. Yao, X., Liu, Y., Lin, G.: Evolutionary programming made faster. IEEE Trans. Evol. Comput. **3**(2), 82–102 (1999)
134. Zhan, Z.-H., Shi, L., Tan, K.C., Zhang, J.: A survey on evolutionary computation for complex continuous optimization. Artif. Intell. Rev. **55**, 59–110 (2022)
135. Zhang, J., Sanderson, A.C.: Jade: adaptive differential evolution with optional external archive. IEEE Trans. Evol. Comput. **13**(5), 945–958 (2009)
136. Zhou, X., Qin, A.K., Gong, M., Tan, K.C.: A survey on evolutionary construction of deep neural networks. IEEE Trans. Evol. Comput. **25**(5), 894–912 (2021)
137. Zhou, Y., Yen, G.G., Yi, Z.: Evolutionary shallowing deep neural networks at block levels. IEEE Trans. Neural Networks Learn. Syst. **33**(9), 4635–4647 (2022)

Chapter 2
Introduction to Coevolutionary Computation

2.1 Coevolutionary Systems: Overview

Coevolutionary Computation collectively refers to a broad family of computational systems that adopt the same principle features in Evolutionary Computation, i.e. these ***coevolutionary systems*** are population-based and employ both evolutionary operators of variation and selection for their search processes. As problem-solving search procedures, both can be viewed within the framework of generate-and-test to realize iterative algorithmic implementations. Indeed, early developments of Coevolutionary Computation share common design principles and employ the same evolutionary operators that have been introduced earlier in EAs. What sets them apart is the specific design of the ***fitness evaluation*** procedure that is used to determine the quality of each candidate solution or individual in the population.

On the one hand, Evolutionary Algorithms (EAs) evaluate the ***objective fitness*** of each candidate with an absolute performance measure. For problems formulated in the context of optimization, a ***fitness function*** f is usually known and constructed in closed form so that the fitness (quality) of a candidate solution is determined by making a direct query on f to obtain a real-valued output. On the other hand, coevolutionary systems evaluate the ***subjective fitness*** of each candidate solution with ***test cases***, which can change throughout the search process. Furthermore, fitness evaluation in coevolution requires an explicit process (e.g. performed through a computational procedure) to determine the quality of a candidate solution with respect to those test cases. In summary, it is crucial to view and apply Coevolutionary Computation in the context of a general problem-solving framework having specific distinguishing features that makes it effective for a variety of problem settings, which are otherwise challenging for its evolutionary counterpart.

In the next Sect. 2.2, we give a categorial view of key developments of coevolutionary systems and then a brief summary on the modern synthesis on natural evolution so as to provide a context with which we can examine the relationship between coevolutionary systems and natural evolution. After that, we will describe

X. Yao, S. Y. Chong, *Coevolutionary Computation and Its Applications*,
Natural Computing Series, https://doi.org/10.1007/978-981-96-2841-4_2

and discuss in more detail the main principle and concepts of coevolution over two sections following next. Section 2.3 will focus on the main principle of coevolution within the generate-and-test framework involving a subjective fitness evaluation procedure. We will discuss the properties of this fitness evaluation and present the argument the conditions describing a successful coevolutionary search. In Sect. 2.4, the two main concepts that govern the operations of coevolution having specific problem-solving capabilities are discussed. These involve population structures and the two types of interactions between agents in coevolution.

2.2 Development of Coevolutionary Computation

We start our review with several key historical developments of coevolutionary systems in Sect. 2.2.1. This is to provide a context with which more detailed discussions will take place in later sections about the process underlying coevolutionary search and the main concepts that govern design of the mechanisms and principal components in coevolutionary systems. Our review of studies that have introduced those developments is not done in a chronological manner. Instead, we consider a more categorical review that focuses on the motivations those studies have sought to address. These motivations can be about verifying a hypothesis with regard to mechanisms in natural coevolutionary processes and also design principles or conceptual frameworks that are implemented as computational systems for real-world problem solving. Indeed, these are the two basic motivations underlying studies of Evolutionary and Coevolutionary Computation: to understand mechanisms in natural evolutionary and coevolutionary processes that then inform the design of computational systems for problem solving. In Sect. 2.2.1, we will briefly introduce the modern synthesis of natural evolution in biology that generally incorporates the theory of population genetics and interpretation of variation within species. This is so that an updated and more modern view of Evolutionary Biology is used as the setting with which the relationship between natural evolution and coevolutionary systems can be discussed.

2.2.1 A Categorical View of Key Developments

Not surprisingly, the problem domain involving situations of strategic decision-making has received early and significant attention in the research and development of coevolutionary systems. In the first place, a formal approach to address these problems in the form of ***Game Theory*** has continued to be developed since its introduction in [84]. Next, we have the general availability of computing machines in various form factors from high-performance computing clusters to mobile devices and across sectors from public to private that are interconnected in the modern world. The underlying research and applied technological developments have culminated in both the pervasiveness and the capability to address these strategic decision-making problems that are the hallmark of many human activities [58]. In the context of Artificial Intelligence, games provide the means with which systems can be developed and evaluated for intelligent behaviours. In particular, games are models of behavioural activities conducted by some agents to which rewards are obtained for certain expenditures of resources.

Moving away from the more formal approaches in Artificial Intelligence, the focus of early Coevolutionary Computation has been motivated to understand how the natural evolution of a population of ***agents***, who are each responsible for making and taking decisions during interactions with the other, gives rise to intelligent decision-making behaviours with little preprogrammed knowledge [12]. A more pertinent challenge to address is whether agents can learn effective decision-making behaviours through coevolutionary processes guided only by their strategic interactions, which can be modelled as game-plays. Consider the simple settings that are involved in the family of ***competitive***, ***zero-sum games*** between two players. Here, the ***payoffs*** that are given to a player are taken away from the other player during any sequence of allowed moves in the game-play. The competitive nature of the game that comes about from the manner with which payoffs are allocated means that there is one that wins and another that loses (a draw is possible, although rules can be designed to break any tie in the final cumulative payoffs). Crucially, information about game outcomes is readily available. They require no further information gathering and associated computations unlike those dealing with specific game properties such as the degree with which a loss of cumulative payoffs is incurred.

In this situation, one could then design the coevolutionary process to selectively favour among the population of game-playing agents those that have a larger number of wins. In any game-play, an agent is only endowed with the ability to take as input the state description and then to output the decision that corresponds to allocation of resources stipulated in the design of the game to complete the move. This behaviour in the form of input-output responses can be modelled by suitable and possibly nonlinear parametric representations such as Artificial Neural Networks (ANNs). This limits the amount of game knowledge that can be incorporated into the decision-making capabilities of agents, save perhaps some form of knowledge required to complete a ***legal*** move according to the rules of the game. Agents are uniquely identified by the parameters in their strategy representation. They can adapt their game-play behaviour through changes to the parameters from the application of the variation operator and their corresponding opportunities for replication with the selection operator throughout the coevolutionary process.

The question remains whether behavioural adaptation through coevolution guided only by the aggregate outcomes of strategic interactions or game-plays between agents would be sufficient to produce those having effective decision-making. Furthermore, each outcome itself is a sum of payoffs corresponding to sequences of decisions that the agent has made. There is no explicit mechanism that assigns more importance to certain patterns of decisions that are deemed crucial by the body of knowledge from the problem domain. Indeed, attempts to coevolve intelligent decision-making agents are noted as early as studies in [6, 68]. However, it would take four decades later whereby with further investigations and availability of faster computing machines that it becomes possible to realize ***coevolutionary learning*** in agents to play competitive, zero-sum board games that start with little preprogrammed expert knowledge [13, 41]. Not only these coevolved agents are able to learn to play these board games to a high competitive level [38], they are also capable of strategic game-plays reminiscent of expert human players [16].

This framework of ***competitive coevolution*** for learning strategic behaviours has been applied to other family of games. One example is the family of ***noncooperative games*** such as the ***nonzero-sum*** game in the form of the ***Iterated Prisoner's Dilemma*** (***IPD***) that has received significant attention. The standard IPD involves two players simultaneously deciding on their individual choices between cooperate and defect over a sequence of moves. The payoffs for all four possible joint-moves made by two players can be represented in a ***normal form*** matrix, which reflects the opposed motives between group rationality to cooperate and individual rationality to defect [2, 3]. Despite having such a simple construction, the IPD is capable of generating complex agent behaviours over sequences of iterated moves when they are allowed to decide based on historical plays [4]. This correspondingly leads to a large number of ***pure strategies*** of fixed responses. The dynamics of coevolution can be complex with different constituents of IPD strategies being present in the population for different configurations of the game (e.g. when more than two choices and noise are allowed in the game) [14, 22]. This coevolutionary framework of IPD games has been further extended to incorporate more involved mechanisms such as indirect interactions (e.g. through reputation) [15, 90] and coalitional formations (e.g. when more than two players can simultaneously engage in the interactions) [73, 74, 88, 89] that are associated with real-world interactions.

Despite the complexity in agent behaviours that can be generated from these coevolutionary processes, the specifications of the games that govern their interactions are actually well-defined. Games are constructed as abstractions of some real-world interactions such as those that can be found in economics and social systems. This makes it easier to perform analysis and identify salient features that make certain strategic behaviours effective. When there is a stronger need to explore or to assess in a what-if manner strategic behaviours that could arise when restrictions placed on agent interactions are relaxed, coevolutionary systems are then designed as a simulation tool. Here, soft ***Artificial Life*** (***ALife***) provides the framework with which both mathematical and computational aspects in modelling and simulations of lifelike behaviours are developed [46]. As an example, ***Cellular Automata*** (***CA***) is typically used as a model of spatial or other forms of localized interactions by way of it defining a specific, grid-like neighbourhood structure for agent interactions in a coevolving finite population [29]. This is in contrast to most selection mechanisms in coevolution whereby some form of random sampling is implemented to choose the agents from the population for interactions.

The seminal work in [76] has considered other aspects of simulating lifelike behaviours that are realistically complex. The interactions still involve two players so that any pair of agents drawn from the population could compete for resources so as to survive and reproduce. However, the settings that are involved in these interactions are realistically complex. In the first place, the environment with which the agents can operate involves a three-dimensional world with simulated physics that includes various effects such as gravity, friction, and collisions. Next, these agents are essentially virtual creatures. Their constructions are highly nontrivial with decision-making capabilities governed by an architecture of neural nodes of various fixed functions, input sensors to detect the environment and output effectors to enable movements and perform tasks. The actual operations of the creatures involve physical locomotions, which are governed by the morphology (relationship) of structures (parts) that constitute their bodies. Both the topologies underlying the morphological structures and neural architectures that control of the body are merged and described via directed graphs. This allows the construction of specific variation operators to physically change the virtual creatures, thus allowing the simulation of agents coevolving their physical morphology and operational behaviours via interactions only.

There is a strong synergy between Coevolutionary Computation and Machine Learning studies. There are studies that make use of principles and mechanisms in Machine Learning to develop and improve coevolutionary systems to learning problems [17, 19, 21]. Other studies have used coevolution to solve various learning problems. Earlier, we have introduced the coevolutionary learning of game-playing strategies, which constitutes a family of problems that is challenging to solve using Machine Learning methodologies without extensive use of domain knowledge and high-performance computational requirements. For the more classical learning problems, as one example, there have been early attempts [51, 63, 65] to coevolve CA rules to perform a binary density classification task. This involves a one-dimensional lattice or array N of cells (the two opposite end cells are connected to then form a circle), each taking on one of the binary states of $\{0, 1\}$. The task is to classify whether an initial configuration describing the array of ones and zeroes is high density (i.e. there are more ones than zeroes in the array) or not. This task is straightforward for a computing machine that is equipped with sufficient memory to just count the number of ones and zeroes and then compare them. However, the real challenge here is that only local computation and very limited memory are afforded. CA rules are used to perform a transformation on each cell by changing its state based on the current states of local neighbourhood of cells for some time steps until a fixed configuration is obtained [57]. The CA rules are deemed to classify the initial configuration as being high density if after the transformations the resulting configuration is all ones, and vice versa.

Even so, coevolutionary learning has been shown to produce high-performing CA rules that correspond to strategies that have been identified earlier for their effectiveness [85]. This is achieved with a two-population, ***predator-prey*** coevolutionary framework, with the population of candidate solutions consisting of CA rules and the other population of test cases consisting of initial configurations of arrays. The fitness of a CA rule is the sum of initial configurations it correctly classifies or solves. The fitness of an initial configuration is the sum of CA rules that do not correctly classify or solve it. The two-population coevolution setup can be further augmented using ***diversity maintenance techniques*** to ensure that the populations are suitably diverse to ensure sustained coevolutionary search progress [18]. A particular mechanism of competitive ***fitness sharing*** is used in [85] that redistribute the contributions or weights associated with solving specific test cases rather than assuming one that is uniform. In the case of the relative fitness calculation for CA rules, initial configurations that are solved by a lower number of CA rules are given higher weights. The same principle applies to the calculations of shared fitness for the evolving initial configurations.

There are other applications of such two-population coevolutionary learning that include grammar induction, gene regulatory inference, and evolutionary robotics. In particular, all these applications are framed in the context of ***systems identification***, whereby the main learning objective is to build a model that is representative of the underlying or internal process of a target system whose external, output behaviour can be observed after being presented with input data [8]. Unlike the situation in most data-driven Machine Learning approaches, data is limited where system identification task is concerned. For example, they might be expensive to procure, and when they are available they might be biased, making it difficult to obtain a representative dataset. The main challenge then in systems identification is that model learning has to be performed through a limited and careful series of experimentation. This challenge is addressed by exploiting the two-population coevolution framework that is reminiscent of an ***active learning*** approach [8]. In particular, the separate population of evolving test cases both circumvents the limitation of dataset availability and generates additional test cases that best differentiate the targeted models in the other population.

Coevolutionary principles have long been used to solve challenging optimization problems. One of the early examples is the seminal work that is reported in [48]. In particular, the study is concerned with searching for a minimal sorting algorithm that transforms a list of unsorted elements to a list that is sorted. The algorithm is called a sorting network because it has descriptive graphical representation of its operations. It takes the form of n number of horizontal lines in parallel that correspond to the n elements in the list. On the left end of these lines is the unsorted list, while on the right end is the sorted list. The operations of the algorithm are represented by directed vertical lines, each indicated by a single arrow that connects two horizontal lines at locations i and j, with $i, j \in \{1, \ldots, n\}$. Each directed vertical line represents a comparison and exchange (swap) operation of the elements that are correspondingly located in the ith and jth horizontal lines. The direction of the arrow represents the direction of the exchange such that the tail will always have the smaller element (e.g. see Figures 1 and 2 in [48]). What the study has demonstrated is that a two-population coevolutionary framework is able to improve on the standard evolutionary optimization approach that uses a fixed set of test cases.

Last but not least is the development of coevolution as a higher-level, general problem-solving framework. The general motivation here is somewhat similar to other frameworks such as ***hyper-heuristics*** in that the focus is about the automated search of lower-level problem-solving approaches to targeted problems. The framework employs coevolutionary principles as the means to search on the space of algorithms or heuristics competitively and other aspects of problem solving that include problem decomposition. Indeed, in the latter setting ***cooperative coevolution*** has been developed early on to solve complex problems whose solution representations admit decompositional structures as components [67]. It is known that not all real-world problems can be decomposed into independent components. There are complex interdependencies between these components. As such, the framework is designed to coadapt these components. For each component, a candidate is sourced from a genetically isolated population (i.e. no crossovers with or migration of candidates from other populations). However, candidates for all components cooperate to form the complete solution to the problem. What remains are fitness evaluations and then applying appropriate selection pressure on these candidates within their respective populations. This can be framed as a credit assignment problem, whereby fixed representatives from other populations are used for fitness evaluations of candidates in the current isolated population.

Controlled empirical studies of such a cooperative coevolution framework have demonstrated its viability in training ANNs for pattern classification tasks. Indeed, various other extensions can be incorporated to cooperative coevolution, which demonstrates the general nature of the framework. In [45], the study extends this framework for ANN training to cooperatively coevolve ***ensemble***. In particular, components now take the form of individual ANNs that can be combined in a soft manner typical of ensemble approaches used in Machine Learning. A multi-objective optimization approach is used in dealing with performance evaluations and selection of coevolved solutions, which now take the form of two separate objectives dealing with classification performance and an ambiguity or correlation measure between outputs of individual ANNs and the constructed ensemble. Other applications of such a cooperative coevolution framework on real-world problem solving are found dealing with optimization and design problems. In [91], the framework is used to solve a particular instance of the Quadratic Assignment Problem (QAP) that takes the form of node (multiprocessor) mappings for scheduled executions of algorithms involving multiple function modules with some underlying dataflow dependencies that are typical in signal processing and multimedia applications. In [55], a parallel implementation of a cooperative coevolution framework is applied to directly solve complex engineering design problems over their entire life cycles, which admit problem decomposition and allow for concurrent optimization tasks to be performed on the smaller, decomposed problems.

This higher-level, general problem-solving coevolutionary framework is not restricted to cooperative settings only. Indeed, the framework also extends to the competitive coevolution settings that have been described earlier. For example, ensembles of game-playing strategies can be coevolved and generated from a single-population competitive setup that is augmented with diversity maintenance [21]. Rigorous quantitative analysis has been carried out to verify that a constructed ensemble possesses higher generalization performance compared to that of individual strategies in the ensemble due to the ***covering*** effect, i.e. each individuals defeat different subsets of opponent strategies [17, 18]. This coevolutionary principle has been used and exploited to develop automated parallel portfolio optimizers. In

[54, 80], a two-population competitive setup is used to coevolve one population of optimizers and another population of generated instances of combinatorial optimization problems. This requires some form of parameterizations of both the optimizers and problem instances. Most heuristic optimizers are designed with parameter settings, which can be exploited together with transformation operators to construct their configuration or parameter spaces. Combinatorial optimization problems such as Travelling Salesman Problems (TSPs) have a certain size measure associated with them, which can be used as a fixed setting of a perturbation process to generate problem instances. The two populations are coevolved in a competitive manner, and this framework has demonstrated success in producing a population of optimizers that collectively through a parallel implementation, to solve new generated problem instances at a performance level near to experts-designed optimizers.

2.2.2 Evolutionary Biology and Its Relationship to Coevolutionary Computation

In the 1997 review paper on Evolutionary Computation [5], the ***neo-Darwinian synthesis*** or ***model*** of natural evolution [49] that combines both theories of Darwinian evolution [23] and Mendelelian genetics is briefly referenced to expound on the basic structures of EAs. In particular, one has in a standard EA a population of individuals undergoing the process of variation and selection. There are many modern EAs that employ significantly more complex and involved methodologies for the evolutionary search process than what is described earlier as a basic EA. However, most general introductions adopted by research studies in Evolutionary Computation (including the review paper [5]) include brief and simplified descriptions of the neo-Darwinian synthesis of evolution. Its original form and the usual introduction of research studies in Evolutionary Biology are significantly more involved and rich that covers concepts and discoveries from a variety of fields that include genetics, ecology, and paleontology among others [49]. Indeed, this ***modern synthesis*** (***evolutionary synthesis***) on the biological evolution process itself is evolving with ongoing debates among experts in the field whether it needs updating [52]. Nevertheless, we provide a brief but more involved introduction of the modern synthesis following the exposition of the Evolutionary Biology text [44]. Our motivation here is both to introduce a more complete exposition of this model of evolution and then to discuss how it relates to some of the more complex systems designed in Evolutionary Computation such as coevolutionary systems.

There are several core concepts that underpin the modern synthesis [44]. First, the basic biological evolutionary process operates at the level of the populations instead of at the level of the individuals. What this means is that biological evolutionary process is conceptualized as changes to the frequency of heritable variations in populations rather than changes to specific individuals. Second, heredity is accomplished via genes that are now generally accepted as DNA or RNA. In particular, the Lamarckian notion of transmission of an individual's experiences to its offspring is considered insofar as a form of cultural inheritance that mediates natural selection. As such and as one example, the sequences of DNA that are transmitted to ***eukaryote*** offsprings via gametes during sexual reproduction are not affected by the parents' experiences. Third, inheritable variations come about from individually infrequent mutations. However, the generally accepted view is that natural selection acts as a stabilizing (purifying) process that enables the maintenance of functional responses to these mutations. There has not been conclusive evidence to support the view that directed mutations resulting in improved functional responses of individuals could occur within biological evolution. The theory of population genetics prescribes mutation to be alteration of heritable variations that can be stably transmitted. Fourth, there are several factors affecting frequencies of heritable variants that include mutation, genetic drift, gene flow, and natural selection. However, natural selection is considered as the sole cause for changes of those frequencies that result in positive adaptations. Here, it is important to take note that natural selection is defined as the consistent difference in offspring productions of reproducing entities that include genes, phenotypically different individuals, species, and populations.

At this point, it is crucial to take note that a large part of the modern synthesis is aimed towards a description of biological evolution that we could observe through various evidence sources of the slow generational change of multicellular eukaryotes that occur in predominantly higher ***taxa*** (e.g. animals). In particular, the large phenotypic changes of such individuals (i.e. those that are physically observable) occur over a very long time through small and incremental changes. Species are considered to be sexually reproducing isolated groups of populations. The evolutionary divergence of species occurs as a result of geographic isolation of their populations. In summary, the modern synthesis [44] is broadly comprised of the framework of population genetics [35, 47, 87] that formalizes the process underpinning changes to the frequency of heritable variations as evolution and the interpretation of variation within species [26]. The core theory of population genetics has accommodated new forms of mutation as they are discovered, e.g. whether mutation acts as single base pair substitutions that each involve swapping a nucleotide with another to insertion of transposable element that constitutes a large sequence of nucleic acid in a regulatory sequence or region. Furthermore, the notion of natural selection itself is nontrivial and can be highly involved. As one example, the ***epistasis*** effects on fitness (rate of reproduction) occurs as a result of interactions between genes. Further adding to the richness in biological evolution is the abundance of species, their specialization in their interactions with other species, and of which some involve processes and outcomes of reciprocal evolutionary change, i.e. coevolution in the form of predators and preys, hosts and parasites, competition, and mutualism [27, 82].

Given such a complex view of the biological evolution, it should not come as a surprise that EAs for which claims are made for being inspired or deriving design principles from processes of natural evolution would be complex as well. Indeed and equipped with a basic view of the modern synthesis of evolution as summarized above, one might now consider that the early EAs such as Genetic Algorithms, Evolutionary Programming, and Evolution Strategies would generally reflect a highly simplified and abstract model of evolution. On the one hand, the review paper [5] and others that adopt a similar presentation would be accurate in their statements that the processes of EAs are analogous to and inspired by those found in evolution under such a restricted setting. On the other hand, there are more complex and involved EAs for search that employ Island models [7, 79], cooperative and competitive coevolution [11, 48, 67, 70] for which there is no notion of absolute fitness on an individual but instead is relative and dynamic that depends and requires interactions between individuals for its computation. The perspective we put forth here is that such systems employ processes that more closely reflect those that are present in biological evolution as they are currently understood by its modern synthesis. Examples include coevolutionary systems that rely on cooperative interactions between individuals that represent potential solutions with respect to subproblems and how their coadaptations could serve to promote the discovery of novel solutions for the complete problem.

In other words, it would appear only natural that the richness of processes in evolution would serve to inspire various mechanisms that can be suitably formulated as computable algorithms in Evolutionary and Coevolutionary Computation. Indeed, a subsequent position text in 2015 [28] has raised a more general point in its discussion on the relationship between natural evolution and artificial evolution. We first note that the term artificial evolution is used in the text to encapsulate those within the paradigm of Evolutionary Computation (in the form of software implementations in computing machines) and Evolutionary Robotics (with processes operating on physical substrates or hardware). In particular, what is more crucial is not so much that EAs are faithful representations of natural evolution but the fact that by having three fundamental components of representation (heredity), variation (mutation), and selection, EAs constitute a form of evolutionary process in the same manner that natural (biological) evolution is. This point can be extended to all other procedural implementations in Coevolutionary Computation since they all possess and operate based on those three components. Given such a perspective that natural evolution and artificial evolution constitute some forms of evolution as a general process that involve those three fundamental components, then there would be linkages between the two that could be exploited in both directions.

2.3 The Main Principle of Coevolution

One of the early motivations for the approach that Coevolutionary Computation takes for fitness evaluations is to address the issue where an absolute objective function that reflects the properties of the problem is not available or is very difficult to formulate. Instead, the fitness of the candidate is evaluated with the use of test cases [48]. This coevolutionary fitness evaluation involves a procedure that takes as input the candidate solution and a set of test cases, and then gives as output a scalar value. The measurements that this procedure carries out should possess two important properties. First, the output of this procedure uniquely reflects the performance of the candidate with respect to the set of test cases. Second, the differences in the output values should correspond to the performance differences of those evaluated candidates with respect to the set of test cases.

Informally, such a coevolutionary performance measurement is said to impose a metric on the set of candidate solutions. This is because one can obtain some understanding about the relationship between any pair of candidate solutions through the difference or distance between their fitness values. This measurement process can be more formally described in the context of ***test-based cooptimization*** [66]. The performance indicating how well a candidate $s \in S$ solves a test case $t \in T$ is described by the distance function or metric $M : S \times T \rightarrow F$ having range over a subset of the set real numbers $F \subset \mathbb{R}_{\geq 0}$. In coevolution, fitness evaluation of a candidate is made with respect to a set of test cases $T' \subseteq T$. The ***quality function*** of coevolved solutions is described by the aggregation of values returned by $M(s, t)$ over the set T' to obtain the quality or fitness value $f(s)$. Note for simplicity of presentation we consider $T' \subseteq T$ as sets rather than multisets or bags (i.e. elements can be repeated) that are quite common in actual practice.

There are various approaches with which one can then aggregate these values. This has been formalized in several studies that consider different ***solution concepts*** (notions of goodness) [30] based on ***generalization*** in Machine Learning [17, 70], ***pareto optimality*** in Multi-objective Optimization [25], and ***Nash equilibrium*** and ***dominance*** in Game Theory [34, 77]. One commonly used solution concept in coevolution is the ***maximum expected performance*** [24], which is related to generalization performance [17, 70]. If one assumes that each test case in the set has equal importance, then one can simply just take the sum $f(s) = \sum_{t \in T'} M(s, t)$ (since the sum and the average is related by a constant multiplier equal to the size of the set of test cases given by $|T'|$). For this solution concept, the best-performing candidate solution is one with $\max_{s \in S} f(s)$. Note as well that the choice of solution concept is crucial because it determines both the interpretation of what represents goodness in the solution but also the form it takes. In the case of maximum expected performance we have defined earlier, the solution can take the form of a single candidate. For other solution concepts, one has a set of solutions to represent the pareto front for pareto optimality [32, 59, 60], and mixtures of solutions or finite approximate proportions of solutions in an infinite population for Nash equilibrium [33].

Although we can now formally define fitness evaluations in coevolution, their actual procedural applications during the search process are more involved. The subjective fitness evaluations of candidate solutions refer to two properties for which fitness values are associated with reproduction success in coevolutionary processes [10]. The fitness of a coevolved candidate is ***relative*** and ***dynamic*** because the set of test cases that is used for the evaluation procedure is evolving as well. To illustrate these two properties in a more practical and actual setting, consider the population (set) of test cases $T'_{t,k}$ that is used at iteration t of a coevolutionary search procedure on the kth run. Note that as with most standard practices in using stochastic algorithms that include EAs and coevolution, the search procedure is repeated $K > 1$ times with individual run indexed $k = 1, 2, 3, \ldots, K$. Suppose we have a distinct coevolved candidate solution $\boldsymbol{s}$ that is present in each iteration and run. Then, its fitness with respect to $\boldsymbol{T}'_{t,k}$ is given by $f(\boldsymbol{s}, \boldsymbol{T}'_{t,k})$. There is the possibility that the fitness value of $\boldsymbol{s}$ differs for different sets of test cases. This is where the notion of relative fitness refers to, i.e. that the fitness values of $\boldsymbol{s}$ depend entirely on $\boldsymbol{T}'_{t,k}$ so that $f(\boldsymbol{s}_i, \boldsymbol{T}'_{t,k}) \neq f(\boldsymbol{s}_i, \boldsymbol{T}'_{u,j})$ whenever $\boldsymbol{T}'_{t,k} \neq \boldsymbol{T}'_{u,j}$. The notion of dynamic fitness refers to the situation within a particular run k that $f(\boldsymbol{s}_i, \boldsymbol{T}'_{t,k}) \neq f(\boldsymbol{s}_i, \boldsymbol{T}'_{u,k})$ whenever $\boldsymbol{T}'_{t,k} \neq \boldsymbol{T}'_{u,k}$ if $t \neq u$.

Note that for this specific solution concept that involves $\max_{\boldsymbol{s} \in \boldsymbol{S}} f(\boldsymbol{s})$, $f(\boldsymbol{s}) = \sum_{\boldsymbol{t} \in \boldsymbol{T}'} M(\boldsymbol{s}, \boldsymbol{t})$, the fitness becomes objective if $\boldsymbol{T}' = \boldsymbol{T}$ is used. In this case, the preferences between any two candidate solutions $\boldsymbol{s}, \boldsymbol{s}' \in \boldsymbol{S}$ are resolved by their fitness values, e.g. as described by the binary relation $f(\boldsymbol{s}, \boldsymbol{T}) \leq f(\boldsymbol{s}', \boldsymbol{T})$. For example, one has the binary relation $\boldsymbol{s} \rightarrow \boldsymbol{s}'$ ($\boldsymbol{s}'$ is preferred over $\boldsymbol{s}$) whenever $f(\boldsymbol{s}, \boldsymbol{T}) < f(\boldsymbol{s}', \boldsymbol{T})$. If one assumes that the two-player game is symmetric, then the complete underlying structure of the pairwise preferences of a finite $\boldsymbol{S}$ is captured by a ***directed graph*** (***digraph***) [20]. If the structure is such that there is a distinct dominant solution $\boldsymbol{s}^*$ whereby $f(\boldsymbol{s}, \boldsymbol{T}) < f(\boldsymbol{s}^*, \boldsymbol{T})$ for all $\boldsymbol{s} \in \boldsymbol{S} \backslash \{\boldsymbol{s}^*\}$, then $\max_{\boldsymbol{s} \in \boldsymbol{S}} f(\boldsymbol{s}) = f(\boldsymbol{s}^*)$ and the solution to this problem is $\boldsymbol{s}^*$.

In practice, $\boldsymbol{T}' = \boldsymbol{T}$ is not used because the computational cost to obtain $f(\boldsymbol{s}, \boldsymbol{T})$, $\boldsymbol{s} \in \boldsymbol{S}$ is too high, e.g. the set $\boldsymbol{T}$ is extremely large. As such, small-sized subsets $\boldsymbol{T}' \subset \boldsymbol{T}$ are used instead to ensure the computational cost for the entire coevolutionary search becomes feasible. In this situation, the relative and dynamic properties of coevolutionary fitness come to play in the search process. To understand in general how these two properties affect coevolutionary search, we continue on with our previous illustration. Suppose that the initial population is seeded with the same set of individual candidate solutions $\{\boldsymbol{s}_i\}$ across all the repeated runs. Note that we drop the indices t, k for simplicity in presentation. However, the set of test cases differs for each run and iterate of the procedure. Assume as well that the fitness value of a solution differs whenever a different set of test cases is used. In this situation, one has the set of fitness values $\{f(\boldsymbol{s}_i, \boldsymbol{T}'_{1,k})\}$ for the initial populations in runs $k = 1, 2, 3, \ldots, K$. The selection of individuals in $\{\boldsymbol{s}_i\}$ for reproduction depends on the fitness values $\{f(\boldsymbol{s}_i, \boldsymbol{T}'_{1,k})\}$. For selection mechanisms that prioritize fitness values such as fitness proportional selection, one can observe that the probability for selection of individual $\mathbb{P}_{\boldsymbol{s}_i}$ differs for each run k. For selection mechanisms that rank candidate solutions according to their fitness values, there is a possibility that changes in the ordering of $\{f(\boldsymbol{s}_i, \boldsymbol{T}'_{1,k})\}$ affect the reproduction rates of $\{\boldsymbol{s}_i\}$. In this specific example, the initial trajectory of the coevolutionary process could differ across each run k even though the same initial population of candidate solutions is seeded.

More importantly, the behaviour or dynamics of the coevolutionary process within a run for a successful search is contingent upon the right sequences of the sets of test cases that are presented to the evolving population of candidate solutions. This is often described as an ***arms race dynamics*** whereby the coevolutionary process drives the search of solutions of increasing quality of measurable overall performance against a more diverse set of test cases [31, 50, 61]. Crucially, this is done in the context of using a small-sized set of test cases that are also changing and improving in terms of presenting challenges to a more diverse set of solutions. Such dynamics is more clearly observed in cases involving a predator-prey coevolution whereby the roles of solutions and test cases are clearly delineated. However, as those studies have clearly noted, such arms race dynamics is by no means automatically realized whenever coevolution is being set up. For example, cyclic dynamics that prevents long-term progress is common and additional mechanisms are required to address it [75].

2.4 Concepts in Coevolution

Earlier, we have briefly discussed about the specific dynamics in the coevolutionary process in the context of an arms race between candidate solutions and test cases that are improving in turn as an indication of a successful coevolutionary search. Such an arms race dynamics is crucial in having to make full use of the subjective fitness evaluation approach that coevolution employs, thereby limiting or addressing inherent issues that come about from employing relative and dynamic fitness values to guide coevolutionary search. A specific coevolutionary setup is used for illustration. However, there are other setups in coevolution that have been designed and applied to solve various types of problems. Here, we will introduce and discuss two important concepts that serve to categorize these various coevolutionary systems. They are about population structures and the types of interactions between members of the coevolving populations. Note that these two concepts are not vacuous and both are crucial towards describing these various complex coevolutionary systems and identifying common features in these systems that allow their categorizations.

2.4.1 Population Structures in Coevolution

Population structures in coevolutionary systems refer to the specific configurations and placements of individuals that are under the action of selection and variation processes. The simplest population structure in coevolutionary systems is those that involve a single panmictic population. In such a ***single-population coevolution***, each member of the population plays the twin roles of being a candidate solution and a test case. As such, a single-population coevolution is associated with competitive coevolution although the reverse is not true. It is also not surprising that most of the single-population coevolutionary systems are applied to solve problems involving game-plays [13], as well as those that are adversarial in nature so as to be modelled as games. The latter includes problems in various domains such as robotics [78] and defence [22].

The more complex population structures involve a multi-population setup. In this case, one has $l \geq 2$ populations in the coevolutionary system. The introduction of such ***multi-population coevolution*** is crucial in allowing for the construction of coevolutionary systems to solve various problems by employing different kinds of mechanisms. In particular, various problem-solving methodologies can now be incorporated into coevolution. Further notions of features in coevolution are required to distinguish and categorize the various coevolutionary systems. One of these notions is about the type of interactions between members of the populations, which will be discussed in the next section in detail. As a very brief introduction, there are coevolutionary systems that are designed with a specific Divide-and-Conquer problem-solving mechanism. In particular, these systems perform problem decomposition requiring different populations whose members represent candidate solutions of the decomposed subproblems. Going back to the discussion about features in coevolution, two important and related aspects in the multi-population setup on population structures that will be further elaborated have to do with the degree of genetic isolation or independence of members of different populations.

We start our discussion with the simple case of a ***two-population coevolution***. Full genetic isolation between members of the two populations can arise because of the roles they play. In most ***predator-prey coevolution***, the candidate solutions and test cases constitute different roles and usually have different parametric representations. In this situation, they are placed in two separate and independent populations with full genetic isolation given that specific variations operators are required to perform the necessary transformation to generate new members in the respective population. For example, this predator-prey setup is used to coevolve a population of solutions in the form of sorting networks with a population of test cases that are lists or sequences of binary numbers to be sorted [48]. In another example, a predator-prey setup is used to coevolve a population of solutions in the form of CA classifiers with a population of test cases that are binary strings to have their density (of 1s or 0s) classified [51, 65]. This is despite in both examples, encoding is used so that the solutions are in the form of binary strings. The different lengths of the binary strings representing the solutions and test cases and crucially the underlying structures that point towards solution optimality and test case challenge rating preclude any meaningful attempt to employ crossovers between the two populations. Note that there are examples of predator-prey setup such as in [50] that coevolves robots having different phenotypic pursue and evade roles (e.g. different specifications such as maximum speed) but for which the underlying genotypic representations are the same, i.e. ANNs as controllers.

Another aspect for genetic isolation between coevolving populations has to do with design features or mechanisms in the system to promote population diversity as the means to have long-term and sustained search performance. As another example, the predator-prey setup can be implemented to coevolve agents for two-player games [72]. Here, the roles that are played by members or agents in both populations are the same. Full genetic isolation is imposed on the populations and that other mechanism for diversity enhancement such as fitness sharing is employed as design features in this coevolutionary system to improve search performance.

This two-population setup can be extended towards a multi-population setup. In [83], nine-population competitive coevolution is developed and shown to have superior search performance against a single-population competitive coevolution for the search of car racing controllers. Genetic isolation is used in the study although it is possible to introduce migration mechanisms in such a multi-population coevolution setup where it is established that the members of the populations all play the same role and thus constitute a valid complete candidate solution to the problem. In the latter situation, the multi-population setup is similar to Island models that are introduced and used for EAs [7, 79].

2.4.2 Interactions in Coevolution

Interactions are important in coevolution because they are the intrinsic means with which fitness values of individuals are derived, obtained, and subsequently interpreted via a defined solution concept and used to guide the search process [25]. So far most of our discussions have focused on interactions in coevolution that are suitably modelled as competitive and noncooperative games given that there are many examples of real-world problems that are adversarial in nature. For the case of a single-population competitive coevolution, the simplest form of interactions is a ***symmetric*** two-player game. There are no distinctions between the two players, i.e. both roles of being a solution and a test case are identical. Any one player can implement a game-playing strategy from the same set of strategies $\boldsymbol{S}$. A symmetric two-player game can be played simultaneously. Simple examples of such games are the IPD [1] and the ***Hawk-Dove*** [34, 39, 40].

For these games that are played ***one shot*** over a single move, complete specifications can be easily made to describe the interactions as well as their outcomes. Each game is specified as the pair $(\boldsymbol{s}_i, \boldsymbol{s}_j) \in \boldsymbol{S} \times \boldsymbol{S}$, with the first player (denoted subscript i) and second player (denoted subscript j) implementing strategies drawn from the same set $\boldsymbol{s}_i, \boldsymbol{s}_j \in \boldsymbol{S}$. The Cartesian product $\boldsymbol{S} \times \boldsymbol{S}$ specifies all the games that can be played over any combination $(\boldsymbol{s}_i, \boldsymbol{s}_j)$ of strategies $\boldsymbol{s}_i, \boldsymbol{s}_j \in \boldsymbol{S}$. For a two-player game with $V = |\boldsymbol{S}|$ number of pure strategies, there are V^2 number of possible games. A one-shot simultaneous game also allows one to specify the outcomes of the game directly in terms of ***payoff***. Then, outcomes from all possible games are stored in a square payoff matrix called the normal form or standard form [58]. For the first player, this is specified by the $\mathbf{A} = (a_{ij} : 1 \leq i, j \leq V)$ with each entry a_{ij} assigned a nonnegative payoff from the outcome of $(\boldsymbol{s}_i, \boldsymbol{s}_j)$. The symmetric game structure allows one to directly obtain the payoff matrix for the second player by taking $\mathbf{B} = \mathbf{A}^\mathsf{T}$, i.e. $(b_{ij} : 1 \leq i, j \leq V) = (a_{ji} : 1 \leq i, j \leq V)$.

Other games that model more complex interactions between two players or roles are possible. The one-shot simultaneous games whose outcomes are described by standard form matrices can be extended to multiple rounds. One classical example is the IPD game that has been extensively studied [12, 17, 19, 36, 37, 43]. Note that in the original Prisoner's Dilemma or other games that can be described in the standard form, immediate analysis can be usually made because the matrices involved are small-sized. This includes determining whether there is a ***dominant*** strategy [58], which is defined as the strategy $\boldsymbol{s}_i \in \boldsymbol{S}$ whose outcomes a_{ij} against any opponent strategy $\boldsymbol{s}_j \in \boldsymbol{S}$ satisfy $a_{ij} \geq a_{kj}$ for any alternative strategy $\boldsymbol{s}_k \in \boldsymbol{S}$. One would immediately observe that extending such games, for example, with multiple rounds or other kinds of extensions such as multiple choices [14] would lead to a combinatorial explosion in the number of pure strategies. This would preclude having immediate analysis on the game unless strong constraints are imposed on the game outcome matrices. More importantly, there are other forms of interactions involving two players beyond simple sequences of moves that are drawn from a fixed set (e.g. cooperate and defect in Prisoner's Dilemma). Board games are played sequentially, with possibly asymmetric roles (white and black players might have different strategies of play), and whose sequences of moves are highly involved, nonlinear, and constrained [16, 38, 41]. What this means is that a game-play has to be realized, by either direct simulation or computation, in order to obtain the outcomes for both players. This is the situation encountered in most practical, real-world applications of coevolution.

However, interactions in coevolution are not restricted to those that are adversarial in nature, and they can take the form of cooperative games. Indeed, in the original study on the framework of cooperative coevolution [67], literature references and discussions have been made on natural and artificial systems in the real world that require some form of joint interactions between multiple agents to form the complete solution. Furthermore and in the language of cooperative game theory, there is the issue of determining whether there are specific ***coalitions*** of these agents that as the combined whole are more optimal or preferred over others. There are few studies such as [64, 86] that have explicitly employed game-theoretic analysis of cooperative coevolutionary systems. These studies are focused on a ***Multi-agent Learning*** setting that involve a team of adaptive agents that have to concurrently learn and jointly interact to solve a task. The challenges posed in such a setting can be naturally addressed through cooperative coevolutionary learning systems. Evolutionary robotics is one such problem domain whereby cooperative coevolutionary systems are used [42]. In addition to the complexity involved in formulating joint interactions to solve coordination tasks, the quality or fitness evaluations in these problem settings are highly nontrivial. The combined solution that is represented by a specific coalition of agents, each of which being a representative of the multiple populations in the system, can be evaluated to determine its quality or fitness. For an individual agent, its quality or fitness might depend on the specific constituents of the remaining part of the team. Thus, it can be challenging or computationally expensive to exactly determine an agent's fitness by way of its marginal contribution to the team's performance. This requires examining through all possible combinations of agents that can be drawn from their populations, and as such an agent's fitness can only be estimated in practice.

Cooperative coevolution has also been applied to other problem-solving settings such as the optimization of stationary objective functions that admit problem decomposition into subproblems [56]. Here, interactions between candidate solutions of their respective subproblems are intimately tied to the nature of problem decomposition. This is because the crucial Divide-and-Conquer strategy in this setting is to identify the independent subproblems or subcomponents and then be able to evaluate candidate solutions for those subcomponents. The specific property that has been studied is that of problem ***separability***. To illustrate this notion of separability, consider a continuous optimization problem with an n-dimensional solution space $\boldsymbol{S} = \mathbb{R}^n$. Assuming a minimization setting, a fully separable optimization problem $f : \boldsymbol{S} \rightarrow \mathbb{R}$ is one that admits

$$\underset{x_1,\ldots,x_n}{\operatorname{argmin}} f(\boldsymbol{x}) = \left(\underset{x_1}{\operatorname{argmin}} f(x_1, \ldots), \ldots, \underset{x_n}{\operatorname{argmin}} f(\ldots, x_n) \right).$$

A fully separable optimization lacks any interdependency between the decision variables x_i. Indeed, a fully separable n-dimensional minimization problem can be reformulated and solved as n independent, one-dimensional minimization problems separately. It can also be solved via cooperative coevolution with a setup of $l = n$ isolated populations. Fitness evaluations could be made straightforward when they are ***completely additively separable*** $f(\boldsymbol{x}) = \sum_i^n f(x_i)$ if representative candidate solutions of the remaining $n - 1$ subcomponents are identified. There are nonseparable problems that are ***additively separable*** in the sense of $f(\boldsymbol{x}) = \sum_i^m f_i(\boldsymbol{x}_i)$, where there are $m \leq n$ independent subcomponents $\boldsymbol{x}_1, \ldots, \boldsymbol{x}_m$. These problems can be solved directly if one is able to identify decision variables that have interdependencies and group them together. Mechanisms for grouping these decision variables range from random approaches [53] to those that employ search [62].

2.5 Concluding Remarks and Notes for Further Reading

In our introduction, we have sought to present Coevolutionary Computation as a general problem-solving framework and in this context provided an account of its historical development from a more categorical perspective. In particular, we have introduced coevolutionary systems as a general, evolutionary-based methodology to solve problems that involve situations of strategic decision-making, which can be abstracted and modelled as games. There is a natural link between Coevolutionary Computation and Game Theory, given that the basis with which both fields of study are framed upon is that of interactions. Our introduction of this framework later progressed from its applications to problems that are framed in the context of Machine Learning and Optimizations, and then to a higher level of problem solving that is akin to metaheuristics and hyper-heuristics. In the latter case and considering a competitive coevolutionary setting, the candidate solutions or agents in such coevolutionary systems take the role of algorithms. Here, coevolutionary principles are used to search over the parameter space of optimizers instead of usual parameter space of decision variables in an optimization problem. Performance measurements of these agents implementing specific optimization algorithms (strategies) are done in a test-based manner against generated instances of some optimization problems. The crucial take-away message about applying Coevolutionary Computation is about how one can frame a specific problem-solving setting such that the evolutionary process for which candidate solutions are subjected to is coupled with the process with which test cases are generated from. This coupling mechanism is achieved through interactions between the two populations of solutions and test cases. Even though the two populations are independent or isolated, their interactions produce fitness or quality measures that are then used within the respective population that favours certain solutions (test cases) for reproduction.

We close this chapter with brief remarks of Coevolutionary Computation and Machine Learning for which there are commonalities between the two fields of study. Our focus here is targeted more towards shared design aspects of individual components from a system's perspective. We will restrict our following discussion at the conceptual level rather than specific implementations. Furthermore, we will consider interactions that are adversarial in nature and can be modelled as competitive games given that in this setting it is more straightforward to look for conceptual similarities. We start this discussion and first consider a more relaxed or general view of coevolutionary principles in the context of agent interactions as the main conceptual link, instead of the specific refinement of relative fitness evaluations arising from interactions against a changing or evolving set (e.g. dynamic fitness landscape view of coevolution [10]). In particular, one of the early studies on competitive coevolution that has made explicit references to mechanisms used in Machine Learning approaches is [70]. The study is mostly credited for introducing two important mechanisms—(1) a diversity maintenance technique through fitness sharing and (2) an external archive called Hall-of-Fame that stores specific test cases (opponent strategies) that is similar to ***taboo search*** mechanism—to promote arms race dynamics in coevolution.

The study employs a predator-prey coevolutionary system to implement a learning procedure for competitive games. Interested readers might note that the same authors have formulated a theoretical framework for competitive learning of games earlier in [69]. Although a large part of this study is theoretical and mostly deals with complexity analysis of their proposed computational learning procedure of competitive games (e.g. establishing the time complexity for discovering the dominant strategy), their theoretical results have shaped and directed the design of specific mechanisms within a two-population competitive coevolutionary learning system in [70]. The complexity analysis of the proposed computational learning algorithm depends on a specific and idealized construction involving two separate sets of learners and teachers. Interactions between members of these two sets are framed as two-player competitive games. The set of learners (learning set) represent the population whose constituents can change with iterative application of the learning algorithm and where the search process for the dominant or optimal strategy of the game resides. The set of teachers (teaching set) constitutes the opponent strategies that are used to evaluate the performance and to distinguish between learners. Iterative algorithm is also applied to the teaching set to ensure the second feature of distinguishing between learners is satisfied. This is achieved whenever any learner in the current learner set is defeated by at least one teacher in the current teaching set. Theoretical results of this procedure have provided insights towards the important requirements of implementing a teaching set (i.e. a small-sized set can be used so long they can distinguish between the learners). These requirements are later encapsulated by the mechanisms of fitness sharing and employing an external archive.

Readers would note that [69, 70] have used various concepts and terminologies from Machine Learning studies. This is not surprising given that the studies have sought to develop a computational learning framework for competitive games while at the same time are influenced and inspired by evolutionary approaches introduced earlier in [48]. In the same manner, we [17, 19] have used the notion of generalization as the means with which performance measurement in competitive coevolution can be made objective and realized. For simplicity in presentation here, we consider interactions in the form of symmetric two-player games. The concept of such performance measure has to do with evaluating the strategy against a sample of test (opponent) strategies, randomly drawn from the strategy space $\boldsymbol{S}$ with an underlying probability distribution $\mathbb{P}_{\boldsymbol{S}}$. Let $i, j \in \boldsymbol{S}$ represent strategies $\boldsymbol{s}_i, \boldsymbol{s}_j \in \boldsymbol{S}$. The ***generalization performance*** G_i of a strategy i is defined as its expected performance $G_i = \mathbb{E}_{\mathbb{P}_{\boldsymbol{S}}(j)}[G_i(J)] = \sum_j^{|\boldsymbol{S}|} \mathbb{P}_{\boldsymbol{S}}(j) G_i(j)$, where the random variable J represents the selection of $j \in \boldsymbol{S}$ with probability $\mathbb{P}_{\boldsymbol{S}}(j)$ and $G_i(j)$ is the game outcome for i between the pair of strategies $\{i, j\}$. Distribution-free estimators for G_i can be easily constructed along with known confidence bounds using a small sample of $N < |\boldsymbol{S}|$ random test strategies $S_N = \{s_1, \ldots, s_N\}$ [17]. The loose confidence bounds can be tightened by exploiting the near-Gaussian nature of the estimate $\hat{G}_i(S_N) = \big(G_i(s_1) + \cdots + G_i(s_N)\big)/N$ via the Central Limit Theorem for the sequence of random samples $\{G_i(s_1), \ldots, G_i(s_N)\}$ [19].

A later study [66] has sought to more explicitly make the conceptual links between ***competitive coevolution***, which is formulated in the context of cooptimization, and supervised learning. This is done by way of mapping the conceptual formalism underlying supervised learning to that of cooptimization. The study begins by defining the problem setting of cooptimization, which we have briefly alluded to when we introduce the notion of fitness evaluation for coevolution in Sect. 2.3. More specifically, competitive coevolution in a ***cooptimization setting*** involves searching over the space of solutions $\boldsymbol{S}$ to with the goal of maximizing the objective $f(\boldsymbol{s}) = \sum_{t \in \boldsymbol{T}} M(\boldsymbol{s}, \boldsymbol{t})$, where $M(\boldsymbol{s}, \boldsymbol{t})$ is a performance metric given by the function $M : \boldsymbol{S} \times \boldsymbol{T} \rightarrow F$, where $F \subset \mathbb{R}_{\geq 0}$. For example, a family of cooptimization problems could involve interactions in the form of symmetric two-player competitive games with win-lose outcomes (e.g. $\{0, 1\} \subset \mathbb{R}_{\geq 0}$ with a win-outcome assigned the real number 1 and a lose-outcome assigned the real number 0). In this setting, the goal is to find the optimal strategy $\boldsymbol{s}^*$ with $\max f(\boldsymbol{s}) = f(\boldsymbol{s}^*)$, whereby $\boldsymbol{T} = \boldsymbol{S}$ and that $\boldsymbol{s}^*$ is a dominant strategy, i.e. $0 < M(\boldsymbol{s}^*, \boldsymbol{t})$ for all $\boldsymbol{t} \in \boldsymbol{T} \backslash \{\boldsymbol{s}^*\}$ [20].

Similarly, the study [66] has described ***supervised learning setting*** within the formalism of cooptimization, choosing the family of binary classification problems for illustration. Here, problem solving in a supervised learning setting involves finding the correct assignment or target function $g : X \rightarrow Y$, where X is the set of objects to be classified and $Y = \{0, 1\}$ is the set of labels. The learning of the target function g is supervised by virtue of the process making use the finite set of labelled examples $\{(x_i, y_i)\} \subset X \times Y$, each example being specified by the pair $(x_i, y_i), i = 1, \ldots, N$ of the object x_i and its correct label y_i. In practice, g is implemented as a classifier in a parametric form c for which its internal parameters are given by $s \in S$. We write $c_s : X \rightarrow Y$ to emphasize function (input-output) response of $s \in S$. In cooptimization formalism, the search process of $s \in S$ is performed using the set of test cases $\{t_i\} \subset T = X$, with the metric $M(s, t) = \mathbf{1}_{\{g(t)\}}\big(c_s(t)\big)$ and having the objective of maximizing $f(s) = \sum_{t \in T} M(s, t)$. Note that the metric $M(s, t)$ is of the type $S \times T \rightarrow F$, where $F = \{0, 1\}$ and is specified by the indicator function of the set $\{g(t)\}$ of correct labels of all classifiable objects $t \in X$. In this case, a classifier s^* is said to be optimal if it classifies all objects in X correctly, with $\max f(s) = |T| = |X|$. In particular, $\mathbf{1}_{\{g(t)\}}\big(c_s(t)\big) = 1$ whenever the classifier agrees with the actual label $c_s(t) = g(t)$ of the test case $t \in T$. The metric can be reformulated as an error using the zero-one loss function given by $L(s, t) = 1 - \mathbf{1}_{\{g(t)\}}\big(c_s(t)\big)$, in which case the objective is then to minimize loss $f(s) = \sum_{t \in T} L(s, t)$.

We close this chapter with a remark comparing competitive coevolutionary learning with ***Temporal Difference (TD)*** learning [9], a class of ***Reinforcement Learning*** approaches, which has been applied successfully to the problem of learning competitive game strategies [81]. There have been studies such as in [71] that compare coevolutionary learning with TD Learning for competitive games. Both methodologies provide the means with which agent behaviours can be adapted while operating in an environment through some bootstrapping mechanisms to address the lack of appropriate target-teacher set or function. In competitive game learning, these bootstrapping mechanisms in principle employ (possibly different versions of the) agent that serves as opponents to generate game-plays that are then used to guide the learning process.

However, this is where the similarity ends since the processes in coevolutionary and TD learning use game-plays differently. On the one hand, coevolution uses outcomes from complete game-plays between various strategies implemented by a population of agents to determine their performance differences as the means to guide the search. On the other hand, TD learning can use the sequences of moves within the game itself through an agent's self-play to guide the search [81]. This principal difference comes about from the interpretation of how problem solving with regard to game-play learning is performed even though similar ANNs are used as solution representations in both methodologies [71]. In coevolutionary learning, ANNs are used as a nonlinear response function that maps (input) game states to (output) legal moves, and thus search the space of game-playing strategies directly. In TD learning, ANN is a function approximator of the value function that maps each game states with a long-term performance measure. This value function is associated with a controller that implements the game-playing strategy (policy) [9].

References

1. Ashlock, D.A.: Cooperation in Prisoner's Dilemma on graphs. In: Proceedings of the IEEE Symposium on Computational Intelligence and Games (CIG 2007), pp. 48–55 (2007)
2. Axelrod, R.: Effective choice in the Prisoner's Dilemma. J. Conflict Resolut. **24**(1), 3–25 (1980)
3. Axelrod, R.: More effective choice in the Prisoner's Dilemma. J. Conflict Resolut. **24**(3), 379–403 (1980)
4. Axelrod, R.: The evolution of strategies in the iterated Prisoner's Dilemma. In: Davis, L.D. (ed.) Genetic Algorithms and Simulated Annealing, chap. 3, pp. 32–41. Morgan Kaufmann, New York (1987)
5. Bäck, T., Hammel, U., Schwefel, H.-P.: Evolutionary computation: comments on the history and current state. IEEE Trans. Evol. Comput. **1**(1), 3–17 (1997)
6. Barricelli, N.A.: Numerical testing of evolution theories: part II: preliminary tests of performance, symbiogenesis and terrestrial life. Acta Biotheor. **16**(3–4), 99–126 (1963)
7. Belding, T.C.: Parallel genetic algorithms for a hypercube. In: Eshelman, L.J. (ed.) Proceedings of the Sixth International Conference on Genetic Algorithms, pp. 114–121. Morgan Kaufmann, San Francisco (1995)
8. Bongard, J.C., Lipson, H.: Nonlinear system identification using coevolution of models and tests. IEEE Trans. Evol. Comput. **9**(4), 361–384 (2005)
9. Bradtke, S.J., Barto, A.G.: Linear least-squares algorithms for temporal difference learning. Mach. Learn. **22**, 33–57 (1996)
10. Bullock, S.: Dynamic fitness landscapes. Technical Report. Research Paper CSRP 350, School of Cognitive & Computing Sciences, University of Sussex (1994)
11. Cartlidge, J., Bullock, S.: Combating coevolutionary disengagement by reducing parasite virulence. Evol. Comput. **12**(2), 193–222 (2004)
12. Chellapilla, K., Fogel, D.B.: Evolution, neural networks, games, and intelligence. Proc. IEEE **87**(9), 1471–1496 (1999)
13. Chellapilla, K., Fogel, D.B.: Evolving an expert checkers playing program without using human expertise. IEEE Trans. Evol. Comput. **5**(4), 422–428 (2001)
14. Chong, S.Y., Yao, X.: Behavioral diversity, choices, and noise in the iterated Prisoner's Dilemma. IEEE Trans. Evol. Comput. **9**(6), 540–551 (2005)
15. Chong, S.Y., Yao, X.: Multiple choices and reputation in multiagent interactions. IEEE Trans. Evol. Comput. **11**(6), 689–711 (2007)
16. Chong, S.Y., Tan, M.K., White, J.D.: Observing the evolution of neural networks learning to play the game of Othello. IEEE Trans. Evol. Comput. **9**(3), 240–251 (2005)
17. Chong, S.Y., Tiňo, P., Yao, X.: Measuring generalization performance in co-evolutionary learning. IEEE Trans. Evol. Comput. **12**(4), 479–505 (2008)
18. Chong, S.Y., Tiňo, P., Yao, X.: Relationship between generalization and diversity in coevolutionary learning. IEEE Trans. Comput. Intell. AI Games **1**(3), 214–232 (2009)
19. Chong, S.Y., Tiňo, P., Ku, D.C., Yao, X.: Improving generalization performance in co-evolutionary learning. IEEE Trans. Evol. Comput. **16**(1), 70–85 (2012)
20. Chong, S.Y., Tiňo, P., He, J., Yao, X.: A new framework for analysis of coevolutionary systems - directed graph representation and random walks. Evol. Comput. **27**(2), 195–228 (2019)
21. Darwen, P.J., Yao, X.: Speciation as automatic categorical modularization. IEEE Trans. Evol. Comput. **1**(2), 101–108 (1997)
22. Darwen, P.J., Yao, X.: Co-evolution in iterated Prisoner's Dilemma with intermediate levels of cooperation: application to missile defense. Int. J. Comput. Intell. Appl. **2**(1), 83–107 (2002)

23. Darwin, C.: On the Origin of Species by Means of Natural Selection, or the Preservation of Favoured Races in the Struggle for Life. John Murray, London (1859)
24. de Jong, E.D.: The maxsolve algorithm for coevolution. In: Proceedings of the 2005 Conference on Genetic and Evolutionary Computation (GECCO'05), New York, pp. 483–489 (2005)
25. de Jong, E.D., Pollack, J.B.: Ideal evaluation from coevolution. Evol. Comput. **12**(2), 159–192 (2004)
26. Dobzhansky, T.: Genetics and the Origin of Species. Columbia University Press, New York (1937)
27. Ehrlich, P.R., Raven, P.H.: Butterflies and plants: a study of coevolution. Evolution **18**(4), 586–608 (1964)
28. Eiben, A.E., Smith, J.: From evolutionary computation to the evolution of things. Nature **521**, 476–482 (2015)
29. Falgueras-Cano, J., Falgueras-Cano, J., Moya, A.: A study of the coevolution of digital organisms with an evolutionary cellular automaton. Biology **10**(11), 1147 (2021)
30. Ficici, S.G.: Solution concepts in coevolutionary algorithms. Ph.D. thesis, Brandeis University, MA (2004)
31. Ficici, S.G., Pollack, J.B.: Challenges in coevolutionary learning: arms-race dynamics, open-endedness, and mediocre stable states. In: Adami, C., Belew, R.K., Kitano, H., Taylor, C. (eds.) Artificial Life VI: Proceedings of the Sixth International Conference on Artificial Life, pp. 238–247. The MIT Press, Cambridge (1998)
32. Ficici, S.G., Pollack, J.B.: Pareto optimality in coevolutionary learning. In: Advances in Artificial Life: 6th European Conference (ECAL'01). Lecture Notes in Computer Science, vol. 2159, pp. 316–325. Springer, Prague (2001)
33. Ficici, S.G., Pollack, J.B.: A game-theoretic memory mechanism for coevolution. In: Proceedings of the 2003 Conference on Genetic and Evolutionary Computation (GECCO'03). Lecture Notes in Computer Science. vol. 2723, pp. 286–297. Springer, Berlin (2003)
34. Ficici, S.G., Melnik, O., Pollack, J.B.: A game-theoretic and dynamical-systems analysis of selection methods in coevolution. IEEE Trans. Evol. Comput. **9**(6), 580–602 (2005)
35. Fisher, R.A.: The Genetical Theory of Natural Evolution. Clarendon Press, Oxford (1930)
36. Fogel, D.B.: Evolving behaviors in the iterated Prisoner's Dilemma. Evol. Comput. **1**(1), 77–97 (1993)
37. Fogel, D.B.: On the relationship between the duration of an encounter and the evolution of cooperation in the iterated Prisoner's Dilemma. Evol. Comput. **3**(3), 349–363 (1996)
38. Fogel, D.B.: Blondie24: Playing at the Edge of AI. Morgan Kaufmann, San Francisco (2002)
39. Fogel, D.B., Fogel, G.B., Andrews, P.C.: On the instability of evolutionary stable strategies. Biosystems **44**, 135–152 (1997)
40. Fogel, D.B., Fogel, G.B.: Simulating natural selection as a culling mechanism on finite populations with the hawk–dove game. Biosystems **104**, 57–62 (2011)
41. Fogel, D.B., Hays, T.J., Hahn, S.L., Quon, J.: A self-learning evolutionary chess program. Proc. IEEE **92**(12), 1947–1954 (2004)
42. Fontbonne, N., Maudet, N., Bredeche, N.: Cooperative co-evolution and adaptive team composition for a multi-rover resource allocation problem. In: European Conference on Genetic Programming (EuroGP 2022). Lecture Notes in Computer Science, vol. 13223, pp. 179–193. Springer, Berlin (2022)
43. Franken, N., Engelbrecht, A.P.: Particle swarm optimization approaches to coevolve strategies for the iterated Prisoner's Dilemma. IEEE Trans. Evol. Comput. **9**(6), 562–579 (2005)
44. Futuyma, D.J.: Evolutionary biology today and the call for an extended synthesis. Interface Focus **7**, 20160145 (2017)
45. García-Pedrajas, N., Hervás-Martínez, C., Ortiz-Boyer, D.: Cooperative coevolution of artificial neural network ensembles for pattern classification. IEEE Trans. Evol. Comput. **9**(3), 271–302 (2005)
46. Gershenson, C., Trianni, V., Werfel, J., Sayama, H.: Self-organization and artificial life. Artif. Life **26**(3), 391–408 (2020)
47. Haldane, J.B.S.: The Causes of Evolution. Longmans, Green and Co., New York (1932)

48. Hillis, W.D.: Co-evolving parasites improve simulated evolution as an optimization procedure. Phys. D **42**, 228–234 (1990)
49. Huxley, J.: Evolution: The Modern Synthesis. Clarendon Press, Oxford (1942)
50. Janssen, R., Nolfi, S., Haselager, P., Sprinkhuizen-Kuyper, I.: Cyclic incrementality in competitive coevolution: evolvability through pseudo-Baldwinian switching-genes. Artif. Life **22**(3), 319–352 (2016)
51. Juillé, H., Pollack, J.B.: Coevolutionary learning: a case study. In: Proceedings for the International Conference on Machine Learning (ICML'98), Madison, WI, pp. 251–259 (1998)
52. Laland, K., Uller, T., Feldman, M., Sterelny, K., Müller, G.B., Moczek, A., Jablonka, E., Odling-Smee, J., Wray, G.A., Hoekstra, H.E., Futuyma, D.J., Lenski, R.E., Mackay, T.F.C., Schluter, D., Strassmann, J.E.: Does evolutionary theory need a rethink? Nature **514**, 161–164 (2014)
53. Li, X., Yao, X.: Cooperatively coevolving particle swarms for large scale optimization. IEEE Trans. Evol. Comput. **16**(2), 210–224 (2012)
54. Liu, S., Tang, K., Yao, X.: Generative adversarial construction of parallel portfolios. IEEE Trans. Cybernet. **52**(2), 784–795 (2022)
55. Lu, X., Menzel, S., Tang, K., Yao, X.: Cooperative co-evolution-based design optimization: a concurrent engineering perspective. IEEE Trans. Evol. Comput. **22**(2), 173–188 (2018)
56. Ma, X., Li, X., Zhang, Q., Tang, K., Liang, Z., Xie, W., Zhu, Z.: A survey on cooperative co-evolutionary algorithms. IEEE Trans. Evol. Comput. **23**(3), 421–441 (2019)
57. Mitchell, M., Crutchfield, J.P., Das, R.: Evolving cellular automata to perform computations: a review of recent work. In: Proceedings of the First International Conference on Evolutionary Computation and its Applications, pp. 42–55. Russian Academy of Sciences, Moscow (1994)
58. Nisan, N., Roughgarden, T., Tardos, E., Vazirani, V.V. (eds.): Algorithmic Game Theory. Cambridge University Press, New York (2007)
59. Noble, J.: Finding robust Texas Hold'em poker strategies using Pareto coevolution and deterministic crowding. In: Proceedings of the 2002 International Conference on Machine Learning and Applications (ICMLA'02), NV, pp. 233–239 (2002)
60. Noble, J., Watson, R.A.: Pareto coevolution: using performance against coevolved opponents in a game as dimensions for Pareto selection. In: Proceedings of the 2001 Genetic and Evolutionary Computation Conference (GECCO'01), CA, pp. 493–500 (2001)
61. Nolfi, S., Floreano, D.: Co-evolving predator and prey robots: do "arms races" arise in artificial evolution? Artif. Life **4**(4), 311–335 (1998)
62. Omidvar, M.N., Li, X., Mei, Y., Yao, X.: Cooperative co-evolution with differential grouping for large scale optimization. IEEE Trans. Evol. Comput. **18**(3), 378–393 (2014)
63. Pagie, L., Mitchell, M.: A comparison of evolutionary and coevolutionary search. Int. J. Comput. Intell. Appl. **2**(1), 53–69 (2002)
64. Panait, L., Luke, S., Wiegand, R.P.: Biasing coevolutionary search for optimal multiagent behaviors. IEEE Trans. Evol. Comput. **10**(6), 629–645 (2006)
65. Paredis, J.: Coevolving cellular automata: Be aware of the red queen! In: Proceedings of the Seventh International Conference on Genetic Algorithms, pp. 393–400. Morgan Kaufmann, East Lansing (1997)
66. Popovici, E.: Bridging supervised learning and test-based co-optimization. J. Mach. Learn. Res. **18**(38), 1–39 (2017)
67. Potter, M.A., Jong, K.A.D.: Cooperative coevolution: an architecture for evolving coadapted subcomponents. Evol. Comput. **8**(1), 1–29 (2000)
68. Reed, J., Toombs, R., Barricelli, N.A.: Simulation of biological evolution and machine learning: I. Selection of self-reproducing numeric patterns by data processing machines, effects of hereditary control, mutation type and crossing. J. Theor. Biol. **17**(3), 319–342 (1967)
69. Rosin, C.D., Belew, R.K.: A competitive approach to game learning. In: COLT'96: Proceedings of the Ninth Annual Conference on Computational Learning Theory, pp. 292–302 (1996)
70. Rosin, C.D., Belew, R.K.: New methods for competitive coevolution. Evol. Comput. **5**(1), 1–29 (1997)

71. Runarsson, T.P., Lucas, S.M.: Coevolution versus self-play temporal difference learning for acquiring position evaluation in small-board go. IEEE Trans. Evol. Comput. **9**(6), 540–551 (2005)
72. Schaul, T., Schmidhuber, J.: A scalable neural network architecture for board games. In: 2008 IEEE Symposium On Computational Intelligence and Games, pp. 357–364 (2008)
73. Seo, Y.G., Cho, S.B., Yao, X.: Emergence of cooperative coalition in NIPD game with localization of interaction and learning. In: Proceedings of the IEEE 1999 Congress on Evolutionary Computation (CEC'99), pp. 877–884. IEEE Press, Piscataway (1999)
74. Seo, Y.G., Cho, S.B., Yao, X.: Exploiting coalition in co-evolutionary learning. In: Proceedings of the IEEE 2000 Congress on Evolutionary Computation (CEC'00), pp. 1268–1275. IEEE Press, Piscataway (2000)
75. Simione, L., Nolfi, S.: Long-term progress and behavior complexification in competitive coevolution. Artif. Life **26**(4), 401–430 (2021)
76. Sims, K.: Evolving 3D morphology and behavior by competition. Artif. Life **1**(4), 353–372 (1994)
77. Smith, J.M.: Evolution and the Theory of Games. Cambridge University Press, Cambridge (1982)
78. Stanley, K.O., Miikkulainen, R.: Competitive coevolution through evolutionary complexification. J. Artif. Intell. Res. **21**, 63–100 (2004)
79. Tanese, R.: Parallel genetic algorithms for a hypercube. In: Grefenstette, J.J. (ed.) Proceedings of the Second International Conference on Genetic Algorithms on Genetic Algorithms and their Application, pp. 177–183. Lawrence Erlbaum, Hillsdale (1987)
80. Tang, K., Liu, S., Yang, P., Yao, X.: Few-shots parallel algorithm portfolio construction via co-evolution. IEEE Trans. Evol. Comput. **25**(3), 595–607 (2021)
81. Tesauro, G.: Temporal difference learning and TD-Gammon. Commun. ACM **38**(3), 58–68 (1995)
82. Thompson, J.N.: The Coevolutionary Process. Chicago University Press, Chicago (1994)
83. Togelius, J., Burrow, P., Lucas, S.M.: Multi-population competitive co-evolution of car racing controllers. In: 2007 IEEE Congress on Evolutionary Computation, pp. 4043–4050 (2007)
84. von Neumann, J., Morgenstern, O. (eds.): Theory of Games and Economic Behavior, 3rd edn. Princeton University Press, Princeton (1953)
85. Werfel, J., Mitchell, M., Crutchfield, J.P.: Resource sharing and coevolution in evolving cellular automata. IEEE Trans. Evol. Comput. **4**(4), 388–393 (2000)
86. Wiegand, R.P., Liles, W.C., Jong, K.A.D.: Analyzing cooperative coevolution with evolutionary game theory. In: Proceedings of the 2002 IEEE Congress on Evolutionary Computation, pp. 1600–1605 (2002)
87. Wright, S.: Evolution in Mendelian populations. Genetics **16**(2), 97–159 (1931)
88. Yao, X.: Evolutionary stability in the n-person iterated Prisoner's Dilemma. BioSystems **37**(3), 189–197 (1996)
89. Yao, X., Darwen, P.J.: An experimental study of n-person iterated Prisoner's Dilemma games. Informatica **18**, 435–450 (1994)
90. Yao, X., Darwen, P.J.: How important is your reputation in a multi-agent environment. In: Proceedings of the IEEE 1999 Conference on Systems, Man, and Cybernetics (SMC'99), Tokyo, pp. 575–580 (1999)
91. Yuan, B., Lu, X., Tang, K., Yao, X.: Cooperative coevolution-based design space exploration for multi-mode dataflow mapping. ACM Trans. Embedded Comput. Syst. **20**(3), 1–25 (2021)

Chapter 3
Principal Approaches of Coevolution: Competitive and Cooperative

3.1 Coevolutionary Algorithms: An Introduction

In Chap. 2, we have used ***coevolutionary systems*** as the general, umbrella term to refer to the various implementations of Coevolutionary Computation. Although a larger number of studies have employed these coevolutionary systems for problem solving in settings such as for learning and optimization problems, these systems are also used for other settings such as simulations. In order to illustrate this further, consider the nature of a broad class of ***noncooperative games*** such as the Iterated Prisoner's Dilemma (IPD) that as a family of games has both competitive and cooperative elements [1, 36]. Here, coevolutionary systems can be used in two general settings: (1) To evaluate the performance of search procedures for game-playing strategies such as ***coevolutionary learning*** that employs ***Coevolutionary Algorithms*** (***CEAs***) when the solution concept is focused on the performance of a game-playing strategy in comparison to other strategies individually (e.g. ***All Defect*** is the ***dominant*** pure strategy) [9, 10]. (2) To study conditions in the adaptation processes that give rise to specific population behaviours through ***coevolutionary simulation*** when the solution concept is focused on the performance of the entire population as a whole (e.g. evolution to mutual cooperation that simultaneously maximizes payoffs of all coevolving agents) [7].

This chapter will focus on the use of coevolutionary systems that are targeted towards those related to the first problem-solving setting that involves an aspect of search for solutions with superior performance. In particular, we will provide descriptions on the key implementation details as well as the mechanisms that are employed to realize these CEAs. Our example-driven presentations will use two problem-solving settings—learning and optimization—to better illustrate algorithmic details of these competitive and cooperative coevolution. Note that in discussing these two problem-solving settings, we consider key aspects in problem challenges that are addressed by the respective coevolutionary approaches within the context of search through the solution space. For example, in the setting of

X. Yao, S. Y. Chong, *Coevolutionary Computation and Its Applications*,
Natural Computing Series, https://doi.org/10.1007/978-981-96-2841-4_3

competitive coevolution for two-player board games, the problem-solving aspect that is addressed is about searching for the optimal game-playing strategies whose solution representation takes a parametric form of board state evaluation functions, e.g. Artificial Neural Networks (ANNs).

These stimulus-response or reflex agents essentially perform a single-ply search with the board state evaluations produced by the parametric functions adopted by the agents [5, 8, 17, 18]. In other words, a large part of the complexity required in reasoning about game-play that requires generating the appropriate sequences of moves is abstracted within the solution representation. This is different compared to the classical Artificial Intelligence problem-solving approach of game-tree search that is about finding paths in the search tree that generates all possible legal sequences of alternating moves between the two players for those that terminate with a win. Other more involved search problems include ***planning*** by ***knowledge-based*** agents that involve procedures for devising a plan or a sequence of actions when operating in an environment based on the use or reasoning with some internal representation of knowledge about the problem [48]. There have been increasing studies on the use of coevolution for these problems, which include among others those in the settings of Evolutionary Robotics and ***Real-Time Strategy*** games [14, 20, 44].

We will describe and discuss in more detail these implementations and mechanisms of Coevolutionary Algorithms (CEAs) for problem solving for the rest of this chapter. Section 3.2 will formulate the main working principles of competitive and cooperative CEAs within the generate-and-test framework involving specific interactions. We will use these competitive and cooperative frameworks as key points in discussions to differentiate between both approaches. Section 3.3 will focus on the use of these coevolutionary approaches for problem solving over two subsections. The first subsection will discuss competitive CEAs for learning problems, while the second subsection will discuss cooperative CEAs for optimization problems. The chapter is closed with remarks and discussions on how both competitive and cooperative coevolutionary approaches can be combined to solve challenging problems.

3.2 The Frameworks of Coevolutionary Algorithms

This section will introduce the general frameworks for both coevolutionary approaches. We start with introducing the framework for competitive CEAs in Sect. 3.2.1, which is then followed by the framework for cooperative CEAs in Sect. 3.2.2. Both frameworks are expanded from the general generate-and-test framework of EAs (see Algorithm 1.1) that has been discussed earlier in Chap. 1.2. One can observe these frameworks as specific extensions that incorporate the principal mechanisms of agent interactions that enable the particular coevolutionary search approaches. They include adversarial interactions in competitive coevolution and Divide-and-Conquer-like strategic interactions in cooperative coevolution.

3.2.1 Competitive Coevolution

The general framework for ***competitive coevolution*** (***competitive CEA***) is described in Algorithm 3.1. Note that in presenting this framework, we consider the more general setting whereby the adversarial interactions between agents in coevolution are abstracted and modelled by asymmetric two-player games. This means that the two players have different roles, with agents playing the role of the first player have access to a different set of strategies compared to agents for the second player. Indeed, for the more specific setting whereby interactions are in the form of a symmetric two-player game, Algorithm 3.1 can be immediately reformulated to use a single population and will procedurally appear more similar to the framework of EAs given by Algorithm 1.1.

However, in the more general setting of asymmetric interactions, the assumption that the sets of strategies differ for the two player roles will point to other design features that are intrinsic to this framework. In particular, one notes the distinguishing feature of having two genetically isolated populations. In the context of cooptimization [41], the population consisting of candidate solutions is represented by $\boldsymbol{X}_{\mathbf{s}} = \big((\boldsymbol{x}_{\mathbf{s}}, f)_i : i = 1, 2, 3, \ldots, n\big)$, while the population of test cases is represented by $\boldsymbol{X}_{\mathbf{t}} = \big((\boldsymbol{x}_{\mathbf{t}}, f)_i : i = 1, 2, 3, \ldots, m\big)$. Note that in the literature of Coevolutionary Computation, other terminologies have been introduced and used. Some are biologically inspired, and they include predator-prey and host-parasite coevolution [4, 24, 46]. However, these coevolutionary systems can be procedurally described within the competitive coevolution framework we introduce and discuss here.

The framework in Algorithm 3.1 describes competitive coevolution as a process on two genetically isolated populations, each undergoing an adaptation process that is performed sequentially or in turn over the other. The general competitive CEA is implemented via the main procedure **CMCEA**, which makes clear the population of candidate solutions $\boldsymbol{X}_{\mathbf{slocal}}$ undergoes adaptation via **EVOLPOPSOL**, while the population of test cases $\boldsymbol{X}_{\mathbf{tlocal}}$ undergoes adaptation via **EVOLPOPTEST**. The genetic isolation of the two populations is crucial because the respective mechanism with which the adaptation process is applied to agents within a particular population differs for the two roles. For example, the specific variation operator that is called and executed within **variationSol** generates individual child candidate solution $(\boldsymbol{x}_{\mathbf{s}})_i$ of type $\mathbb{R}^D$. The procedure **variationSol** copies these generated individuals into a partition of the full population $\boldsymbol{X}_{\mathbf{slocal}}$ before returning the new full population $\boldsymbol{X}_{\mathbf{slocal}}$. We have **variationTest** operating on $\boldsymbol{X}_{\mathbf{tlocal}}$ by generating and copying over individual child test cases $(\boldsymbol{x}_{\mathbf{t}})_i \in \mathbb{R}^E$. The variation operators in **variationSol** and **variationTest** differ at the very least in their specifications given that $\mathbb{R}^D \neq \mathbb{R}^E$. However, both variation operators can employ similar mechanisms (e.g. in this situation the family of real-valued variation operators as introduced in 1.2.3).

Likewise, the need for genetic isolation between the two populations can be ascertained for the selection operators. On the one hand, **selectionSol** as a function of type $\mathcal{X} \to \mathcal{X}$ operates on the population configuration $\boldsymbol{X}_{\mathbf{slocal}}$ that as a whole

Algorithm 3.1 A general competitive CEA

Input:

Y_s	Population of candidate solutions
Y_t	Population of test cases
s_{term}	Termination criterion selection
csol	Termination criterion selection for evolutionary cycle of candidate solutions
ctest	Termination criterion selection for evolutionary cycle of test cases

Output:

X_s	Population of evolved candidate solutions

```
procedure CMCEA(Ys, Yt, sterm, csol, ctest)
    t := 0                                         ▷ Initialize time-stamp
    bterm := true                                  ▷ Initialize loop
    Xs := initializeSol(Ys)                        ▷ Initialize population of parents in Xs
    Xt := initializeTest(Yt)                       ▷ Initialize population of parents in Xt
    while bterm do
        Xs := EVOLPOPSOL(Xs, Xt, csol)             ▷ Evolving candidate solutions
        Xt := EVOLPOPTEST(Xt, Xs, ctest)           ▷ Evolving test cases
        t := t + 1                                 ▷ Increment coevolutionary cycle count
        bterm := termination(Xs, t, sterm)         ▷ Check for termination
    end while
    return Xs
end procedure

procedure EVOLPOPSOL(Xs, Xt, csol)
    cs := 0                                        ▷ Initialize cycle-time-stamp
    csolterm := true                               ▷ Initialize loop
    Xslocal := seedSol(Xs)                         ▷ Seed population of parents in Xslocal
    while csolterm do
        Xslocal := variationSol(Xslocal)           ▷ Apply variation operator on Xslocal
        evaluationSol(Xslocal, Xt)                 ▷ Fitness Evaluation of Xslocal using Xt
        Xslocal := selectionSol(Xslocal)           ▷ Apply selection operator on Xslocal
        cs := cs + 1                               ▷ Increment evolutionary cycle count
        csolterm := terminationSol(Xslocal, cs, csol)   ▷ Check for termination
    end while
    return Xslocal
end procedure

procedure EVOLPOPTEST(Xt, Xs, ctest)
    ct := 0                                        ▷ Initialize cycle-time-stamp
    ctestterm := true                              ▷ Initialize loop
    Xtlocal := seedTest(Xt)                        ▷ Seed population of parents in Xtlocal
    while csolterm do
        Xtlocal := variationTest(Xtlocal)          ▷ Apply variation operator on Xtlocal
        evaluationTest(Xtlocal, Xs)                ▷ Fitness Evaluation of Xtlocal using Xs
        Xtlocal := selectionTest(Xtlocal)          ▷ Apply selection operator on Xtlocal
        ct := ct + 1                               ▷ Increment evolutionary cycle count
        ctestterm := terminationTest(Xtlocal, ct, ctest)   ▷ Check for termination
    end while
    return Xtlocal
end procedure
```

is of type $\mathcal{X}$. This operation involves taking as input a population configuration, changing its configuration by placing selected individuals in the top part of the population (the partition that by addressing refers to the parents), and producing as output a new population configuration $\boldsymbol{X}_{\mathbf{slocal}} = X_j \in \mathcal{X}$ for the next iteration. On the other hand, **selectionTest** acts on $\mathcal{Y}$ in the same manner. It is a function of type $\mathcal{Y} \rightarrow \mathcal{Y}$, taking as input a population configuration, performing a change on it, and producing as output the population configuration $\boldsymbol{X}_{\mathbf{tlocal}} = Y_j \in \mathcal{Y}$. Earlier, we have specified that the population of candidate solutions and test cases to be $\big((\boldsymbol{x}_{\mathbf{s}}, f)_i : i = 1, 2, 3, \ldots, n\big)$ and $\big((\boldsymbol{x}_{\mathbf{t}}, f)_i : i = 1, 2, 3, \ldots, m\big)$, respectively. As such, the selection operators in **selectionSol** and **selectionTest** differ at the very least by their specification given that the spaces of configurations differ (i.e. $\mathcal{X} \neq \mathcal{Y}$ if one has $n \neq m$) even though one can use selection operators that employ the same mechanisms (e.g. tournament selection as described earlier in 1.2.4).

We use the variable names $\boldsymbol{X}_{\mathbf{slocal}}$ in **EVOLPOPSOL** and $\boldsymbol{X}_{\mathbf{tlocal}}$ in **EVOLPOPTEST** to emphasize that the adaptation processes are applied locally to the populations of candidate solutions and test cases and thus maintain genetic isolation. However, the two populations interact through procedures that evaluate the fitnesses or performances of population members. For example, **evaluationSol** will use $\boldsymbol{X}_{\mathbf{tlocal}}$ to perform updates on $\boldsymbol{X}_{\mathbf{slocal}}$, specifically all individual fitness values in $(\boldsymbol{x}_{\mathbf{s}}, f)_i$, $i = 1, 2, 3, \ldots, n$ that are given the variable name f and addressed via $\boldsymbol{X}_{\mathbf{slocal}}[i].f$. In the same manner, **evaluationTest** makes use of $\boldsymbol{X}_{\mathbf{slocal}}$ to update fitness values in $\boldsymbol{X}_{\mathbf{tlocal}}$. In considering both **EVOLPOPSOL** and **EVOLPOPTEST** as employing evolutionary adaptation processes, one procedurally describes **CMCEA** as a coevolutionary system with ***coupled*** fitness evaluations.

The parameters csol and ctest control the duration of the adaptation processes in **EVOLPOPSOL** and **EVOLPOPTEST**, respectively. Any one of these two parameters determines how many repeated evolutionary cycles of the application of variation and selection operators (including the fitness evaluation of population members) are made in the respective procedure of **EVOLPOPSOL** or **EVOLPOPTEST** on the current population before termination. They are tracked by the variables c_s and c_t, respectively, for **EVOLPOPSOL** or **EVOLPOPTEST**. Then, the main procedure **CMCEA** will call and execute the evolutionary adaptation process on the other population. These sequences of repeated calls and executions of **EVOLPOPSOL** and **EVOLPOPTEST** are made until the termination criterion that is specified by $\mathrm{s_{term}}$ triggers exiting the loop in the main procedure **CMCEA** upon a change in the flag $\mathrm{b_{term}}$, thus ending the computational simulation of the coevolutionary process. Termination criterion can be a fixed number of iterations for the loop, or some other more elaborate schemes, each of which is written down in **termination** and chosen for execution by $\mathrm{s_{term}}$. Note that a complete ***coevolutionary cycle*** refers to completed executions of **EVOLPOPSOL** and **EVOLPOPTEST** in the main loop of **CMCEA** and it is tracked by the variable t.

3.2.2 Cooperative Coevolution

We present the general framework for ***cooperative coevolution*** (***cooperative CEA***) that is described in Algorithm 3.2. In this framework, the assumption made is that the general setting involves the family of additively separable problems that admit a

Algorithm 3.2 A general cooperative CEA

Input:

D	Dimensionality of the problem
N	Size of individual population
$\mathrm{f_{sel}}$	Selection of problem from library of additively separable problems
$\mathrm{s_{term}}$	Termination criterion selection
csol	Termination criterion selection for evolutionary cycle of an individual population

Output:

$\boldsymbol{x}$	Evolved candidate solution

```
procedure CoCEA(f_sel, D, N, s_term, csol)
    t := 0                                        ▷ Initialize time-stamp
    b_term := true                                ▷ Initialize loop
    F := problemDecom(D, f_sel)                   ▷ Initialize structure F from problem decomposition
    b := initContext(D, F)                        ▷ Initialize context vector b randomly
    f_b := evaluate(b)                            ▷ Set f_b = f(b)
    for i = 0 : F.m do
        X_i := initializeSubPop(D, N, F, i)       ▷ Initialize populations for F.m subcomponents
    end for
    while b_term do
        b := EVOLPOPSOL(f_sel, D, N, X_i, F, i, b, f_b, csol)   ▷ Evolving population of
                                                                ▷ i-th subcomponent
        t := t + 1                                ▷ Increment coevolutionary cycle count
        b_term := termination(b, t, s_term)       ▷ Check for termination
    end while
    x := b
    return x
end procedure

procedure EVOLPOPSOL(f_sel, D, N, X_i, F, i, b, f_b, csol)
    c_s := 0                                      ▷ Initialize cycle-time-stamp
    csol_term := true                             ▷ Initialize loop
    X_slocal := seedSol(X_i)                      ▷ Seed population of parents in X_slocal
    while csol_term do
        X_slocal := variationSol(N, X_slocal, F, i)        ▷ Apply variation operator on X_slocal
        evaluationSol(f_sel, N, X_slocal, F, i, b)         ▷ Fitness Evaluation of X_slocal
        b := updateContext(D, N, X_slocal, F, i, b, f_b)   ▷ Update context vector b
        f_b := evaluate(b)                                 ▷ Update f_b = f(b)
        X_slocal := selectionSol(N, X_slocal)              ▷ Apply selection operator on X_slocal
        c_s := c_s + 1                                     ▷ Increment evolutionary cycle count
        csol_term := terminationSol(X_slocal, c_s, csol)   ▷ Check for termination
    end while
    return b
end procedure
```

decomposition in the sense that the D-dimensional problem's objective function can be formulated as $f(\boldsymbol{x}) = \sum_{i=1}^{m} f_i(\boldsymbol{x}_i)$ with $m \leq D$ independent subcomponents $\boldsymbol{x}_1, \ldots, \boldsymbol{x}_m$. Our framework is similar to that presented in the recent survey paper of cooperative CEAs in [30], which is based on the cooperative coevolution framework that is introduced in [43]. Unlike the case where they have presented one in which the problem is decomposed into D one-dimensional subproblems (i.e. the case of $m = D$), we have made general the framework and present one that can be applied to cases of $m \leq D$. Indeed, this requirement is made explicit whereby the main procedure **CoCEA** takes as one of the inputs the selection for the problem $\mathrm{f}_{\mathrm{sel}}$ from a library of additively separable problems.

The general cooperative CEA is implemented via the main procedure **CoCEA**, which after initializations proceeds to call the procedure **problemDecom** that performs a decomposition on the selected problem. The latter returns as its output a data structure $\boldsymbol{F}$ that captures and represents the problem's decomposition involving nonseparable decision variables being grouped into subcomponents $\boldsymbol{x}_1, \ldots, \boldsymbol{x}_m$. This is achieved by capturing and then grouping the relevant index numbers from the decision variable vector $\boldsymbol{x} = (x_1, \ldots, x_D)$, as values to be stored by the associated subcomponents. Let $\mathbb{Z}_{1 \leq z \leq D} = \{1, 2, 3, \ldots, D\}$ be the set of integers that index $\boldsymbol{x}$. The data structure $\boldsymbol{F}$ can be described as follows:

$$\begin{aligned} \boldsymbol{F} &= (m, s, \mathbf{IDX}), \\ m &\in \mathbb{Z}_{1 \leq z \leq D}, \\ s &= (s_i : i = 1, \ldots, m), \;\; s_i \in \mathbb{Z}_{1 \leq z \leq D}, \\ \mathbf{IDX} &= (\mathbf{idx}_i : i = 1, \ldots, m), \\ \mathbf{idx}_i &= (\mathrm{index}_{i,j} : j = 1, \ldots, s_i), \mathrm{index}_{i,j} \in \mathbb{Z}_{1 \leq z \leq D}. \end{aligned} \tag{3.1}$$

In (3.1), $\boldsymbol{F}.m$ refers to the number of disjoint subcomponents following problem decomposition. $\boldsymbol{F}.s[i]$ refers to the dimensionality of the vector representing the ith subcomponent $\boldsymbol{x}_i$ and is given by a positive integer s_i in the range from 1 to D. $\boldsymbol{F}.(\mathbf{IDX}[i]) = (\mathrm{index}_{i,1}, \ldots, \mathrm{index}_{i,s_i})$ is a vector of dimension s_i, whereby the distinct positive integers $\mathrm{index}_{i,j},\; j = 1, \ldots, s_i$ are taken from the set $\mathbb{Z}_{1 \leq z \leq D}$. Essentially it is a list of positive integers, each representing the index numbers from $\boldsymbol{x} = (x_z : z = 1, 2, 3, \ldots, D)$ associated with the ith subcomponent $\boldsymbol{x}_i$. There are two properties or invariants associated with this problem decomposition. First, one has $\sum_{i=1}^{m} s_i = D$. Second, the union of the lists of distinct positive integers $\boldsymbol{F}.(\mathbf{IDX}[i])$s that is given by $\mathrm{index}_{i,j},\; i = 1, \ldots, m,\; j = 1, \ldots, s_i$ is simply a permutation of $1, \ldots, D$. Implementation wise, $\boldsymbol{F}$ is used by all other procedures as indices for the subcomponents $i = 1, \ldots, m$ and the associated decision variables therein.

Such a decomposition can be illustrated with the following example. Consider an ***additively separable*** problem with decision vector $\boldsymbol{x} = (x_1, x_2, x_3)$ that admits a decomposition whereby $f(\boldsymbol{x}) = \sum_{i=1}^{2} f_i(\boldsymbol{x}_i)$. In particular, $\boldsymbol{x}_1 = (x_2)$ and $\boldsymbol{x}_2 = (x_1, x_3)$ (note the slight departure from standard notation we have used to denote a vector as a list of variables). The procedure **problemDecom** returns the following:

$$\boldsymbol{F} = (m, s, \mathbf{IDX}),$$

$$m = 2,$$

$$s = (s_1, s_2) = (1, 2),$$

$$\mathbf{IDX} = (\mathbf{idx}_1, \mathbf{idx}_2),\ \mathbf{idx}_1 = (2),\ \mathbf{idx}_2 = (1, 3).$$

Algorithm 3.2 can exploit this decomposition structure of the problem that is captured and represented by $\boldsymbol{F}$ and subsequently solve the optimization of additively separable problems in a cooperative manner. This is achieved with the use of the context vector $\boldsymbol{b}$ and adaptation processes of $2 \leq m \leq D$ genetically isolated population implemented sequentially. The context vector $\boldsymbol{b}$ performs two important roles in this cooperative CEA framework: (1) It represents the current complete solution that is assembled $\boldsymbol{b} = (\boldsymbol{b}_i : i = 1, \ldots, m)$ based on the selection of an individual from each ith population (i.e. $\boldsymbol{b}_i \in \boldsymbol{X}_i.\boldsymbol{x}$), upon completion of evolutionary adaptation processes for all the m populations via the procedure **EVOLPOPSOL**. (2) It provides the means to evaluate the fitness or performance of each individual candidate solution representing the ith subcomponent $\boldsymbol{x}_i$ during the evolutionary adaptation process of the respective ith population. This is because the fitness function of each subcomponent f_i is typically unknown (e.g. in Blackbox optimization).

The first role of the context vector $\boldsymbol{b}$ is associated with the procedure **evaluate** that computes the fitness of the current complete solution via $f(\boldsymbol{b})$. At the start of Algorithm 3.2, the complete solution represented by $\boldsymbol{b}$ is initialized via **initContext**. This can be performed in the usual manner via some random sampling of $\mathbb{R}^D$ to more involved initialization procedures that employ seeding of known solutions. More importantly, the context vector $\boldsymbol{b}$ is updated via **updateContext** for each ith subcomponent in a sequential manner $i = 1, \ldots, m$ within the evolutionary adaptation process **EVOLPOPSOL** of the ith population $\boldsymbol{X}_i$. Consider a minimization problem that is additively separable with objective $f(\star)$, the correct problem decomposition $\boldsymbol{F}$, and with known subcomponent fitness functions f_i, $i = 1, \ldots, m$. The new or updated selection of the ith subcomponent is $\boldsymbol{b}_i = \operatorname{argmin}_{\boldsymbol{x}_{i,k} \in \boldsymbol{X}_i.\boldsymbol{x}} f_i(\boldsymbol{x}_{i,k}) = \operatorname{argmin}_{\boldsymbol{x}_{i,k} \in \boldsymbol{X}_i.\boldsymbol{x}} \left(f_i(\boldsymbol{x}_{i,k}) + C\right)$. During an update for the ith subcomponent, $C = \sum_{j \in \{1,\ldots,m\} \setminus \{i\}} f_j(\boldsymbol{b}_j)$ is a fixed value since all other subcomponents $\{\boldsymbol{b}_1, \ldots, \boldsymbol{b}_m\} \setminus \{\boldsymbol{b}_i\}$ are fixed or remain as they are selected previously. Then, in effect one has $\boldsymbol{b}_i = \operatorname{argmin}_{\boldsymbol{x}_{i,k} \in \boldsymbol{X}_i.\boldsymbol{x}} f_i(\boldsymbol{x}_{i,k})$. This cannot be implemented in practice because f_i is unknown. Actual implementation to select the appropriate $\boldsymbol{b}_i$ in **updateContext** requires comparison of $f(\boldsymbol{x}_{i,k} \cup \boldsymbol{b}^c)$ values with respect to a priori selected individuals of all other subcomponents $\boldsymbol{b}^c$. These $f(\boldsymbol{x}_{i,k} \cup \boldsymbol{b}^c)$ values act as surrogate to determine which of the candidate solutions of the ith subcomponent $\boldsymbol{x}_{i,k}$ satisfy $\min_{\boldsymbol{x}_{i,k} \in \boldsymbol{X}_i.\boldsymbol{x}} \left(f_i(\boldsymbol{x}_{i,k}) + C\right)$ [30]. The notation

$x_{i,k} \cup b^c$ refers to the complete solution vector. This process to compute the fitness of $f_i(x_{i,k})$ via $f(x_{i,k} \cup b^c)$ is called ***complementation***.

The procedure **EVOLPOPSOL** implements the evolutionary adaptation processes for each ith population $X_i = \big((x_i, f)_k : k = 1, \ldots, N\big)$ consisting of individual candidate solutions $x_{i,k}$ for the ith subcomponent. The process is the same as with any EA, and the parameter csol controls the duration of this process or evolutionary cycle (e.g. how many times the evolutionary operators are repeatedly called), which is tracked by the variable c_s. The only exception is that the subcomponents x_i, $i = 1, \ldots, m$ might differ in their dimensionalities s_i, $i = 1, \ldots, m$ so that the variation operator **variationSol** to generate child candidates requires information of problem decomposition structure F. The selection operator **selectionSol** to choose parents for the next generation requires the basic information of the population size N. This parameter is fixed for all populations although one can choose different population sizes N_i for each ith subcomponent. The fitness evaluation of each candidate solution $x_{i,k}$ in the population is implemented via the procedure **evaluationSol**. The procedure requires the current context vector b with the fitness of the candidate solution $x_{i,k}$ being calculated as $X_i[k].f = f(x_{i,k} \cup b^c)$.

Lastly, note that **EVOLPOPSOL** implements **updateContext** to select the best candidate solution for the ith subcomponent in the context vector. Note that $b_i = \text{argmin}_{x_{i,k} \in X_i.x} \big(f_i(x_{i,k}) + C\big) = \text{argmin}_{x_{i,k} \in X_i.x} f_i(x_{i,k})$ for additively separable problem with the correct problem decomposition F. Then, after the coevolutionary cycle that corresponds to the sequence of calls of the procedure **EVOLPOPSOL** for $i = 1, \ldots, m$ is completed, one obtains the complete candidate solution to the original problem given by $b = \big(\text{argmin}_{x_{1,k} \in X_1.x} f_1(x_{1,k}), \ldots, \text{argmin}_{x_{m,k} \in X_m.x} f_m(x_{m,k})\big)$. The termination criterion for the cooperative CEA is specified by $\mathrm{s_{term}}$. This could involve a fix number of iterations or coevolutionary cycles tracked by the variable t to more sophisticated mechanisms such as those that detect (premature) convergence of the populations. Crucially, the final updated context vector b is one with the best performance, and upon termination, it is returned as the final solution by the main procedure **COCEA**.

3.3 Coevolutionary Problem Solving

In Chap. 2, we have provided an overview on the literature of Coevolutionary Computation where we have discussed the development of coevolutionary systems for various real-world problem solving. In particular, the two general settings where coevolutionary systems are focused on and have been studied extensively are those formulated as learning and optimization. Section 2.4.2 has shown that the use of competitive coevolution for learning and cooperative coevolution for optimization are based on specific formulations of population interactions within the two coevolutionary approaches. These formulations are exploited and subsequently formulated as the principal mechanisms for performance or fitness evaluations in the algorithmic frameworks of competitive and cooperative CEAs that are introduced in the previous Sect. 3.2. Here, we will discuss in more detail how these general algorithmic frameworks of coevolution can be applied for the two problem-solving settings of learning and optimization. Our presentation will be example-driven, and we will elaborate on specific issues that have been raised in the literature on the design aspects of those CEAs to address inherent challenges in those problem-solving settings.

3.3.1 Competitive Coevolution for Learning

We start the discussion with the single-population competitive CEA for learning problems of games whereby any agent in the population can play both roles as candidate solution and test case. Indeed, it is only in competitive CEA that such a single-population configuration is common. Here, such ***competitive coevolutionary learning*** provides an alternative approach to search for game-playing strategies. For this setting, it is either known that the game is symmetric so that the set of game-playing strategies is the same for both players or that one can make the assumption that the game is near symmetric. Such an assumption has been made for board games involving sequential moves between two players such as Checkers [5], Othello [8], Chess [18], and Go [47], among other games. One could even make a more involved assumption that the coevolved strategies have to learn to play both colours (i.e. as the first and second players), and formulate the fitness evaluation accordingly so as to take into account performances playing as the first and second players against any test opponent strategy.

We begin with the competitive coevolution of learning strategies of two-player, symmetric IPD games [3, 6, 11, 15, 16, 19]. For this setting, the competitive CEA is similar to that of a general EA (as described in Algorithm 1.1) with the main principal difference being the fitness evaluation of game-playing agents in the population that is based on game-plays. The process one goes through in the design of EAs for specific problem-solving settings applies to CEAs as well. In particular, the design choice of solution or strategy representation has a significant impact on the evolutionary operators and especially on the variation operators. A two-player IPD game involves a sequence of joint-moves that are made by the two players simultaneously for a fixed number of rounds. In each round, a player can choose between cooperate (C) or defect (D), the move to play. Memory-based reactive strategies that use history of past moves made by oneself, the opponent, or the joint-moves of oneself and the opponent to decide the move to play have been well-studied. For example, the seminal study [3] has restricted the coevolutionary learning process as a search on the space of memory-three, reactive IPD strategies.

Such a strategy can be represented as a look-up table of past three joint-moves consisting of $(2^2)^3$ entries. (Note: There are 2^2 possible joint-moves in one past round $(C, C), (C, D), (D, C), (D, D)$.) Each entry itself consists of the move to be played, i.e. cooperate (C) or defect (D), which is represented by

p_{addr}. Each entry location of the look-up table is addressed using information of all the past three joint-moves. The complete list of addresses $\boldsymbol{A}$ consists of $|\{\text{C}, \text{D}\}|^{|\{1,2\}\times\{1,2,3\}|} = 2^6 = 64$ elements. Each address addr $= \big((m \in \{\text{C}, \text{D}\}, i, r) : i = 1, 2,\ r = 1, 2, 3\big)$ is written down as a list of six 3-tuples, where index $i = 1, 2$ refers to the player who made the move (1 for self, 2 for opponent) and index $r = 1, 2, 3$ refers to the previous rth round. A historical play of three joint-moves involves six moves with the indices i, r in addr being used to inform a strategy, which player and round a move refers to. For example, when both players cooperated in all the past three rounds, the first player processes the historical play as addr $= \big((\text{C}, 1, 1), (\text{C}, 2, 1), (\text{C}, 1, 2), (\text{C}, 2, 2), (\text{C}, 1, 3), (\text{C}, 2, 3)\big)$. Then, the strategy takes the form of a map $\boldsymbol{T} : \boldsymbol{A} \to \{\text{C}, \text{D}\}$ with $\boldsymbol{T} = (p_{\text{addr}} \in \{\text{C}, \text{D}\} : \text{addr} \in \boldsymbol{A})$ whose output being a value either C or D is obtained from the corresponding entry in the look-up table $\boldsymbol{T}(\text{addr}) = p_{\text{addr}} \in \{\text{C}, \text{D}\}$.

Note that for the first three rounds, there are not enough historical joint-moves for which one can access the look-up table $\boldsymbol{T}$. As such, any memory-based reactive strategy requires an additional mechanism to initiate moves whenever complete historical joint-moves are not available. In [3], this involves an assumed initial history of three rounds for each strategy. As such, the strategy can now start to play the game by processing these assumed historical plays to address the specific entry location in the look-up table and obtain as output $\boldsymbol{T}(\text{addr})$. This also completes the specification of the memory-three reactive IPD strategies. The study further employs a genetic (binary) encoding scheme for the strategy representation. Since there are $(2^2)^3 = 64$ elements (entries) in $\boldsymbol{T}$ and a further $(2 + 2 + 2) = 6$ elements for the assumed historical plays, then the encoding scheme produces a binary string of length 70. Each bit is associated to a particular element, and one can arbitrarily assign a binary value of 0 for C and 1 for D.

The genetic encoding scheme for the strategy representation allows the use of a Genetic Algorithm (GA) to implement this competitive CEA to search for IPD strategies. In [3], a single population of 20 agents is used. The variation operators consist of crossover such as one point with bit-flip mutation that is set to occur at significantly lower probability. The selection operator used is frequency proportional. Fitness evaluation for each agent in the population is performed by taking the sum of the IPD game outcomes obtained from all game-plays. Each agent competes with all other 19 agents (i.e. self-play is excluded) in the population. For each game-play, the outcome for an agent implementing a particular IPD strategy is the cumulative IPD game payoffs as a result of the joint-moves it makes with the opponent over 151 rounds.

The IPD game payoffs for the first player are specified by the matrix $\mathbf{G} = (g_{ij} : i, j = \text{C}, \text{D})$, with the index i representing the (row) first player's move and j representing the (column) second player's move. This payoff matrix $\mathbf{G}$ for any two-player, two-choice IPD game consists of four nonnegative real numbers $(R, S, T, P) = (g_{\text{CC}}, g_{\text{CD}}, g_{\text{DC}}, g_{\text{DD}}),\ R, S, T, P \in \mathbb{R}_{\geq 0}$. The four variables R, S, T, P have specific game interpretations: reward (R) for mutually cooperating, taking a sucker's (S) payoff when the other player defected, having the temptation (T) to defect, and punishment (P) for mutually defecting [2]. They must satisfy the following conditions: (1) $T > R$ and $P > S$ (defection pays more), (2) $R > P$ (mutual cooperation pays more than mutual defection), (3) $R > (T + S)/2$ (alternating does not pay compared to mutual cooperation). The study in [3] uses $(R, S, T, P) = (3, 0, 5, 1)$, although other values can be used as long as they satisfy the IPD game conditions. Regardless, suitable normalizations over the number of rounds and games will ensure that the fitness of any agent in the population to range over $[S, T]$ for IPD game outcomes defined as cumulative payoffs.

Note that the general principles involved in the design of single-population competitive CEAs apply to other two-player games with assumed symmetric strategy set. Most of the specific details are concentrated on the design of the parameterized representation of the game-playing strategies and the associated variation operators. In particular, ANNs have been used to provide a parametric representation for game-playing strategies based on board state evaluation functions in various two-player board games [5, 8, 18, 47]. In most studies, real-valued variation operators are designed and applied to generate new game-playing strategies represented as ANNs. For example, the competitive CEA for learning Go game-playing strategies in [47] considers weighted linear functions for board state evaluations. This is represented as a single-node ANN with a non-linear hyperbolic tangent function output. The competitive CEAs that have been investigated include both the elitist $(1 + \lambda)$-ES and nonelitist $(1, \lambda)$-ES (i.e. the single parent is always replaced by the best child in the population). Furthermore, the variation operators that are studied involve two nonisotropic mutation operators, one of which employs fixed σ_i and another employs self-adaptation on the σ_i (one can refer to Sect. 1.2.3 for more details). More complex ANN architectures have been used such as ***Multilayer Perceptrons*** (***MLPs***) in [5, 8, 18]. However, one can certainly design more sophisticated combination of strategy representation and variation operators that evolve representation structures (e.g. ANN topologies) or employ game domain knowledge and implement user-designed fixed structures (e.g. three ANNs are used in [18] that focus on different areas of the board for the game of Chess).

Earlier we have briefly discussed that for two-player sequential board games, one could make a more sophisticated argument for the application of competitive CEA for agents to simultaneously learn both first player and second player game-playing strategies. For example, the study [47] employs a commonly used round-robin tournament selection operator, i.e. an agent will engage in game-plays will all other agents in the population but exclude self-play. Furthermore, the fitness evaluation considers both player roles where in any pairwise encounter between two agents, two sets of games are played with each agent taking turns being the first and second players of the game. Game outcomes are defined as (win, lose, draw) = $(1, -2, 0)$ where rather than the usual ***zero-sum*** formulation, the agent that loses the game is penalized twice the amount of what a win gives. Such an arbitrary specification is not unusual in competitive coevolutionary learning setting [5].

However, there are other games such as Nim and 3D tic-tac-toe whereby the roles of the first and second players are different. For these two games, there are inherent advantages to being the first player, for example, by enabling a win with optimal game-plays (Nim) or exploiting some inherent advantages in the game design (3D tic-tac-toe). This points to different strategy sets for the first and second players. As such, one could employ a two-population competitive CEA and formulate coevolutionary learning for first player strategies instead. In such a setting, the population of second player agents will take on the role of test strategies. The study [46] has demonstrated the utility of such a two-population competitive CEA with a predator-prey setup in comparison to using a fixed sample of second player game-playing agents that are randomly generated. In particular, the competitive CEA incorporated advanced coevolutionary design features such as fitness sharing to encourage arms race dynamics between the two coevolving but genetically isolated populations of first and second player agents.

In the context of competitive learning, such dynamics refers to coevolution providing a ***pedagogical*** sequence of test strategies throughout evolutionary time that the (first player) agents can learn to compete and defeat. This is in contrast to the use of fixed sample of random test strategies whereby the system now performs evolutionary learning. The study [46] has argued that for the more complex game of Nim, the fixed sample of test strategies appears to the evolutionary learning system analogously as a rough fitness landscape that lacks gradients and is more akin to a ***Needle in a Haystack***. There are insufficient information associated with defeating the other remaining test strategies in the fixed sample for evolutionary learning to discover first player strategies to overcome those few difficult test strategies. Instead, coevolutionary learning is able to search for those necessary intermediate test strategies that are not found by random sampling when the system is equipped with additional mechanisms to differentiate and maintain (in external archives) new and unique test strategies.

3.3.2 Cooperative Coevolution for Optimization

Cooperative coevolution solves optimization problems via a ***Divide-and-Conquer*** problem-solving approach that combines ***problem decomposition*** with a multi-population coevolutionary search. This requires optimization problems to admit problem decompositions that result in simpler problem subcomponents, each of which is individually targeted by a specific, genetically isolated and evolving population of candidate solutions. Instead of searching for the optimal solutions for independent subcomponents individually in a systematic manner, these populations interact to enable a cooperative search so that after each iterate candidate solutions for the subcomponents are improving in performance in tandem. This also means that cooperative CEAs involve multiple evolving populations as described in the algorithmic framework of 3.2. This assumes the ideal scenario that there is an exact procedure with which additively separable optimization problems are correctly decomposed. In particular, the procedure outputs subcomponents that correctly group interacting variables together. If the procedure **problemDecom** cannot decompose and break the optimization problem into simpler subcomponents but as a result return the original problem, then the cooperative CEA will lose its distinctive search characteristics and operate as a single-population EA.

However, one might have no prior knowledge of problem structures (i.e. they are ***Blackbox*** optimization problems) in real-world implementations of such cooperative CEAs even for additively decomposable optimization problems. In this situation, cooperative CEAs with incorrect problem decomposition could be trapped in pseudominima that come about as a result of the combination of interactions between subcomponents and the sequence with which the associated populations evolve that cause the bias of cooperative search towards those pseudominima. The recent survey [30] has provided a simple example to illustrate how such a pseudominimum is formed and searched by cooperative CEAs in the supplementary document. In addition, one can refer to the text about fitness landscapes in [12] that discuss an analogous situation on the creation of local minima as a result of specific binary encoding scheme with single bit-flip mutations for a continuous optimization problem. In both of these problem settings, the appropriate choice would be to use a representation that considers these interacting variables as a group and design variation operators that make changes to them as a whole. Other problem settings involve complex and higher-order interactions (e.g. those beyond pairwise interdependencies) between decision variables underlying high-dimensional optimization problems that make problem decomposition challenging [32, 40, 55].

As a consequence, significant research has been made to develop strategies that can effectively group interacting variables together within their respective subcomponents as an alternative to problem decompositions. This is crucial for the effective and efficient operation of cooperative coevolutionary search for such Blackbox additively separable optimization problems. As such, our discussion in this section of real-world implementations of cooperative CEAs will focus on this problem decomposition aspect. These ***variable grouping strategies*** include those employing mechanisms that are static, random, linkage learning-based, overlap and hierarchical-based, and domain-knowledge-based [30]. We will discuss in more detail the first three strategies that employ mechanisms of increasing sophistication as the means to showcase the issues and challenges on their implementations in cooperative CEAs.

We start our discussion with the simple ***static variable grouping strategy***. The various mechanisms in this category do not employ any systematic or learning procedure to discover the underlying decomposition structures of the optimization problems. Instead, they use fixed and arbitrary assignments of decision variables into groups to form the subcomponents. The study [42] first proposes the simple static variable grouping whereby a D-dimensional optimization problem is decomposed into D number of one-dimensional subcomponents. Each subcomponent is associated with a distinct decision variable of the original problem. The cooperative CEA that is developed in the study can be described by Algorithm 3.2, employing D number of coevolving populations. GA is used for the procedure **EvolPopSol**, and the resulting cooperative coevolutionary GA is more effective than baseline GA in optimizing fully separable benchmark problems. It is less effective on problems with interdependencies between the decision variables. However, it is still superior to the baseline GA at the expense of increased computational overheads for employing the more sophisticated cooperative search mechanism.

Other static variable grouping strategies have been developed. For example, the study [52] has proposed a generalized mechanism that statically decomposes D-dimensional optimization problems into m subcomponents, each associated with s number of distinct decision variables and satisfying $D = m \times s$. Various combinations of m number of subcomponents of size s for the static decomposition can be used although their effectiveness would depend on the degree of interdependencies of the decision variables in the original problem. More importantly, the study has employed PSO for the procedure **EVOLPOPSOL** to construct a cooperative coevolutionary PSO. The seminal study [42] has used GA, while elsewhere in [28] EP has been used to implement **EVOLPOPSOL** for evolutionary search on each individual population of candidate solutions for the respective subcomponent. This demonstrates that cooperative CEA embodies a higher-level problem-solving approach as indicated by its framework in the form of Algorithm 3.2. As such, any EA can be used if it is suitably prepared to search for the subcomponent solutions given their parameterized representational form. Nearly independent of this, various variable grouping strategies can be designed and assessed for their effectiveness with cooperative coevolutionary search.

The next category with increased sophistication in mechanism design is the ***random variable grouping strategy***. Here, the mechanisms are designed to address two related issues that are associated with the earlier category. In particular, the grouping of decision variables is fixed both in terms of the subcomponent constituents during the coevolutionary cycle and the size of each subcomponent. The general idea of mechanism design of random variable grouping strategy is to move and then make the decomposition process as part of the coevolutionary search cycle. In the context of Algorithm 3.2, this requires moving the procedure **problemDecom** and other associated operations (e.g. initializations or seeding of the populations of subcomponents) into the main loop of the main procedure **COCEA** and then makes the appropriate changes to the procedure **EVOLPOPSOL** that is used to perform evolution search on the candidate solutions of the respective subcomponent (e.g. in the framework of Algorithm 3.2 this involves passing of associated parameters to the procedure).

One of the more straightforward mechanisms that has been proposed for random variable grouping strategy involves decomposing the original problem into m subcomponents, each of size s. In [25, 53], the problem decomposition is performed at the start of each coevolutionary cycle. The number of subcomponents m and their sizes s are fixed throughout the coevolutionary cycle. However, the s number of distinct decision variables that are selected from those representing the original problem $\boldsymbol{x} = (x_1, \ldots, x_D)$ and then grouped into their respective subcomponents $\boldsymbol{x}_1, \ldots, \boldsymbol{x}_m$ is performed in a random manner. Procedurally, one can simply permute the list $x_1, \ldots, x_D$ and then group every s number of decision variables left-to-right to produce m subcomponents. The probability that any decision vector x_i is selected into a subcomponent $\boldsymbol{x}_k$ will be uniform. Crucially, the study [54] has used a combination argument and then provides examples that there is a high probability associated with selecting a group of decision variables together over several coevolutionary cycles. This suggests that cooperative CEAs using such a random variable grouping strategy would spend enough cycles evolving interdependent decision variables together. Further extensions that have been studied include implementing variable sizes for the subcomponents that can be changed throughout coevolutionary cycles. In [26], a systematic approach is introduced that involves choosing a variable size s from the set $\{i \in \mathbb{Z}_{>0} : D \mod s = 0 : s\} \backslash \{s^*\}$ with uniform probability that is different to current size s^* if there are no quantifiable fitness improvements. Another example is introduced in [51] that considers an adaptive approach to simultaneously change the subcomponent size s_i and associated population size N_i during coevolutionary cycles. Selection of a configuration involves a comparison phase that implements a greedy search over various configurations (s_i, N_i), $i = 1, \ldots, q$ to determine the one configuration (s_j, N_j) that provides the best fitness improvements.

The next category of variable grouping strategies involves procedures that detect variable interactions in the problem structures to determine if and where an optimization problem can be decomposed. This is an active area of research and development, with the survey study [30] naming these strategies as interaction learning and listing a further five subcategories. We list them here along with some of the associated literature studies: perturbation [34, 35, 55], statistical model [37, 45], distribution model (e.g. EDAs) [32, 33], approximate model [27], and linkage adaptation [49]. Our brief discussion here will focus on ***perturbation-based mechanisms*** to better highlight the issue of variable interactions that are present in additively separable optimization problems. We start with pairwise interactions that might be present in these problems although the mechanisms have been improved to handle higher-order interactions [55]. These interactions can be revealed through specific perturbations that are introduced to the decision variables.

Let $f(\boldsymbol{x}) = f(x_1, x_2, x_3, \ldots, x_D)$ be the objective or fitness function of a D-dimensional additively separable optimization problem. Let the positive integers $i, j \in \{1, 2, 3, \ldots, D\}$ satisfy $i < j$. The forward difference $\Delta_{\delta, x_i} f(\boldsymbol{x})$ of f with respect to the decision variable x_i and the interval $\delta > 0$ is defined as

$$\Delta_{\delta, x_i} f(\boldsymbol{x}) = f(\ldots, x_i + \delta, \ldots) - f(\ldots, x_i, \ldots).$$

Note that only the explicitly written out decision variable x_i is changed, where on the left of the difference of the two terms for $f(\star)$ one has $x_i + \delta$ and on the right x_i. The horizontal ellipsis symbol "…" is used to indicate that all other decision variables remain unchanged. In addition, the forward difference $\Delta_{\delta, x_i} f(\boldsymbol{x})$ having substituted values $x_i = a$ and $x_j = b$ is written as $\Delta_{\delta, x_i} f(\boldsymbol{x})|_{x_i = a, x_j = b}$.

The study [38] has established a theoretical result for continuous additively separable optimization problems f (i.e. having partial derivatives $\delta f / \delta x_i$ with respect to any x_i). In particular, the ***nonseparable condition*** is proved whereby any pair of decision variables x_i and x_j is nonseparable if the condition $\Delta_{\delta, x_i} f(\boldsymbol{x})|_{x_i = a, x_j = b_1} \neq$

$\Delta_{\delta,x_i} f(\boldsymbol{x})|_{x_i=a,x_j=b_2}$ holds for all reals $a, b_1 \neq b_2, \delta > 0$. The proof arguments actually establish the contrapositive of the statement. This is more useful in practice since this implies that x_i has no interaction with x_j (i.e. the pair x_i, x_j is separable) whenever the forward differences $\Delta_{\delta,x_i} f(\boldsymbol{x})|_{x_i=a,x_j=b_1}$ and $\Delta_{\delta,x_i} f(\boldsymbol{x})|_{x_i=a,x_j=b_2}$ induced by the perturbation δ on x_i result in similar values when they are evaluated with different values b_1 and b_2 substituted into x_j, respectively. In particular, grouping mechanisms based on perturbation can be designed based on detections of pairwise interactions. This includes the Differential Grouping (DG) [38] and Linkage Identification by Nonlinearity Check (LINC) [35] mechanisms. For example, one could set the parameter $\epsilon > 0$ and check if $\Delta_{\delta,x_i} f(\boldsymbol{x})|_{x_i=a,x_j=b_1}$ and $\Delta_{\delta,x_i} f(\boldsymbol{x})|_{x_i=a,x_j=b_2}$ differ to within ϵ to determine pairwise interaction of x_i, x_j.

We will close this section with a brief discussion on ***collaboration selection strategies*** in the framework of cooperative CEA. There are two aspects for which collaborations of candidate solutions from populations representing other subcomponents are crucial. The first is concerned with how candidate solutions with respect to each subcomponent are chosen so as to form the complete solution. The second is concerned with fitness evaluations of candidate solutions with respect to a subcomponent that are based on the choice of those made for other subcomponents. Various mechanisms have been developed for collaboration selection strategies, with the survey study [30] noting the two associated main factors that differentiate their operations: (i) Collaborator selection pressure that is associated with the level or degree of greediness in choosing a candidate solution with respect to a subcomponent population to form the complete solution. (ii) Collaborator selection size that corresponds to the number of complete solutions used to evaluate the fitness of a candidate solution of a subcomponent population.

We highlight three related mechanisms that are named best, random, and elite collaborator selection. The first mechanism on ***best collaborator selection strategy*** [38, 42] is already formulated in Algorithm 3.2 whereby a greedy approach is used to evaluate the fitness of candidate solutions for each subcomponent as well as choosing the one as the representative subcomponent solution to form the complete solutions. Note that the two aspects on collaboration selection for forming the complete solution and fitness evaluation of subcomponent candidate solutions are mutually resolved for additively separable optimization problems with the correct decomposition. There are other problems whereby the right decomposition is not made with interdependencies between the decomposed subcomponents. The greedy nature of the best collaboration selection strategy is not appropriate in these settings. The ***random collaboration selection strategy*** [13, 39] could be an alternative. The mechanism randomly chooses a candidate solution from the respective population of each subcomponent to form the complete solution. For the ***elite collaborator selection strategy*** [21], the mechanism will consider K candidate solutions from each population of size N in the collaborator pool and then randomly choose candidates from the pool to form the complete solutions. The parameter K controls the level of greediness in the selection of candidate solutions from each population and it ranges $[1, N]$. At $K = 1$, then the strategy operates as the best collaborator selection strategy with maximum level of greediness since the best candidate solution in the population is always chosen. At $K = N$, the strategy operates in the same way as the random collaborator selection strategy.

3.4 Concluding Remarks and Notes for Further Reading

We have provided two algorithmic formulations of coevolutionary principles that are tasked to search for solutions with superior performance and discussed in more detail their specific implementations on some problem examples. The first involves the competitive CEA framework to solve problems in a learning setting. The second is about the cooperative CEA framework on optimization problem setting. In the case of competitive CEA on learning problems, interactions take the form of a competitive two-player game with candidate solutions and test cases taking on first and second player roles, respectively. The main coevolutionary principle that governs design aspects of a competitive CEA to search for solutions (i.e. learn first player game-playing strategies) would leverage on two genetically isolated populations of candidate solutions and test cases, each undergoing an evolutionary search process guided by the interaction (game) outcomes with members of the opposing population. A successful implementation of competitive coevolution couples both evolving populations to promote arms race dynamics between the two populations in general succession of coevolutionary cycles repeatedly.

The case of cooperative CEA to solve optimization problems in a Divide-and-Conquer manner is more involved in terms of both algorithmic design and its specific implementation with respect to the problem at hand. Our presentation has focused on the class of additively separable optimization problems that admit decomposition into smaller and simpler subcomponents. Here, a cooperative CEA consists of multiple genetically isolated populations with each population consisting of candidate solutions for a particular subcomponent. Cooperative coevolution occurs when each of these populations undergoes evolutionary process to search for its own subcomponent candidate solutions that is guided by outcomes of interactions with those from the other populations to form the complete solution. In particular, the main coevolutionary principle that governs design aspects of a cooperative CEA involves mechanisms for collaborator selections to combine candidate solutions of subcomponents to form the complete solution as well as the individual fitness assignments to these candidate solutions required for evolutionary search processes in their respective populations. This is crucial towards a successful cooperative search so that these multiple populations are improving in performance with each cycle of coevolution.

We close this chapter with brief remarks of coevolutionary systems that combine both competitive and cooperative search aspects in problem solving. Indeed, subsequent chapters in Parts II and III will present in more detail studies we have done to develop [29, 50] or analyse [9, 10] such complex coevolutionary systems. However, the idea of combining both competitive and cooperative coevolutionary search is not new. There have been early studies [22, 23] that have proposed a framework of ***competitive-cooperative CEA*** for solving optimization problems. The proposed framework could be considered as an extension to the existing cooperative CEA framework. In particular, the Divide-and-Conquer problem-solving aspect is still being addressed via a cooperative coevolutionary search. Instead of designing fixed or the more elaborate learning mechanism that is implemented to enable decision variable groupings required for problem decomposition, the motivation of the proposed competitive-cooperative CEA framework is to exploit the natural evolutionary process so that decision variable groupings come about as its emergent property.

In [22], the competitive-cooperative coevolution framework has been applied to solve the more complex multi-objective optimization problems. However, the framework can be easily adapted to solve the more simple single-objective optimization problems. This is because fitness evaluation and selection of individuals are accomplished through separate procedures **evaluationSol** and **selectionSol**. The relevant mechanisms within both procedures can be changed so that Pareto ranks are used in solving a multi-objective optimization problem. For the single-objective optimization problem setting, the ranking is reduced to comparison and ordering of real-valued numbers associated with the fitness of individuals. Regardless, the cooperative coevolution aspect in [22] optimizes a single variable for each of the genetically isolated populations. In this manner, the core of the competitive-cooperative coevolution framework is based on the original cooperative coevolution framework of [43] that decomposes D-dimensional problems into one-dimensional subproblems. However, instead of assigning a fixed population to represent candidate solutions for a specific subcomponent, the framework incorporates mechanisms for different and multiple genetically isolated populations competing to offer candidate solutions for the role of any particular one-dimensional subcomponent or decision variable.

The benchmark problems that the study [22] has used to evaluate the competitive-cooperative CEA involve continuous functions with real-valued input or decision variables. In this competitive-cooperative coevolution framework, the use of multiple genetically isolated populations with each population consisting of members that are able to represent candidate solutions for one-dimensional decision variable provides two features. First, the members within any population are compatible in terms of their specification with members of other populations since they are all of the same type $\mathbb{R}$. This type compatibility simplifies the coevolution framework so that competitive and cooperative search mechanisms can be applied to all the populations. Second, the genetic isolation of the populations in this coevolutionary setting serves to maintain diversity instead of the usual type protection in other cooperative coevolutionary search settings. In most other existing cooperative settings, populations represent candidate solutions of subcomponents with different types (i.e. different dimensionalities because the subcomponents consist of groups with a different number of decision variables) and require genetic isolation for this reason. Here, procedures **variationSol** with the same type $\mathbb{R} \to \mathbb{R}$ can be applied to any of the populations to produce new child candidates from the parents. Crucially, the various and different populations competing to offer to represent the role of specific subcomponents that are then combined to provide the full solution give rise to emergent interdependencies between the subcomponents.

In [22], the two mechanisms are applied in sequence with a fixed frequency cycle each. For example, competitive coevolution is activated for some fixed number of coevolutionary cycles that is then followed by cooperative coevolution with the same frequency, and so forth until termination. The study has employed a binary encoding scheme so that crossovers and bit-flip mutation operators can be used. The study in [23] has implemented the same competitive-cooperative coevolution framework but with the baseline adaptation process on candidate solutions being changed to that of a PSO. As such, the studies have demonstrated again that the framework of competitive-cooperative coevolution operates at higher level of problem solving. This indicates that other suitable population-based, stochastic search algorithms can be used to search for the individual candidate solutions.

Finally, there have been studies that have investigated what these competitive-cooperative coevolutionary systems entail in terms of their inherent complexity. The example study we highlight here is from [31] that has used a linguistic prediction game with solution representation in the form of ***Finite State Automata*** (***FSA***). As in the case of the IPD game we have discussed earlier, the interaction or game-play of this linguistic prediction game involves two players that simultaneously output symbols drawn from a fixed set of alphabets over a fixed number of rounds. Unlike the case of the IPD game, different payoff matrices are used depending on the type of pairwise interactions involved. In [31], two interaction types are considered, i.e. cooperative and competitive. In the cooperative game, any two agents that interact will receive the same payoff if their symbols match in each round, otherwise a zero payoff is obtained. This stands in contrast in the competitive game where zero payoff is given to both agents if the same symbols are produced in each round, and otherwise a payoff of one is obtained for only one of the agents depending on the specific pair of symbols that are played.

In order to facilitate having both types of interactions or games, the current population of agents is configured as a network with the type of edge connection specifying the interaction. Furthermore, the linguistic prediction game is chosen because one is able to measure the complexity of an agent's behaviour by the smallest equivalent FSA that generates the same game-playing strategy. The controlled empirical study in [31] has demonstrated that when both types of competitive and cooperative interactions are present, the agents coevolved to higher level of behavioural complexity as measured by their minimal FSA representation compared to cases where only one type of interaction is present. Note that each agent represents the full candidate solution for the problem at hand, for example, which is to learn to play both the competitive and cooperative games. Furthermore, the specific use of FSA as the game-playing strategy representation can grow in complexity of behaviours as indicated by the specific sequences of symbols generated as a response to opponents' plays. In such a perspective, one could argue that coevolutionary systems that employ both competitive and cooperative search aspects would be more able to solve problems that require a higher level of complexity in the representation of the solutions.

References

1. Axelrod, R.: More effective choice in the Prisoner's Dilemma. J. Conflict Resolut. **24**(3), 379–403 (1980)
2. Axelrod, R.: The Evolution of Cooperation. Basic Books, New York (1984)
3. Axelrod, R.: The evolution of strategies in the iterated Prisoner's Dilemma. In: Davis, L.D. (ed.) Genetic Algorithms and Simulated Annealing, chap. 3, pp. 32–41. Morgan Kaufmann, New York (1987)
4. Cartlidge, J., Bullock, S.: Combating coevolutionary disengagement by reducing parasite virulence. Evol. Comput. **12**(2), 193–222 (2004)
5. Chellapilla, K., Fogel, D.B.: Evolution, neural networks, games, and intelligence. Proc. IEEE **87**(9), 1471–1496 (1999)
6. Chong, S.Y., Yao, X.: Behavioral diversity, choices, and noise in the iterated prisoner's dilemma. IEEE Trans. Evol. Comput. **9**(6), 540–551 (2005)
7. Chong, S.Y., Yao, X.: Multiple choices and reputation in multiagent interactions. IEEE Trans. Evol. Comput. **11**(6), 689–711 (2007)
8. Chong, S.Y., Tan, M.K., White, J.D.: Observing the evolution of neural networks learning to play the game of Othello. IEEE Trans. Evol. Comput. **9**(3), 240–251 (2005)
9. Chong, S.Y., Tiňo, P., Yao, X.: Measuring generalization performance in co-evolutionary learning. IEEE Trans. Evol. Comput. **12**(4), 479–505 (2008)
10. Chong, S.Y., Tiňo, P., Ku, D.C., Yao, X.: Improving generalization performance in co-evolutionary learning. IEEE Trans. Evol. Comput. **16**(1), 70–85 (2012)
11. Darwen, P.J., Yao, X.: Co-evolution in iterated Prisoner's Dilemma with intermediate levels of cooperation: application to missile defense. Int. J. Comput. Intell. Appl. **2**(1), 83–107 (2002)
12. De Jong, K.A.: Fitness landscapes. In: Burke, E.K., Kendall, G. (eds.) Search Methodologies: Introductory Tutorials in Discrete Optimization and Decision Support Techniques, pp. 681–705. Springer, New York (2014)
13. de Oliveira, F.B., Enayatifar, R., Sadaei, H.J., Guimarães, F.G., Potvin, J.-Y.: A cooperative coevolutionary algorithm for the multi-depot vehicle routing problem. Expert Syst. Appl. **43**, 117–130 (2016)
14. Elfeky, E.Z., Elsayed, S., Marsh, L., Essam, D., Cochrane, M., Sims, B., Sarker, R.: A systematic review of coevolution in real-time strategy games. IEEE Access **9**, 136647–136665 (2021)
15. Fogel, D.B.: Evolving behaviors in the iterated Prisoner's Dilemma. Evol. Comput. **1**(1), 77–97 (1993)
16. Fogel, D.B.: On the relationship between the duration of an encounter and the evolution of cooperation in the iterated Prisoner's Dilemma. Evol. Comput. **3**(3), 349–363 (1996)
17. Fogel, D.B.: Blondie24: Playing at the Edge of AI. Morgan Kaufmann, San Francisco (2002)
18. Fogel, D.B., Hays, T.J., Hahn, S.L., Quon, J.: A self-learning evolutionary chess program. Proc. IEEE **92**(12), 1947–1954 (2004)
19. Franken, N., Engelbrecht, A.P.: Particle swarm optimization approaches to coevolve strategies for the iterated Prisoner's Dilemma. IEEE Trans. Evol. Comput. **9**(6), 562–579 (2005)
20. García, E., Villar, J.R., Tan, Q., Sedano, J., Chira, C.: An efficient multi-robot path planning solution using A^* and coevolutionary algorithms. Integr. Comput.-Aided Eng. **30**, 41–52 (2023)
21. Glorieux, E., Svensson, B., Danielsson, F., Lennartson, B.: Improved constructive cooperative coevolutionary differential evolution for large-scale optimisation. In: Proceedings of the IEEE 2015 Symposium Series on Computational Intelligence (SSCI'15), pp. 1703–1710. IEEE Press, Piscataway (2015)
22. Goh, C.-K., Tan, K.C.: A competitive-cooperative coevolutionary paradigm for dynamic multiobjective optimization. IEEE Trans. Evol. Comput. **13**(1), 103–127 (2009)
23. Goh, C.-K., Tan, K.C., Liu, D.S., Chiam, S.C.: A competitive and cooperative co-evolutionary approach to multi-objective particle swarm optimization algorithm design. Euro. J. Oper. Res. **202**, 42–54 (2010)

24. Hillis, W.D.: Co-evolving parasites improve simulated evolution as an optimization procedure. Physica D **42**, 228–234 (1990)
25. Li, X., Yao, X.: Tackling high dimensional nonseparable optimization problems by cooperatively coevolving particle swarms. In: Proceedings of the IEEE 2009 Congress on Evolutionary Computation (CEC'09), pp. 1546–1553. IEEE Press, Piscataway (2009)
26. Li, X., Yao, X.: Cooperatively coevolving particle swarms for large scale optimization. IEEE Trans. Evol. Comput. **16**(2), 210–224 (2012)
27. Li, G., Rosenthal, C., Rabitz, H.: High dimensional model representations. J. Phys. Chem. A **105**(33), 7765–7777 (2001)
28. Liu, Y., Yao, X., Zhao, Q., Higuchi, T.: Scaling up fast evolutionary programming with cooperative coevolution. In: Proceedings of the IEEE 2001 Congress on Evolutionary Computation (CEC'01), pp. 1101–1108. IEEE Press, Piscataway (2001)
29. Liu, S., Tang, K., Yao, X.: Generative adversarial construction of parallel portfolios. IEEE Trans. Cybern. **52**(2), 784–795 (2022)
30. Ma, X., Li, X., Zhang, Q., Tang, K., Liang, Z., Xie, W., Zhu, Z.: A survey on cooperative co-evolutionary algorithms. IEEE Trans. Evol. Comput. **23**(3), 421–441 (2019)
31. Moran, N., Pollack, J.: Effects of cooperative and competitive coevolution on complexity in a linguistic prediction game. In: Proceedings of the 14th European Conference on Artificial Life (ECAL 2017), pp. 298–305 (2017)
32. Mühlenbein, H., Mahnig, T.: FDA – a scalable evolutionary algorithm for the optimization of additively decomposed functions. Evol. Comput. **7**(4), 353–376 (1999)
33. Mühlenbein, H., Paaβ, G.: From recombination of genes to the estimation of distributions i. binary parameters. In: Parallel Problem Solving from Nature PPSN IV. Lecture Notes in Computer Science, vol. 1141, pp. 178–187. Springer, Berlin (1996)
34. Munetomo, M.: Linkage identification based on epistasis measures to realize efficient genetic algorithms. In: Proceedings of the IEEE 2002 Congress on Evolutionary Computation (CEC'02), pp. 1332–1337. IEEE Press, Piscataway (2002)
35. Munetomo, M., Goldberg, D.E.: Linkage identification by non-monotonicity detection for overlapping functions. Evol. Comput. **7**(4), 377–398 (1999)
36. Nisan, N., Roughgarden, T., Tardos, E., Vazirani, V.V. (eds.): Algorithmic Game Theory. Cambridge University Press, New York (2007)
37. Omidvar, M.N., Yao, X.: Cooperative co-evolution with delta grouping for large scale nonseparable function optimization. In: Proceedings of the IEEE 2010 Congress on Evolutionary Computation (CEC'10), pp. 1762–1769. IEEE Press, Piscataway (2010)
38. Omidvar, M.N., Li, X., Mei, Y., Yao, X.: Cooperative co-evolution with differential grouping for large scale optimization. IEEE Trans. Evol. Comput. **18**(3), 378–393 (2014)
39. Panait, L., Luke, S., Harrison, J.F.: Archive-based cooperative coevolutionary algorithms. In: Proceedings of the 2006 Genetic and Evolutionary Computation Conference (GECCO'06), pp. 345–352 (2006)
40. Pelikan, M., Goldberg, D.E., Cantú-Paz, E.: Boa: The Bayesian optimization algorithm. In: Proceedings of the 1999 Genetic and Evolutionary Computation Conference (GECCO'99), pp. 525–532 (1999)
41. Popovici, E.: Bridging supervised learning and test-based co-optimization. J. Mach. Learn. Res. **18**(38), 1–39 (2017)
42. Potter, M.A., De Jong, K.A.: A cooperative coevolutionary approach to function optimization. In: International Conference on Parallel Problem Solving from Nature PPSN III. Lecture Notes in Computer Science, vol. 866, pp. 249–257. Springer, Berlin (1994)
43. Potter, M.A., Jong, K.A.D.: Cooperative coevolution: an architecture for evolving coadapted subcomponents. Evol. Comput. **8**(1), 1–29 (2000)
44. Qu, H., Xing, K., Alexander, T.: An improved genetic algorithm with co-evolutionary strategy for global path planning of multiple mobile robots. Neurocomputing **120**, 509–517 (2013)
45. Ray, T., Yao, X.: A cooperative coevolutionary algorithm with correlation based adaptive variable partitioning. In: Proceedings of the IEEE 2009 Congress on Evolutionary Computation (CEC'09), pp. 983–989. IEEE Press, Piscataway (2009)

46. Rosin, C.D., Belew, R.K.: New methods for competitive coevolution. Evol. Comput. **5**(1), 1–29 (1997)
47. Runarsson, T.P., Lucas, S.M.: Coevolution versus self-play temporal difference learning for acquiring position evaluation in small-board go. IEEE Trans. Evol. Comput. **9**(6), 540–551 (2005)
48. Russell, S., Norvig, P.: Artificial Intelligence: A Modern Approach, 4th edn. Pearson, New Jersey (2021)
49. Smith, J., Fogarty, T.C.: Recombination strategy adaptation via evolution of gene linkage. In: Proceedings of the IEEE 1996 International Conference on Evolutionary Computation, pp. 826–831. IEEE Press, Piscataway (1996)
50. Tang, K., Liu, S., Yang, P., Yao, X.: Few-shots parallel algorithm portfolio construction via co-evolution. IEEE Trans. Evol. Comput. **25**(3), 595–607 (2021)
51. Trunfio, G.A., Topa, P., Wąs, J.: A new algorithm for adapting the configuration of subcomponents in large-scale optimization with cooperative coevolution. Inf. Sci. **372**, 773–795 (2016)
52. van den Bergh, F., Engelbrecht, A.P.: A cooperative approach to particle swarm optimization. IEEE Trans. Evol. Comput. **8**(3), 225–239 (2004)
53. Yang, Z., Tang, K., Yao, X.: Differential evolution for high-dimensional function optimization. In: Proceedings of the IEEE 2007 Congress on Evolutionary Computation (CEC'07), pp. 3523–3530. IEEE Press, Piscataway (2007)
54. Yang, Z., Tang, K., Yao, X.: Large scale evolutionary optimization using cooperative coevolution. Inf. Sci. **178**(15), 2986–2999 (2008)
55. Zhong, R., Zhang, E., Munetomo, M.: Cooperative coevolutionary differential evolution with linkage measurement minimization for large-scale optimization problems in noisy environments. Complex Intell. Syst. (2023). https://doi.org/10.1007/s40747-022-00957-6

Chapter 4
Applications of Coevolution in Real-World Problem Solving

4.1 General Introduction

This chapter will showcase how the principles of coevolution can be exploited for real-world problem solving. This follows from and links with the two preceding chapters. In particular, Chap. 2 has focused on the main principle of coevolution that involves populations of candidate solutions and test cases whose evolutionary adaptations are guided by their mutual strategic interactions. Under some conditions that promote these interacting populations to coevolve with arms race dynamics, one would observe candidate solutions evolve towards increasingly higher performance levels and test cases evolve towards presenting increasingly more challenges. Meanwhile, Chap. 3 has focused on the two principal approaches of coevolution in problem solving. Competitive coevolution is primarily concerned with formulating a problem-solving approach in which the lack of an objective fitness measure to evaluate solution performance and appropriate samples of test cases can be circumvented. Instead, a coevolutionary system can be designed such that the evolving population of candidate solutions compete against the evolving population of test cases. Cooperative coevolution incorporates Divide-and-Conquer problem-solving methodology to decompose the original complex problems into simpler subcomponents that are then amenable to iterative search approaches such as Evolutionary Algorithms. In this chapter, we review various studies and highlight how they have specifically made use of the principles and concepts of competitive or cooperative coevolution to solve those real-world problems.

We consider four general problem domains: (1) Business, Economics, and Public Policy. (2) Entertainment Games. (3) Engineering, Logistics, and Transportation. (4) Software, Systems, and Services. They are mostly selected as additional problem domains compared to those covered in Parts II and III of this book. For example, in Chap. 10 we will present our work that directly applies to problem-solving settings in the domain of politics and serious games. The selection of review of studies that have been carried out by others in Sect. 4.2 on economics and

X. Yao, S. Y. Chong, *Coevolutionary Computation and Its Applications*,
Natural Computing Series, https://doi.org/10.1007/978-981-96-2841-4_4

Sect. 4.3 on entertainment games would complement our study in Chap. 10. Most problem domains such as logistics and transportation as an example include a variety of problem settings. In Chap. 6, we present our work specifically about solving logistics problems that are formulated as the Travelling Salesman Problem and Vehicle Routing Problem. However, in Sect. 4.4 of this chapter, we would present a study that has applied a cooperative Coevolutionary Algorithm (CEA) to solve a specific transportation problem setting on vehicle platooning. We have also conducted a few studies on the use of coevolutionary principles in software development applications. These studies are directly incorporated as part of the presentation in Sect. 4.5 on the wider problem domain of software, systems, and services.

Despite the variety of specific problem settings that the selected studies have chosen to solve, in general they have been abstracted and then formulated as optimization and learning problems. As such, one can apply cooperative or competitive CEAs as the appropriate iterative solvers for those formulated problems. Our review is directed towards these individual studies. In each section that focuses on a large problem domain, we consider several studies that have investigated the use of coevolutionary principles in optimizers or learning systems to solve specific real-world problems. For each of these studies, we cover more details than what a systematic review would do. In particular, we would describe the actual problem setting in its real-world context that would allow us to more clearly highlight the practical impact coevolved solutions would provide. After that, we will proceed to briefly detail the specific problem abstraction and subsequent formulation. Although our presentation would be less formal and technical, we would emphasize and clarify details on those abstraction approaches. Our aim is so that one could still develop intuitive understanding of the problem formulations. Once a problem has been appropriately formulated as an optimization or learning problem with specific characteristics, then it would be a matter of choosing the appropriate coevolutionary problem-solving approach (i.e. cooperative or competitive) and deciding on design choices that provide good search performance (e.g. population structures, solution representation, variation, and selection operators).

Given the number of studies that apply cooperative or competitive coevolution across various problem domains, a systematic review that focuses on detecting general patterns on how coevolutionary systems are designed for different problems might gloss over those important problem abstraction and formulation details. Instead, our aim is to provide sufficient details for readers to obtain general ideas how these selected studies have solved those real-world problems via coevolution. In addition, both practitioners and designers of coevolutionary systems might be able to derive some general patterns with which the principles and concepts of coevolution are exploited after reading through several example studies that we have presented in this chapter. For the practitioners, one might want to read through the entire section related to the problem domain areas. For the designers, one might want to select one or two examples from each section. Nevertheless, in the concluding Sect. 4.6, we would discuss some general aspects and patterns arising from the selected studies in applying coevolutionary principles and concepts.

4.2 Business, Economics, and Public Policy

In this section, we first present the applications of coevolution for problem solving in the domains of business and economics that are of great practical importance in the real world. Our selected literature review would focus on several studies that have applied coevolutionary principles for modelling and analysis of several different markets that include real, financial, and energy (electricity) markets. After that, we close this subsection with a discussion of the potential applications of coevolution in the broad and general domain of public policy. Notwithstanding the case that details of coevolutionary principles are specifically provided for problem solving in applications of business and economics from prior studies and be indicated for in applications of public policy, the main conceptual view that is emphasized is that ***coevolutionary simulation*** constitutes a methodology within the broad approach of ***Agent-based Modelling*** (***ABM***) or ***Agent-based Simulation*** (***ABS***). As such, our discussion on potential applications in public policy will be based on the review of selected studies that have already made arguments about the use of ABM/ABS.

For our first discussion on the application of coevolution for problem solving in business, we choose two representative and related studies [42, 44] in the problem setting of strategic interactions of agents in real markets. Here, these decision-making agents participate in a real market and are conceived as representing brand managers that could implement a set of alternative marketing strategies to affect their sales performance. In particular, the competitive setting is based on the retail coffee market model that has been proposed and developed earlier in [17]. This particular model exhibits four properties or aspects crucial in a real market, of which two are related to the asymmetric nature of competitions found in real markets and the other two we highlight them next. One property has to do with the situation that brands differ or have unique responses of marketing actions to consumer responses. In other words, some brands can obtain more sale returns for similar allocation of advertising. Another property is about the large and rapid changes in volume over time when viewed at the store or chain level (e.g. via supermarket scanner data of sales) that can come about as a result of certain strategic market-mix decisions of agents. A one-shot (single-shot) market simulator called Competitive Analysis System for Promotional Effectiveness Research (CASPER) has been developed that combines the asymmetric market-share model and the category volume model. Associated with the simulator is a dataset of the actual retail coffee market of eight brands competing over 52 weeks [17]. As such, both evolutionary and coevolutionary simulations can be conducted, with the former making use of the dataset while the latter employing the market simulator.

In particular, CASPER is able to take the marketing actions of agents (brand managers) in terms of the prices and promotional activities of its brands and then return profits via estimates of markets shares and sales volume. This is done for each single week to reflect the outcome from joint actions (interactions) of agents. This can be iterated over a finite number of weeks (rounds of interactions). Note that the subsequent studies [42, 44] have not employed CASPER to perform online calculations of agent profits due to high computational costs. Instead, these studies derive payoff matrices with profits for each agent as a result of joint actions that can occur from their weekly market participations. Four actions (price levels) are considered: *High* price to approximate cooperative/collusive price, *high* price to approximate the two-brand coalition price, *low* price to approximate the noncooperative or Nash-Cournot price, and *Low* price to approximate the envious price. In effect, CASPER-games are constructed for which the derived payoff matrices are saved and used as look-up routines to obtain the payoff (profit) of the one-shot interaction of each agent. Crucially, the simulator has been developed to allow for the construction of n-player CASPER-games with the studies [42, 44] employing reduced three-player CASPER-game for computational cost savings reason. For these reduced CASPER-games, the remaining brands are represented by nonstrategic price taking players that do not change their implemented marketing strategies.

Both studies [42, 44] have employed a ***multi-population coevolution*** setup whereby multiple genetically isolated populations are used. The primary motivation for using a multi-population competitive coevolution is that the underlying real market the two studies are investigating consists of asymmetric players. There are two sources of such asymmetric competition for the real market setting considered. One has to do with stable, cross-competitive effects that as one example reflects the situation whereby sales from some brands in one category cannot be countered by sales of brands in another category (e.g. national versus regional brands). The other has to do with temporal effects in the sense that within the same category, sales from a brand would produce more returns if it is made on its own instead of with sales of other brands. Given that the retail coffee market as modelled by the CASPER-game is multiplayer with asymmetric competitions, then a particular population is used to represent a player (brand). Stimulus-response agents are used in each population with bit-string encoding similar to studies that have used competitive coevolution of the Iterated Prisoner's Dilemma games [6] (the latter has been briefly detailed in Chap. 3.3.1). As such, agents representing a strategic brand manager (player) can change their marketing actions based on historical joint actions with other players. The fitness evaluation of an agent representing a brand in a particular population is based on CASPER-game interactions with agents representing different brands in other populations. In other words, the required agents are drawn from their respective populations so as to play a multiplayer CASPER-game with payoffs given by the CASPER-game-derived payoff matrices to take into account asymmetric competitions. In this manner, an agent's fitness can take the form of average CASPER-game outcomes from different combinations of agents representing various brands. The studies [42, 44] have found that such a multi-population competitive coevolution can learn marketing strategies for specific brands that are superior to those developed by human experts and those found through an evolutionary learning approach based on the dataset.

In our next review discussion, we focus on studies that have used coevolutionary simulation for problem solving in the domain of economics. In particular, the coevolutionary simulations of electricity and financial markets in these studies represent a specific form of ***Agent-based Computational Economics*** (***ACE***) model [15]. ACE is used as a computational methodology to study economic systems whereby processes are modelled as dynamical systems of interacting agents [59]. In ACE, each participant of the market is modelled by an independent decision-making agent. These agents interact with each other and are capable of learning so as to adapt the decision-making behaviours. Indeed, one would note the correspondence between such computer-based simulation approaches in ACE with that of a coevolutionary system. Game-theoretic approaches have modelled such markets via the Standard Cournot model whereby agents employ quantity strategies for one-shot game (trading interaction) with the market and the Pareto Improvement Model whereby agents are allowed to collude for repeated games with the market.

The study in [15] has introduced a coevolutionary framework for analysis of an oligopolistic electricity market through multi-agent simulation. Both one-shot and repeated games are considered for agent interactions. In the framework, each agent represents an electricity power producer that employs quantity decision-making as its trade strategy. All agents submit their respective quantities to be generated to the Power Exchange, which would use the appropriate market rules and demand characteristics based on assumed interactions (i.e. Standard Cournot and Pareto Improvement for one-shot and repeated games, respectively). The multi-population coevolution setup is used consisting of multiple genetically isolated populations, with each population corresponding to an electricity power producer. As such, the members of a population represent the candidate solution (trading strategies) for that specific producer. Fitness evaluation for an individual member (agent) can be performed when it is used with representatives of the other producers are chosen to form a market of agents (e.g. either randomly or using a greedy approach to choose the best representative for each producer). For agent interactions with the market modelled as one-shot games, the profit for the agent can be calculated directly from the Standard Cournot model and is used as its fitness. In the case that the market allows for repeated trading submissions, the coevolutionary framework for the market simulation now takes the form of a ***Multi-objective Optimization Problem*** (***MOP***). This is because in the case of infinitely repeated ***noncooperative games***, there are many colluding groups of agents that can be formed, each of which constitute a noncooperative equilibrium. Here, the Pareto optimal solutions are these noncooperative equilibria. ***NSGA-II*** is used as the underlying ***Multi-objective EA*** (***MOEA***) to search for these Pareto optimal solutions [19]. In both cases, the multi-population CEA is able to find the Cournot-Nash and Pareto Optimal solutions, respectively.

The authors later apply the same coevolutionary framework to analyse the case whereby market clearance is based on the Supply Function Equilibrium (SFE) model [14]. For this market model formulation, the solution that takes the form of an equilibrium of all agents maximizing their profits consists of a set of supply functions, with a supply function representing the trading strategy of a producer. These supply functions can take the form of a specific affine or piecewise affine function. As such, the multi-population CEA searches for the group of producers in market equilibrium, with each producer represented by the best agent within a genetically isolated population having specific parameters associated with the supply function used for trading strategy. As before in [15], a greedy approach is used to choose the best member in each population to be the representative agent implementing the trading strategy for a producer. Results of computational studies in [14] have demonstrated that the multi-population CEA is able to find the solution in the form of supply functions in equilibrium in all the cases investigated. This coevolutionary framework has been extended and adapted in two different studies. In [68], the coevolutionary framework is successfully applied to solve the more complex electricity market that incorporates a forward (contracts) market in addition to the existing spot (trading) market as a two-stage game. In [27], the coevolutionary framework is used for analysis and to find the solution for a complex electricity market involving active distribution whereby producers (agents) adopt heterogeneous strategies to represent microvirtual power plants in the modern setting.

We remark here that the coevolutionary framework used in these reviewed studies [14, 15, 27, 68] is a particular type of competitive CEA that is formulated to solve noncooperative games. In game theory, a basic noncooperative two-player game refers to a setting whereby each player has access to a set of strategies it can implement and for which payoffs are defined for any encounter involving any pair of strategies from the set [46]. Most of the reviews we have done so far in Chap. 3 and later in Sect. 4.4 have predominantly focused on competitive CEAs of agent interactions modelled as two-player games. However, the interactions described in those studies on electricity market are actually multi-player games. In particular, the number of players corresponds to the number of electricity producers that are considered. One such solution concept for a multi-player game involves the Nash equilibrium that constitutes the set of players implementing specific strategies such that each player cannot individually deviate (by implementing a different strategy) to obtain a better payoff. The reason for implementing multiple genetically isolated populations in their competitive CEA is so that the learning process or search for strategies can be independently performed for each player in the setting of multi-player games. Indeed, this design choice that the original study [15] has later been demonstrated in [27] to be appropriate when there is a need to consider heterogeneous strategy representations as opposed to a homogeneous strategy representation considered in many studies that apply competitive CEAs for games.

Next, we review other studies such as in [43] that has developed a coevolutionary simulation of a financial market. In particular, a software system called Coevolved Heterogeneous Artificial Stock Market (CHASM) has been developed that allows users to conduct simulations of stock markets with artificial traders in various settings as the means to assess and analyse different aspects that could influence a simulated financial market. The stock market model and mechanisms follow an agent-based market model proposed earlier in [29], which itself constitutes an extension of the Santa Fe artificial market introduced in [4]. The base agents employ inductive reasoning and are boundedly rational in the sense that their strategic decision-making are based on limited past information of their actions. Each agent is equipped with a finite set of pure (fixed) strategies that process as inputs available trading information and then output a decision (i.e. buy stocks, sell stocks, or do nothing). The study in [26] that considers a simpler model of the Santa Fe artificial market uses a single-population competitive CEA and that each agent in this population employs a simple autoregressive predictor.

In [43] that considers the more complex market model, heterogeneous agents are considered. There are three types of agents: noise, value, and technical traders. A noise trader makes a decision (buy, sell, or do nothing) with specific probabilities. A value trader based its decision on price differences between the current value given by the market and a fundamental value the trader adopts. A technical trader's decision-making is significantly more sophisticated in that it is capable of performing analysis on the past trading information given by the market for decision-making. In particular, the strategic decision-making of technical traders is represented in the form of trees structures and evolved via ***Genetic Programming*** (***GP***). Under such setting, the study has employed a multi-population competitive CEA. Each trader (agent) in coevolution consists of a genetically isolated population of GP trees so that the learning process of each trading agent is formulated as an evolutionary search process on the corresponding population of GP trees. These populations coevolve through interactions of their respective evolved agents through stock prices in the changing market. Such a multi-population coevolutionary simulation is sufficient to generate a time series of stock returns (prices) having stylized facts (statistical properties) reminiscent of those observed (measured) in real-world stock markets.

A more recent study on the application of coevolution of trading strategies in [45] has investigated both cooperative and competitive CEAs for the search of trading strategies that are compared against an EA. Instead of simulating an artificial stock market, datasets of a real stock market are used to train the agents. As with other studies such as [43], agents are technical traders whose decision-making rules are represented as GP trees. The study has employed a fitness function to measure the performance of agents as trading strategies. This fitness function measures the difference between the compounded excess returns of trading strategies with the return of a baseline buy-and-hold strategy, which are employed over the given dataset. As such, an EA can be designed to evolve GP trees that are evaluated based on this fitness function. The cooperative CEA involves two genetically isolated populations representing buying and selling strategies. As such, a member from each of the two populations is required to construct the full solution capable of buying and selling stocks or shares. The same fitness function is used to evaluate the complete solution (strategy), with a credit assignment mechanism used to apportion fitness values for the respective buying or selling agents. For the example of a buying agent, this involves taking the sum of fitness values, each of which is obtained by pairing with a selling agent. In the multi-population competitive CEA, each genetically isolated population represents a specific agent. The relative fitness value for each member in the current population is calculated by means of aggregating interaction outcomes with all members from other populations. An interaction involves a pairwise comparison of absolute fitness values between the two individual members, which are obtained using the fitness function employed in the EA setup. The study [45] has found out that the multi-population competitive CEA is able to search for robust trading strategies in comparison to the EA approach and cooperative CEA.

We close this section with a brief discussion of potential applications of coevolutionary principles in problem-solving settings in the very important domain of public policy. One handbook [34] on this subject matter has generally described public policy to include both elements of construction and analysis of policies and associated programmes that are used by officers of the state (e.g. the governing arm of a nation) to rule. The real-world issues that public policy aims to address are wide-ranging but at the core of most if not all those issues is the matter of humans living and thus having to consume various resources from a finite world. One seminal paper is Hardin's exposition about the ***Tragedy of the Commons*** [35], which includes among several examples one about general pollution that is not often readily appreciated by the general public as an issue of consumption of a finite resource. This is because the planet has a limited even though initially a vast capacity of mediating pollutants that are created from human activities. Regardless of whether one is developing or analysing public policy, there is this crucial aspect of persuasion because at the heart of an effective public policy is one that can be successfully implemented. This means one that carries the general support of the public [34]. Given the substantial amount of literature that is available for such a topic, we have aimed to select several articles in more specific areas of public policy that have incorporated or emphasized agent-based methodologies in ABM for review and discussion.

We reiterate that coevolutionary systems constitute a methodology within ABM that has as a main feature and property the adaptation of interacting agents. This crucial aspect is one that might appeal to the technocrat policy-makers as they attempt to develop an understanding of large-scale, complex, and adaptive systems that underpin the real-world issues that policies and programmes have to address. We start with the review paper [39] that has sought to demonstrate how such an understanding of ***Complex Adaptive Systems*** (***CAS***) can be made in a manner that provide insights towards the management of those systems underlying the world people live in and rely upon. On the one hand, there is the economic system that arises from human activities. On the other hand, there is the ecological system that comes about from processes of the physical and biological world. Both economic and ecological systems are individually CAS, interlinked and causally interacting as lower-level (micro) components of a higher-level (macro) CAS. It is this formation of the macro-coupled CAS with an underlying coevolutionary process between two micro-agents representing economic and ecological systems that are of interest and in the remit of public policy.

Here, the coupled economic-ecological CAS provides an exemplar of the world affected by modern human activities as an addition to other physical and biological activities. One is now able to model behaviours of this coupled CAS in the context of nonlinear dynamical systems. The coupled CAS would be sufficiently complex so as to realize a variety of dynamical regimes that include the existence of multiple fixed points representing steady states, some of which are associated with local basins of attraction, while others could be saddle points with locally stable and unstable dynamics. This has the crucial real-world implication in terms of the management of real-world systems because of the interpretation of what human intervention means [39]. For example, human intervention could reinitialize the system to a new initial condition, one preferably close to a desired steady state and then thereafter allow the dynamics of the coupled system to evolve to that steady state. In another example, alternative and desired steady states could include a saddle point that could be reached via a trajectory representing an optimal control process realizable by human intervention. To give a real-world interpretation what the behaviours of the coupled CAS mean in terms of public policy, consider the problem of agricultural (e.g. fertilizer) runoff that could shift the state of a lake between being oligotrophic (healthy) to eutrophic (unhealthy) as a result of algae growth. Human interventions could be a one-time cleanup that results in state reinitialization while policies and programmes to regulate fertilizer usage as the optimal control process to alter the dynamics of the lake's health towards an oligotrophic state [39]. At this point, we would emphasize the potential of applying coevolutionary simulation to studies of most coupled CAS for public policy because one still has to rely on numerical means via computer-generated trajectories to reveal CAS dynamics despite having some qualitative insights from prior mathematical analysis if the latter is available.

Despite theoretical appeal of more formal methodologies reviewed in [39], technocratic policy-makers could adopt that enable them to better understand the real-world problem they have to solve by developing an effective suite of public policy, there is the issue of persuading the general public to support the associated programmes for which the latter's participation is crucial for successful implementation. ABM can help to bridge this gap by incorporating knowledge from other disciplines particularly those in the social sciences. The paper in [51] has pushed forward this perspective with respect to the important issue of climate policy. Not only that more formal methodologies could lack appeal to the general public during communication phases to garner support, perhaps more crucially they might not take into account what typically occurs in the real world when associated programmes of a policy are implemented. There could be the narrow operationalization of agents in more formal models as a result of adopting the standard assumption of rationality, and the limited understanding and as a consequence of models that do not incorporate agents with vested interests, among others. Such biases in the model could turn away policy-makers from certain criteria when it comes to evaluation of policy impacts, and thus policy could be developed that eventually lack social and political support [51].

However, the framework of ABM is general, flexible, and can accommodate the incorporation of elements from other disciplines. In the context of climate policy, a literature review has been provided in [51] whereby three pairs of disciplines have been integrated into ABM in existing studies. First, elements of behavioural and management sciences can be integrated into ABM so that more realistic agent representation of the various actors in climate policy is enabled. In particular, real-world actors (e.g. firms and households) are not perfectly informed utility maximizers but instead are oriented towards goal satisfaction (e.g. via organization routines and individual habits) and capable of adaptations in light of access to new information or feedbacks for performance improvements. Second, elements of sociology and psychology can be integrated into ABM so as to more realistically capture the nature and consequences of interactions between agents. As one example, real-world individuals do not have stable preferences and are socially isolated. Instead, their preferences are multi-faceted and can be affected by their social networks. Third, elements from political and economic sciences can be integrated into ABM to take into account that real-world agents are heterogeneous in their feature identifications such as political views and income. For example, the evaluation of policy acceptance could be made better if the formation and development of political coalition and social movement are taken into account as part of the process involving repeated interactions between policy-makers and the general public for policy development.

In all three integration examples, there is scope and potential to apply coevolutionary principles into those ABM methodologies to better assess public policy design and public support, which in real-world settings are interdependent with causal relationship. Indeed, this is one discussion point that is raised in the position paper [51]. Furthermore and elsewhere in [9], the same discussion point is expanded and amplified within the setting of communication policy for governance underpinning the management of nations and other large organizations of people. However, the notion of governance that is considered is more broad and beyond the usual scope of governments through the mechanisms of law, regulation, and other traditional instruments. Instead, both social and technological mechanisms that could shape the operations and outcomes of the communication system are considered in governance. This is crucial given the pervasiveness of modern technologies in human activities, of which one aspect is the proliferation of modern digital communication media (e.g. the Internet). The framework of coevolution is argued for in [9] as a more effective conceptual means with which the multi-agent and multi-level CAS that underpin communication governance could be studied given that the individual actors such as media consumers and content providers have interdependent relationships at multiple levels that include consumption and regulation as examples.

Our final discussion is on the paper [11] about incorporating and applying ABM in empirical social research since the integration of various disciplines that cover physical, natural, and social sciences has the potential impact towards improved public policy development and support. The first motivation that is raised in [11] has been discussed elsewhere by other studies we have introduced above. Interdependencies between microcomponents of a macrosystem and feedbacks between the two give rise to non-linear micro-macro relationship, which can be observed as emergent macropatterns that constrain behaviours of individual microcomponents. The second motivation is particularly relevant to addressing two related issues that come about from data-driven approaches that are prevalent in empirical social research. Here, one might have access to a large amount of data but is unsure which statistical model to use, and consequently make decisions specified by an adopted model without knowing whether those decisions have the right impact.

Another issue is the availability of rich concepts and ideas about human behaviours compared to what available data could represent. ABM could address these issues by guiding the specifications for statistical models and data collection efforts through the use of simulations that allows for comparison of macropatterns with observed data and provide what-if evaluations that come about from different assumptions of individual microbehaviours [11].

One crucial aspect applying ABM in empirical social research is to determine the appropriate level of complexity in the model for the policy-making task at hand. The review paper [11] has discussed a range of approaches of varying model complexities that have been employed. They include on one end empirically validated ABMs to function as simulations or virtual laboratories [22], and to other end abstract models that are used to uncover current understanding about theoretical underpinnings and mechanisms of the system under study [5, 52]. There are also approaches that fall somewhere in the middle whereby ABMs that better reflect properties of real-world actors are incorporated into traditional models that operate at the macrolevel (e.g. aggregates arising from interactions of many individual agents) that are computationally less costly and thus faster in providing predictions. One example is the study on epidemiological prediction of disease spread in a geographic population that combines a traditional low-dimensional dynamical systems approach operating at a single variable of interest (e.g. infection level of the population) with ABM that operates at the level of individual agents [23]. In this setting, one has coupled dynamics of disease and fear of contracting the disease that together give rise to more complex dynamics of infection levels in the population reminiscent of those observed in the real world (e.g. multiple waves of infections over a longer period). For such investigation settings, there is the potential of applying coevolutionary principles to better model how individual agents adapt their behaviours and thus provide new analytical tools for policy-making. Studies can be made to conduct what-if simulations of government interventions through policy and programmes that are aimed to promote adaptation towards certain agent behaviours so as to guide the variable of interest at the macrolevel of the system towards a desired value. These are done through coevolutionary simulations with different interventions to investigate changes to external or environmental conditions that are conducive for different behaviours adopted by the majority of individual adaptive agents.

4.3 Entertainment Games

Unlike the case with other problem domains, most designers and practitioners of CEAs would have readily come across applications of coevolution in games. In particular, competitive games have long constitute a relevant and useful problem domain with which deep understanding of coevolutionary systems can be made [12, 13, 16, 18, 25]. Indeed, all three chapters in this book in Part III cover research we have conducted over the years studying, designing, and applying coevolutionary systems to solve the problems of games. For example, the final two chapters have focused on the use of the basic two-player competitive games with deterministic outcomes with which deep theories could be developed to perform both qualitative and quantitative characterizations of coevolutionary problems and search processes. These theories are developed by modelling interactions in ***competitive CEA*** as game-plays. However, our focus in this section is on the applications of coevolutionary systems to solve the problems of games, and in particular, on entertainment games rather than the mathematical abstractions of games that are considered in those chapters. We will provide some detailed descriptions of these games and the challenges involved as they are not just for entertainment but actually involve some abstractions of other real-world problems.

Our brief survey of coevolution on entertainments games is based on selected literature reviews, one of which is the recent review paper that has performed a systematic review whereby coevolutionary systems are applied to ***Real-Time Strategy Games*** (***RTSs***), which is a particular subgenre of ***video games*** [21]. Indeed, the importance of video games in its global industry could not be understated, with a USD 110 billion market worldwide in 2017, and between then to 2020, represent the industry having the highest annual revenue and revenue growth within the category that include all other forms of media entertainment. There are five genres of video games: (1) traditional games, (2) simulation games, (3) strategy video games, (4) action video games, and (5) fantasy games. In the case of RTS, it is categorized together with ***Turn-based Games*** (***TBSs***) under strategy video games. The main difference between these two subgenres has to do with the implementation of strategic actions in the game by the players, where in RTS those actions have to be performed in real time by all players, while in TBS players are each allocated a dedicated time to complete actions, while all other players have to wait for their turns in sequence. Note that the review paper [21] considered RTS game in a more general sense, i.e. games having properties, rules, and strategic decision-making aspects of players similar to what an RTS video game as a simplified simulation of military warfare would have been considered [49].

Nevertheless, RTS games such as StarCraft and Warcraft series produced by the company Blizzard involve different settings of game-play. Classically these games start with a single human player against the computer (AI player) within a storyline backdrop but then would go on to include multiple-players setting (human- and AI players) in a local area network (LAN) all the way to a massively multiplayer setting on the Internet. It involves a two-dimensional map with specific terrain properties that include resources at specific locations. Buildings and units can be placed within this map satisfying certain conditions, which are predicated by matching properties of the terrain, buildings, and units. Resources are limited by their types and amount. The collection and expenditure of these resources constitute the basic economy for each player. Buildings are usually immoveable structures, while moveable units are typically of two types, i.e. resource collection and for engaging in warfare. A player in an RTS game is given a specific task to achieve, for example, in a standard match setting between two players, this means a player has to defeat the opponent by destroying all of its buildings and units. Decision-making actions that each player takes can be categorized in two levels, i.e. macromanagement (strategic) and micromanagement (tactical) [21, 49]. The first involves longer-term and higher-level planning with corresponding actions to maintain economy so that resource expenditures can be directed towards production of units and technological upgrades that confer some advantages, e.g. by way of modifying unit properties and enabling production of new unit types. The second involves short-term actions of warfare by the player's units, e.g. against the opponent's units and buildings, and medium-term actions that involve tactical coordination of groups of units for offence and also in conjunction with specific buildings for defence.

In the context of applications of coevolution to general RTS games [21], three areas have been addressed in the literature: (1) Evolving game-playing strategies. (2) Evolving ***Non-player Characters*** (NPCs) for game AI. (3) Evolving contents. The difference in the first two areas is on the targeted application of the game, in which case the second area is on producing AI players to compete with human players. Regardless, these two areas constitute atypical application for coevolutionary search methodologies. One example involves the formulation of the problem of winning a skirmish or encounter involving multiple units between two players as an optimization task [49]. Unlike the more challenging macromanagement task, such a micromanagement and well-defined task might admit optimal solutions and be searched effectively with heuristics. Here, a multi-objective cooperative CEA [1] has been proposed to search for such tactical strategic actions for a group of units consisting of different types, e.g. melee and ranged units. Note that this is a more complex problem compared to the case of using a group of homogeneous units (all of the same type). A heterogeneous or mixture of units with different properties would open up the possibility of other tactics, which induces a larger search space. Here, a cooperative CEA is employed to provide an effective Divide-and-Conquer approach to solve this problem. The problem decomposition employed is to divide the problem of tactical actions of a mixed group of units in the context of their types, i.e. tactics of melee units as one subcomponent and tactics of ranged units as another subcomponent. The MOP formulation is to take into account aspects that are critical in such a skirmish, i.e. maximizing damage to enemy units while minimizing damage from friendly fire. The multi-objective cooperative CEA is able to search for tactics involving such a mixed group of melee and ranged units with good performance against the default tactics in RTS game of StarCraft II. Note that there are other studies such as in [32] that investigate a cooperative coevolutionary formulation for control of multi-type agents in the more general problem setting of Evolutionary Robotics.

Another example involves the more challenging tasks associated with the macromanagement aspects in RTS games. Again, it is possible to focus on a specific task such as the planning of which units to be built and the order they are built to make the problem tractable to search methodologies. Indeed, such a problem can be formulated as a specific optimization task in the form of a scheduling problem whereby the solution to the task is represented by the specific build order [7]. The objective function typically involves resource expenditure, the number of opponent units and buildings that are destroyed, with the latter being given higher weights.

Performance evaluation of a build order requires actual simulation of the game whereby one main aspect of the player's strategy is the build order implemented. Simplification and abstraction of a StarCraft-like RTS game called Watercraft are required to speed up performance evaluation through reduced simulation whereby players implement their strategies restricted to specific aspects of game-playing in a controlled manner. In [7], this is done in a two-step process whereby each build order of the two competing players is first executed to produce the resulting warfare capable units, and then the outcome of a skirmish is simulated upon a player initiating an attack with all available units from both players involved. Although such a reduced setting makes it possible to employ a greedy search approach, nevertheless the study [7] has adapted a competitive CEA introduced by [50] that employs fitness sharing and a Hall-of-Fame archive to perform a coevolutionary learning search process of build orders. The coevolutionary approach is able to search for more robust build orders that perform well against a variety of other build orders compared to other heuristics approaches such as a Genetic Algorithms that employ a fixed set of opponent build orders to perform an evolutionary learning process and a Hill Climbing algorithm that is essentially a $(1 + 1)$ version of the competitive CEA.

Lastly for this section, we briefly discuss other applications of coevolution for RTS games that has considered generating new contents. In [47], contents of RTS games are restricted to the game maps to make the problem tractable. The RTS game considered is Planet Wars of the 2010 Google AI Challenge international contest. The specific mechanisms for the search of game maps fall under the area of ***Procedural Content Generation*** (***PCG***) of games that is concerned with various automatic and computer-assisted methodologies for generation of contents [54]. The study in [47] uses a predator-prey coevolutionary approach [36] due to the asymmetric nature of agents that implement game-playing strategies and those that represent game-playing environments in the form of maps. This means having two genetically isolated populations, one for agents implementing game-playing strategies and another for agents implementing maps. Hall-of-Fame archives are employed as a diversity maintenance technique to improve coevolutionary search performance and also function for performance evaluations of agents for both roles. The performance evaluation of agents has to be specifically formulated with respect to the roles they play. For the agent implementing a game-playing strategy, its performance incorporates individual encounters with a set of fixed game-playing strategies that is called the Fixed Rival Bot (FRB) agent, over the Hall-of-Fame archive of maps. For the agent implementing a map, its performance incorporates FRB agent performance against the game-playing agent and also a penalty term. The latter is employed as the means for evolutionary pressure to act to control a feature in the map (i.e. the number of planets) so that the coevolutionary process would not produce degenerate maps that could not be defeated by game-playing strategies. The study has demonstrated that such a predator-prey coevolutionary approach can be applied to both generate game-playing strategies and also content in the form of maps being game-playing environments.

4.4 Engineering, Logistics, and Transportation

The general domain of engineering provides a large number of design problems with different degrees of complexity. They all would have practical significance because their respective solutions are representative of product or systems to be constructed. In this section, we review several studies on the application of coevolution to ***Engineering Design problems***. We start with the study in [58] that has developed a dual-system variable-grain optimization framework that is based on cooperative coevolution for the purpose of solving layout design of complex engineering systems. In particular, the study has considered a ***Satellite-Module Layout Design*** (***SMLD***) problem. This involves placement of objects (e.g. devices and equipment) inside or outside of the satellite module in a manner that optimizes design objectives, being further subject to spatial and performance constraints. In the context of mathematical optimization, the SMLD problem as a three-dimensional layout optimization problem can be abstracted as a constrained combinatorial optimization problem. The formulated objective function is made single objective from multi-objective via a penalty function approach. The characteristics or properties of formulated objective function are typically nonlinear, discontinuous, and multimodal. As such, they pose a large challenge to standard EAs whereby premature convergence would be a typical outcome leading to low success rate.

As such, the study [58] has developed a ***Dual-System Variable-Grain Cooperative Coevolutionary Genetic Algorithm*** (***DVGCGA***) to address such challenging constrained combinatorial optimization problems. There are three main features in DVGCGA. The first feature is that a coevolutionary Divide-and-Conquer approach in the form of a cooperative CEA is employed to perform problem decomposition of the original optimization problem into several subcomponents (subsystems). In the case of DVGCGA, the cooperative CEA implementation is based on the original framework proposed in [48]. The evolutionary search process for the individual population of candidate subsystem solution is done via a GA. However, DVGCGA is different compared to a conventional cooperative CEA due to its second feature that incorporates a dual-system problem-solving approach. In general, dual-system refers to the application of two search approaches in parallel with communication between the two. In [58], this involves having copies of the decomposed problems in the first and second systems. The second system will implement a cooperative CEA. In the first system, the subsystems are treated virtually for the purpose of communication (via migration of candidate subsystem solutions) with the second system. The first system will implement an evolutionary search on the whole, original problem. The third feature involves a variable coarse graining, which in the context of a Divide-and-Conquer search of DVGCGA means that a mechanism that controls the number of subcomponents is implemented. The coarse graining mechanism in DVGCGA exploits the fact that mass of objects is a useful surrogate or indicator. Computational study in [58] has demonstrated significant effectiveness of DVGCGA compared to other approaches that include standard cooperative CEA and island-based EA.

Next, we review the study [63] that has developed a particular implementation of cooperative CEA for the design of digital ***Infinite-Impulse Response*** (***IIR***) filter. This is of practical significance in the modern world as it is possible to design both digital hardware and software implementations of such filters, the latter due to availability of computing power in modern machines. However, a traditional digital IIR filter design actually involves design in the analog hardware space with the added requirement of an analog-to-digital transformation. There are several challenges in such a traditional design approach. They include single-solution output, inefficiencies in the transformation, quantization errors of the filter coefficients, among others. The introduction of modern optimizers such as EAs has made it possible for effective design of digital IIR filters directly in the digital space. In particular, the IIR filter is specified by its transfer function, which has a cascade representation that takes the form of products of transfer functions of first- and second-order filters with their respective coefficients [63]. The performance of these digital IIR filters can be characterized by their magnitude response, their approximately linear phase response, and the orders of the filters used. As such, the design of digital IIR filters can be formulated as an MOP.

The study [63] has developed a cooperative CEA to solve this MOP formulation of the digital IIR filter design. The cooperative coevolutionary principle is used here for a Divide-and-Conquer problem-solving approach. In particular, problem decomposition is done along the line of IIR filter's structure and coefficients. As such, two genetically isolated populations are used, whereby one population consists of candidates representing the structure or control on filter designs (e.g. fixed numbers of n-order filter units for a bounded positive integer $n < N$) and another population consists of candidates representing the coefficients of the filters. Given that IIR filter control structures are largely responsible for meeting the three objectives, the population of IIR filter controls take the primary role to produce the Pareto solution that is formed when candidate coefficients are taken from the other population. This necessitates having to use an MOEA on the population of IIR filter controls. In particular, NSGA-II is used [19]. For the second population, Simulated Annealing (SA) with heat reservation and reheating strategies is employed to search for these filter coefficients. The algorithm design aspect here is that a diverse population of coefficients is required in the event that the Pareto solution consists of various IIR filter designs with various control structures. The proposed cooperative coevolutionary approach to IIR filter design is comparable to the effective Hierarchical GA approach [57] but with the further performance advantage of producing lower-order filter designs.

The next study [24] that we review is about the development of a competitive CEA for the purpose of optimizing a ***Fuzzy Logic Controller*** (***FLC***) for the Antilock Braking System (ABS) under varying road conditions. The ABS is one particular noteworthy and major active safety advancements in vehicular technology. The purpose of the ABS is to ensure that the vehicle equipped with such a system has reduced braking distances and is manoeuvrable during emergency situations by preventing the wheels from locking or slipping under braking. Modern controllers are able to apply continuous braking pressure on the wheels while avoiding locking situations, of which FLCs or systems [38, 41] that made use of ***Fuzzy Logic*** [64] represent a widespread application. The main reasons FLCs are used in ABS are because they are able to meet real-time computation requirements and being adapted to a variety of vehicles and road conditions. A FLC has a parallel structure and low computation requirements compared to other systems based on adaptive control laws and complex mathematical computations. Another reason that Fuzzy Logic is used in these controllers as opposed to employing Artificial Neural Networks is because the latter constitute blackbox models as opposed to the former that constitute whitebox models for which one can still derive understanding of their computations.

Despite these inherent advantages, FLCs in ABS require optimization of their control logics because of the varying operational conditions that are imposed on ABS in real-world scenarios. In this aspect, previous studies rarely consider full vehicle dynamics as well. As such, the study [24] performs the optimization of FLCs in ABS of a full multibody model of a two-wheeled vehicle in simulations that take into account varying operational conditions and transients on the tire slip-adhesion model. In particular, slip and road conditions are estimated using an extended Kalman filter as part of the simulation to be fed to the FLC. In this manner, the task of the FLC is to regulate brake pressure as its output given inputs that measure the slip error (difference between current slip and optimum slip) and the rate it changes, as well as the road conditions. The study uses the predator-prey competitive coevolution [36] in this optimization task. One main reason for the use of competitive coevolutionary principle is to address the need to reduce high computational costs from using full vehicle simulations through rigorous testings of FLC's ABS performance. Instead, the idea is to engineer an arms race dynamics in the predator-prey (FLC-test-suites) interactions. In this setting, two genetically isolated populations are used, whereby one represents candidate FLCs and the other represents test suites (i.e. road conditions) to which an evolved FLC is subjected to determine its ABS performance over varying operational conditions under vehicle dynamics simulations. The predator-prey competitive CEA is able to search for robust FLCs over varying road conditions compared to conventional controllers through systematic computational studies.

Our next review focuses on the application of coevolutionary principles for the purpose of kinematic calibrations of parallel manipulators in a manufacturing setting [67]. In general, the structure of a parallel robotic manipulator consists of two or more independent motion chains instead of one serial link. Although such parallel manipulators have inherent advantages over serial manipulators in terms of payload, stiffness, accuracy, and other operational aspects, they are also of much higher complexity. For applications of parallel robotic manipulators in an industrial manufacturing setting, accuracy is an important factor. Here, kinematic calibrations are required because deflections can occur at the joint variables of these complex parallel manipulators. Kinematic calibrations ensure that the nominal values are approximately equivalent to actual values for the pose of end-effectors under different inputs of actuation variables. There are multiple sources that contribute to errors resulting in deflections at joints. They include errors from manufacturing and assembly, thermal, and nonlinear stiffness in the robotic manipulators. These multiple errors can be accounted for from a single hypothetical error source generated in some random fashion based on the pseudoerror theory from the field of measurement and signal processing [67].

Given this problem formulation in place, the study [67] employs a neural-network-based approach to develop the complex, nonlinear association between information generated from a pseudoerror source and measured parameters, which is then used in conjunction with an inverse kinematic model of the parallel manipulator. Optimal kinematic calibration can then be obtained by tuning the parameters of the neural network. This tuning process is done using a cooperative CEA. In particular, the coevolutionary Divide-and-Conquer approach is applied on the neural network. This is achieved by decomposing the connection weights associated with the multilayer neural network into respective subcomponents. For example, the connection weights between the first and the second layers form a subcomponent. Multiple genetically isolated populations are used, with each population used in the evolution of candidate solutions for the subcomponent. In this manner, the optimization process on the entire neural network can be done in a cooperative coevolutionary manner. The computational study carried out in [67] has demonstrated the effectiveness of the cooperative coevolutionary neural network approach compared to three other neural-network-based approaches, i.e. a backpropagation neural network using the more advanced Levenberg-Marquardt algorithm, a general regression neural network, and an evolutionary neural network.

For the last two reviews, we focus on operational problems that have been addressed primarily in the area of ***Operations Research*** (***OR***). The first involves a ***scheduling problem*** in the specific setting on hydrothermal power generation [40]. Here, a hydrothermal power generator consists of two power generating units: hydro and thermal. The hydropower generating units are used to offset the additional load during peak demand period over the thermal power generating units due to high-fuel costs. Water reservoir of hydropower generating units is refilled during off-peak load periods. However, operations of such a hydrothermal power generating unit are more complex and require good schedules. This is due to the presence of various constraints that include system balance, the hydro and thermal unit specifications, limited water resource constraint, and others. In particular, the hydrothermal generation scheduling problem has been formulated as a mixed-integer optimization problem that is non-linear and nonconvex. In general, the formulation captures the scheduling of hydro units and the dispatching of thermal units requiring system constraint satisfaction of generation-load balance and spinning reserve of both unit types [40].

The study [40] used a specific ***two-population competitive coevolution*** approach to solve this hydrothermal generation scheduling problem that is first introduced in [56]. The approach is called the ***Coevolutionary Augmented Lagrangian Method*** (***CEALM***), and it has been introduced as a general method to solve nonconvex constrained optimization problems. This is first achieved through the construction of the dual formulation of the original (primal) constrained optimization problem that now includes both the decision variables and the Lagrange multipliers. The augmented Lagrangian methods convexify the primal objective function with quadratic penalty terms that are associated with the constraints to ensure that the solution in the dual problem corresponds to that of the primal problem. The solution that includes optimal values for the decision variables and Lagrange multipliers is constructed in such a manner that locally (over some defined neighbourhood) it corresponds to a saddle-point solution (e.g. minimizing over decision variables and maximizing over multipliers). What is novel in CEALM is that the search for such a saddle-point solution formulated as a particular two-player zero-sum game can be achieved with a competitive CEA.

In [56], CEALM involves two genetically isolated populations whereby one corresponds to candidate solutions for the decision variables and another corresponds to candidate solutions for the Lagrange multipliers. The evolutionary search procedure for any one of the two populations is based on Evolution Strategies with self-adaptation for continuous optimization problems [10]. Interactions between the pairs of candidates drawn from the respective competing populations form the matrix game of finite dimension. This matrix constitutes a finite approximation of the zero-sum game. The matrix is updated as the individual evolutionary process is applied to the respective minimizing and maximizing populations such that they then form a dense sample around the saddle-point solution. The zero-sum game formulation of the dual problem obtained from augmented Langragian ensures that the security strategy of the respective player (minimizer or maximizer) satisfies the inequality between the minimizing and maximizing objective functions. This means that the coevolution of these two competing populations would converge to the saddle-point solution. This solution is the global optimum if the problem has only one optimum, or a local optimum if the problem has many optima. In [40], the CEALM framework is applied to solve the hydrothermal generation scheduling problem that constitutes combinatorial optimization. The zero-sum game formulation now involves strategy representation in the form of a binary string. As such, the baseline evolutionary search process is based on a GA. The computational study demonstrates that CEALM can search for superior solutions compared to a GA and a SA.

Lastly, we review the applications of coevolution to solving operational problems in the large domain involving logistics and transportation. The following Chap. 6 presents research works we have carried out that made use of coevolutionary principles at the higher problem-solving level to develop a new framework of parallel portfolio of optimizers to search for solutions of logistics problems that can be formulated as the Travelling Salesman Problem and the more complex Vehicle Routing Problem. For this subsection, we review a representative study by Deng et al. [20] that focuses on a specific ***transportation problem*** involving cooperative driving of ***Connected and Automated Vehicles*** (***CAVs***). Here, cooperative driving of CAVs holds the key to modern transportations that could optimize traffic flow and hence among others leads to minimizing of energy usage. One particular area that is of increasing research interests is that of vehicle platooning whereby CAVs at specific regions equipped with the relevant facilities could automatically engage in cooperative merging and eco-driving, coordination at intersections without traffic or signal lights, and speed harmonization in highways or freeways. There are three phases involved in vehicle platooning: (1) the formation phase (where vehicles joined to form a group by coordinated merging and speed adjustment), (2) the maintaining phase (where vehicles maintain the same speed and gap), and (3) the separation phase (where vehicles leave the group).

The study in [20] has focused on the more challenging and less investigated formation phase of vehicle platooning. In particular, the problem setting involves vehicles merging to form a coordinated group in a particular road segment. Assuming that other objectives could be satisfied (e.g. safety through appropriate speed and gap of CAVs), then the main objective of such a coordination of CAVs within the control region is to minimize energy usage. The CAVs are assumed to be electric vehicles, in which case energy usage is a function of velocity, acceleration, and roadway grade and as such can be estimated. In this manner, the optimization task involves getting the CAVs to form a cooperative platoon formation with fixed vehicle speed and spacing between CAVs that result in minimization of energy usage. This involves optimization of a CAV trajectory, which can be illustrated as a specific function plot on a graph of vehicle spacing over time [28]. Then, the formulation of a numerical optimization problem that captures the salient aspect of the cooperative platoon formation requires a compact parameterized representation of a CAV trajectory. Instead of the many via points that describe a CAV trajectory, the study [20] employs a cubic spline representation. Here, several via points called knots that are evenly distributed in the vertical axis representing vehicle spacing are sufficient to approximate the vehicle trajectory through cubic spline interpolation. Crucially, a numerical optimization formulation of the cooperative platoon formation can be made for which numerical solvers can be applied to find the optimal solution.

In [20], a ***cooperative CEA*** is used as the solver for this numerical optimization problem. The specific formulation that the study has developed allows for a coevolutionary Divide-and-Conquer problem-solving approach. The numerical optimization problem has a natural problem decomposition in which the objective function can be decomposed along the individual vehicle trajectory representation, i.e. each subcomponent constitutes an individual CAV trajectory in the platoon. Multiple genetically isolated populations are used, whereby each population corresponds to an evolving sample of candidate solutions for a specific subcomponent. The numerical optimization for the cooperative platoon formation involves constrained optimization. However, the actual formulation is done via unconstrained optimization with penalty terms to reflect those constraints. The latter is exploited in the cooperative coevolutionary search. In particular, the evolutionary search process of individual population is done via local fitness function that only captures energy usage of the individual vehicle. The collaborative element in the cooperative platoon formation is reflected by the penalty terms. Instead of designing specific collaboration selection strategy that requires the combination of candidate solutions of subcomponents to form the complete solution, the study [20] directly exploits the penalty terms. In particular, the subcomponent population can be evolved individually, but that the cooperative coevolutionary process involves a local search of the current best full solution that minimizes the penalty terms. Computational study has demonstrated that the proposed cooperative CEA can generate feasible solutions and that the overall performance is superior compared to human-designed heuristics and EAs.

4.5 Software, Systems, and Services

In this section, we would provide a brief review of selected applications of coevolution in software, computer and network systems, and services. Indeed, these problem domains constitute an active area of research in the design and development of coevolutionary systems to provide effective solutions. We start the review with the domain of software development in the specific area of ***Automatic Programming*** (***Automatic Refinement***) that is concerned with the automatic generation of programs (e.g. codes). The generation of programs that constitute low-level implementation in some target programming languages from their high-level formal specifications is a very challenging task. We have introduced a conceptual framework based on coevolution to address this task [2, 3]. The framework is based on the ***predator-prey coevolution*** [36] of two genetically isolated populations, one of which consists of programs and another of unit tests. The framework is first developed for the simpler task of automated bug fixing whereby coevolution is used to search for refined programs from initial program with bugs [2]. Later, the framework is extended to a restricted setting of automatic generation of programs from a set of fixed primitives that define a nontrivial programming language [3].

In both settings, formal specifications (preconditions and postconditions) of the intended program are exploited so as to construct unit tests that are employed in the manner of software testing as part of the procedure to detect faults in the evolved programs. This is different compared to a rigorous determination of a program correctness. Software testing of programs does not guarantee correctness even if there is zero fault. Nevertheless, a quantitative statement can be made about the reliability of those programs. More specifically, a program $g : \boldsymbol{X} \rightarrow \boldsymbol{Y}$ is viewed as a transformation of the input variables $\mathbf{x} \in \boldsymbol{X}$ to the output variables $\mathbf{y} \in \boldsymbol{Y}$. A unit test is viewed as the pair $(\mathbf{x}, c)$ of input variables $\mathbf{x} \in \boldsymbol{X}$ for the program and an associated heuristic cost function $c : \boldsymbol{X} \times \boldsymbol{Y} \rightarrow \mathbb{R}_{\geq 0}$. The cost function $c(\mathbf{x}, \mathbf{y})$, $g(\mathbf{x}) = \mathbf{y}$ outputs a nonnegative real value, where 0 implies that the program output is equivalent to the correct output $\mathbf{y} \equiv \mathbf{y}^*$ or positive real values for other outputs. This cost function c is a heuristic in the sense that the program output is tested against the program specification, i.e. of satisfying preconditions and postconditions (note: the true correct output is unknown) and that distances have to be defined for outputs that do not meet the specification. The evolutionary process of the programs is based on GP, specifically the Strongly Type GP, while the evolutionary process of the unit tests requires the construction of a set of predicates from which program specifications and associated distance functions are constructed [3]. Here, the predator-prey competitive coevolutionary process is used so that the evolutionary refinement of initial programs that do not fully meet specifications can be made with a pedagogical evolutionary sequence of unit tests that are informative towards refinement of programs.

There are other studies that feature principles and mechanisms of coevolution as part of complex systems that are designed and developed for real-world problem solving. For example, the study in [8] has developed a system based on cooperative CEA and GP to solve the common search-and-replace text processing task that constitute a basic feature or functionality in modern text processing software and word processors (e.g. Microsoft Word), document preparation (e.g. LaTeX), as well as integrated development environments (IDEs) that are used for modern software developments. Indeed, the study has argued that software with modern text processing capabilities are already able to handle user input ***regular expressions***, which constitute a specific and standardized textual syntax for representing text matching patterns, even though most users other than those with specific training (e.g. programmers) are not familiar with them.

As such, the study [8] develops an automatic system for ***search-and-replacement task*** that actually learns from user-provided examples written in normal texts. These examples contain the search pattern s and replacement expression r that the system has to discover so as to perform the correct search-and-replacement task in a later user-provided texts. GP is naturally used to construct the pair (s, r) of search pattern and replacement expression in the form of regular expression for the former and include back references for the latter. For example, the regular expression `(\d+)-(\d+)-(\d+)` is used to express a search pattern involving three capturing groups, each for repeated digits $\{0, 1, \ldots, 9\}$, for dates that are normally written in a text in the form of three dash-separated numbers with replacement expression with back references `$2-$1-$3` that represents a date interpreted as month-day-year. Crucially, the study [8] has improved on the initial GP-based system by employing a conceptually simpler coevolutionary framework for search. In particular, cooperative CEA is used to decompose the problem of searching for the pair (s, r) of search pattern and replacement expression. Two genetically isolated populations are used, one for representing search patterns and the other for replacement expressions. The fitness of the complete solution takes the form of an MOP having three minimization objectives. The cooperative coevolutionary GP system is found to be significantly superior in performance compared to the initial GP-based system on various benchmarks.

We now turn our discussion to applications of coevolution in mobile systems, which constitutes an area of increasing importance given that mobile devices now serve as the primary platform with which various daily and work tasks are performed for many users. In particular, the study [53] has noted increasing security concerns on mobile devices as a result of larger volume of Android mobile devices and interests in malware developers targeting these devices. Evolutionary Computation studies for both ***malwares*** and ***anti-malwares*** as programs have used GP given that tree-based representation is a natural choice to construct and evolve programs. In the case of malware evolution, methodologies have been developed in which executable application files on Android can be converted into their representative source codes or class files conforming to Android's Java Virtual Machine (VM) implementation. Then, a GP tree can be constructed from these codes or files to represent these malwares. Malwares are evolved through the application of evolutionary processes on their GP trees, upon which new executable application files are recreated from evolved GP trees. In the case of anti-malware evolution, the ***malware detection*** program is one based on static analysis of application files for which a representative GP tree could be constructed. This GP tree essentially functions as a decision tree, whereby static features extracted from a potential application file with malware can be analysed based on features related to the application programming interface (API) calls and permissions. Although studies have been made to malware and anti-malware individually, the study [53] implements a competitive coevolution whereby the populations of malwares and anti-malwares are now both under evolutionary pressure from their adversarial interactions to improve. Indeed, the study has made a strong claim that such a predator-prey coevolution setup is unique. Their computational study has revealed coevolved malwares that are more evasive with respect to standard anti-malware programs, and coevolved anti-malwares that have more robust detection performance of new malwares.

Next, we review two representative studies on the applications of coevolutionary principles to develop a framework for performance-security trade-off optimization in ***Network Control System*** (***NCS***). In particular, an NCS involves both elements of communication as a networked system and control of entities that form the network [65]. Distributed NCS (D-NCS) constitutes a spatially distributed system with the control loops being closed through a real-time network. For example, a D-NCS implementation would include a distributed set of sensors, actuators, and controllers that are interfaced via a communication network [66]. The issue raised in [65] is that NCS is being employed in various critical infrastructures in energy, water, manufacturing, and transportation sectors to name a few, for which security is one major issue. Although security system such as those based on the use of cryptography is employed to ensure safe operations, there is also a competing performance issue in NCS. As such, there is a performance-security trade-off because system resources are inherently limited. In [65], the study has investigated this trade-off optimization in the context of MOP for a simulated networked DC motor as a model of NCS. Here, the performance aspect is related to the operations of the DC motor, while the security aspect is related to the need to ensure sufficient secured bandwidth via encryption for communication channels of sensor to controller and controller to actuator. In particular, a cooperative CEA is designed to search for the set of nondominated solutions (pareto front) whereby one population is evolved for performance-related decision variables, while another population is evolved for security-related decision variables for this problem. The study has also verified experimentally that a particular nondominated solution corresponds to the Nash equilibrium. In general, the cooperative CEA is able to find solutions that are superior in terms of performance and security compared to two EAs, one of which is based on swarm optimization [37].

In [66], cooperative coevolutionary principles are applied for a more complex D-NCS simulation setting involving multiple agents in the form of ***Unmanned Ground***

Vehicles (***UGV***). The simulation setting used is iSpace, which constitutes a network-based integrated navigation system to guide multiple UGVs via navigational information from remote controllers. These UGVs constitute the distributed network of sensors and actuators of the D-NCS. The added complexity of this setting compared to the earlier study [65] involving a networked DC motor comes about from the distributed and multiple agents of the same type in the form of UGVs. The cooperative CEA has to be extended with multiple populations representing performance-related decision variables for the UGVs, while a single population represents the security-related decision variables for the communication network that are secured via encryption as before. The cooperative CEA is able to converge to the solution that corresponds to the Nash equilibrium. Furthermore, the cooperative CEA is also able to deal with situations that would have impaired the D-NCS performance through the need to ensure increased security requirements in the form of added time delays in communication. In particular, the obtained solutions are able to maintain increased security requirements while still providing optimal performance that leads to good tracking performance on the UGVs [66].

We close this section of review with a representative application of coevolution on services. In particular, the study [62] has utilized coevolutionary principles as part of a complex approach for ***Dynamic Multi-objective Optimization Problem*** (***DMOP***) formulation for ***Internet of Things*** (***IoTs***) services. An IoT refers to a complex heterogeneous network of devices equipped with sensors, software, and various technologies for communication and exchange of data. One main goal of the IoT is to enable connection and interaction between physical devices and the digital world without human intervention so that seamless services can be provided. These services need not occur through the World Wide Web, but instead there are various other communication and network protocols for global Internet connectivity. In this manner, extensive services can be enabled and provided in important areas related to manufacturing, smart (energy) grids, security, healthcare, among others [61]. IoT services have been categorized to four general classes on identify-related services, information aggregation services, collaborative-aware services, and ubiquitous services [30]. The study in [62] is focused on the last class on ubiquitous services that are particularly challenging to optimize as a result of the nature of the underlying IoT having heterogeneous and dynamic links that are further compounded by huge uncertainties.

More specifically, a real-world implementation of a large-scale IoT consists of thousands of distributed entities (that include devices) where upon detection of service requests, multiple optimal services from a large candidate set have to be provided immediately. There are various tasks that the IoT must enable. These include providing services for dynamic concurrent requests, reducing energy consumption and service time, and improving information accuracy. Crucially, IoT operates in resource-constrained environments whereby service availability, bandwidth allocation, and reliability all need to be taken into consideration. In [62], the optimization problem involved in such an IoT is formulated as a DMOP. The MOP formulation part of the problem involves two competing objectives that have to be minimized, i.e. total energy consumption (by way of distance-related transmission costs incurred for communication and service-related energy consumption for providing services upon requests) and service time (by way of the maximum completion time involved in providing all services by all service providers). The dynamic optimization formulation part of the problem involves a varying number of service providers and requests with time, as well as time-varying distances between providers and devices that impact transmission costs and ultimately energy consumption.

As a result, the Pareto solutions change dynamically with time. There is a need for a sophisticated optimization algorithm that is able to quickly find Pareto solutions that would change throughout the operation of the IoT. The approach used in [62] is to incorporate a multi-population ***competitive-cooperative CEA*** that employs both competitive and cooperative mechanisms. The competitive-cooperative coevolutionary framework has been introduced earlier in [31], which extends the original cooperative CEA of [48] for the single-objective optimization problem. Here, problem decomposition is performed to produce one-dimension subcomponents and as such does not require type protection (in other words, members from any population can be compatibly used for the subcomponents because the latter are all one-dimension decision variable). In [62], a similar problem decomposition principle is employed whereby decomposition is applied on the Pareto solution. Each population would search and produce a subset of the Pareto solution. These Pareto optimal sets can be combined to form the entire candidate Pareto solution. As such, the problem-solving issue of having to rapidly discover a Pareto solution can be achieved through a cooperative coevolutionary search. The search process within each population to obtain nondominated members is achieved via a particular multi-objective search algorithm, specifically the ***Nondominated Immune Algorithm*** (***NNIA***) [33].

The coevolving populations that are individually under NNIA search process constitute a particular layer of a multilayer framework inspired by the human nervous, endocrine, and immune (NEI) systems. In particular, the proposed bio-inspired self-learning CEA (BSCA) employs a hierarchical search process inspired by the mechanisms of the human NEI systems. A regulation mechanism is designed to adapt the migration rate between the coevolving populations situated lower at Layer 3. The middle Layer 2 is tasked with cloning each of the Pareto optimal set obtained from Layer 3 as whole individual and then performs an adaptation or learning process in order to produce new Pareto optimal sets. The learning rate in this layer is regulated as well. Lastly, the upper Layer 1 is tasked with performing local search on the solutions produced in Layer 2 as the means to track moving Pareto solutions that constitute a characteristic of a DMOP. An agricultural IoT setting has been formulated, and the computational study in [62] demonstrates that the proposed BSCA is superior to other algorithms adapted for DMOP to track and search for moving Pareto solutions (e.g. those based on MOEAs and multi-objective IAs).

4.6 Concluding Remarks and Notes for Further Reading

Here, we have provided a literature review that is directed at individual studies whereby the principles of coevolution have been applied for real-world problem solving. We have discussed various studies applying coevolutionary problem solving (including those having potential applications) in four general problem domains of business, economics and public policy, entertainment games, engineering, logistics and transportation, and software, systems, and services. These problem domains have been selected to complement those we have addressed in studies we will present later in Parts II and III to develop coevolutionary systems for solutions. Despite the variety of problem settings and applications of coevolutionary systems, there are commonalities among all these studies. In particular, these real-world problems are abstracted and formulated as either optimization or learning problems. In general, one can then apply the appropriate coevolutionary principles for problem solving such as competitive CEAs for learning problems and cooperative CEAs for optimization problems. We have already discussed these two general coevolutionary approaches to problem solving in Chap. 4. There are specific problem settings whereby coevolutionary principles are used other than these two general approaches, which we would note and discuss. Nevertheless, we will discuss the general aspects and patterns of applications of coevolutionary principles and concepts of the selected reviewed studies towards problem solving.

We start first with competitive CEAs and specifically the applications of predator-prey competitive coevolution in real-world problem solving. In the seminal study [36], this principle of competitive coevolution has been proposed for problem settings whereby one is faced with difficulties in constructing useful objective functions and obtaining samples of test cases so that iterative procedures such as EAs can be applied to search for the required solutions. Furthermore, these two asymmetric roles that are played between solutions and test cases necessitate the use of at least two genetically isolated populations. Interactions between these two populations occur only at the phenotypic level via fitness evaluations and selection process. This distinguishes such predator-prey competitive CEAs against those that use single-population competitive CEAs. In the latter case, interactions can occur both at the genotypic and phenotypic levels. Among the collection of studies we have reviewed in this chapter, four problem settings have been addressed through predator-prey competitive CEAs. Even though it is possible to design test cases to some extent with the use of appropriate experts and problem domain knowledge, almost all of these studies [2, 3, 24, 47, 53] have noted the need for evolving solutions to successfully interact and solve a variety of test cases in order to obtain robust solutions with high-performance levels.

There are two main motivations why predator-prey competitive coevolution is used. The main motivation is that it provides automatic approach to obtain samples of test cases required for the evolutionary search of solutions. Here, the main challenge is that it is very difficult to construct objective functions with which solution performance can be evaluated. Note that this notion of solution performance can apply to the problem setting of solvability through the means of specifying the appropriate solution concept. For example, our study in [3] on automatic search or generation of software or computer programs based on formal specifications through coevolution exploits the earlier notion of refining initial software with bugs in the codes or faults in their implementations [2]. In both related problem settings, coevolution is employed for automatic construction of unit tests that could reveal implementation faults of those evolved software and specify in some quantitative manner how far are those tested software away from one that is reliable and fault free. Here, it is a challenging and nontrivial task to construct such unit tests. Significant domain experts knowledge and time resource would have been required to produce such a set of unit tests. However, such applications of coevolutionary principle to construct test cases apply even to problem settings whereby the actual need is of a less serious nature. For entertainment games, predator-prey competitive coevolution has been combined with PCG for the generation of game maps [47].

Another motivation being that coevolution can provide the means for an automatic approach to obtain diverse samples of test cases that can be used in the evolutionary search processes of robust solutions. In Sect. 4.4, the generated FLCs for ABS system are coevolved against a variety of road conditions as test cases because effective applications of continuous braking pressures on the brakes in vehicles equipped with the ABS systems differ with changing road conditions [24]. The adaptive FLCs can learn these different braking applications when they are presented with such a variety of road conditions through coevolution. Crucially, the predator-prey competitive coevolutionary process provides an automated approach with which road testing conditions are varied and adapted to present more challenges for these FLCs to cope with. In Sect. 4.5, the study [53] has applied predator-prey competitive coevolution to search for more robust malware detection programs against malwares that are also evolving to elude detections.

At this point of discussion, one could consider the fundamental differences on the nature of what diverse samples of test cases constitute across various problem settings. In particular, this diversity of test cases could manifest in different aspects throughout the actual operational lifetime of the solution systems. On the one end, one could assume a more passive aspect with which different test cases would present themselves to the solution system. In the context of evolving FLC for ABS systems, there are optimal brake application controls for different road conditions. The expectation of actual operations of the ABS systems is vehicles having to encounter varying but a fixed number of road conditions [24]. In such a setting, the solution has to be able to deal with and respond against a fixed number of test cases that could be discovered in advance (i.e. prior to actual lifetime operations). On the other end, one could assume a more active aspect with which test cases present themselves to the solution system. In the problem setting investigated by Sen et al. [53], the expectation of actual lifetime operation of a malware detection program is one that the malwares are actively changing to elude detection (e.g. either automatically through some preprogrammed procedures or by human-assisted alterations). In response, the malware detection programs also have to be changed (e.g. automatically through some procedures).

Most of the problem settings above that involve the use of predator-prey competitive coevolution to provide solutions are formulated in the context of learning problems that are analogous to active learning within the framework of supervised learning. However, competitive coevolutionary systems are used in other problem settings such as simulations within the framework of ***multi-agent system***. In Sect. 4.2 that has focused on the problem domain of economics, most problem settings involving a group of interacting decision-making agents can be abstracted as multiplayer games. One can then formulate specific notions that describe optimal decision-making for the individual and group of agents such as those involving the Nash equilibrium and Pareto optimality. For example, the studies in [14, 15, 27, 68] have formulated multiplayer games to describe the interaction of power producing agents in energy markets. Others in [42, 44] have considered multiplayer games for real markets (e.g. retail coffee markets). Unlike the interaction setting within a predator-prey competitive CEA that can be formulated as two-player competitive games, these multiplayer game formulations involve more than two players. Game payoffs and subsequent outcomes can be resolved after all players have implemented their strategic decisions required for the game that is played to completion (e.g. whether single shot or repeated). It is not surprising that the competitive coevolutionary systems that are used for these problem settings have employed a multi-population setup with each population representing the collection of candidate strategies that are implemented for each player (agent).

The study [43] has considered a more complex problem setting involving a financial or stock market with the multiple agents now able to implement heterogeneous strategies as opposed to homogeneous strategies in [14, 15, 27, 68]. These heterogeneous strategies include technical trading strategies that are able to vary the amount of stocks to be traded based on their analysis of the market compared to value trading strategies with linear decision responses towards the amount of stocks to be traded based on simple indicator such as price. Unlike most optimization and learning problems whereby specifications and parameters in the problem formulations are fixed, MAS simulations involve studying optimal or realized outcomes of the MAS when problem specifications and parameters are varied. Coevolutionary simulation is central to these problem settings because of the need to model and provide for the means with which adaptive agents can learn behaviours based on their interactions with other adaptive agents. In [43] that study simulated artificial stock market systems, established observations have been made in real-world financial systems whereby human traders are capable of adapting their trading strategies in response to varying market conditions that are attributed to changes in other trading agents' behaviours.

Note that single-population competitive CEAs are used for real-world problem solving in the appropriate setting even though most of the selected studies for review in this chapter have employed multi-population competitive CEAs. For example, the study [45] has compared three different coevolutionary systems that include multi-population and single-population competitive CEAs and a cooperative CEA. This comparison of different coevolutionary systems are made possible because the problem setting of searching for optimal individual trading agents in the context of a financial market for which the agents have no influence over its operations is akin to a one-player game against the environment [55, 60]. This is a particular optimization problem setting. Here, the fitness of a trading strategy can be estimated based on performance difference against an established baseline strategy, which is used as the basis to construct an objective fitness function. Crucially, relative fitness can be constructed on the basis of pairwise comparison of fitness values associated with any two implemented strategies. Similarly in [7], the application of single-population competitive CEA to search for macromanagement strategies in an RTS game that are abstracted and then formulated as build orders in the context of scheduling problems is made possible through the use of pairwise comparison of any two implemented strategies. A build order strategy is ultimately tasked with producing warfare units to engage in skirmishes. The pairwise comparison is the result of the skirmish outcome of units produced by two opposing build order strategies. The study has found that coevolutionary search is able to produce more robust build order strategies than evolutionary search. As such, it is crucial to note that for the same real-world problem, it is possible to formulate the problem settings as both optimization and learning.

We now turn our attention to most of the remaining reviewed studies that have employed cooperative CEAs for real-world problem solving. In particular, the problem settings they have considered are formulated as optimization problems. Cooperative CEAs are employed because the formulated problem settings admit problem decomposition so that Divide-and-Conquer methods can be used. As such, they vary in terms of the problem characteristics, for example, whether the problem is single objective or multi-objective. In Sect. 4.4, two studies [20, 67] have addressed optimization problems that are single objective given that the components involved in assessing the performance of a candidate solution can be aggregated. For example, the study [67] has addressed the inverse kinematic calibration of a robotic parallel manipulator that is subjected to pose deflections from various error sources through a multilayer Artificial Neural Network (ANN). Cooperative CEA is used to optimize the topology and connection weights of the ANN. Decomposition is done along the hidden layers of the ANN, with different populations used to represent candidate solutions for different hidden layers. Here, the single-objective optimization formulation comes about from having various error sources that cause deflections on the manipulator's pose aggregated to a single error source that is to be minimized by the ANN. Another study [58] in Sect. 4.4 has reformulated the problem from multi-objective to single objective via the use of penalty terms.

Most of the other problem settings across different domains involved MOP formulation. This is not surprising given the complexity involved in most real-world problem solving. Furthermore, such an MOP formulation comes about whether one is able to concretely formulate the problem in full detail (e.g. the search-and-replacement task in text processing [8] and digital IIR filter design [63]) or that one has to first employ a high level of problem abstraction (e.g. the micromanagement of a mixed group of units for skirmish in RTS games [1]). Despite the added complexity that comes about from the multi-objective nature of these optimization problems, once they have been appropriately formulated, it is straightforward to identify the point where the problem can be decomposed to subcomponents. For example, the study [8] has improved on the initial GP-based system and employed a conceptually simpler coevolutionary framework to search for the optimal pair (s, r) of search pattern and replacement expression in a text search-and-replacement task. Since the MOP formulation involves three objectives, all of which have to be minimized, then problem decomposition can be done along the point of search pattern s and replacement expression r. In [1] on micromanagement of a mixed group of units for skirmish in RTS games, the MOP formulation involves one objective to be maximized and another to be minimized. However, since both objectives involve functions that relate configurations of the mixed group of units as the input with damages as the output, then problem decomposition can be done along the point of the types of units, i.e. a subgroup of all melee units and another subgroup of all ranged units.

In closing this chapter, we remark that there are specific problem settings that require more unique applications of coevolutionary principles beyond general usage of cooperative CEAs for optimization problems and competitive CEAs for learning problems. For example, the most complex problem-solving setting in terms of problem characteristics is addressed in [62]. Here, the problem setting involves an IoT operation that is formulated as a DMOP. The MOP formulation part of the problem involves two competing objectives on total energy consumption and service time that have to be minimized. The dynamic optimization formulation part of the problem involves a varying number of service providers and requests with time, as well as time-varying distances between communicating devices that ultimately impact energy consumption. The approach used in the study [62] to solve such a DMOP employs cooperative coevolution to find the Pareto optimal set with multiple populations setup, whereby each population constitutes specific subsets of a Pareto optimal set. Such a problem decomposition strategy is used because of the need to be able to rapidly find a new Pareto optimal set when the constituents of the communicating devices change with time. In another example, the study [56] constructed a specific two-population competitive CEA to solve nonconvex constrained optimization problems. The problem is first appropriately transformed via augmented Lagrange method to obtain the dual problem formulation that is unconstrained and then made convex through the use of appropriate penalty terms to the primal or original problem. Crucially, the search for such a saddle-point solution involving minimizer over decision variables and maximizer over Lagrange multipliers can be reformulated as solving a specific two-player zero-sum game, the latter for which a two-population competitive CEA can be used.

References

1. Adhikari, N.K., Louis, S.J., Liu, S.: Multi-objective cooperative co-evolution of micro for RTS games. In: Proceeding of the IEEE 2019 Congress on Evolutionary Computation (CEC'19), pp. 482–489. IEEE Press, Piscataway (2019)

2. Arcuri, A., Yao, X.: A novel co-evolutionary approach to automatic software bug fixing. In: 2008 IEEE Congress on Evolutionary Computation (IEEE World Congress on Computational Intelligence), pp. 162–168. IEEE, Piscataway (2008)
3. Arcuri, A., Yao, X.: Co-evolutionary automatic programming for software development. Inf. Sci. **259**, 412–432 (2014)
4. Arthur, W.B.: Inductive reasoning and bounded rationality. Am. Econ. Rev. **84**(2), 406–411 (1994)
5. Axelrod, R.: The Evolution of Cooperation. Basic Books, New York (1984)
6. Axelrod, R.: The evolution of strategies in the iterated Prisoner's Dilemma. In: Davis, L.D. (ed.) Genetic Algorithms and Simulated Annealing, chap. 3, pp. 32–41. Morgan Kaufmann, New York (1987)
7. Ballinger, C., Louis, S.J., Liu, S.: Coevolving robust build-order iterative lists for real-time strategy games. IEEE Trans. Comput. Intell. AI Games **8**(4), 363–376 (2016)
8. Bartoli, A., De Lorenzo, A., Medvet, E., Tarlao, F.: Automatic search-and-replace from examples with coevolutionary genetic programming. IEEE Trans. Cybernet. **51**(5), 2612–2624 (2021)
9. Bauer, J.M., Latzer, M.: Co-evolution: applications and implications for governance research in communication studies. In: Puppis, M., van den Bulck, H., Mansell, R. (eds.) Handbook of Media and Communication Governance. Elsevier, Amsterdam (2023). https://ssrn.com/abstract=4433276
10. Beyer, H.-G., Schwefel, H.-P.: Evolution strategies: a comprehensive introduction. Nat. Comput. **1**, 3–52 (2002)
11. Bruch, E., Atwell, J.: Agent-based models in empirical social research. Sociological Methods Res. **44**(2), 186–221 (2015)
12. Burgin, G.H.: On playing two-person zero-sum games against nonminimax players. IEEE Trans. Syst. Sci. Cybernet. **5**(4), 369–370 (1969)
13. Chellapilla, K., Fogel, D.B.: Evolution, neural networks, games, and intelligence. Proc. IEEE **87**(9), 1471–1496 (1999)
14. Chen, H., Wong, K.P., Chung, C.Y., Nguyen, D.H.M.: A coevolutionary approach to analyzing supply function equilibrium model. IEEE Trans. Power Syst. **21**(3), 1019–1028 (2006)
15. Chen, H., Wong, K.P., Nguyen, D.H.M., Chung, C.Y.: Analyzing oligopolistic electricity market using coevolutionary computation. IEEE Trans. Power Syst. **21**(1), 143–152 (2006)
16. Chong, S.Y., Tiňo, P., Yao, X.: Measuring generalization performance in co-evolutionary learning. IEEE Trans. Evol. Comput. **12**(4), 479–505 (2008)
17. Cooper, L.G., Nakanishi, M.: Market-Share Analysis: Evaluating Competitive Marketing Effectiveness. Kluwer, Boston (1988)
18. Darwen, P.J., Yao, X.: Speciation as automatic categorical modularization. IEEE Trans. Evol. Comput. **1**(2), 101–108 (1997)
19. Deb, K., Pratap, A., Agarwal, S., Meyarivan, T.: A fast and elitist multiobjective genetic algorithm: NSGA-II. IEEE Trans. Evol. Comput. **6**(2), 182–197 (2002)
20. Deng, Z., Fan, J., Shi, Y., Shen, W.: A coevolutionary algorithm for cooperative platoon formation of connected and automated vehicles. IEEE Trans. Veh. Technol. **71**(12), 12461–12474 (2022)
21. Elfeky, E.Z., Elsayed, S., Marsh, L., Essam, D., Cochrane, M., Sims, B., Sarker, R.: A systematic review of coevolution in real-time strategy games. IEEE Access **9**, 136647–136665 (2021)
22. Epstein, J.M.: Modeling to contain pandemics. Nature **460**, 686–687 (2009)
23. Epstein, J.M., Parker, J., Cummings, D., Hammond, R.A.: Coupled contagion dynamics of fear and disease: mathematical and computational explorations. PLoS ONE **3**, E3955 (2008)
24. Fernández, J.P., Vargas, M.A., Velasco García, J.M., Cabrera Carrillo, J.A., Castillo Aguilar, J.J.: Coevolutionary optimization of a fuzzy logic controller for antilock braking systems under changing road conditions. IEEE Trans. Veh. Technol. **70**(2), 1255–1268 (2021)
25. Fogel, D.B.: Blondie24: Playing at the Edge of AI. Morgan Kaufmann, San Francisco (2002)

26. Fogel, D.B., Chellapilla, K., Angeline, P.J.: Inductive reasoning and bounded rationality reconsidered. IEEE Trans. Evol. Comput. **3**(2), 142–146 (1999)
27. Fu, H., Zhang, W.: Market equilibrium in active distribution system with μVPPs: a coevolutionary approach. IEEE Access **5**, 8194–8204 (2017)
28. Ghiasi, A., Li, X., Ma, J.: A mixed traffic speed harmonization model with connected autonomous vehicles. Transport. Res. Part C: Emerg. Technol. **104**, 210–233 (2019)
29. Giardina, I., Bouchaud, J.: Bubbles, crashes and intermittency in agent based market models. Euro. Phys. J. B Condens. Matter Complex Syst. **31**, 421–437 (2003)
30. Gigli, M., Koo, S.: Internet of things: services and applications categorization. Adv. Internet Things **1**(2), 27–31 (2011)
31. Goh, C.-K., Tan, K.C.: A competitive-cooperative coevolutionary paradigm for dynamic multiobjective optimization. IEEE Trans. Evol. Comput. **13**(1), 103–127 (2009)
32. Gomes, J., Mariano, P., Christensen, A.L.: Dynamic team heterogeneity in cooperative coevolutionary algorithms. IEEE Trans. Evol. Comput. **22**(6), 934–948 (2018)
33. Gong, M., Jiao, L., Du, H., Bo, L.: Multiobjective immune algorithm with nondominated neighbor-based selection. Evol. Comput. **16**(2), 225–255 (2008)
34. Goodin, R., Moran, M., Rein, M. (eds.): The Oxford Handbook of Public Policy. Oxford University Press, Oxford (2008)
35. Hardin, G.: The tragedy of the commons. Science **162**(3859), 1243–1248 (1968)
36. Hillis, W.D.: Co-evolving parasites improve simulated evolution as an optimization procedure. Physica D **42**, 228–234 (1990)
37. Kennedy, J., Eberhart, R.: Particle swarm optimization. In: Proceedings of ICNN'95 - International Conference on Neural Networks, pp. 1942–1948. Perth (1995)
38. Khatun, P., Bingham, C.M., Schofield, N., Mellor, P.H.: Application of fuzzy control algorithms for electric vehicle antilock braking/traction control systems. IEEE Trans. Ind. Electron. **52**(5), 1356–1364 (2003)
39. Levin, S., Xepapadeas, A.: On the coevolution of economic and ecological systems. Ann. Rev. Resour. Econ. **13**, 355–377 (2021)
40. Liang, R., Ke, M., Chen, Y.: Coevolutionary algorithm based on Lagrangian method for hydrothermal generation scheduling. IEEE Trans. Power Syst. **24**(2), 499–507 (2009)
41. Lin, C., Li, H.: Intelligent hybrid control system design for antilock braking systems using self-organizing function-link fuzzy cerebellar model articulation controller. IEEE Trans. Fuzzy Syst. **21**(6), 1044–1055 (2013)
42. Marks, R.E., Midgley, D.F., Cooper, L.G., Shiraz, G.M.: Coevolution with the genetic algorithm: repeated differentiated oligopolies. In: Proceedings of the 1st Annual Conference on Genetic and Evolutionary Computation (GECCO'99), Orlando, pp. 1609–1615 (1999)
43. Martinez-Jaramillo, S., Tsang, E.P.K.: An hcterogeneous, endogenous and coevolutionary GP-based financial market. IEEE Trans. Evol. Comput. **13**(1), 33–55 (2009)
44. Midgley, D.F., Marks, R.E., Cooper, L.G.: Breeding competitive strategies. Manage. Sci. **43**(3), 257–275 (1997)
45. Nicholls, J.F., Engelbrecht, A.P.: Co-evolved genetic programs for stock market trading. Intell. Syst. Accounting, Finance Manage. **26**(3), 117–136 (2019)
46. Nisan, N., Roughgarden, T., Tardos, E., Vazirani, V.V. (eds.): Algorithmic Game Theory. Cambridge University Press, New York (2007)
47. Nogueira-Collazo, M., Porras, C.C., Fernández-Leiva, J.A.: Competitive algorithms for coevolving both game content and AI. A case study: planet wars. IEEE Trans. Comput. Intell. AI Games **8**(4), 325–337 (2016)
48. Potter, M.A., Jong, K.A.D.: Cooperative coevolution: an architecture for evolving coadapted subcomponents. Evol. Comput. **8**(1), 1–29 (2000)
49. Robertson, G., Watson, I.: A review of real-time strategy game AI. AI Magaz. **35**(4), 75–104 (2014)
50. Rosin, C.D., Belew, R.K.: New methods for competitive coevolution. Evol. Comput. **5**(1), 1–29 (1997)

51. Savin, I., Creutzig, F., Filatova, T., Foramitti, J., Konc, T., Niamir, L., Safarzynska, K., van den Bergh, J.: Agent-based modeling to integrate elements from different disciplines for ambitious climate policy. Wiley Interdiscip. Rev. Clim. Change **14**(2), E811 (2022)
52. Schelling, T.C.: Micromotives and Macrobehavior. W. W. Norton & Company, New York (1978)
53. Sen, S., Aydogan, E., Aysan, A.I.: Coevolution of mobile malware and anti-malware. IEEE Trans. Inf. Forensics Secur. **13**(10), 2563–2574 (2018)
54. Shaker, N., Togelius, J., Nelsonl, M.J.: Procedural Content Generation in Games. Springer, Cham (2016)
55. Tadelis, S. (ed.): Game Theory: An Introduction. Princeton University Press, New Jersey (2013)
56. Takh, M., Sun, B.: Coevolutionary augmented Lagrangian methods for constrained optimization. IEEE Trans. Evol. Comput. **4**(2), 114–124 (2000)
57. Tang, K., Man, K., Kwong, S., Liu, Z.: Design and optimization of IIR filter structure using hierarchical genetic algorithms. IEEE Trans. Ind. Electron. **45**(3), 481–487 (1998)
58. Teng, H., Chen, Y., Zeng, W., Shi, Y., Hu, Q.: A dual-system variable-grain cooperative coevolutionary algorithm: satellite-module layout design. IEEE Trans. Evol. Comput. **14**(3), 438–455 (2010)
59. Tesfatsion, L.: Agent-based computational economics: a constructive approach to economic theory. In: Handbook of Computational Economics, vol. 2, pp. 831–880. Elsevier, Amsterdam (2006)
60. Volker, B., Udo, B., Wilfried, S.: Strategic option pricing. Econ. Bus. Rev. **6**(3), 118–129 (2020)
61. Want, R., Schilit, B.N., Jenson, S.: Enabling the internet of things. Computer **48**(1), 28–35 (2015)
62. Yang, Z., Jin, Y., Hao, K.: A bio-inspired self-learning coevolutionary dynamic multiobjective optimization algorithm for internet of things services. IEEE Trans. Evol. Comput. **23**(4), 675–688 (2019)
63. Yu, Y., Xinjie, Y.: Cooperative coevolutionary genetic algorithm for digital IIR filter design. IEEE Trans. Ind. Electron. **54**(3), 1311–1318 (2007)
64. Zadeh, L.A.: Fuzzy sets. Inf. Control **8**(3), 338–353 (1965)
65. Zeng, W., Chow, M.: Optimal tradeoff between performance and security in networked control systems based on coevolutionary algorithms. IEEE Trans. Ind. Electron. **59**(7), 3016–3025 (2012)
66. Zeng, W., Chow, M.: Modeling and optimizing the performance-security tradeoff on D-NCS using the coevolutionary paradigm. IEEE Trans. Ind. Inf. **9**(1), 394–402 (2013)
67. Zhang, D., Gao, Z.: Optimal kinematic calibration of parallel manipulators with pseudoerror theory and cooperative coevolutionary network. IEEE Trans. Ind. Electron. **59**(8), 3221–3231 (2012)
68. Zhang, S.X., Chung, C.Y., Wong, K.P., Chen, H.: Analyzing two-settlement electricity market equilibrium by coevolutionary computation approach. IEEE Trans. Power Syst. **24**(3), 1155–1164 (2009)

Chapter 5
Pathologies and Analysis in Coevolutionary Search

5.1 General Challenges in Coevolution

In Chap. 3, we have introduced and discussed the two frameworks of competitive and cooperative coevolution. They are presented in the form of Algorithms 3.1 and 3.2, respectively. In our discussion of competitive and cooperative Coevolutionary Algorithms (CEAs) particularly about the fundamental procedural components such as the variation and selection operators, we have made frequent references of them having similar design considerations with equivalent procedural components in Evolutionary Algorithms (EAs) as given by Algorithm 1.1. This means that most of the challenges that one encounters in having to design effective components in CEAs would be the same as in the case of designing EAs with high performance. In particular, one would have to consider among the available design choices and specifications for the solution representation, variation, and selection operators, those that are most appropriate for the problem at hand to bring about effective and efficient operations of the (co)evolutionary search.

Chapter 1 has presented the general details and challenges associated with these basic components of EAs, which are relevant and apply to those in CEAs as well. We have introduced the various form of solution representations that are used in EAs and briefly discussed their respective design considerations due to their specific constructions in 1.2.2. Most of the discussions on cooperative CEAs in Chap. 3 that are applied to additively separable optimization problems would have straightforward implementation of solution representations, particularly in the setting of continuous optimization problems involving known benchmark problems [41, 63, 66]. However, Chap. 4 has introduced more challenging problem-solving settings involving more sophisticated use of solution representation with CEAs. For example, the study [60] on the use of cooperative CEA with Genetic Programming (GP) tree representation for Cloud Computing resource allocation rules. In the application of competitive CEAs for problems involving games, various studies

X. Yao, S. Y. Chong, *Coevolutionary Computation and Its Applications*,
Natural Computing Series, https://doi.org/10.1007/978-981-96-2841-4_5

have investigated the uses of complex Artificial Neural Networks (ANNs) to provide game-playing strategies [13, 16, 37, 40].

The use of any solution representation is closely tied with the specific variation operator. This is because the main responsibility of the variation operator is to generate new child candidate solutions based on the parent candidate solutions in the current population. Indeed, earlier in 1.2.3 we have introduced and discussed how the form of solution representation and the variation operator together specify the neighbourhood structure of the search space. As such, there is a strong connection between design considerations of both solution representations and the associated variation operators for EAs. In the same manner, one can find similar research and development of solution representation and variation operators in CEAs. As an example, the use of complex ANNs with multiple layers would necessitate employing sophisticated variation operators with mechanisms for parameter control such as self-adaptation [13, 16] to allow for effective coevolutionary search of strategies for two-player board games. However, if one is able to discover and exploit useful invariants and other important structures that enable more effective and compact representations of solutions as a result of domain knowledge, then one can use simpler variation operators that would still allow for effective coevolutionary search [46, 48].

Given that the mechanisms of selection operators in both CEAs and EAs make use of fitness values associated with candidate solutions to decide the rate they are to be reproduced in the following generation, then it would stand to reason that the design of selection operators would have an impact on the search performance of (co)evolutionary processes. As before, the same selection operators that have been designed for use in EAs (see 1.2.4) are also implemented in CEAs. However, the main conceptual difference between the frameworks of CEAs and EAs is in the fitness evaluation of generated candidate solutions. As such, this chapter will focus on the main dynamical behaviours that are intrinsic to coevolutionary processes because of the principal use of relative and dynamic fitness measures of candidate solutions. The motivation of our presentation here is to unravel why and how these pathological behaviours come about in coevolutionary processes that have been observed to negatively impact search performance of CEAs.

The remainder of this chapter is organized into two main sections. Section 5.2 will first introduce and informally describe these coevolutionary pathologies. We will make use of a competitive coevolution setting in our discussion so as to more clearly present how these pathologies affect coevolutionary search of complete solutions. Section 5.3 will introduce recent studies that have developed new frameworks that capture specific structures in problems and processes of coevolution. This leads to subsequent development of tools to analyse coevolution formally, and as a result, they more precisely describe some of the coevolutionary pathologies. The chapter closes with remarks and discussions on how these pathologies can be addressed and have their negative effects on coevolutionary search processes mitigated.

5.2 Pathologies in Coevolutionary Search

In this section, we will introduce ***coevolutionary pathologies*** that collectively refer to a number of characteristic behaviours or dynamics observed in the coevolutionary process that affect the search performance of a coevolutionary system [6, 8, 11, 14, 22, 30, 55]. The first two subsections will introduce and discuss coevolutionary pathologies that are generally well-understood in the literature and for which both theoretical and empirical methodologies have been developed to study them. One key advantage of employing theoretical methodologies despite having to work on simplified and more abstract coevolutionary systems is that they allow for more precise descriptions of these pathologies. Section 5.2.1 describes the first group of pathologies on overspecialization, focusing, and intransitivity that can be understood by studying structures in the problems coevolution is applied to solve. Section 5.2.2 describes the second group of pathologies on cycling and forgetting that can be understood by studying the structures underlying coevolutionary processes. There are some linkages between cycling dynamics of coevolutionary processes and intransitivity in the relations of the space of solutions for coevolutionary problems. Section 5.2.3 will introduce and discuss the pathologies of disengagement and mediocre stable states for which most of the relevant studies in the literature have carried out empirical investigations.

5.2.1 *Overspecialization, Focusing, and Intransitivity*

In coevolutionary systems, ***overspecialization*** refers to the situation where the candidate solutions in the population are ***focusing*** on some specific properties of the problem instead of addressing all the relevant properties and thus solve the problem in its entirety. This particular coevolutionary pathology has been extensively studied and is generally well-understood. There are two perspectives that one can adopt to better understand the issue of overspecialization in coevolution. In a competitive coevolution setting, one perspective is to consider the issue at the level of pairwise interactions between candidate solutions as two-player competitive games. In other words, the problem coevolution that aims to solve—***coevolutionary problem***—is two-player competitive games. As a solution concept, one might consider dominance of a strategy over all others in the set of strategies in a pairwise manner. A two-player competitive game is said to have a dominant strategy if it has the property that there is a unique strategy that defeats the remaining strategies in the set [51]. However, games with a very large strategy set might have complex pairwise or binary relations over the set of strategies. In particular, the existence of a unique strategy that is dominant does not preclude that for the subset of the remaining dominated strategies, there has to be a strategy that dominates all other strategies within this subset.

Indeed, it is now known that these games can admit binary relation structures that form an ordered sequence of components (subsets of strategies). In particular, components that are ordered higher up dominate those that are ordered below them thereby forming a transitive chain of dominating components. However, there is no single dominant strategy within each component [19, 50]. Here, one has the situation whereby any strategy specializes in defeating only some other rather than all strategies [24] within that component. This comes about as a result of ***intransitivity*** in the binary relation of strategies in the component. Given a set of strategies $\boldsymbol{S}$, a strategy $s_i \in \boldsymbol{S}$ that ***dominates*** (defeats) another strategy $s_j \in \boldsymbol{S}$ is expressed as $s_i \leftarrow s_j$. The binary relation A on the set of strategies S is transitive if for all $s_i, s_j, s_k \in \boldsymbol{S}$, one has whenever $s_i \leftarrow s_j$ and $s_j \leftarrow s_k$, then $s_i \leftarrow s_k$. Intransitivity occurs where there are $s_i, s_j, s_k \in \boldsymbol{S}$ such that $s_i \leftarrow s_j$, $s_j \leftarrow s_k$, and $s_i \rightarrow s_k$. Informally, the triple $s_i, s_j, s_k \in \boldsymbol{S}$ forms a ***cycle*** as in the example game of only three strategies of ***Rock, Paper, and Scissors*** instead of a transitive triple. We will revisit this in more detail later in Sect. 5.3 where we will present a theoretical tool for analysis for such problem structures in coevolution that is based on Directed Graph Theory.

Another perspective is to consider the issue within the interaction or the game-play. In particular, the study [65] has introduced the ***numbers games*** that provide two useful features towards the study of coevolutionary systems. First, these games can illustrate the nature of intransitivity in the pairwise comparison (relation) of the set of objects with which one can only determine which objects are deemed superior by way of comparison. Second, the geometric structure that one can obtain through some specific means of organizing objects in the set [8] can be exploited to specifically highlight the focusing issue in coevolutionary systems [9]. In general, the numbers game involves a set of n-tuples of natural numbers $\mathbb{N}$. A particular numbers game is specified by the manner in which the outcome of a comparison between any pair of n-tuples is calculated as well as the choice of $n \geq 2$. Note that if one considers $n = 1$ and no self-play by disallowing comparison of similar numbers, then the binary relation $<$ on $\mathbb{N}$ is transitive and thus the problem represented by this numbers game becomes trivial. Most studies [9, 65] have considered two-player numbers games involving the set of 2-tuples (i.e. a pair of numbers) given that these games are already able to elicit coevolutionary pathologies such as overspecialization of agents that focus on solving (defeating) only some aspects of their opponents' game-playing strategies.

In a symmetric two-player numbers game, both players can choose to implement strategies from the set $\boldsymbol{S} \subset \mathbb{N}^2$ specified by the 2-tuple $(x, y) \in \mathbb{N}^2$. Let the first player implement the strategy (x_1, y_1) and the second player implement the strategy (x_2, y_2). For a two-player numbers game called the ***intransitive game***, the following gives the payoff function for the first player:

$$p_{\mathrm{IG}}\big((x_1, y_1), (x_2, y_2)\big) = \begin{cases} 1 & \text{if } \delta_x > \delta_y \wedge y_1 > y_2 \\ 1 & \text{if } \delta_y > \delta_x \wedge x_1 > x_2 \\ 0 & \text{otherwise,} \end{cases}$$

where $\delta_x = |x_1 - x_2|$, $\delta_y = |y_1 - y_2|$ [9]. For simplicity, one can restrict this game to only win-lose outcome with the player having the bigger payoff value winning the game, and ensuring that no draw outcome is obtained when both players have the same payoff values. The later can be achieved by restricting the pair of strategies drawn from $\boldsymbol{S}$ so that: (i) $x_1 \neq x_2$ and $y_1 \neq y_2$ (i.e. no self-play) and (ii) $\delta_x \neq \delta_y$. Following the example provided in [65], the three strategies $s_1 = (1, 6), s_2 = (4, 5), s_3 = (2, 4)$ form a cycle. The cycle is obtained based on the following results by direct calculation: (i) $s_1 \leftarrow s_2$ with $p_{\text{IG}}(s_1, s_2) = 1$ since $\delta_x > \delta_y \wedge y_1 > y_2$ and $p_{\text{IG}}(s_2, s_1) = 0$ since $\delta_x > \delta_y \wedge y_1 \not> y_2$. (ii) $s_2 \leftarrow s_3$ with $p_{\text{IG}}(s_2, s_3) = 1$ since $\delta_x > \delta_y \wedge y_1 > y_2$ and $p_{\text{IG}}(s_3, s_2) = 0$ since $\delta_x > \delta_y \wedge y_1 \not> y_2$. (iii) $s_1 \rightarrow s_3$ with $p_{\text{IG}}(s_1, s_3) = 0$ since $\delta_y > \delta_x \wedge x_1 \not> x_2$ and $p_{\text{IG}}(s_3, s_1) = 1$ since $\delta_j > \delta_i \wedge i_1 > i_2$.

The geometric organization that comes about from the structure of the pairwise relationships underlying the set of strategies $\boldsymbol{S}$ for the numbers game can be suitably exploited to enable some form of analysis with visualizations [8, 9]. A ***partially ordered set*** (***poset***) $(\boldsymbol{S}, \preceq)$ compares elements of $\boldsymbol{S}$ and orders them according to the binary relation $\preceq$. (Note: One has a reduced poset $(\boldsymbol{S}, \prec)$ by disallowing self-comparison, e.g. no self-play for the relation specified by a two-player game). Consider an Intransitive Game with $\boldsymbol{S} = \{s_1, s_2, s_3\}$. If all games are played on the triple s_1, s_2, s_3 and that the dominance relation $\leftarrow$ satisfies the transitive property, then the poset $(\boldsymbol{S}, \leftarrow)$ is ***totally ordered*** or ***linearly ordered*** with relations on $\boldsymbol{S}$ monotonically expressed as $s_1 \leftarrow s_2 \leftarrow s_3$ since by transitivity $s_1 \leftarrow s_3$. Informally and visually, the elements of $\boldsymbol{S}$ can then be placed as distinct points in a one-dimensional line with their ordering specified by their placements (e.g. as $s_3 \rightarrow s_2 \rightarrow s_1$ in an ascending order). For the previous example of $\boldsymbol{S}$ whereby $s_1 = (1, 6), s_2 = (4, 5), s_3 = (2, 4)$, the triple forms a cycle with incomparable elements due to intransitivity. The triple cannot be placed in a one-dimensional line and be linearly ordered.

However, this $\boldsymbol{S}$ can be placed or embedded into $\mathbb{N}^2$, which itself can be embedded into (and hence visualized in) the two-dimensional space $\mathbb{R}^2$ [9]. For a strategy $s_i = (x, y)$, let the value along each dimension be addressed as $s_i.x$ and $s_i.y$, respectively. In the current example, $(\boldsymbol{S}, \leftarrow)$ has the linear order $s_1 \rightarrow s_3 \rightarrow s_2$ based on $s.x$-dimension and another linear order $s_3 \rightarrow s_2 \rightarrow s_1$ based on $s.y$-dimension. The fact that $(\boldsymbol{S}, \leftarrow)$ is embedded in a two-dimensional space with two linear orderings implies there are cycles in pairwise relations on $\boldsymbol{S}$. Assume now that one adds to the current set of strategies a new strategy $s_4 = (s_1.x + 10, s_1.y + 10) = (11, 16)$. For this new set of strategies $\boldsymbol{S} \cup \{s_4\}$, one could then visualize the cluster given by the subset $\{s_1, s_2, s_3\}$ as before and then located in the top-right one has $\{s_4\}$. Indeed, one can work out that the strategy s_4 dominates each of the strategies in $\{s_1, s_2, s_3\}$ by direct calculation. However, one could also make use of the geometric organization of the cluster $\{s_1, s_2, s_3\}$ that forms a cycle and is embedded in $\mathbb{R}^2$, and the knowledge about the game that a strategy s_4 dominates the subset of strategies $\{s_1, s_2, s_3\}$ by having larger values in each of the two dimensions, i.e. $s_4.x > \max\{s_1.x, s_2.x, s_3.x\}$ and $s_4.y > \max\{s_1.y, s_2.y, s_3.y\}$. In particular, s_4 does not focus in having a larger value on one of the two dimensions but employs larger values in both dimensions to dominate the remaining strategies. This is reflected by the geometric location of s_4 with respect to the cluster of $\{s_1, s_2, s_3\}$ in the direction of the diagonal away from the origin of $\mathbb{R}^2$.

5.2.2 Cycling and Forgetting

Thus far in our current discussion, we have used the term cycle to refer to the intransitivity in the binary relation on the triple $s_i, s_j, s_k \in \mathbf{S}$ in contrast to one that satisfies the transitivity property. There are studies on ***population dynamics*** in coevolutionary systems. Here, the focus is on how the configuration of the population could change as a result of repeated applications of variation and selection operators. The main pathological issue of interest is ***cycling population dynamics*** or simply ***cycling*** [11, 14, 22, 30, 55]. Earlier in 1.2 of Chap. 1, we have described a discrete-time dynamical system $\mathcal{F} : \mathcal{X} \rightarrow \mathcal{X}$ whereby the variation $\mathcal{V}$ and selection $\mathcal{S}$ operators that constitute the coevolutionary process $\mathcal{F} = \mathcal{S} \circ \mathcal{V}$ act on the space of population configurations $\mathcal{X}$. This approach of mathematical abstraction and modelling has been applied to coevolution as well [35, 62]. Such ***coevolutionary dynamical systems*** allow for analysis of complex coevolutionary population dynamics via trajectories in $\mathcal{X}$ as iterates $\mathbf{X}_1, \mathbf{X}_2, \mathbf{X}_3, \ldots$ generated by the abstract coevolutionary model $\mathcal{F}$. In particular, the notion of regimes of ***periodic*** cycles has been used to more precisely describe cycling population dynamics observed in coevolution. Here, a periodic cycle of period $n \geq 1$ is described as the trajectory or sequence of population configurations or states $\mathbf{X}_1, \mathbf{X}_2, \mathbf{X}_3, \ldots, \mathbf{X}_n$ that repeat, e.g. $\mathbf{X}_n = \mathbf{X}_1$. Later in Sect. 5.3, we will give the precise theoretical formulation of such a coevolutionary dynamical system that allows for the rigorous analysis of population dynamics. For the rest of this section, we first introduce and discuss the relevant dynamical regimes arising from coevolutionary processes that are attributed to pathological search.

It is crucial that one distinguishes between cycle structures in population dynamics as a result of action of coevolution $\mathcal{F}$ on the space of population configuration $\mathcal{X}$ and cycle structures in the pairwise relation $\leftarrow$ on the set of strategies $\mathbf{S}$. They refer to different aspects of a coevolutionary system with the former describing structures associated with the coevolutionary process while the latter describing structures associated with the coevolutionary problem. Despite this, there are links between these two cycle structures because the selection mechanism in a coevolutionary process operates by taking interaction outcomes of strategies implemented by the population of agents. In turn, these outcomes are governed by the underlying structures of the pairwise relations in the set of strategies [19]. For example, the ***phase portrait*** (***phase diagram***) of the abstract coevolutionary dynamical system for the game of Rock, Paper, and Scissors with intransitivity would indicate cyclic dynamics. For example, the study [30] has produced such phase portraits in the form of line plots of trajectories that live in the two-dimensional standard ***Simplex*** $\mathbb{S}^2$ embedded in $\mathbb{R}^{2+1}$. (Informally, the reduction by one dimension comes from the fact proportions sum to one, with knowledge of proportions for two strategies automatically gives the proportion for the remaining strategy.) The phase space or space of population configuration $\mathcal{X}$ is given by $\mathbb{S}^2$ and consists of specific mixtures or proportions of the population implementing the three strategies Rock, Paper, and Scissors.

Periodic cycling dynamics come about due to the presence of ***fixed points*** in the phase space representing specific population configurations in an ***equilibrium*** that are ***unstable*** and act as ***repellors***. The dynamics are such that trajectories of points in the neighbourhood around those repelling fixed points (e.g. as a result of any small perturbation such as the action of mutation operator that changes the population configuration) would move away. Coevolutionary dynamical systems with three pure strategies for agents to choose to implement can have fixed points that represent ***monomorphic*** and ***polymorphic*** equilibrium representing population configuration of only one strategy and those with at least two strategies implemented, respectively. For the Rock, Paper, and Scissors game, the coevolutionary dynamical system has an unstable polymorphic equilibrium with equal proportions of Rock, Paper, and Scissors strategies and three monomorphic equilibria, one for each of the all Rock, Paper, and Scissors strategies. Visually, the Simplex $\mathbb{S}^2$ is an equilateral triangle with each corner containing one of the saddle points and at its centroid the unstable polymorphic equilibrium. The centroid also corresponds to the ***Nash equilibrium*** given that in a population of agents implementing equal proportions of Rock, Paper, and Scissors strategies, they will all receive the same game payoffs under ***complete mixing***, i.e. each agent play with all agents (including self-play).

The three ***saddle points*** at each corner are due to intransitivity of relations in Rock, Paper, and Scissors strategies. Saddle points have both attracting and repelling components in their dynamics. If one restricts dynamics along the sides of the Simplex $\mathbb{S}^2$ where any non-corner point corresponds to a population of agents implementing only two of the three available strategies for some proportions, then it is attracted to one corner point and repelled by the opposite corner point. For example, any non-corner point being attracted to the corner point representing all Paper and repelled away from the corner point representing all Rock strategies. As such, for trajectories that start from a non-equilibrium interior point of the Simplex $\mathbb{S}^2$, they would cycle in that part of the phase space. In general, the presence of multiple unstable fixed points with repelling dynamics gives rise to various cycling dynamics including those that are periodic. It is thus not surprising that cyclic dynamics can arise even for some coevolutionary dynamical systems with two pure strategies such as the Hawk-Dove because the structures of the games are such that the polymorphic and monomorphic equilibria are all unstable [35]. Furthermore cyclic dynamics of coevolution involving more complex games, e.g. those involving the Iterated Prisoner's Dilemma (IPD) with more choices have been observed via single-valued measurements $\mathbb{S}^n \rightarrow \mathbb{R}_{\geq 0}$ of the evolving population configurations with domain knowledge about the game [15, 25].

The pathological issue of ***forgetting*** in coevolutionary dynamics refers to the sequence of changing population configurations whereby agents exhibiting some specific behavioural traits that are present earlier are lost and then later rediscovered. The study [30] has listed three situations in which a trait is lost during coevolution.

The first situation involves a trait being selected against during coevolution because agents that implement strategies with the trait would receive lower fitness values. The next two situations involve the trait being selected for. However, selection pressure can be low in which case drift could set in due to the action of the variation operator, or in the more extreme case, the mechanism in the variation operator has inherent bias against the trait that precludes it from being present in generated offsprings. In the discussion of behavioural traits, one could consider the more straightforward setting of pure strategies representing those traits in their entirety. One example is the Rock, Paper, and Scissors game. Coevolutionary dynamical systems can be constructed having specific perturbations arising from the action of the variation operator to demonstrate cyclic dynamics exhibiting forgetting pathology. This involves perturbed trajectories that cycle through the sides of the Simplex $\mathbb{S}^2$ so that periodically one of three strategies is not implemented by the coevolving population. However, behavioural traits could also refer to some aspects of game-play that a strategy or a subset of strategies would employ against the opponent strategy. One simple example is the numbers game where a trait refers to a particular dimension. In another example such as IPD games, behavioural trait could refer to the propensity of strategies to employ specific choices for game-plays such as to cooperate, defect, or some other more elaborate game-plays involving intermediate cooperation levels and a combination of cooperation and defection. As such, cyclic dynamics could coincide with the issue of forgetting of behavioural traits for the coevolution of IPD game strategies with intermediate levels of cooperation.

5.2.3 Disengagement and Mediocre Stable States

In this last subsection, we introduce and discuss two other remaining pathologies with respect to coevolutionary processes for which most of the relevant studies in the literature have carried out empirical investigations. Furthermore, most of these studies have investigated the two pathologies that are typically observed in the more complex setting of a two-population competitive coevolution. This is unlike the case with the earlier two groups of pathologies for which studies could be made for the generally simpler setting of single-population competitive coevolution. Despite this, controlled empirical studies can be applied to reveal this last group of coevolutionary pathologies and isolate the settings that contribute them. Some of the explanations that have been offered for these pathologies are derived from other fields of studies such as those from Evolutionary Biology.

The pathology of ***disengagement*** refers to the situation in a two-population competitive coevolution with asymmetric interactions (e.g. predator-prey coevolution) whereby one of the population mostly outperform the other population [11]. When this situation occurs, then the use of relative fitness evaluation is no longer useful because the opposing population is unable to discriminate the performance of the population at hand. One typical example is the predator-prey coevolution that underlies the adaptations of robots under pursuit-evasion interactions in Evolutionary Robotics as simulations of those that occur in the natural world [21]. Here, the ***Red Queen Effect*** has been proposed in [64] to explain continued arms race between two coevolving populations that take turn in evolutionary time to erode and nullify the discovered trait that gives adaptive advantage in the opposing population through responsive adaptations in the current population to counter the aforementioned trait. Such dynamics between two coupled populations could persist if there are sufficient diversity present in the populations during the phase of adaptation in one population and then counter-adaptation in the other population.

However, it is possible a particular adaptive trait with inherent advantages against opposing population is discovered by the current population members and then rapidly proliferates through reproduction over the entire population within a short evolutionary cycle. This will then lead to the onset of disengagement because most of the members in the current population outperform those in the opposing population. When this occurs, relative fitness no longer is able to distinguish between the population members for selection and then they start to genetically drift (e.g. via mutation operator) [11]. Likewise, the same dynamic could occur in the opposing population as well. Here, the population members are mostly outperformed and obtained equally poor fitness values that could not be used to distinguish them. As such, the opposing population also start to drift. When this occurs, the two coevolving populations are said to suffer from ***full disengagement***. There are two other degrees of disengagement that have been identified in [11].

Asymmetrical disengagement is said to occur when one population reaches global optimality, while the opposing population is left to drift. In one example on the host-parasite competitive coevolution for sorting networks [44], the study [11] has illustrated a scenario of asymmetrical disengagement at work that impairs search performance. Note that the host-parasite competitive coevolution can have in addition to the primary objective of searching for the optimal sorting networks, the secondary objective of searching for the best test lists. Assume that the problem setting has asymmetric performance as well such that the best sorting network is able to sort the most difficult test list. If the population of sorting networks evolves to the point of discovering the global optimal solution, then asymmetric disengagement would set in. The global optimal solution would start to proliferate, while in the opposing population of test lists genetic drift could set in and stop search for more difficult test cases. The coevolutionary system in this scenario is able to achieve the primary objective but then fails to achieve the secondary objective. Furthermore, if the coevolutionary system continues to operate, then ongoing genetic drift could introduce deleterious mutations to the population of sorting networks that might cause them to loose even the optimal solutions if the selection operator has some elements of nonelitist mechanism.

Collusive disengagement occurs between two coevolving populations when one of them enters into a specific stable configuration or equilibrium of members having suboptimal performances. However, the population as a whole is said to ***collude*** because the said population members have similar fitness values. The study [11] has illustrated one example involving the ***coin toss game*** for a two-population coevolutionary system with a population of generators and predictors. Here, the optimal solutions are generators that randomly produce heads:tails at 50:50 ratio and predictors that guess these two possible states of the game each at equal probability. However, one could have the situation whereby the population of predictors consists of equal proportions of always guessing heads and always guessing tails. Although these predictors are specialists having on average similar fitness values or performances individually, as an entire population they collude and operate as a mixture of strategies with optimal play that is effective against a population of diverse generators. The latter population would then start to genetically drift due to asymmetric disengagement. The effectiveness of such population configuration would then mean that it would be difficult for evolutionary process to search for the actual optimal individual predictors. There would be a strong selection pressure to maintain such a population configuration or state, and in this situation it is also referred as the pathological issue of ***mediocre stable states*** [31]. Indeed, the population of generators could also enter into such a state having equal proportions of always tossing heads and always tossing tails. Both populations of generators and predictors are said to collude as well.

5.3 Analysis of Coevolutionary Search

In this section, we will introduce two formal frameworks that have been developed for the analysis of competitive coevolutionary systems. These theoretical frameworks allow for precise descriptions of the salient characteristics with which coevolutionary systems operate to conduct search. They capture the main structures that contribute to several coevolutionary pathologies that one could encounter when applying coevolutionary systems for problem solving. Crucially, the two frameworks provide the means with which tools can be constructed for rigorous analysis of coevolutionary systems. Section 5.3.1 will first present the formal framework for analysis of competitive coevolutionary problem structures based Directed Graph Theory. Next, Sect. 5.3.2 will present the formal framework for analysis of competitive coevolutionary processes based on Dynamical Systems Theory.

5.3.1 Cycle Structures in Coevolutionary Problems

Recently, we [19] have introduced a framework that uses ***directed graphs*** (***digraphs***) to capture the entire structure in any coevolutionary problem taking the form of two-player strategic games with deterministic outcomes, typically encountered in the competitive coevolutionary setting [1, 13, 15, 25, 36, 40]. This ***coevolutionary problem*** is represented as the ***coevolutionary digraph*** $D_C = (V_S, A_R)$. The vertex set represents the solution (strategy) set $V_S = S$ where each distinct ***vertex*** $v \in V_S$ corresponds to a single unique solution (i.e. pure strategy). The arc set captures the binary relation on S that is given by $A_R = R$ where each ***arc*** (directed edge) in one direction $u \leftarrow v \vee u \rightarrow v$ or both directions $u \leftrightarrows v$ corresponds to the pairwise preference relationship between the pair of solutions $u, v \in V_S$. In this manner, a wide range of coevolutionary problems as two-player strategic games can be represented by these digraphs that completely capture their pairwise relation (dominance) structures on the corresponding strategy sets.

The game between a pair of strategies is captured by the edge $\{u, v\}$ connecting the two vertices $u, v \in V_S$ in the digraph. Note that the underlying graph of D_C is ***complete*** so that there is an edge connecting any pair of vertices, i.e. there is a game-play between any pair of strategies. Its win-lose outcome is captured by the ***orientation*** of this edge, e.g. the arc $u \leftarrow v$ indicates that u ***dominates*** (defeats) v. In this manner, a ***coevolutionary tournament*** represents a coevolutionary problem modelled by a two-player game having win-lose game outcomes for a specific strategy set with all strategies competing with one another (except self-play). If the outcome includes a draw $u \leftrightarrows v$, then all such coevolutionary problems given by games with win-lose-draw outcomes are included in the family of ***coevolutionary semicomplete digraphs***. For the remaining of this subsection, we will focus our discussion on the family of coevolutionary tournaments that are sufficient to present digraph theoretic notions of cycle structures crucial towards the complete characterization of problem structures in coevolution.

We present a few more notions and properties in digraphs that are used for later discussion. The notion of domination can be extended to subsets of V_S, whereby a vertex is a subset of V_S consisting of one element. For two disjoint subsets V_S^1 and V_S^2 where $V_S^1 \cap V_S^2 = \emptyset$, $V_S^2 \Leftarrow V_S^1$ denotes that all vertices $v \in V_S^2$ dominate all vertices $u \in V_S^1$. A ***walk*** in the digraph $D_C = (V_S, A_R)$ is the alternating sequence of vertices and arcs $v_1 a_1 v_2 a_2 v_3 \ldots a_{k-1} v_k$, where $v_1, \ldots, v_k \in V_S$ and $a_1, \ldots, a_{k-1} \in A_R$. It is common to simply write this walk as the sequence $v_1 v_2 v_3 \ldots v_k$. The length of a walk is given by the number of arcs in it. In the previous definition this is simply k. A walk is ***closed*** if $v_k = v_1$. Next, one can then define a (v_1, v_k)-***path*** as a walk where all vertices in it are distinct. Then, a ***k-cycle*** is a closed (v_1, v_k)-walk of length $k \geq 3$ on the (v_1, v_{k-1})-path. The digraph is ***strongly connected*** (or simply ***strong***) if for each pair of distinct vertices $u, v \in V_S$ there exist both (u, v)-path and (v, u)-path. A ***reducible*** digraph is one where its vertex set V_S admits a ***vertex partition*** into two disjoint, nonempty subsets V_S^1 and V_S^2 satisfying $V_S^1 \cap V_S^2 = \emptyset$ and $V_S^1 \cup V_S^2 = V_S$, with the further property that the two subsets are oriented in one direction, i.e. $V_S^1 \Rightarrow V_S^2 \vee V_S^1 \Leftarrow V_S^2$. Otherwise, the digraph is ***irreducible***. Indeed, all irreducible digraphs are strongly connected by definition. Note that vertex partitioning of a reducible digraph D induces subdigraphs, e.g. D^1 and D^2. We denote the vertex set $V(D^{(i)})$ and arc set $A(D^{(i)})$ of a subdigraph $D^{(i)}$.

We can now use the language of digraph theory to precisely describe specific properties of games that have been used to illustrate cycle structures associated with coevolutionary pathologies that in the past have been informally explained.

For example, the structures in the game of Rock, Paper, and Scissors can be captured by the coevolutionary digraph $D_C = (V_S, A_R)$ where $V_S = \{v_1, v_2, v_3\} = \{\text{Rock, Paper, Scissors}\}$ and $A_R = \{v_1 \to v_2, v_2 \to v_3, v_3 \to v_1\}$. For this game of only three strategies, the triple also constitutes the strategy set. ***Intransitivity*** in the dominance relations of this triple is precisely captured by the 3-cycle $v_1v_2v_3v_1$ on the (v_1, v_3)-path. Since the closed cycle and path include all distinct strategies in V_S, they are both said to be ***Hamilton***.

In another example of the one-dimensional numbers game with $V_S = \{v_1, v_2, v_3\} = \{1, 2, 3\}$ and $A_R = \{v_1 \to v_2, v_2 \to v_3, v_1 \to v_3\}$, the triple satisfies the transitivity property. Here, the coevolutionary digraph $D_C = (V_S, A_R)$ is ***transitive***. Let $v_1, v_2, v_3, \ldots, v_n$ be an ordering of V_S such that for all arcs $v_i \to v_j \in A_R$ one has $i < j$. Such a digraph is said to have an ***acyclic ordering*** and furthermore acyclic digraph has no cycles. Indeed, it is known that a tournament is transitive if and only if it is acyclic [5], since one can express all the dominance relations monotonically as $v_1 \to v_2 \to v_3 \to \cdots \to v_n$. The Rock, Paper, and Scissors game and the one-dimensional numbers game of only three strategies would seem to be trivial examples now that one is equipped with the language of digraph theory. Nevertheless, the notions of cycle structures and acyclic ordering as the means with which tournaments can be characterized can be generalized to larger tournaments. In particular, the theoretical results to obtain those characterizations have exploited the fact that non-Hamilton lower-order cycles can be packed into subsets of V_S as strongly connected components, and that the dominance relations on these strong components are monotonic.

Indeed, the motivation for using a directed graph representation is not only that it naturally captures all relevant structures underlying coevolutionary problems, but all the various machineries and powerful results in digraph theory and specifically tournament theory [50] can be used to rigorously analyse them. In particular, we [19] have linked the digraph theoretic notion of ***irreducibility*** and game theoretic notion of ***solvability*** in qualitatively characterizing these coevolutionary tournaments. Indeed, one main result is that coevolutionary tournaments are either solvable (i.e. there exists a dominant subset of strategies) or not. A ***solvable*** coevolutionary tournament $T_C = (V_S, A_R)$ is ***reducible***, i.e. the strategy set can be vertex-partitioned with a dominant subset, e.g. $V_S^1 \Rightarrow V_S^2$. The cycle structures in a coevolutionary tournament determine whether it is solvable (reducible) or not.

One of the main technical results is to demonstrate that a reducible coevolutionary tournament T can be ***strongly decomposed*** into subdigraphs $T^{(1)}, T^{(2)}, T^{(3)}, \ldots, T^{(l)}$ where $\cup_{i=1}^{l} V(T^{(i)}) = V(T)$ and $V(T^{(i)}) \cap V(T^{(j)}) = \emptyset$ when $i \neq j$. Furthermore, each subdigraph is a strong component and that they form a chain of dominance relation that can be expressed monotonically $V(T^{(1)}) \Rightarrow V(T^{(2)}) \Rightarrow V(T^{(3)}) \Rightarrow \cdots \Rightarrow V(T^{(l)})$. Another main technical result is that an irreducible coevolutionary tournament T is strongly connected and furthermore is ***vertex pancyclic***, meaning that for every $v \in V(T)$ and every $k \in \{3, 4, 5, \ldots, n\}$, there is a k-cycle containing v in the tournament T. As such, we are able to make use of digraph theoretic analysis to reveal all cycle structures that arise in coevolutionary problems. Our analysis is able to reveal that these cycle structures are more complex than what prior literature studies have shown. One such example is that lower-order cycles are contained in higher-order cycles in various degrees, which even in the simple setting of one-population coevolution that are modelled as Markov chains can affect search performance [19]. They also corroborate with other theoretical studies such as in [49] that specifically examines the setting whereby an appropriate objective measure having crucial properties can be constructed in CEA leads to an EA-like search behaviour. For example, a coevolutionary problem with win-lose outcomes (internal subjective performance measure) for which solutions form a transitive chain has a rank equivalent to external objective performance measure given by the score sequence of solutions, where a solution's score is the number of wins.

We close this subsection with example descriptions of coevolutionary tournaments to highlight key properties one can obtain about these cycle structures. One can easily obtain a vertex pancyclic coevolutionary tournament by a single arc reversal of a transitive tournament. For the transitive tournament with vertices labelled according to the acyclic ordering $v_1, v_2, v_3, \ldots, v_n$, one has on one end v_1 representing the worst performing strategy with no wins and then on the other end v_n representing the best-performing strategy with the maximum number of wins $n-1$. The triple v_1, v_2, v_n has the dominance relations $v_1 \rightarrow v_2, v_2 \rightarrow v_n, v_1 \rightarrow v_n$. It suffices to reverse the arc $v_1 \rightarrow v_n$ to $v_1 \leftarrow v_n$ to produce a vertex pancyclic tournament T. Indeed, one now has for T a Hamilton cycle $v_1v_2v_3 \ldots v_nv_1$ on the Hamilton path $v_1v_2v_3 \ldots v_n$. Such a vertex pancyclic tournament has the least number of 3-cycles in them [10]. A ***regular*** tournament such that every vertex has the same number of incoming arcs (representing games it wins) and outgoing arcs (representing games it loses) has the most number of 3-cycles. Only tournaments with odd number of vertices can be regular and that it forms a generalization of the smallest regular tournament of only three vertices as informally illustrated by the Rock, Paper, and Scissors game. Informally, it takes more time for coevolution to escape from a regular subcomponent that has the most number of 3-cycles compared to one that has the least number of 3-cycles [19].

5.3.2 *Population Dynamics of Coevolutionary Processes*

We will first introduce the specific formulation of competitive coevolution as discrete-time dynamical systems and then within such mathematical setting discuss the intricate cycle structures that underlie population dynamics generated from coevolutionary processes. The study [35] is the first to formalize such dynamical systems for coevolution although there are others such as the earlier study in [38] as well as a much later sequel in [39] that have empirically investigated population dynamics of coevolution through computer-generated trajectories. We have already introduced the notion of population dynamics earlier in Sect. 5.2.2 as a collection of trajectories, each being the particular forward sequence $\{\boldsymbol{X}_n : n \in \mathbb{N}_{\geq 1}\} =$

$\{\mathcal{F}^0(\boldsymbol{X}_1), \mathcal{F}^1(\boldsymbol{X}_1), \mathcal{F}^2(\boldsymbol{X}_1), \ldots\}$ of population configurations generated under the coevolutionary process $\mathcal{F} : \mathcal{X} \rightarrow \mathcal{X}$ starting from $\boldsymbol{X}_1 \in \mathcal{X}$.

The are two motivations for using dynamical systems as abstract mathematical models of coevolution: (1) The mechanisms of coevolutionary processes can be suitably formalized under some settings. (2) The behaviours or ***regimes*** of those trajectories whether simple or complex can be qualitatively characterized and precisely described. However, the population dynamics generated by dynamical systems $\mathcal{F}$ acting on a high-dimensional metric space $\mathcal{X} = (\mathbb{R}^D, d)$ for competitive coevolution of interactions between agents being able to choose from the strategy set having a large number $(D - 1)$ of pure strategies are very rich. Indeed, we have earlier highlighted that even in the case where there are only two pure strategies, coevolution can generate complex population dynamics. Nevertheless, it is still possible to provide qualitative and quantitative characterizations of such systems that are interpretable. As such, the majority of studies [35, 38, 39, 61, 62] have considered coevolution involving two-player, two-pure-strategy games such as the Hawk-Dove. Note that dynamical systems analysis has been formulated and applied for cooperative coevolution setting as well [53, 54], although it is more complex and as such mostly relied on computer-generated trajectories as source for analysis.

For the remainder of this subsection, our discussion will focus on such ***coevolutionary dynamical systems*** involving two-player, two-pure-strategy games that we formally describe next. They belong to the family of one-dimensional discrete-time dynamical systems that are specified by the function $f : \mathcal{X} \rightarrow \mathcal{X}$ on a metric space with distance measure d given by the compact interval of the real line $([0, 1], d)$. Subsequently, we will use the simplified notation $f : [0, 1] \rightarrow [0, 1]$. The actions of the map f on $[0, 1]$ generate an orbit (trajectory) described by the difference equation $x_{n+1} = f(x_n) \in [0, 1], n \in \mathbb{N}_{\geq 1}$, starting from a given initial condition $x_1 \in [0, 1]$. The true orbit of x_1 is the sequence $\{x_n : n \in \mathbb{N}_{\geq 1}\} = \{x_1, f(x_1), f^2(x_1), \ldots\}$, where the iterates f^m, $m \in \mathbb{N}$ of f on x_1 are defined inductively with $f^0(x_1) = x_1$ and $f^n(x_1) = f\big(f^{n-1}(x_1)\big)$.

Under such a setting, one main contribution of [35] is to identify the primary modelling interface via ***Evolutionary Game Theory*** (***EGT***) [58] between competitive coevolution and dynamical systems. The model for the coevolutionary process assumes: (1) An infinite population of agents (each agent implements one of the two pure strategies for game-plays). (2) Every agent interacts with all other agents in the population (i.e. ***complete mixing***) and accumulates payoffs from game outcomes. (3) Agents reproduce asexually in proportion to their cumulative payoffs (producing only clones). In this manner, different specifications in games and selection mechanisms among others lead to different constructions of the ***replicator maps*** $f : [0, 1] \rightarrow [0, 1]$ [45, 57].

We consider interactions of the form of a two-player, two-pure-strategy symmetric games that are simultaneous and one shot so that outcomes are specified by the normal form payoff matrix, e.g. $\mathbf{A} = (a_{ij} : 1 \leq i, j \leq 2)$ with each entry a_{ij} assigned a nonnegative payoff for the first player. Let $p_1, p_2 \in [0, 1]$ represent the proportions of the two strategies adopted by the infinite population of agents. Since strategies are inheritable and in any generation (iterate) t all agents compete for payoffs that decide their average reproductive success, the replicator map f governs how the population state observed as p_1, p_2 change with time t. To formulate f, collect the proportions of strategies in a state (column) vector $\mathbf{p}(t) = \big(p_1(t), p_2(t)\big)^{\mathsf{T}}$ and the associated strategy cumulative payoffs in vector $\mathbf{w}(t) = \big(w_1(t), w_2(t)\big)^{\mathsf{T}} = \mathbf{Ap}(t)$, at every iterate t. Let $p_1 = p$ and $p_2 = 1 - p_1 = 1 - p$. Then,

$$w_1 = p_1 a_{11} + p_2 a_{12} = p a_{11} + (1-p) a_{12}$$
$$w_2 = p_1 a_{21} + p_2 a_{22} = p a_{21} + (1-p) a_{22}.$$

The replicator $f(p_i) = p_i w_i / \sum_{i=1}^{2} p_i w_i$ uses fitness proportional selection. For p_1,

$$f(p) = \frac{p \cdot w_1}{p \cdot w_1 + (1-p) \cdot w_2}.$$

Note that the coevolutionary dynamical system is one-dimensional because the dynamics given by f evolves on the one-dimensional standard Simplex $\mathbb{S}$ embedded in $\mathbb{R}^2$ with $\sum_{i=1}^{2} p_i(t) = 1$ for all $t = 0, 1, 2, \ldots$. The dynamics described by $\mathbf{p}(t+1) = f\big(\mathbf{p}(t)\big)$ is preserved when one instead uses $p(t+1) = f\big(p(t)\big)$, i.e. observing changes to $p = p_1$ in time since the line segment associated with the compact interval $[0, 1]$ for the coordinate axis representing p_1 of $\mathbb{R}^2$ is a projection of $\mathbb{S}$. Other replicator maps that model different selection mechanisms have been constructed and studied [35, 62]. Informally, coevolutionary dynamical systems are best viewed as models to study long-term population dynamics under the actions of selection. The actions of variation (e.g. mutation) can be considered in the context of specific finite perturbations on the trajectories in addition to other contributing effects that change the population state outside of selection such as finite-population effects [61, 62].

More importantly, now that one is able to formulate these coevolutionary dynamical systems as specific replicator maps $f : [0, 1] \to [0, 1]$, one can then use the extensive and expressive language of Dynamical Systems Theory to precisely study and describe population dynamics generated from these maps. Let $O_f = \{x_n : n \in \mathbb{N}\}$ be the orbit generated from f. There are various dynamical behaviours or ***regimes*** associated with orbits generated from replicator maps f of coevolution that we will introduce here. One such simple regime involves the ***fixed point*** $f(q) = q \in [0, 1]$ of f, and we use $O_f = \{q\}$ to denote orbits of fixed points of f for emphasis. This notion can be extended to fixed points of f^k to describe ***periodic orbits of period*** k, which is defined as the sequence of k distinct points $O_{f^k} = \{q_1, q_2, q_3, \ldots, q_k\} = \big\{f^0(q_1), f^1(q_1), f^2(q_1), \ldots, f^{k-1}(q_1)\big\}$. Note that this assumes a minimality condition whereby for each $q \in O_{f^k}$, $f^j(q) \neq q, 1 \leq j \leq k-1$. In particular, O_{f^k} is said to be a ***cycle*** if $f^k(q_i) = q_i \in (0, 1)$ for all $q_i \in O_{f^k}$. One important property to note for a cycle O_{f^k} is that if the system starts at a ***periodic point*** $q_i \in O_{fk}$ of period k, then those orbits stay fixed in O_{fk} (cycling through $O_{fk} = \{q_1, q_2, q_3, \ldots, q_k\}$ indefinitely for $k \geq 2$).

Another notion that Dynamical Systems Theory precisely describes is that of an orbit of p ***accumulating*** to a fixed point [7, 28]. One such observation that is often made through measurements of real-world systems or computational simulations is the long-run behaviour of orbits that enter and remain in a fixed point. In particular, the periodic orbit O_{fk} of period k is said to be ***attracting*** if orbits of p in the neighbourhood of the periodic points accumulate to it, i.e. $f^j(p) \to O_{fk}$ as $j \to \infty$ [28]. As such, attracting periodic orbits of period $k \geq 2$ are also called ***limit cycles*** of period k, and they have been reported in studies of competitive coevolution [35, 38]. One typical characteristic of such limit cycles is the observation of trajectories of population states (whether produced by measurements or computer-generated) that initially traverse through the state space $[0, 1]$ and then quickly settles into periodicity because they enter into one of the periodic points.

In addition to these regimes of attracting fixed point and periodic orbits that are well documented in studies of coevolutionary dynamics, there are more complex behaviour observed in population dynamics of coevolution. They have been attributed to ***chaotic dynamics*** [35, 38]. However, they are yet to be precisely described in the current literature in coevolutionary dynamics. Part of the issue is that most studies of coevolutionary dynamics are based on computer-generated trajectories and performed over a short time-scale. In [35], population dynamics produced over 10^4 iterates that are attributed to being chaotic actually consist of computer-generated ***pseudo-orbits*** (***pseudo-trajectories***) that are subjected to finite-machine computing errors. But chaotic dynamics are known to feature a very ***sensitive dependence*** on initial condition such that two orbits of nearby points can ***exponentially separate*** (e.g. starting close by at 10^{-14}, a typical doubling error rate would mean it takes around 50 iterates to have a single unit error of $\approx 2^{50} \times 10^{-14}$) [43].

But there are studies in dynamical systems for which it can be rigorously shown that computer-generated pseudo-orbits from chaotic maps can be tracked over 10^5 iterates [20]. We [61, 62] have started studying chaotic dynamics in coevolution via continuous replicator maps f that might feature ***hyperbolicity***. This is crucial because hyperbolic maps are ***structurally stable*** in the sense that their orbit structures are rich and robust. Despite having sensitive dependence on initial condition, hyperbolic maps possess the ***shadowing*** property so that remarkably any pseudo-orbit obtained as a result of finite perturbation at every iterate can be closely followed by a true orbit. As with the use of digraph theory to study cycle structures in coevolutionary problems, the motivation for using more of the methodologies in Dynamical Systems Theory is to be able to rigorously analyse such intricate structures underlying the more complex and chaotic dynamics that can arise in coevolutionary systems and for which under specific circumstances (e.g. shadowing dynamics) that computer-assisted tools are useful.

Indeed, these two seemingly opposed features in chaotic dynamics actually allow the study of dynamical systems that exhibit very complex behaviours. In addition to the seemingly irregular and unpredictable manner with which chaotic orbits go through the population state space (e.g. the unit interval [0, 1] for coevolutionary dynamical systems that are chaotic), they also return nearby their starting points. The latter is a result of chaotic maps being ***topologically transitive*** in the sense that for any pair of open nonempty subsets $U, V \subset X$ there exists a $n > 0$ such that $f^n(U) \cap V \neq \emptyset$ [29]. One informal approach to understand the richness of orbits in a chaotic $f : [0, 1] \to [0, 1]$ is to consider the variety of regimes of the orbits and their density. For illustration, remove all the disjoint periodic orbits of all periods that live in [0, 1], orbits of points that are ***eventually periodic*** (within finite iterates of f), and those whose orbits are ***asymptotically periodic*** [47]. Then, one still has points that belong to the set U and whose orbits are not even asymptotically periodic but ***recurrent*** in the sense that $f^n(U) \cap U \neq \emptyset$ for some $n > 0$.

5.4 Concluding Remarks and Notes for Further Reading

In this chapter, we have provided a summary of a very rich area of research in coevolution on the various pathological issues that are known to negatively affect coevolutionary search performance. We have started with a review of these coevolutionary pathologies and grouped them into three categories. The first has to do with pathologies of overspecialization, focusing, and intransitivity that are related by way of cycle structures in coevolutionary problems and for which some deep and theoretical understanding have been made. These structures come about from the binary relations underlying the set of solutions whereby one makes use of pairwise comparisons to establish performance of those solutions within a competitive coevolution setting for problem solving. Although we have [19] discussed how Digraph Theory has been used to provide complete characterization of structures in coevolutionary problems, we have also referenced other studies [8, 9] that have used Order Theory and provide similar descriptions. This is not surprising given that acyclic graphs connecting strong components correspond to partial ordering of finite sets [2–4].

Cycle structures in competitive coevolution can be described as a spectrum of coevolutionary digraphs, where on the one end resides regular digraphs with the maximum number of 3-cycles and towards the other end transitive digraphs with no cycles. Between these two extremes, one has a large region of the spectrum of irreducible digraphs that are differentiated by the number of 3-cycles in them and then another large region of reducible digraphs that all admit strong decompositions into components that form a transitive chain of dominance relations. When a competitive coevolution setting is viewed in the context of a two-player strategic game, the binary relations on the solution set are now expressed as dominance relations on strategy set. This provides link to Game Theory [51] that many past studies in competitive coevolution have used. In addition to current qualitative understanding that coevolutionary problems (digraphs) can be categorized as either solvable (reducible) or not (irreducible), there are also quantitative characterizations on these solvable coevolutionary problems in the context how fast coevolution can search for the dominant solutions [19].

The remaining two groups of pathologies are predominantly about structures on the dynamics of coevolutionary processes. In particular, the primitive object that is under consideration is the population configuration (state), each consisting of the specific solutions that are present as well as how many they are in the population. One can easily ascertain that the primitive object in question is significantly more involved now and as such the field understanding is partially filled in with controlled empirical studies employing notions from the biological domain. For example, the last group of pathologies that we have categorized that include disengagement and mediocre stable states refer to pathologies observed in the even more complex two-population competitive coevolution with asymmetric interactions.

Nevertheless, some deep and theoretical understanding on the group of pathologies that include cycling and forgetting are obtained through the use of Dynamical Systems Theory. Here, Game Theory and specifically EGT play a prominent role as the modelling interface allowing for the construction of coevolutionary dynamical systems in the form of replicator maps [35]. Most observations on cycling and forgetting can be expressed as periodic trajectories generated from these coevolutionary dynamical systems. However, there are other significantly more complex and chaotic dynamics in which more formal studies have begun [61, 62]. Note thus far these formal studies on coevolution have considered ***deterministic dynamics***. There are other studies that have considered ***stochastic dynamics*** for which the coevolutionary processes directly incorporate finite-population effects and other stochasticity involving evolutionary processes of selection and variation, e.g. through the use of ***Markov process*** to investigate population dynamics in coevolution [33] as well as CEA search performance for different problem structures [18, 19].

We provide a brief summary by way of a literature review on several key studies and focus our discussion on the methodology that aims to address various pathological issues that could arise as a result of certain properties (e.g. intransitivity) in coevolutionary problems and directly operates in the selection process. Note that there are other mechanisms and heuristic approaches, e.g. those that employs external archives to affect selection [26, 34, 56] as one example. Here, we focus on those change or reformulate fitness values of solutions directly. These approaches have been collectively referred to as ***diversity maintenance techniques*** [17] as a result of previous claims and observations that their impact on coevolutionary search performance is due to their effects of introducing and maintaining diversity in the population. We start with one particular family of approaches that involve ***fitness sharing***. The notion of fitness sharing is not new, having been introduced and employed earlier in GA to better solve multimodal optimization problems [42]. In the case of competitive CEAs, fitness sharing is employed to provide a mechanism in the selection process to change fitness values assigned to competing agents in a manner that introduce and maintain a diverse population of test cases that collectively would present different challenges to the population of solutions to overcome. As such, the use of fitness sharing might prevent the onset of overspecialization and focusing in competitive coevolution. In the following, we introduce and briefly discuss two different implementations of fitness sharing in coevolution.

The first involves the ***competitive fitness sharing*** that is developed for a predator-prey competitive coevolution setting in [56]. The implementation can be easily adopted for a single competitive coevolution setting although in our following description the notation reflects a two-population competitive coevolution setting as introduced in [56]. Each evolved solution agent $\boldsymbol{s} \in \boldsymbol{S}$ interacts with a random subset of opponent or test agents $\boldsymbol{T}'_s \subset \boldsymbol{T}$ if a tournament setting is considered instead of a round robin that will involve the full population $\boldsymbol{T}$. Unlike the usual case of interactions modelled as a win-lose game outcome that associates the same value for a win against any opponent, in competitive fitness sharing, different values can be associated depending on the opponent. This is achieved by assigning a particular weight w_t for a win against an opponent $\boldsymbol{t} \in \boldsymbol{T}'_s$ that is given by $1/n_t$, where n_t is the number of solution agents in its full population that have defeated $\boldsymbol{t}$. Let the size of this full population be $|\boldsymbol{S}|$. Consider only a win-lose game outcome and that draws (including those from self-play) result in a lost for both players. Then, one has the range $w_t \in [1/(|\boldsymbol{S}| - 1), 1]$. The shared fitness for $\boldsymbol{s}$ is calculated as $f_{\text{sh}}(\boldsymbol{s}) = \sum_{\boldsymbol{t} \in \boldsymbol{T}'_s} w_t \cdot M(\boldsymbol{s}, \boldsymbol{t})$, where $M(\boldsymbol{s}, \boldsymbol{t})$ is the usual assignment function with 1 if $\boldsymbol{s}$ defeats $\boldsymbol{t}$, and 0 otherwise. Essentially, the approach encourages the selection of unique solutions in terms of those that solves specific test cases few other competing solutions could achieve. The value of $1/n_t$ is reciprocally proportionate to n_t, i.e. it takes on a larger value when n_t is smaller.

Implicit fitness sharing [24] employs an algorithmic procedure that approximates the shared fitness value for the solution agent $\boldsymbol{s} \in \boldsymbol{S}$ [59] and thus has the advantage of not having to design a sharing function that can be problem specific [23]. We will introduce the approach here in a single-population competitive coevolution setting. Let $g(\boldsymbol{s}, \boldsymbol{t})$ be the game payoff that strategy $\boldsymbol{s}$ obtains in a two-player game against $\boldsymbol{t}$. (Likewise, $g(\boldsymbol{t}, \boldsymbol{s})$ is the game payoff that strategy $\boldsymbol{t}$ receives.) For each agent in the population $\boldsymbol{s} \in \boldsymbol{S}$, random subsets of test opponents $\boldsymbol{T}'_c$, $c = 1, \ldots, C$ are drawn from the population where $\boldsymbol{T}'_c \subset \boldsymbol{S}$. Note that each $\boldsymbol{T}'_c$ excludes $\boldsymbol{s}$ to disallow self-play. For each random subset $\boldsymbol{T}'_c$, the procedure assigns a fitness value equivalent to the game payoff to the opponent that achieves the largest game payoff in the two-player game against $\boldsymbol{s}$, which is given by $\max_{\boldsymbol{t} \in \boldsymbol{T}'_c} g(\boldsymbol{t}, \boldsymbol{s})$. In the event of ties, the opponents will share the fitness equally (the game payoff divided by the number of opponents in such ties). Note that this is used for interactions modelled as nonzero-sum games where solution concept or notion of optimality is about maximizing accumulated payoffs. For zero-sum games, one could use the largest winning margin, i.e. $\max_{\boldsymbol{t} \in \boldsymbol{T}'_c} \delta_{\text{win}}(\boldsymbol{t}, \boldsymbol{s})$, where $\delta_{\text{win}}(\boldsymbol{t}, \boldsymbol{s}) = g(\boldsymbol{t}, \boldsymbol{s}) - g(\boldsymbol{s}, \boldsymbol{t})$ if $\boldsymbol{t}$ wins against $\boldsymbol{s}$, and 0 otherwise.

Another approach that directly works to change fitness values to influence selection of agents is ***reducing virulence*** [11, 12]. This nature-inspired approach is based on the observations that reduced virulence could address the issue of disengagement in biological coevolutionary systems. Highly virulent parasites could incapacitate the hosts and subsequently reduce their reproduction chances to infect new hosts. In comparison, strains with moderate virulence could reproduce faster and thus gain an evolutionary advantage. The argument for artificial competitive coevolutionary systems to employ reduced virulence then discourages the onset of disengagement from rapid proliferation of high-performing agents in the population with respect to test cases so as to promote longer-term and sustained coevolutionary search. Assume that fitness values $f(\boldsymbol{s})$ for the agent $\boldsymbol{s}$ are proportional to performance against solution test cases, e.g. $f(\boldsymbol{s}) = \sum_{\boldsymbol{t} \in T'} M(\boldsymbol{s}, \boldsymbol{t})$. The implementation of reduced virulence rescales $f(\boldsymbol{s})$ via $f(\boldsymbol{s}, v) = 2f(\boldsymbol{s})/v - \left(f(\boldsymbol{s})/v\right)^2$ [12], where the parameter $v \in [0.5, 1.0]$ controls the virulence level that represents tendency to maximize $f(\boldsymbol{s})$. For example, standard coevolution operates at maximum virulence level with $v = 1.0$. The function with domain over $f(\boldsymbol{s})$ and range over $f(\boldsymbol{s}, 1.0)$ is injective (one to one), thereby encouraging agents to maximize their fitness values. At $v = 0.75$, the function is surjective (many to one) where some high values of $f(\boldsymbol{s})$ are mapped to lower values $f(\boldsymbol{s}, 0.75)$, thereby penalizing higher-performing agents in the population to discourage search from focussing on solutions that maximize relative fitness values.

Lastly, we introduce the ***pareto coevolution*** approach that is based on the Multi-objective EA framework [27, 32, 52]. In competitive coevolution, one could argue against the use of relative fitness values based on a simple weighted sum of values such as $f(\boldsymbol{s}) = \sum_{\boldsymbol{t} \in T'} M(\boldsymbol{s}, \boldsymbol{t})$. This is because outcomes against some test cases could be more important, which requires nonuniform weights. More importantly, the set of weights could change for different population configurations generated by coevolution. Pareto coevolution addresses this issue directly by employing test cases to represent different, underlying objectives of the problem that a solution agent must address. The selection process is based on the ***pareto dominance*** relation of the competing agents in the population with respect to a set of test cases.

For a two-population competitive coevolution of solutions $\boldsymbol{s}_i$, $i = 1, \ldots, S$ and test cases $\boldsymbol{t}_k$, $k = 1, \ldots, T$ [27], this can be expressed as $\boldsymbol{s}_i \succ \boldsymbol{s}_j \iff \forall \boldsymbol{t}_k : g(\boldsymbol{s}_i, \boldsymbol{t}_k) \geq g(\boldsymbol{s}_j, \boldsymbol{t}_k) \wedge \exists \boldsymbol{t}_k : g(\boldsymbol{s}_i, \boldsymbol{t}_k) > g(\boldsymbol{s}_j, \boldsymbol{t}_k)$. In words, $\boldsymbol{s}_i$ pareto dominates $\boldsymbol{s}_j$ when $\boldsymbol{s}_i$ has at least the same game payoff as $\boldsymbol{s}_j$ for every one of the T objectives (each represented by the test case $\boldsymbol{t}_k$), and there is at least one objective for which $\boldsymbol{s}_i$ has a better game payoff than $\boldsymbol{s}_j$. They are ***mutually nondominating*** when $\boldsymbol{s}_i$ is better than $\boldsymbol{s}_j$ on at least one objective and $\boldsymbol{s}_j$ is better than $\boldsymbol{s}_i$ on at least another objective, with both being equivalent on the remaining objectives. In general, the mechanism will seek to rank agents according to the pareto layers obtained from their pareto dominance relations. Mutually nondominating agents in the ***pareto front*** would be the most preferable for selection. Subsequent pareto layers can be obtained systematically by removing mutually nondominating agents ranked in higher pareto layers [32]. Note that in a single-population competitive setting, such pareto dominance ranking would produce an acyclic dominance chain of strong components that correspond to the pareto layers [19].

References

1. Axelrod, R.: The evolution of strategies in the iterated Prisoner's Dilemma. In: Davis, L.D. (ed.) Genetic Algorithms and Simulated Annealing, chap. 3, pp. 32–41. Morgan Kaufmann, New York (1987)
2. Backhouse, R.: An analysis of repeated graph search. In: Mathematics of Program Construction, pp. 298–328. Springer, Berlin (2019)
3. Backhouse, R., Doornbos, H., Glück, R., van der Woude, J.: Components and acyclicity of graphs. An exercise in combining precision with concision. J. Logical Algebraic Methods Program. **124**, 100730 (2022)
4. Backhouse, R., Doornbos, H., Glück, R., van der Woude, J.: Elements of Algorithmic Graph Theory. An Exercise in Combining Precision with Concision. Unpublished monograph (2022). https://www.cs.nott.ac.uk/~psarb2/MPC/BasicGraphTheory.pdf
5. Bang-Jensen, J., Gutin, G.Z.: Digraphs Theory, Algorithms and Applications. Springer, London (2009)
6. Bari, A.G.: Interactive fitness domains in competitive coevolutionary algorithm. Ph.D. thesis, University of South Florida, FL (2019)
7. Barwell, A.D.: ω-limit sets of discrete dynamical systems. Ph.D. thesis, University of Birmingham, Birmingham (2010)
8. Bucci, A.: Emergent geometric organization and informative dimensions in coevolutionary algorithms. Ph.D. thesis, Brandeis University, MA (2007)
9. Bucci, A., Pollack, J.B.: Focusing versus intransitivity geometrical aspects of co-evolution. In: Proceedings of the 2003 Genetic and Evolutionary Computation Conference (GECCO'03), pp. 250–261 (2003)
10. Burzio, M., Pelant, J.: On complete strongly connected digraphs with the least number of 3-cycles. Discrete Math. **155**, 27–30 (1996)
11. Cartlidge, J.: Rules of engagement: Competitive coevolutionary dynamics in computational systems. Ph.D. thesis, University of Leeds, Leeds (2004)
12. Cartlidge, J., Bullock, S.: Combating coevolutionary disengagement by reducing parasite virulence. Evol. Comput. **12**(2), 193–222 (2004)
13. Chellapilla, K., Fogel, D.B.: Evolution, neural networks, games, and intelligence. Proc. IEEE **87**(9), 1471–1496 (1999)
14. Chong, S.Y.: Generalization and diversity in co-evolutionary learning. Ph.D. thesis, University of Birmingham, Birmingham (2007)
15. Chong, S.Y., Yao, X.: Behavioral diversity, choices, and noise in the iterated Prisoner's Dilemma. IEEE Trans. Evol. Comput. **9**(6), 540–551 (2005)
16. Chong, S.Y., Tan, M.K., White, J.D.: Observing the evolution of neural networks learning to play the game of Othello. IEEE Trans. Evol. Comput. **9**(3), 240–251 (2005)
17. Chong, S.Y., Tiňo, P., Yao, X.: Relationship between generalization and diversity in coevolutionary learning. IEEE Trans. Comput. Intell. AI Games **1**(3), 214–232 (2009)
18. Chong, S.Y., Tiňo, P., He, J.: Coevolutionary systems and pagerank. Artif. Intell. **277**, 103164 (2019)
19. Chong, S.Y., Tiňo, P., He, J., Yao, X.: A new framework for analysis of coevolutionary systems - directed graph representation and random walks. Evol. Comput. **27**(2), 195–228 (2019)
20. Chow, S., Palmer, K.J.: On the numerical computation of orbits of dynamical systems: the one-dimensional case. J. Dyn. Differ. Equ. **3**(3), 361–379 (1991)
21. Cliff, D., Miller, G.F.: Tracking the red queen: measurements of adaptive progress in co-evolutionary simulations. In: Advances in Artificial Life: Proceedings of the Third European Conference on Artificial Life. Lecture Notes in Computer Science, vol. 929, pp. 200–218. Springer, Berlin (1995)
22. Darwen, P.J.: Co-evolutionary learning by automatic modularization with speciation. Ph.D. thesis, University of New South Wales, Sydney (1996)

23. Darwen, P.J., Yao, X.: A dilemma for fitness sharing with a scaling function. In: Proceedings of the 1995 IEEE Conference of Evolutionary Computation (ICEC'95), Perth, pp. 166–171 (1995)
24. Darwen, P.J., Yao, X.: Speciation as automatic categorical modularization. IEEE Trans. Evol. Comput. **1**(2), 101–108 (1997)
25. Darwen, P.J., Yao, X.: Co-evolution in iterated prisoner's dilemma with intermediate levels of cooperation: application to missile defense. Int. J. Comput. Intell. Appl. **2**(1), 83–107 (2002)
26. de Jong, E.D.: The incremental pareto-coevolution archive. In: Proceedings of the 2004 Conference on Genetic and Evolutionary Computation (GECCO'04), pp. 525–536 (2004)
27. de Jong, E.D., Pollack, J.B.: Ideal evaluation from coevolution. Evol. Comput. **12**(2), 159–192 (2004)
28. de Melo, W., van Strien, S.: One-Dimensional Dynamics. Springer, Berlin (1993)
29. Devaney, R.L.: An Introduction to Chaotic Dynamical Systems, 2nd edn. Addison-Wesley, Boston (1989)
30. Ficici, S.G.: Solution concepts in coevolutionary algorithms. Ph.D. thesis, Brandeis University, MA (2004)
31. Ficici, S.G., Pollack, J.B.: Challenges in coevolutionary learning: arms-race dynamics, open-endedness, and medicocre stable states. In: Proceedings of the Sixth International Conference on Artificial Life (ALIFE), pp. 238–247 (1998)
32. Ficici, S.G., Pollack, J.B.: Pareto optimality in coevolutionary learning. In: Advances in Artificial Life: 6th European Conference (ECAL'01). Lecture Notes in Computer Science, vol. 2159, pp. 316–325. Springer, Prague (2001)
33. Ficici, S.G., Pollack, J.B.: Evolutionary dynamics of finite populations in games with polymorphic fitness equilibria. J. Theor. Biol. **247**, 426–441 (2007)
34. Ficici, S.G., Pollack, J.B.: A game-theoretic memory mechanism for coevolution. In: Proceedings of the 2003 Conference on Genetic and Evolutionary Computation (GECCO'03). Lecture Notes in Computer Science, vol. 2723, pp. 286–297. Springer, Berlin (2003)
35. Ficici, S.G., Melnik, O., Pollack, J.B.: A game-theoretic and dynamical-systems analysis of selection methods in coevolution. IEEE Trans. Evol. Comput. **9**(6), 580–602 (2005)
36. Fogel, D.B.: Evolving behaviors in the iterated prisoner's dilemma. Evol. Comput. **1**(1), 77–97 (1993)
37. Fogel, D.B.: Blondie24: Playing at the Edge of AI. Morgan Kaufmann, San Francisco (2002)
38. Fogel, D.B., Fogel, G.B., Andrews, P.C.: On the instability of evolutionary stable strategies. Biosystems **44**, 135–152 (1997)
39. Fogel, D.B., Fogel, G.B.: Simulating natural selection as a culling mechanism on finite populations with the hawk–dove game. Biosystems **104**, 57–62 (2011)
40. Fogel, D.B., Hays, T.J., Hahn, S.L., Quon, J.: A self-learning evolutionary chess program. Proc. IEEE **92**(12), 1947–1954 (2004)
41. Goh, C.-K., Tan, K.C.: A competitive-cooperative coevolutionary paradigm for dynamic multiobjective optimization. IEEE Trans. Evol. Comput. **13**(1), 103–127 (2009)
42. Goldberg, D.E., Richardson, J.: Genetic algorithms with sharing for multimodal function optimization. In: Proceedings of the Second International Conference on Genetic Algorithms and Their Applications, pp. 41–49 (1987)
43. Hammel, S.M., Yorke, J.A., Grebogi, C.: Do numerical orbits of chaotic dynamical processes represent true orbits? J. Complexity **3**(2), 136–145 (1987)
44. Hillis, W.D.: Co-evolving parasites improve simulated evolution as an optimization procedure. Physica D **42**, 228–234 (1990)
45. Hofbauer, J., Sigmund, K.: Evolutionary game dynamics. Bull. Am. Math. Soc. **40**(4), 479–519 (2003)
46. Krawiec, K., Szubert, M.G.: Learning n-tuple networks for Othello by coevolutionary gradient search. In: Proceedings of the 2011 Genetic and Evolutionary Computation Conference (GECCO'11), pp. 355–362 (2011)
47. Li, T., Yorke, J.A.: Period 3 implies chaos. Am. Math. Mon. **82**(10), 985–992 (1975)

48. Lucas, S.: Learning to play othello with n-tuple systems. Aust. J. Intell. Inf. Proces. Special Issue on Game Technology **9**(4), 1–20 (2007)
49. Luke, S., Wiegand, R.P.: When coevolutionary algorithms exhibit evolutionary dynamics. In: Barry, A. (ed.) Workshop on Understanding Coevolution: Theory and Analysis of Coevolutionary Algorithms (GECCO'02), pp. 236–241 (2002)
50. Moon, J.W.: Topics on Tournament. Holt, Rinehart and Winston, New York (1968)
51. Nisan, N., Roughgarden, T., Tardos, E., Vazirani, V.V. (eds.): Algorithmic Game Theory. Cambridge University Press, New York (2007)
52. Noble, J., Watson, R.A.: Pareto coevolution: using performance against coevolved opponents in a game as dimensions for pareto selection. In: Proceedings of the 2001 Genetic and Evolutionary Computation Conference (GECCO'01), CA, pp. 493–500 (2001)
53. Popovici, E., De Jong, K.: The dynamics of the best individuals in co-evolution. Nat. Comput. **5**, 229–255 (2006)
54. Popovici, E., Jong, K.D.: Relationships between internal and external metrics in co-evolution. In: Proceedings of the IEEE 2005 Congress on Evolutionary Computation (CEC'05), Edinburgh, pp. 2800–2807 (2005)
55. Rosin, C.D.: Coevolutionary search among adversaries. Ph.D. thesis, University of California, San Diego, CA (1997)
56. Rosin, C.D., Belew, R.K.: New methods for competitive coevolution. Evol. Comput. **5**(1), 1–29 (1997)
57. Sigmund, K., Nowak, M.A.: Evolutionary game theory. Curr. Biol. **9**, R503–R505 (1999)
58. Smith, J.M.: Evolution and the Theory of Games. Cambridge University Press, Cambridge (1982)
59. Smith, R.E., Forrest, S., Perelson, A.S.: Searching for diverse, cooperative populations with genetic algorithms. Evol. Comput. **1**(2), 127–149 (1992)
60. Tan, B., Ma, H., Mei, Y., Zhang, M.: A cooperative coevolution genetic programming hyperheuristics approach for on-line resource allocation in container-based clouds. IEEE Trans. Cloud Comput. **10**(3), 1500–1514 (2022)
61. Tino, P., Chong, S.Y., Yao, X.: On reliability of simulations of complex co-evolutionary process. In: Proceedings of the 24th European Conference on Modelling and Simulation (ECMS 2010), pp. 258–264. European Council for Modeling and Simulation, Kuala Lumpur (2010)
62. Tiňo, P., Chong, S.Y., Yao, X.: Complex co-evolutionary dynamics - structural stability and finite population effects. IEEE Trans. Evol. Comput. **17**(2), 155–164 (2013)
63. van den Bergh, F., Engelbrecht, A.P.: A cooperative approach to particle swarm optimization. IEEE Trans. Evol. Comput. **8**(3), 225–239 (2004)
64. van Halen, L.: A new evolutionary law. Evol. Theory **1**, 1–30 (1973)
65. Watson, R.A., Pollack, J.B.: Coevolutionary dynamics in a minimal substrate. In: Proceedings of the 2001 Genetic and Evolutionary Computation Conference (GECCO'01), pp. 702–709 (2001)
66. Yang, Z., Tang, K., Yao, X.: Large scale evolutionary optimization using cooperative coevolution. Inf. Sci. **178**(15), 2986–2999 (2008)

Part II
Coevolutionary Computation on Optimization Problems

Chapter 6
Coevolutionary Construction of Parallel Algorithm Portfolio Optimization

6.1 Introduction to Parallel Portfolio Optimization

Iterative search methods have become major de facto approaches for tackling various computationally hard problems. These established iterative search methods come in two broad categories, i.e. exact methods such as mixed integer programming solvers in ***Linear Programming*** to heuristic methods that include local search and metaheuristics. The latter also involve a spectrum of approaches. On the one hand, there are specialized heuristic algorithms that are design-tailored for a particular class of problems, e.g. the Lin-Kernighan (LK) heuristic for the ***Traveling Salesman Problem*** (***TSP***) [52]. On the other hand, there are general algorithmic frameworks, e.g. Evolutionary Algorithms (EAs) [7, 27, 93] (see Chap. 1 for a brief introduction) that are applicable over a variety of problem classes across different domains. Despite the variety of mechanisms employed in these heuristic approaches to perform iterative search, they all share a common feature, i.e. they are parameterized algorithms. This means that they involve parameters that must be configured by users before the algorithm is applied to solve a problem.

There is now a collection of powerful frameworks for which rigorous theoretical analysis for many parameterized algorithms can be performed that indicate worst or average bounds on their performances and for which optimal parameter settings can be obtained [24]. Likewise, actual performances of EAs in real-world problem solving are in many situations highly sensitive to the settings of parameters [26, 39, 42, 46]. There is the assumption that an ***optimal configuration*** (or setting) of parameters exists, which might be found and in such a process requires knowledge of both the algorithm and the problem to solve. However, manual tuning (by the designer or practitioner) cannot be readily performed. Applications of such parameterized algorithms often involve a large number of free parameters that need to be carefully tuned so that they can achieve their best performance [20, 55, 60, 63]. In many cases, the manual tuning is done in an ad hoc manner, requiring considerable human effort, and is often limited to the exploration of few parameter settings

X. Yao, S. Y. Chong, *Coevolutionary Computation and Its Applications*,
Natural Computing Series, https://doi.org/10.1007/978-981-96-2841-4_6

[41, 43, 56, 58]. The performances of parameterized algorithms that are manually tuned might be far from the optimal.

As such, substantial efforts have been made to automate this important procedure that is crucial towards realizing the potential of the algorithm in solving a problem. These automated procedures are called ***Automatic Parameter Tuning*** [25, 39] when the algorithms have relatively few parameters with mostly real-valued domains, or ***Automatic Algorithm Configuration*** [3, 35] when the algorithms have more types (e.g. ordinal and categorical) of parameters. Several high-performance algorithm configurators have been developed and demonstrated to be effective: ParamILS [41], GGA [2], irace [58], and SMAC [42]. Here, a ***configuration*** of a parameterized algorithm refers to a complete setting of the parameters of the algorithm such that the algorithm's behaviour on a given problem instance is completely specified (up to randomization of the algorithm itself). Methods that include Automatic Algorithm Configuration essentially involve a high-level iterative generate-and-test procedure. The key idea is to parameterize many aspects of the algorithms and thus come up with a large space of algorithms as the ***configuration space***, from which effective ***algorithm configurators*** are used to search for high-performance configurations or solvers with respect to a targeted problem. Then, different configurations are iteratively generated and tested on the instance set from the target problem class. For example, ParamILS and GGA utilize direct search methods on the configuration spaces, i.e. an iterated local search algorithm in ParamILS and a specific GA in GGA, while SMAC and irace both rely on built meta-models to guide the sampling of the configuration spaces. The procedure outputs the configuration that performs the best on the instance set upon termination. Note that since a configuration fully instantiates a parameterized algorithm, we will also use the term configuration to directly denote the resulting solver specified by the procedure for the rest of this chapter.

Later studies consider more complicated structures of solvers that are crucial for automatic portfolio construction where the targeted object is now a set or portfolio of solvers that are chosen from a configuration space. Such a setting essentially means the search space with respect to a portfolio of solvers is generally much larger than the case with respect to sequential solvers, thus providing more degree of freedom on the portfolios that are searched and possibly leading to better performance. One main enabler for increasing interests in both developing and applying these methods is the increasing availability and decreasing cost of high-performance computing systems. Indeed, such approaches have been demonstrated to be both practical and effective in cases of constructing sequential solvers [11, 49], sequential portfolios (i.e. algorithm portfolios with selectors/scheduling) [45, 75, 76, 91, 92], and parallel portfolios [53, 54]. Note that a parallel portfolio could be easily converted into a sequential portfolio by using algorithm selection methods [78] to build selectors on the solvers in the portfolio. For portfolios with selectors, there are two representative approaches: Hydra [92] and ISAC [45]. Hydra constructs a portfolio iteratively by finding a configuration in each iteration that maximizes marginal contribution to the current portfolio. ISAC clusters the training instances based on features and independently runs an algorithm configurator on each cluster. They are later adapted to develop parallel portfolios, e.g. PARHYDRA and CLUSTERING [53] and PCIT [54]. There are other competing approaches that are motivated to increase performance via adaptive solvers such as adaptive parameter control [5, 31, 37, 46], reactive search [10, 47], and hyper-heuristics [19, 69, 74] that all dynamically determine the best solver setting or configuration while solving a problem instance.

There are two main research issues that need to be addressed to allow for the automated construction of effective solvers. The first issue has to do with the requirement of these methods that a training instance set (consisting of instances of the problem domain of interest) is available for the process of constructing the solvers, particularly to evaluate the quality of solvers so that they can be differentiated and a decision can be made which solvers to choose. Much like the setting of supervised learning, there is an indispensable assumption that the training instance set is a good representative of the target use cases [36] so that trained solvers can generalize well to the instances outside of the training set. In practice given a specific application, the trained solver has encountered and solved instances during training that are similar to the new problem instance. Such an assumption would be valid on some well-investigated computationally hard problems in planning [76] and those abstracted as the ***Boolean Satisfiability Problem*** (***SAT***) [45, 49, 53, 54, 92], and TSP [54] where availability of benchmark suites allows for the usual training-testing split. However, there might arise two situations the training set might not be sufficiently representative and as a consequence causes a major impact on the applicability of the constructed solvers in practice: (1) Only a limited number of instances are accumulated, and they might not cover the whole possible instances of a target problem. (2) The accumulated instances are outdated and could not reflect the properties of current instances well. These two situations are not rare and have been discussed in different areas of the literature. In combinatorial optimization, commonly used benchmark instances are deemed not challenging enough [34], narrowly defined [79], and not reflective of real-world problem instances [72]. In real-world domain such as logistics, there are also concerns that problem instances proposed decades ago might not represent current real-world problems whereby big cities are growing larger [83, 88].

An obvious but intuitive general approach to handle this issue is to find means to augment the training instance set for which generating additional instances appears to be an alternative. Various instance-generation methods have been proposed to problem domains such as TSP [28, 61, 66, 80, 81, 85], SAT [59, 85], job shop scheduling problems [18], constraint satisfaction problems [64, 85], graph-colouring problems [79], and bin-packing problems [1]. However, developing procedures that generate good training set or data in practice is not trivial. Note that the main goal for having a representative training instance set is to achieve good generalization of the constructed solvers. Thus, what is considered as a representative training set very much depends on the specific solvers that are considered. However, constructed solvers themselves are dependent on the training instance set. In other words, it is very difficult to obtain a concrete definition of representativeness in advance, which is crucial for evaluating a given training set and thereby generating a representative one as well. This difficulty could be alleviated by allowing some redundancy in the training instance set, since in the extreme case, one could obtain perfect generalization if all possible target instances are included in the training set. However this could lead to overwhelming cost to obtain the training instance set as well as for constructing the solvers.

The second issue deals with improving the generalization ability of these automatically constructed solvers. Much in the same way that motivates the introduction and development of ensemble methods in Machine Learning, one can develop methods for automatic construction of ***Parallel Algorithm Portfolios*** (***PAPs***) [29, 40, 54, 70] that seek to identify a set of component solvers or configurations to form a PAP. To solve a problem instance, all the component solvers are run independently and typically in parallel to obtain multiple solutions. Then, the best solution will be taken as the output of the PAP. Although a PAP would consume much more computational resources than a single-configuration solver, it has two important advantages. First, the performance of a PAP on any given problem instance is the best performance achieved among its component solvers, thus achieving a much stronger overall performance than any of its component solver on the problem instance in isolation. Second, there has been significant development of parallel computing architectures [6] (e.g. multi-core CPUs) and that exploiting parallelism has become very important as well as the new normal to tackle computationally hard problems. PAPs employ parallel solution strategies and thus exploit modern high-performance computing facilities in a simple way.

Note that these two issues also overlap. The motivation for developing problem-solving systems that are robust is so that they address the whole target problem class rather than a few specific problem instances. Here, a PAP construction method is expected to identify a PAP that generalizes well. In practice, one is unlikely to know in advance the exact problem instances that a problem-solving system will encounter. The need for generalization requires the training instance set used for PAP construction to be sufficiently large so that it consists of good representatives of all instances of the target problem class. Unfortunately, in a real-world scenario PAP construction is very likely to face the few-shots challenge, i.e. the available training instance set is not only of small size but also may not well represent the target problem class. For example, the widely studied TSP benchmark suites (i.e. TSPlib [73]) consist of a few hundred TSP instances, while there could be millions of possibilities for concrete TSP instances even with respect to a fixed number of cities. What this imply is that if one has a more powerful PAP construction method, there is a higher risk that the obtained PAP will over-fit to the problem instances involved during the construction process.

In this chapter, we will demonstrate how both issues can be addressed through our studies in [57, 84] where we have developed two approaches for the automatic construction of parallel portfolios. Unlike previous approaches that generate additional training instances to construct a PAP, our approaches employ a dynamically changing training instance set. In Sect. 6.2, we introduce the first approach that is called ***Generative Adversarial Solver Trainer*** (***GAST***) [57] that puts instance generation and portfolio construction in an adversarial game. The aim here is that the adversarial process would drive the portfolio to satisfactorily solve more problem instances, thus improving its generalization performance, while the training instance set is allowed to dynamically change in response to improving portfolio performance. Our computational study will demonstrate the effectiveness of GAST against other portfolio construction methods and state-of-the-art manually designed parallel solvers in data-scarce and data-biased scenarios involving TSP and SAT. The next Sect. 6.3 will start with a theoretical analysis to show that the generalization ability of PAP can be enhanced by training against an augmented set of problem instances that are of increasing difficulty. This motivates the formulation of a new PAP construction framework, namely ***Coevolution of Parameterized Search*** (***CEPS***) [84]. The generalizable PAPs are modelled as a coevolutionary system with two genetically isolated populations, where one represents the configurations that as a whole form the PAP and another represents the population of problem instances. Computational study is carried out whereby procedural implementations of CEPS are applied on two hard combinatorial optimization problems, i.e. TSP and ***Vehicle Routing Problem with Simultaneous Pickup-Delivery and Time Windows*** (***VRPSPDTW***) [86]. We note that CEPS framework on coevolving algorithms and problem instances is independent of domain-specific knowledge with potential applications beyond optimization problems, e.g. to planning and learning problems. Domain-specific knowledge is only employed to develop procedures to parameterize and then vary algorithms and problem instances.

6.2 Construction of Parallel Algorithm Portfolio Through a Generative Adversarial Approach

This section presents the study [57] we have carried out to develop an approach for automatic construction of PAPs called GAST. The general principle underlying GAST is similar to that of Generative Adversarial Networks [30] that have been applied to problems in vision-related domains [22, 62, 95], whereby an adversarial game-like mechanism is used to train a discriminator to distinguish real samples from fake samples synthesized by a generator, and that the generator in turn is also trained to deceive the discriminator by producing ever more realistic samples. Here, our focus is on problem-solving domains such as planning and optimization. GAST culminates from our study to address two main and interrelated challenges: (1) How to generate useful instances for portfolio construction. (2) How to appropriately integrate the instance generation into portfolio construction process that allows increasing generalization performance of constructed portfolio. Section 6.2.1 will introduce these PAPs and discuss important aspects such as their operations, performance evaluation, parameterization, and finally the formulation of a search on PAPs. The following Sect. 6.2.2 will introduce and describe in detail the procedural implementation of GAST with an analysis of its time complexity in comparison with current PAP construction approaches. Section 6.2.3 will detail the specific implementations of GAST on TSP and SAT as well as the problem-specific settings used in the experiments. Lastly, Sect. 6.2.4 will present the computational study we have carried out to verify the effectiveness of GAST against other state-of-the-art approaches on two well-established combinatorial optimization problems of TSP and SAT.

6.2.1 Parallel Algorithm Portfolios

A PAP with K component solvers is denoted as a K-tuple $c_{1:K} = (c_1, \ldots, c_K)$, where c_k represents the kth component solver of $c_{1:K}$. Note that a PAP solving a problem instance will execute or run all component solvers $c_1, \ldots, c_k$ independently in parallel until some termination condition is met. Termination condition might vary according to the problem domains considered and the performance metrics of interest. In a decision problem (e.g. SAT), all component solvers will be terminated if one of them outputs an answer to the problem instance (e.g. SATISFIABLE or UNSATISFIABLE). Here, the runtime needed by the PAP $c_{1:K}$ to solve the problem instance is the runtime needed by the best component solver for solving this problem instance. In an optimization problem (e.g. TSP), the termination conditions are different according to the performance metrics of interest. One metric considered is the runtime needed to find a good enough solution of accepted quality level (e.g. within a predefined gap to the optimum). Here, termination condition is when any of the component solvers finds such a solution. Another metric considered is the quality of the best solution found within a time budget. Termination condition includes the case when the time budget is exhausted, where each component solver will be terminated and the best solution among the ones found by the component solvers will be returned as the output of $c_{1:K}$. In practice, a ***cutoff time*** or maximum runtime will be introduced for any problem-solving setting to prevent the problem-solving process from being run for a prohibitively long time when none of component solvers could solve the problem instance.

Consider a minimization setting whereby lower values on a performance metric or measure indicate better performances. Then, the performance $P(c_{1:K}, s)$ of a PAP $c_{1:K}$ on a problem instance s is defined as the best performance achieved among $c_1, \ldots, c_K$ on s and is given by

$$P(c_{1:K}, s) = \min_{k \in \{1,\ldots,K\}} m(c_k, s), \tag{6.1}$$

where $m(c_k, s)$ is the performance of c_k on s according to a performance metric m (e.g. runtime or solution quality). Note that in practice, optimal solutions with respect to instances of an optimization problem might be unknown. In this situation, the runtime metric would not be measurable since it is not possible to determine whether solutions obtained by the component solvers are of accepted quality levels, in which case the procedure will terminate the running of all component solvers.

We can extend the performance of a PAP $c_{1:K}$ on a training instance set T as aggregation of $P(c_{1:K}, s)$ values with respect to all instances $s \in T$. As such, we define $P(c_{1:K}, T)$ as the weighted average performance

$$P(c_{1:K}, T) = \frac{1}{|T|} \sum_{s \in I} w_s \cdot P(c_{1:K}, s), \tag{6.2}$$

where $|T|$ is the number of the instances in T and the set of weights $\{w_s\}_{s=1}^{|T|}$ is introduced to handle different scales of the performances on different instances (and usually used when m is related to solution quality).

When constructing a PAP $c_{1:K}$, each component solver is an individual configuration $c_1, \ldots, c_K \in C$ selected from a configuration space C. The space C is induced by a set of parameterized base solvers Bs. This is illustrated in Fig. 6.1. If there is only one base solver, the configuration space is exactly the base solver's parameter space. Otherwise, the configuration space takes each base solver's parameter space as a subspace and would include an additional top-level parameter as an index to decide which subspace (base solver) would be used. The full configuration space of a PAP $c_{1:K}$ with each component solver corresponding to a particular kth base solver is given by the Cartesian product $C^K = \prod_{i=1}^{K} C_k$, where the Cartesian product of two configuration spaces A_1 and A_2 is $A_1 \times A_2 = \{(a_1, a_2) \mid a_1 \in A_1 \wedge a_2 \in A_2\}$. Then, the size of the full configuration space for $c_{1:K}$ is $|C_1| \cdot |C_2| \cdot \cdots \cdot |C_K|$. If we further assume the use of the same set of paramaterized base solvers Bs that induce the configuration space C to construct a PAP $c_{1:K}$, then we have that $|C_k| = |C|$ for all $k = 1, \ldots, K$ and that the size of the full configuration space is $|C|^K$.

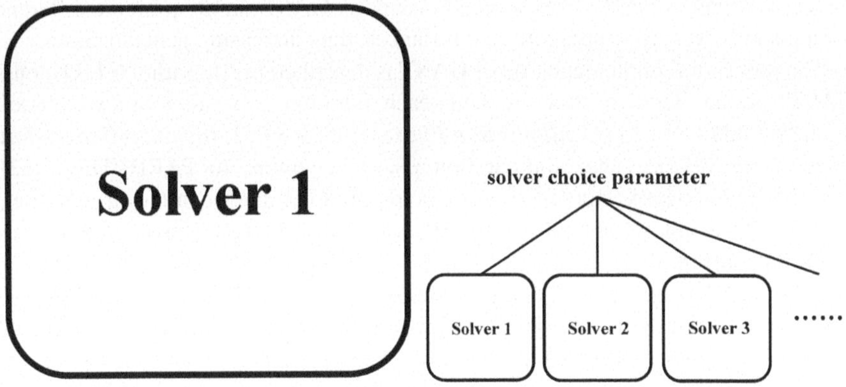

(a) C in the case of $|B| = 1$ (b) C in the case of $|B| > 1$

Fig. 6.1 Configuration space C induced by a set of parameterized solvers B in cases of $|B| = 1$, i.e. $B = \{\text{Solver1}\}$ in (**a**), and $|B| > 1$, i.e. $B = \{\text{Solver1, Solver2, Solver3}, \ldots\}$ in (**b**), respectively. Each rounded rectangle in this figure represents the parameter space of the corresponding base solver [57]

We can now use the definitions above to formulate PAP construction problem. Consider a possibly non-representative training instance set $T \subset \mathbf{\Omega}$, a performance metric m, and K-tuples of parameterized base solvers Bs taken from the induced configuration space C^k. The PAP construction problem can be formulated as a search for the set of configurations $c_{1:K} = (c_1, \ldots, c_K) \in C^K$ such that $c_{1:K}$ has good generalization performance $P(c_{1:K}, \mathbf{\Omega})$ on the target instance set $\mathbf{\Omega}$. Note that the target instance set $\mathbf{\Omega}$ of a problem domain is often impossible to be enumerated in advance, e.g. they can be very large and in some other problem-solving settings they can change over time.

6.2.2 Generative Adversarial Solver Trainer (GAST)

Here we introduce and describe in detail our automatic PAP construction approach called GAST. There are two key design principles for GAST. The first is concerned with the process of generating useful and representative training instances that cover target cases comprehensively throughout the lifetime running of GAST. For example, the instances that are out of the training set but represent target cases that are already be solved by the component solvers would not help the construction procedure to improve the generalization of solvers. Desirable generated instances are those not present in or withhold from the training set and challenging for constructed solvers. The second principle is concerned with the complementarity among the component solvers [45, 53, 54, 70, 92], which is crucial for the effectiveness of any PAP. The performance of a PAP on an instance depends on the best-performing component solver on that instance. A unique component solver is unlikely to perform best on all instances. GAST should promote different component solvers to handle different instances, thus increasing generalization.

The procedural implementation of **GAST** is described in Algorithm 6.1. Overall, **GAST** has an iterative structure with each iteration consisting of two phases implemented in turn: (1) ***Configuration Phase*** (lines 3–8). (2) ***Instance-Generation Phase*** (lines 9–31). The configuration phase is similar to PARHYDRA [53] whereby component solvers of $c_{1:K}$ are configured iteratively. In the ith iteration, **GAST** calls an algorithm configurator **AC** in with time budget t_C to configure c_k to add to the current portfolio $c_{1:k-1} = (c_1, \ldots, c_{k-1})$ such that the performance of the resulting portfolio $c_{1:k} = (c_1, \ldots, c_k)$ on instance set T is optimized (line 4). Note that one can define the object with a data structure to store performances of the portfolio against individual problem instances and them as a whole, e.g. object $\boldsymbol{P}(c_{1:k}, T)$ contains those performance values $P(c_{1:k}, T)$ where optimizers $c_{1:k}$ and problem instances T are interpreted as addressing lines when passed to $\boldsymbol{P}(\,)$ (line 5). During the configuration process of c_k, **GAST** runs the entire portfolio on the considered instances T with the existing portfolio $c_{1:k-1} = (c_1, \ldots, c_{k-1})$ being fixed. The aim here is to find a configuration c_k that maximizes the marginal performance contribution across the configurations identified in the previous iterations $c_{1:k-1}$. Note that most generic algorithm configurators implement some randomized

Algorithm 6.1 GAST [57]

Input:

B	Base solver B with configuration space C
K	Number of component solvers
T	Training instance set
m	Performance metric setting
n	Number of independent configurator runs
t_C	Time budgets for configuration
t_V	Time budgets for validation
t_K	Time budgets for instance generation

Output:

$c_{1:K}$	PAP

procedure GAST$(B, K, T, m, n, t_C, t_V, t_K)$
  for $k := 1 : K$ **do**
    for $j := 1 : n$ **do** ▷ **Configuration Phase** Start
      $c^j_{1:k} :=$ **AC**$(c^j_{1:k-1}, B, m, t_C)$ ▷ Algorithm Configurator procedure **AC** operates ▷ on $(c_{1:k-1}) \times \{c \mid c \in C\}$ to generate c_k
      $P(c^j_{1:k}, T) :=$ **validate**$(c^j_{1:k}, T, m, t_V)$ ▷ Object $P(c^j_{1:k}, T)$ with container
    end for ▷ (data structure) $P(c^j_{1:k}, T)$ addressed by $c^j_{1:k}, T$
    $j_{\text{best}} :=$ **argminP**$(P(c^j_{1:k}, T), n)$ ▷ $j_{\text{best}} := \text{argmin}_{j \in \{1,\ldots,n\}} P(c^j_{1:k}, T)$
    $c_{1:k} := c^{j_{\text{best}}}_{1:k}$
    ▷ **Configuration Phase** End
    if $k = K$ **then** ▷ **Instance-Generation Phase** Start
      break ▷ Skip instance generation as PAP $c_{1:K}$ constructed
    end if
    $L :=$ **sizeOf**(T) ▷ $L := |T|$
    initialize$(q_{s_1:s_L})$
    for $l := 1 : L$ **do**
      $q_{s_l} :=$ **quality**$(s_l, P(c_{1:k}, s_l))$ ▷ $s_l \in T$ given $q_{s_l} = w_{s_l} \cdot P(c_{1:k}, s_l)$
    end for ▷ $q_{s_1:s_L} = (q_{s_1}, \ldots, q_{s_L})$
    $T_b := T$ ▷ Backup T
    while within time budget t_K **do**
      initInstance(T_{new}) ▷ Initialize $T_{\text{new}} := \emptyset$
      $L :=$ **sizeOf**(T)
      for $l := 1 : L$ **do**
        refset := **randSample**(**delete**$(I, \{s_l\})$) ▷ **delete**$(I, \{s_l\})$ returns $I \backslash \{s_l\}$
        $s_{\text{new}} :=$ **variation**(s_l, refset)
        $T_{\text{new}} :=$ **combine**$(T_{\text{new}}, \{s_{\text{new}}\})$ ▷ $T_{\text{new}} := T_{\text{new}} \cup \{s_{\text{new}}\}$
      end for
      $H :=$ **sizeOf**(T_{new}) ▷ $H := |T_{\text{new}}|$
      initialize$(p_{s_1:s_H})$
      for $h := 1 : H$ **do**
        $p_{s_h} :=$ **quality**$(s_h, P(c_{1:k}, s_h))$ ▷ $s_h \in T_{\text{new}}$ given $p_{s_h} = w_{s_h} \cdot P(c_{1:k}, s_h)$
      end for ▷ $p_{s_1:s_H} = (p_{s_1}, \ldots, p_{s_H})$
      $T :=$ **btourSel**$(T, q_{s_1:s_L}, T_{\text{new}}, p_{s_1:s_H})$ ▷ Binary tournament selection on $T \cup T_{\text{new}}$
    end while
    $T :=$ **combine**(T, T_b) ▷ **Instance-Generation Phase** End
  end for
  return $c_{1:K}$
end procedure

methods. **GAST** performs n independent runs of **AC** while configuring c_k to obtain n different portfolios $c^1_{1:k}, \ldots, c^n_{1:k}$. This is to ensure the reliability of the outputs of **AC** [41, 42]. These portfolios are tested via **validate** on T with time budget t_V (line 5), and the one that achieves the best validation performance is retained (lines 7–8).

The instance-generation phase begins once the configuration phase finishes. Note in the last iteration (i.e. the kth iteration), the instance generation is skipped (line 10) with a manual **break** since $c_{1:K}$ has been completely constructed and there is no need to generate new instances. The general idea of instance generation in **GAST** is to iteratively incorporate new instances into the instance set that are more challenging to the PAP. Procedurally, a backup T_b of the current training set T is made (line 16) and then restored back to form the new training set at the end of this phase (line 31). **GAST** employs an iterative process to: (1) generate new instances based on current training set T (lines 19–25), (2) test these new instances with current portfolio $c_{1:k}$ (lines 26–30), and (3) use them to update the instance set T (line 31). This process is carried out until the total time spent reaches the time budget t_T (line 18).

The procedure to generate a new instance s_{new} is described specifically as follows. The procedure **randSample** outputs a random set of instances refset obtained from $T \backslash \{s\}$ (line 22). This is so that a new instance s_{new} can be generated via the procedure **variation** that uses and obtains an existing instance $s \in T$ as the base instance, and then modifies s with random perturbation and insertion of structures (components) extracted from the reference refset (line 23). This procedure is repeated until a set of new instances T_{new} is generated. Instances generated in this manner are expected to differ but preserved some characteristics of the existing instances. This allows the appropriate exploration of the instance space that is crucial towards the process of constructing portfolios with improving generalization. Moreover, the diversity in T_{new} is expected to be enhanced since each new instance differs from the base instance from T. Note that the development of the procedure **variation** depends on the specific problem domain considered. Later in Sect. 6.2.3 we will describe in more detail the procedure **variation** for TSP and SAT.

The two remaining aspects of the instance-generation phase are the instance evaluation and the instance selection. As we have mentioned earlier, only instances that cannot be solved well by the current portfolio $c_{1:k}$ is valuable for the process **GAST** employs to make changes to the portfolio and improve generalization. Since the portfolio is formulated in a minimization setting, then the opposite (maximization) setting is applied for the instance. In particular, each instance s is assigned with a quality score $w_s \cdot P(c_{1:k}, s)$ given by the performance of the current portfolio on it (line 15 for each $s \in T$ and line 29 for each $s_{\text{new}} \in T_{\text{new}}$), with the interpretation that s achieves a higher score for lower performances of $c_{1:k}$. Note that w_s is just the normalization factor to handle different scales of the performances. In practice, $P(c_{1:k}, s)$ could represent the mean of repeated application of $c_{1:k}$ to solve s given that component solvers in $c_{1:k}$ are randomized solvers. Lastly, the procedure **btourSel** incorporates T_{new} and T to form a population of instances and then applies binary tournament selection on the population [8], in which two random instances are compared and the one with lower quality score being removed from the population, repeatedly (line 31). This ensures that the size of the instance set remained the same.

GAST iteratively runs the configuration phase and instance-generation phase in turn so as to always promote configuring the component solver in the current iteration to handle the newly generated instances that are different compared to those from the previous iterations so that the complementarity among the component solvers of the constructed portfolio would be enhanced. The mechanism in GAST that provides such a problem-solving ability is made more explicit in this following example. In particular, the objective of the PAP construction problem is to find $c_{1:K} = (c_1, \ldots, c_K)$ from configuration space C^K having maximum coverage on the target instance space $\boldsymbol{\Omega}$. Recall that instance coverage here means the portfolio $c_{1:K}$ solving the instance. In general, the problem is NP-hard and can be approximated within $1 - \frac{1}{e} + o(1) \approx 0.632$, being achieved by the generic greedy method [68]. More specifically, this is an iterative method that starts with an empty portfolio and at each iteration selects a configuration $c \in C$ that covers the largest number of uncovered instances in $\boldsymbol{\Omega}$ to add to the portfolio. The iterative framework of GAST (the outermost loop in Algorithm 6.1) also employs this greedy method but in addition has an instance-generation phase in each iteration. The assumption here is that at the start, a training set T that is non-representative of $\boldsymbol{\Omega}$ is provided. During the process of portfolio construction, it is unclear which instances in $\boldsymbol{\Omega}$ are not covered by the current portfolio. Thus, it is necessary to first identify those uncovered instances in $\boldsymbol{\Omega}$ so that subsequently the portfolio's coverage on $\boldsymbol{\Omega}$ can be enlarged. This is what the mechanism of adversarial instance generation provides. In comparison, existing approaches do not involve such a mechanism, and hence they could only optimize and possibly overtune the portfolio's coverage on the training set T.

We will finish our discussion with a time complexity analysis comparing those of GAST with existing PAP construction approaches. Note that the most time-consuming part of GAST is those runs of the component solvers on the problem instances for which the incurred computational costs account for the vast majority of the total computational costs of GAST. Therefore, we analyse the time complexity of GAST in terms of the total number of the runs of the solvers. In Algorithm 6.1, the solvers are invoked in three places, i.e. configuration, validation, and instance generation. Recall that in the kth iteration of GAST, there are k component solvers in the portfolio, and they are always executed in parallel. Let N_C, N_V, and N_T denote the number of the runs of each component solver during configuration, validation, and instance generation, respectively. The total number of runs of solvers in the kth iteration of GAST is $k \cdot \big(n \cdot (N_C + N_V) + N_T\big)$, where n is the number of independent configurator runs. Since the instance-generation phase is skipped in the last iteration of GAST, the time complexity of GAST in terms of the number of the runs of the solvers is $O\big(\sum_{k=1}^{K} k \cdot n \cdot (N_C + N_V) + \sum_{k=1}^{K-1} i \cdot N_T\big) = O\big(K^2 n(N_C + N_V) + K^2 N_T\big)$ [57]. Similarly, we could obtain the time complexity of existing parallel portfolio construction approaches. For PARHYDRA, GLOBAL, and PCIT, they are

$O(K^2 n(N_C + N_V))$ [53], $O(Kn(N_C + N_V))$ [53], and $O(Kn(N_C + N_V))$ [54], respectively.

Note that for a specific portfolio construction approach, the values of N_C, N_V and N_T depend on the predefined time budgets t_C, t_V, and t_K, respectively. For different approaches, these time budgets could be set differently. For time budgets t_C, t_V and t_K, the total CPU time consumed by GAST is $\sum_{k=1}^{K} k \cdot n \cdot (t_C + t_V) + \sum_{k=1}^{K-1} k \cdot t_T$. The n independent runs of **AC** and the validation processes can be performed in parallel if n machines (with each having K-cores) are available, in which case GAST will require around $K \cdot (t_C + t_V) + (K-1) \cdot t_T$ wall-clock time to complete [57]. These CPU times are listed in Table 6.1 for GAST, PARHYDRA, GLOBAL, and PCIT, which will be referenced later in Sect. 6.2.3.

6.2.3 *GAST on TSP and SAT*

We start by introducing details underlying the procedure **variation** that depends on the domain knowledge as represented by the target problems of TSP and SAT. Their specific development leads to GAST-TSP and GAST-SAT. For the TSP, we consider the ***symmetric TSP***, i.e. the distance between two cities is the same in each opposite direction, with distances taken in a two-dimensional Euclidean space. Each TSP instance is represented by a list of coordinates $\{(x, y)\}_i^N$, each distinct coordinate (x_i, y_i) corresponding to a city. We extend the variation strategy used in [85] that requires the base instances and the reference instances to have the same size N (i.e. the number of the cities) to allow the use of instances of different sizes. GAST-TSP requires only one reference instance in refset, whereby the procedure **variation** applies a variable-length crossover and a uniform mutation to base instance s and reference instance s^*. Let $|s|$ and $|s^*|$ be the lengths of the coordinates lists s and s^*, respectively. First, crossover randomly selects $\min\{|s|, |s^*|\} - 1$ split points in both lists and then constructs a new coordinates list (instance) s_{new} in a sequential manner by choosing each segment from either of the two lists with equal probability. Second, the new list is then subjected to the mutation operator that replaces each coordinate in the list with a random coordinate chosen uniformly within the range of the corresponding x or y coordinate, bounded by the minimum and maximum

Table 6.1 Computational costs for existing parallel portfolio construction approaches. t_C, t_V, t_T are time budgets for configuration, validation, and instance generation, respectively. K is the portfolio size. n is the number of independent runs of the used algorithm configurator. Note that for different approaches t_C, t_V could be set to different values [57]

	CPU time
GAST	$\sum_{k=1}^{K} k \cdot n \cdot (t_C + t_V) + \sum_{k=1}^{K-1} k \cdot t_T$
PARHYDRA	$\sum_{k=1}^{K} k \cdot n \cdot (t_C + t_V)$
GLOBAL	$n \cdot K \cdot (t_C + t_V)$
PCIT	$n \cdot K \cdot (t_C + t_V)$

values from the lists of s and s^*, with a probability (mutation rate) of $1/|s_{\text{new}}|^{\frac{1}{2}}$ [57].

The SAT involves finding the appropriate assignments into the variables that satisfy a propositional logic expression, with solutions typically in standard Conjunctive Normal Form. The procedure **variation** in GAST-SAT utilizes the ***spig*** technique proposed in [59], which iteratively removes particular components (bounded together through a core variable) from the base instance s and then inserts such structures extracted from the reference instances into s. The generated instance will only be accepted by *spig* if all of its features are within σ standard deviations of the mean value across all instances (including s and the reference instances). Instead of setting a small and fixed value of σ [59] that would lead to the generation of similar instances with small variations, we set σ as a random variable whose value is randomly sampled from $[3, 300]$ for each acceptance check to introduce more randomness in the generated instances. The size of the reference instance set is set at $|\text{refset}| = \lceil |T|^{\frac{1}{2}} \rceil$, where $|T|$ is the size of the training set to prevent the runtime of *spig* from being too long.

Next, we detail the ***evaluation instance sets*** used to compare performances of PAPs obtained by GAST and other automatic parallel portfolio construction approaches. Since we focus on the scenarios where the available training instances are non-representative, it is very important to decide an appropriate way to choose the instances used for evaluations of algorithm outputs. We consider two different methods to obtain these instances, thus dividing our experiments into two parts. In the first part, we obtain instances through instance generators to evaluate GAST-TSP because for TSP there exist generators that could generate instances with diverse characteristics. Specifically, we use the ***portgen*** and ***portcgen*** generators from the 8th DIMACS Implementation Challenge [44] to generate 150 uniform-instances (in which the cities are randomly distributed) and 150 clustering-instances (in which the cities are distributed around different central points) to form a set of 300 instances denoted as $\text{TSP}_{\text{whole}}$. The problem sizes of all these generated instances are within $[400, 600]$.

However, the instance-generation method has two potential issues: (1) The generated instances might not reflect real-world cases, thus making their use as evaluation instance sets to lack practical significance. (2) GAST employs instance generation, and there might be a possibility the mechanism used in GAST is similar to the instance generators used to obtain the evaluation sets. As such, we obtain the instances from the industrial benchmark suites to evaluate GAST-SAT to avoid these issues in the second part. Specifically, we use two industrial benchmarks, IBM Hardware Verification (HV) benchmark, and Bounded Model Checking (BMC) benchmark, both from the Algorithm Configuration Library (AClib) [43]. We randomly select 150 instances from each of the two sets to form a set of 300 instances denoted as $\text{SAT}_{\text{whole}}$.

Following the common scheme, in the experiments the portfolio construction approaches use a training set, and then they are evaluated on an unseen test set. We only describe how we split $\text{TSP}_{\text{whole}}$ here. The same splitting procedure is applied to SAT_{whole}. In particular, we split TSP_{whole} into training sets and test sets in two different approaches to simulate two different cases. The first case SMALL means the available training set contains only a small number of instances. We randomly select 1/6 instances (50 in total) from TSP_{whole} as training instances and use the remaining instances (250 in total) as test instances. The second case BIAS means the training instances are biased to narrowly defined cases. We randomly select 1/3 instances from TSP_{whole} corresponding to one of the two types of instances (50 in total; recall that there are 150 uniform-instances and 150 clustering-instances) as training instances, with the remaining being test instances (250 in total).

We repeat the split procedure for 4 times for each of SMALL and BIAS cases to ensure the reliability of our experiments given that the split procedure is randomized and the specific realizations or choices of training/test instances used would affect the performances of portfolio construction approaches [57]. In total, we have eight different experimental scenarios, whereby each scenario corresponds to a unique pair of training set and test set, for the TSP and SAT domains. We denote these scenarios as TSP-SMALL/BIAS-1/2/3/4 and SAT-SMALL/BIAS-1/2/3/4 to denote these scenarios. For the subsequent presentation, we use the term TSP-SMALL to denote a set of four scenarios corresponding to {TSP-SMALL-1/2/3/4}. Likewise, we use the terms TSP-BIAS, SAT-SMALL, and SAT-BIAS to denote the respective sets, each of which corresponds to the 1/2/3/4 realizations obtained from the associated randomized splitting procedure.

Table 6.2 summarizes the instance sets, the cutoff time, and the base solvers used in different scenarios. The base solver used in TSP-SMALL/BIAS is the Helsgaun's Lin-Kernighan Heuristic (LKH) version 2.0.7 [32] (with 23 parameters), one of the

Table 6.2 Summary of the instance sets, the cutoff time, and the base solver in each scenario [57]

	Instance sets	Cutoff time	Base solvers
TSP-SMALL	Randomly select 50 instances from TSP_{whole} as training instances and the remaining 250 instances are used as test instances	1s	LKH version 2.0.7 [32] with 23 parameters
TSP-BIAS	Randomly select 50 instances from 150 uniform-instances or 150 clustering-instances in TSP_{whole} as training instances and the remaining 250 instances are used as test instances		
SAT-SMALL	Randomly select 50 instances from SAT_{whole} as training instances and the remaining 250 instances are used as test instances	150s	Lingeling-ala [12] with 118 parameters
SAT-BIAS	Randomly select 50 instances from 150 IBM-HV-instances or 150 BMC-instances in SAT_{whole} as training instances and the remaining 250 instances are used as test instances		

state-of-the-art inexact solver for TSP. The base solver used in SAT-SMALL/BIAS was lingeling-ala [12] (with 118 parameters), the gold medal winning solver in the application track of the 2011 SAT Competition.

6.2.4 *Computational Study of GAST on TSP and SAT*

Here, we present our computational study we have carried out in [57] to demonstrate the effectiveness of GAST with regard to two target problems represented by TSP and SAT. We start with initial discussion on the experimental settings that we have used. For the portfolio size, we set the number of component solvers K to 4 since 4-core machines are widely available. For the performance metric, we consider the runtime needed by a solver to solve the problem instances for SAT and to find the optima of the problem instances for TSP. In particular, we set m to ***Penalized Average Runtime-10*** (***PAR-10***) [41], which is the average runtime over all the test runs with unsuccessful runs (i.e. unable to solve the given instance within the cutoff time) counted as 10 times the cutoff time. Note for PAR-10 the weight w_s in Eq. ((6.2)) is set to 1. The optimal solutions for TSP instances are obtained using Concorde [4], an exact TSP solver.

We will compare GAST against the state-of-the-art automatic parallel portfolio construction approaches: GLOBAL, PARHYDRA [53], and PCIT [54]. Note that all the considered approaches use SMAC version 2.10.07 [42] for the algorithm configurator **AC**. Since the performance of SMAC could be often improved when used with the instance features, we have provided SMAC access to the 126 SAT features and the 114 TSP features used in [54]. The detailed settings of the time budgets for each approach are given in Table 6.3. In general, the experiments involving GAST would consume around 10% more CPU time than other approaches for the construction of PAPs. All the experiments have been conducted on a cluster of 3 Intel Xeon machines with 128 GB RAM and 24 cores each (2.20 GHz, 30 MB

Table 6.3 Detailed time budget in terms of hours of CPU time for each approach in each scenario. "TSP" represents eight scenarios TSP-SMALL/BIAS-1/2/3/4 and "SAT" represents eight scenarios SAT-SMALL/BIAS-1/2/3/4. In the experiments the number of independent runs of algorithm configurator n for all approaches was set to 10. t_C, t_V, t_T are time budgets for configuration, validation, and instance generation, respectively. See Table 6.1 for how to estimate the needed CPU time for each approach [57]

	TSP				SAT			
	t_C	t_V	t_T	Total	t_C	t_V	t_T	Total
GAST	1.5 h	0.5 h	5 h	230 h	8 h	4 h	40 h	1440 h
GLOBAL	5 h	0.5 h	–	220 h	30 h	4 h	–	1360 h
PCIT	5 h	0.5 h	–	220 h	30 h	4 h	–	1360 h
PARHYDRA	1.5 h	0.5 h	–	230 h	8 h	4 h	–	1200 h

Cache), running Centos 7.5. The entire experiments have taken almost 2.5 months to complete.

For each experimental scenario, we have tested each obtained portfolio by running it on the test set 50 times for TSP and 5 times for SAT [57]. The mean $\pm$ stddev of the test performances (PAR-10 score over the test instances) across these runs and the total number of timeouts (#TOs) are presented in the $PAR-10^{\dagger}$ columns and the #TOs columns, respectively, in Table 6.4. The validation performances of the portfolios constructed by PARHYDRA, GLOBAL, and PCIT (except for GAST since it procedurally changes the training set during operation) on the training sets are also reported in the $PAR-10^{\star}$ columns in Table 6.4. We have performed a Wilcoxon signed-rank test (with significance level $p = 0.05$) to determine whether the differences between the test performances (i.e. the $PAR-10^{\dagger}$ columns) are significant in each scenario. A PAR-10 score is indicated in boldface if it was not statistically significantly different from the best test PAR-10 score with respect to the scenario.

Results in Table 6.4 indicate that GAST is the overall best-performing PAP construction approach whereby it constructed significantly and substantially better portfolios than other approaches in most cases. We note that PARHYDRA could be considered as a variant of GAST without the instance-generation mechanism. The superior performances of GAST over PARHYDRA indicate the effectiveness of the instance generation for improving the portfolio's generalization. We further note that we have used generated instances for TSP and industrial instances for SAT. The consistent strong performances of GAST on both domains indicate that the mechanism employed in GAST for generating new instances via recombination of existing instances and random perturbation (taking into account domain knowledge of problem representation in constructing the operators in **variation**) is a robust and effective way for training data augmentation. Another important observation from Table 6.4 is that the gaps between the validation performances and the test performances are usually very large for existing approaches. This strongly suggests that the training sets are non-representative, thus indicating the need for instance generation that we have incorporated in GAST [57].

We have carried out a controlled experiment to systematically compare GAST against PARHYDRA in the context of improvement in generalization from the addition of solvers to the constructed portfolio [57]. This is because both GAST and PARHYDRA iteratively add one component solver to the portfolio per iteration. We have run GAST and PARHYDRA for eight iterations in four scenarios TSP/SAT-SMALL/BIAS-1. Let GAST_i and PARHYDRA_i denote the resulting portfolios at the end of the ith iteration of GAST and PARHYDRA, respectively, and $i = 1, \ldots, 8$. In each scenario, we have tested the corresponding GAST_i and PARHYDRA_i on the test instances. Let $P[\text{portfolio}, \text{scenario}]$ be the average test result in terms of PAR-10 scores. For example, $P[\text{GAST}_2, \text{TSP} - \text{SMALL} - 1]$ is the average PAR-10 score of the test result of GAST_2, i.e. the resulting portfolio at the end of the second iteration of GAST on the test instances of TSP-SMALL-1. The results are plotted along the number of iterations in Fig. 6.2, which provide three observations: (1) Test performances for both GAST and PARHYDRA

Table 6.4 Results of validation and testing in the 16 experimental scenarios. Validation performances in terms of PAR-10 scores over the training set are presented in the PAR-10* columns. Test performances in terms of mean ± stddev of the PAR-10 scores across the 50 runs (for TSP) and the 5 runs (for SAT) over the test set are presented in the PAR-10† columns. The total number of timeouts (#TOs) in testing is presented in the #TOs columns. The name of the construction approach is used to denote the portfolios constructed by it. The test PAR-10 score of a portfolio is shown in boldface if it was not significantly different from the best test performance in the scenario (according to a Wilcoxon signed-rank test with $p = 0.05$) [57]

	GAST		GLOBAL			PARHYDRA			PCIT		
	PAR-10†	#TOs	PAR-10*	PAR-10†	#TOs	PAR-10*	PAR-10†	#TOs	PAR-10*	PAR-10†	#TOs
TSP-SMALL -1/2/3/4	**0.48 ± 0.09**	327	0.23	1.02 ± 0.24	934	0.18	**0.67 ± 0.18**	565	0.18	0.58 ± 0.12	388
	0.48 ± 0.08	282	0.24	0.71 ± 0.16	521	0.16	0.50 ± 0.13	354	0.19	**0.56 ± 0.12**	390
	0.47 ± 0.11	277	0.29	0.98 ± 0.22	868	0.18	0.50 ± 0.12	356	0.19	**0.48 ± 0.09**	292
	0.51 ± 0.12	319	0.28	1.01 ± 0.29	918	0.16	0.51 ± 0.15	346	0.18	**0.50 ± 0.11**	326
TSP-BIAS -1/2/3/4	**0.68 ± 0.18**	550	0.17	1.04 ± 0.20	925	0.15	0.82 ± 0.18	715	0.18	1.26 ± 0.21	1196
	0.50 ± 0.14	340	0.20	0.72 ± 0.18	590	0.14	0.59 ± 0.16	471	0.20	0.55 ± 0.14	369
	0.62 ± 0.14	434	0.26	0.90 ± 0.20	767	0.13	**0.86 ± 0.20**	753	0.17	0.75 ± 0.16	599
	0.51 ± 0.13	307	0.23	0.76 ± 0.18	604	0.18	**0.60 ± 0.18**	442	0.18	0.50 ± 0.13	333
SAT-SMALL -1/2/3/4	**317 ± 4.93**	258	280	351 ± 15.3	285	255	333 ± 5.66	271	251	327 ± 5.66	266
	283 ± 4.16	230	403	317 ± 7.66	258	398	321 ± 9.39	261	402	322 ± 15.0	262
	335 ± 8.77	274	324	355 ± 9.88	290	307	**334 ± 9.58**	272	310	**374 ± 2.49**	306
	295 ± 4.07	240	279	346 ± 9.65	281	285	336 ± 8.21	273	276	356 ± 10.4	291
SAT-BIAS -1/2/3/4	**299 ± 4.79**	243	377	383 ± 17.9	313	337	353 ± 3.25	288	366	369 ± 16.7	302
	272 ± 6.53	220	517	350 ± 7.42	285	504	299 ± 11.0	241	512	287 ± 11.1	233
	331 ± 4.36	271	250	394 ± 11.2	322	245	355 ± 6.77	290	248	387 ± 8.19	318
	324 ± 5.58	264	393	348 ± 10.2	284	388	**310 ± 7.48**	250	364	334 ± 12.1	273

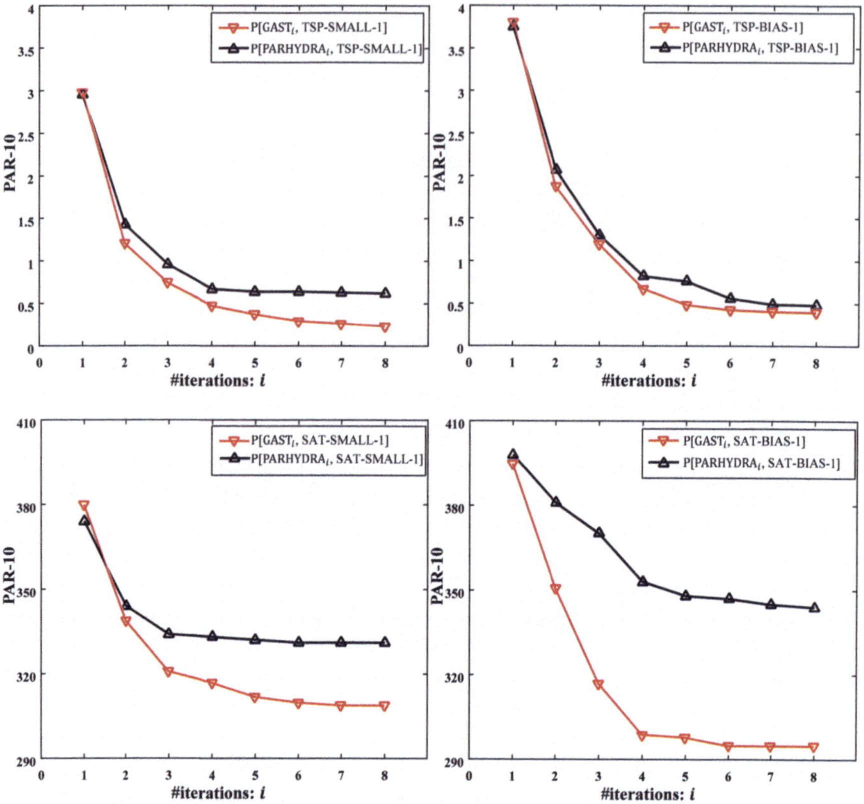

Fig. 6.2 Test performance (in terms of average PAR-10 scores) progress when $K = 8$ for GAST and PARHYDRA in four scenarios (TSP/SAT-SMALL/BIAS-1) along the number of iterations. GAST$_i$ and PARHYDRA$_i$ are the resultant portfolios at the end of the ith iteration of GAST and PARHYDRA, respectively. P[portfolio, scenario] is the average test result in terms of PAR-10 scores on the test set in the scenario [57]

monotonically improve with each subsequent iteration. The addition of a new component solver would lead to a new portfolio with better performance (or at least similar performance with those obtained in the previous iterates) given that portfolio performance is defined by the best solver performance among the set of component solvers. (2) The rate of performance increase with respect to the iterations gradually decreases for both GAST and PARHYDRA (e.g. around $i = 5$). The performance of the portfolio is lower-bounded by the best theoretical performance of the base solver, while in practice the limit can be affected by what the algorithm configurator can achieve. (3) In general, GAST achieves better performance improvements than that PARHYDRA (e.g. significantly so for SAT-BIAS-1) because GAST could address the limitation of being given a non-representative training data by augmenting it via instance-generation mechanism [57].

Next, we have carried out a controlled experiment to systematically compare GAST against PARHYDRA with augmented training set (PARHYDRA-A) [57]. However, PARHYDRA-A only uses the instance-generation procedure in GAST to obtain an augmented training set. In particular, it is not equipped with the specific adversarial mechanism used in GAST. This will allow us to answer more conclusively whether augmentation of training set that adds more training instances (e.g. $s_{\text{new}} := \textbf{variation}(s_l, \text{refset})$) is sufficient, or one that adds better quality training instances, e.g. via an adversarial process whereby newly generated instances have to compete with existing ones (e.g. the additional selection process $T := \textbf{btourSel}(T, q_{s_1:s_L}, T_{\text{new}}, p_{s_1:s_H})$ is required for handling data-scarce/biased scenarios. In each of the eight SAT scenarios (SAT-SMALL/BIAS-1/2/3/4), PARHYDRA-A generates a larger training set of 300 instances, i.e. 5 times of that used in PARHYDRA. The results are summarized in Table 6.5. They indicate that PARHYDRA-A still could not reach the performance levels of GAST, and over specific scenario actually has a drop in actual performance. On the one hand, our results here strongly suggest that the adversarial mechanism used to generate more challenging instances against an improving portfolio is key to the success of GAST. On the other hand, treating data augmentation and portfolio construction as two sequential and independent phases does not lead to construction of portfolio with improving generalization performance [57].

Lastly, we have compared portfolios constructed by GAST against the state-of-the-art manually designed parallel solvers on SAT [57]. Specifically, we have tested each of the eight portfolios constructed by GAST in SAT-SMALL/BIAS-1/2/3/4 on the entire SAT instance set $\text{SAT}_{\text{whole}}$ and reported the best, the worst, and the median performance (in terms of PAR-10) achieved among these portfolios in Table 6.6. For manually designed solvers, we have selected Plingeling-ala [12] (which is the official parallel version of lingeling-ala, the latter being the base solver in all the SAT scenarios in our experiments), pfolioUZK [89] (which is the

Table 6.5 Test performances in terms of mean ± stddev of the PAR-10 scores across the five runs over the test set in the 8 SAT scenarios, i.e. SAT-SMALL-BIAS-1/2/3/4. The name of the construction approach is used to denote the portfolios constructed by it. PARHYDRA-A refers to PARHYDRA configuring based on augmented training sets. A PAR-10 score is shown in boldface if it was not significantly different from the best test performance in the scenario (according to a Wilcoxon signed-rank test with $p = 0.05$) [57]

	GAST	PARHYDRA	PARHYDRA-A
SAT-SMALL-1/2/3/4	**317 ± 4.93**	333 ± 5.66	**325 ± 4.89**
	283 ± 4.16	321 ± 9.39	300 ± 8.90
	335 ± 8.77	**334 ± 9.58**	349 ± 6.76
	295 ± 4.07	336 ± 8.21	329 ± 10.4
SAT-BIAS-1/2/3/4	**299 ± 4.79**	353 ± 3.25	331 ± 13.4
	272 ± 6.53	299 ± 11.0	323 ± 12.5
	331 ± 4.36	355 ± 6.77	345 ± 12.8
	324 ± 5.58	**310 ± 7.48**	333 ± 12.4

Table 6.6 Test performances in terms of mean ± stddev of the PAR-10 scores across the 5 runs over SAT_{whole}. UZK refers to pfolioUZK. ALA refers to Plingeling-ala. BBC refers to Plingeling-bbc. GAST-B, GAST-W, and GAST-M refer to the best, the worst, and the median performance achieved among the eight portfolios constructed by GAST in SAT-SMALL/BIAS-1/2/3/4. A PAR-10 score is shown in boldface if it was not significantly different from the best test performance (according to a Wilcoxon signed-rank test with $p = 0.05$) [57]

	UZK	ALA	BBC	GAST-B	GAST-W	GAST-M
SAT_{whole}	317 ± 7.48	310 ± 9.24	**274 ± 4.03**	**282 ± 5.18**	313 ± 5.79	301 ± 8.21

gold medal winning solver of the parallel track of the SAT'12 Challenge), and Plingeling-bbc [13] (which is the gold medal winning solver of the parallel track of the SAT'16 Competition). Note that all the manually designed solvers considered here have implemented far more advanced parallel-solving strategies (e.g. clause sharing) than only independently running component solvers in parallel. The default settings of these solvers are used, and all of them are tested on SAT_{whole}. Their test performances are presented in Table 6.6 [57].

In general, our results in Table 6.6 have indicated that GAST is competitive against some of state-of-the-art manually designed parallel solvers. The portfolios constructed by GAST perform better than pfolioUZK and in most cases perform better than Plingeling-ala. It is impressive to note that in the best case, the portfolio constructed by GAST (regardless of its simple parallel-solving strategy) could reach the performance level of the more state-of-the-art Plingeling-bbc. Moreover, the performance difference between them is not statistically significant. These results indicate that GAST could identify powerful parallel portfolios, with little human effort involved. It is expected that given more state-of-the-art base solvers, e.g. lingeling-bbc, GAST could deliver parallel portfolios with even better performance [57].

6.3 Coevolutionary Framework of Parallel Algorithm Portfolio Construction

Here, we will present our study [84] to develop the framework of coevolution for PAP construction. Note that the main underlying motivation of our study is more general in that the framework is for the coevolution of algorithm configurations and problem instances. In this two-population competitive coevolutionary system, one population represents the algorithm configurations and another population represents problem instances. Both populations undergo a coevolutionary process that involves an evolutionary adaptation process (of variation and selection) in any one of the two populations, which is guided by the respective population members' strategic interactions with members of the other population. The emphasis here is not just on the specific competitive process involved in assessing the quality of population members (whether they play the role of solvers or problem instances) within a population and the adversarial competition between both populations, but the other two processes of variation and selection of those population members. Note as well that the coevolutionary framework is independent of domain-specific knowledge and thus has the potential to be applied to other problem-solving settings such as planning and learning. As such in Sect. 6.3.1, we perform a theoretical analysis on the generalization performance of PAP and demonstrate that the PAPs constructed by augmented instance set (in the sense of being more representative than current instance set) will have better generalization. The next Sect. 6.3.2 will introduce a procedural implementation of our framework that is called Coevolution of Parameterized Search (CEPS). Then in Sect. 6.3.3, we will demonstrate how the incorporation of domain-specific knowledge with regard to two target problems to the higher-level problem-solving aspect of CEPS leads to two instantiations, i.e. CEPS-TSP and CEPS-VRPSPDTW. Lastly, Sect. 6.3.4 will present the computational study we have carried out to verify the effectiveness of CEPS against other state-of-the-art approaches on two well-established combinatorial optimization problems of TSP and VRPSPDTW. More importantly, CEPS-TSP and CEPS-VRPSPDTW demonstrate that the coevolutionary framework is applicable in few-shots setting with scarce availability of initial training instances.

6.3.1 *Parameterized Solvers Made Generalizable*

In the following performance analysis we have carried out in [84], assume a PAP is built for a problem class (e.g. TSP). We will first introduce our notations and definitions. Let $s \in \mathbf{\Omega}$ be an instance of the problem class and $\mathbf{\Omega}$ be the set of all possible and distinct instances s. Given a parameterized algorithm, each component solver of the PAP is a configuration (full instantiation) of the algorithm. In general, the parameterized algorithm can take the form of any concrete computational process, e.g. a traditional heuristic search process such as the LK Heuristic for TSP or even an ANN [21, 50, 67] that outputs a solution for a given instance of the target problem class. Let θ be a configuration. Then, we have $\Theta = \{\theta_k : k = 1, \ldots, K\} = \{\theta_1, \ldots, \theta_K\}$ denoting a PAP that contains K different configurations (component solvers). The quality or performance of the solver θ on the instance s is denoted as $f(s, \theta)$. The performance indicator or measure could concern many aspects, e.g. the quality of the obtained solution [58], the CPU time required to achieve a solution above a given quality threshold [41], or even be stated in a multi-objective form [15]. As before, the performance $f(s, \Theta)$ of a PAP Θ on an instance s is defined as the best performance achieved among its component solvers $\theta_1, \ldots, \theta_k$ on s (where assuming a minimization context)

$$f(s, \Theta) = \min\{f(s, \theta_1), \ldots, f(s, \theta_K)\}. \tag{6.3}$$

Following the above definitions, we can state optimizing the generalization performance of a PAP as

$$\min_{\Theta} J(\Theta) = \int_{s \in \boldsymbol{\Omega}} f(s, \Theta) p(s) ds, \tag{6.4}$$

where $p(s)$ stands for the prior (probability) distribution of s. In practice, the prior distribution is usually unknown, and a uniform distribution can be assumed without loss of generality [84]. Equation (6.4) can be simplified as follows (omitting a normalization constant):

$$\min_{\Theta} J(\Theta) = \int_{s \in \boldsymbol{\Omega}} f(s, \Theta) ds. \tag{6.5}$$

In both formulations, one should take note of that within the integral the function $f(s, \Theta)$ takes on a minimization term that is expressed as $\min\{f(s, \theta_1), \ldots, f(s, \theta_K)\}$, which we have formulated as Eq. (6.3) earlier.

In practice, the challenge associated by the optimization setting as expressed by Eqs. (6.4) and (6.5) is that direct optimization is not possible since the set $\boldsymbol{\Omega}$ is generally unavailable. Instead, only a set of training instances, i.e. a subset $T \subset \boldsymbol{\Omega}$, is given for the purpose of constructing Θ. Indeed, the so-called ***over-tuning phenomenon*** [14, 56], which is analogous to the ***over-fitting phenomenon*** in Machine Learning, has been observed when the size of the training instance set is rather limited. It is this latter setting that is referred to as the ***few-shots challenge***. Here, the test (generalization) performance of the obtained configurations is arbitrarily bad even if their performance on the training set is excellent. More importantly, given a training set T collected from the real world, it is nontrivial to verify whether it is a good representative of $\boldsymbol{\Omega}$. In the situation where the training instance set is too small, or is not a good representative of the whole problem class, then the best PAP obtained with such a training instance set would fail to generalize.

A natural idea to tackle the few-shots challenge is to augment T with a set of synthetic instances T' such that the PAP obtained with $T \cup T'$ would generalize better than that obtained with T. This idea can be realized in principle because as T' continues to grow in size, $T \cup T'$ eventually approaches $\boldsymbol{\Omega}$. Hence, the key question for practical implementation is how a generalizable PAP could be obtained with a sufficiently small T'. More specifically in the context of allowing for the implementation of automatic PAP construction: How to generate synthetic training instances, such that the generalization of the obtained PAP could be improved as much as possible with a T' of (say predefined) small size.

For a parameterized algorithm, suppose the best-performing PAP Θ with respect to training instance set T has been obtained. Then, a synthetic instance set T' is generated with the aim that a new PAP Θ' obtained with $T \cup T'$ would outperform the current Θ in terms of generalization as much as possible. Assume that $T \cap T' = \emptyset$ (i.e. the subset $T' \subset \boldsymbol{\Omega}$ is non-overlapping with $T \subset \boldsymbol{\Omega}$) and disregard the inner optimization or tuning process with which Θ' and Θ are obtained for the moment.

The issue of generating a high-quality or informative T' could be formally stated as the following optimization problem [84]:

$$\min_{T'} \{J(\Theta') - J(\Theta)\} = \int_{s\in\Omega} f\left(s, \Theta'\right) ds - \int_{s\in\Omega} f\left(s, \Theta\right) ds$$

$$= \left[\sum_{s\in T} f(s, \Theta') + \sum_{s\in T'} f(s, \Theta') + \int_{s\in\Omega\setminus(T\cup T')} f(s, \Theta') ds\right]$$

$$- \left[\sum_{s\in T} f(s, \Theta) + \sum_{s\in T'} f(s, \Theta) + \int_{s\in\Omega\setminus(T\cup T')} f(s, \Theta) ds\right]. \tag{6.6}$$

Equation (6.6) expresses the optimization aim of achieving the largest improvement over $J(\Theta)$, which is taken as a constant since Θ is obtained with respect to T. Since Θ' is obtained with $T \cup T'$, we further assume that $f(s, \Theta') \leq f(s, \Theta)$ for any $s \in \boldsymbol{\Omega}$. Although this might seem a rather restrictive assumption, it will be shown later that it could be fulfilled when Θ is a subset of Θ'. Applying this assumption to the right-hand side of Eq. (6.6), we have the following inequalities [84]:

$$\sum_{s\in T} \left(f\left(s, \Theta'\right) - f(s, \Theta)\right) \leq 0,$$

$$\sum_{s\in T'} \left(f\left(s, \Theta'\right) - f(s, \Theta)\right) \leq 0,$$

$$\int_{s\in\Omega\setminus(T\cup T')} \left(f\left(s, \Theta'\right) - f(s, \Theta)\right) ds \leq 0. \tag{6.7}$$

We exploit the fact that in the right-hand side of Eq. (6.6), the value associated with the $\boldsymbol{\Omega}\setminus(T \cup T')$ terms is unknown but a constant. As such, we can discard the terms regarding $\boldsymbol{\Omega}\setminus(T \cup T')$ and retain the ones regarding T and T'. Given that the inequalities expressed by (6.7) all deal with nonpositive values, we can discard terms for which we have a fixed unknown to obtain the following upper bound [84]:

$$J\left(\Theta'\right) - J(\Theta) \leq \sum_{s\in T\cup T'} \left(f\left(s, \Theta'\right) - f(s, \Theta)\right). \tag{6.8}$$

We note that the inequality expressed by (6.8) gives an upper bound of $J(\Theta') - J(\Theta)$ that depends on the current instance set T, the target instance set T', the current PAP Θ, and the new PAP Θ'. Under the assumption of $f(s, \Theta') \leq f(s, \Theta)$ for any $s \in \boldsymbol{\Omega}$ with which the inequalities are constructed, Eq. 6.8 guarantees that the new PAP Θ' has better generalization than the current PAP Θ. More importantly, both $J\left(\Theta'\right)$ and $J(\Theta)$ cannot be exactly measured in practice. As such, the upper bound expressed in Eq. (6.8) provides a measurable surrogate for minimizing $J(\Theta') - J(\Theta)$. Note that since our constructions involve upper bounds, larger performance improvements due to actual smaller values of $J(\Theta') - J(\Theta)$s could be achieved although they are not measurable. Therefore, for a PAP Θ obtained with a training instance set T, an improved PAP Θ' (in terms of generalization performance) could be obtained with a two-step strategy that seeks to minimize the upper bound given in Eq. (6.8) [84]:

1. Identify the T' that maximizes $\sum_{s \in T \cup T'} f(s, \Theta)$ (this is equivalent to maximizing $\sum_{s \in T'} f(s, \Theta)$ since $\sum_{s \in T} f(s, \Theta)$ is a constant given that T and Θ are fixed).
2. Identify the Θ' that minimizes $\sum_{s \in T \cup T'} f\left(s, \Theta'\right)$ (once T' is generated, the term $\sum_{s \in T \cup T'} f(s, \Theta)$ is a constant and can be omitted).

This two-step strategy naturally serves as the core building block of an iterative process that gradually seeks PAPs with better generalization performance. Recall that this two-step strategy is derived from the assumption that $f(s, \Theta') \leq f(s, \Theta)$ for any $s \in \mathbf{\Omega}$. This assumption holds if $\Theta \subseteq \Theta'$ because by definition the performance of PAP as defined by Eq. (6.3), we have that $f(s, \Theta') = \min\{f(s, \Theta), \min_{\theta \in \Theta' \setminus \Theta} f(s, \theta)\} \leq f(s, \Theta)$. There could be many ways to design such an iterative process. Earlier in Sect. 6.2, we have presented GAST that employs an adversarial process as the mechanism for constructing PAPs that are improving in response to increasingly more challenging training instance set. Here, we will employ two-population competitive coevolution to provide a readily available and general framework. Following this, one could design the evolution of the PAP (the configuration population) as identifying new configurations to insert into the current PAP Θ, such that the new PAP $\Theta' \supseteq \Theta$ minimizes $\sum_{s \in T \cup T'} f\left(s, \Theta'\right)$, while the evolution of the training instance set is enhanced via more representative $T \cup T'$ satisfying $T \cap T' = \emptyset$ [84].

In practice however, such a mechanism would suffer from the PAP-size issue whereby the number of the component solvers in the PAP will keep increasing with each coevolutionary cycle. Note that a PAP runs its component solvers in parallel, and thus its size is limited by the available computational resources (e.g. the number of available CPU cores) and cannot grow indefinitely. One approach to possibly address this issue is to first remove some configurations from the PAP Θ, resulting in a temporary PAP $\bar{\Theta}$, and then identify new configurations to insert into $\bar{\Theta}$, such that the final PAP Θ' is of the same size as Θ. However, this approach no longer guarantees the validity of the above assumption and as a consequence Θ' may generalize worse than Θ. This issue can be mitigated by increasing redundancy in the evolution of the PAP. More specifically, one could repeat the configuration-removal-insertion procedure to Θ for n times: (1) Removal phase to obtain n temporary PAPs $\bar{\Theta}_1, \ldots, \bar{\Theta}_n$. (2) Insertion phase where new configurations are identified and inserted to obtain n new PAPs $\Theta'_1, \ldots, \Theta'_n$, each being the same size as Θ. Then, one selects from $\Theta'_1, \ldots, \Theta'_n$ the PAP that performs best against $T \cup T'$ [84].

6.3.2 The Framework of Coevolution of Parameterized Search

We have incorporated the procedure we have described at the end of the previous Sect. 6.3.1 into a coevolutionary search process to formulate the CEPS framework. In turn, the framework is procedurally implemented as **CEPS** in the form Algorithm 6.2 [84]. In general, CEPS consists of two major phases: an initialization phase (lines 2–10) and a coevolution phase (lines 11–34). We start out description of CEPS with the initialization phase. Given an initial training instance set T, a simple greedy strategy is adopted to initialize a configuration population that represents the PAP Θ of size K. First, a set of candidate configurations S_{config} is obtained whereby each configuration is randomly sampled from the configuration space via **rconfigS**(B) (line 2) and tested on the training set T via **VALIDATET**(Θ, T) (line 3). Then, Θ is constructed (lines 4–10) via an iterative process. At each iteration k, the configuration $\theta_k \in S_{\text{config}} \backslash \{\theta_{k-1}, \ldots, \theta_1\}$ is identified and selected via **FINDC** (line 6). After that, the configuration θ_k is inserted into the current PAP of size $k-1$ to produce the new PAP $\Theta \cup \{\theta_k\}$ of size k whereby the former's inclusion into the current PAP leads to the largest performance improvement (line 7). The process terminates when K configurations have been selected.

The coevolution phase is further subdivided into two alternating evolutionary phases for the configuration population (representing the PAP) (lines 12–21) and the instance population (representing the training instances) (lines 22–33). A single, complete coevolutionary cycle consists of a complete evolutionary cycle for the configuration population and then a complete evolutionary cycle for the instance population. The number of coevolutionary cycles or iterations is specified via t_M in Algorithm 6.2. The Θ **Evolutionary Phase** starts with a given configuration population Θ, for which then n temporary PAPs $\bar{\Theta}_j$ $j = 1, \ldots, n$ are generated whereby each $\bar{\Theta}_j$ is a copy of Θ with a random configuration $\theta \in \Theta$ removed (lines 14–15). After that, an existing automatic algorithm configuration tool is used on each $\bar{\Theta}_j$ to search in the configuration space a new configuration θ' with the target that the resulting PAP $\Theta'_j := \bar{\Theta}_j \cup \theta'$ improves its performance on the training set T by way of minimizing $\frac{1}{|T|} \sum_{s \in T} f(s, \Theta \cup \theta')$ (line 17). Here in particular, we use **SMAC** [42]. Lastly, the best-performing PAP among $\Theta'_1, \ldots, \Theta'_n$ is selected to replace Θ via **bestPAP** (line 21). This evolutionary process can be viewed as a $(1, n)$-ES with **SMAC** employed as the mutation operator.

In the T **Evolutionary Phase**, **CEPS** starts by creating a copy of the training instance set $T' := T$ that subsequent processes will operate on (line 25). The aim of the evolution of the instance population is to identify the set T' that is hard for Θ, i.e. maximizing $\sum_{s \in T'} f(s, \Theta)$. As such, each instance $s_l \in T'$, $l = 1, \ldots, |T'|$ is assigned the performance $f(s_l, \Theta)$ of Θ on s_l as the fitness value $q_{s_l} = f(s_l, \Theta)$. The procedure **initFit** will perform the associated computation and assignment process (line 27). In each generation of the evolution of the instance population, **CEPS** applies the procedure **mutate** on a random instance $s \in T'$ to generate an offspring s' (line 29). Then, s' competes with another random instance $s^* \in T'$ and only replaces s^* if the latter has a lower fitness value $f(s^*, \Theta) < f(s', \Theta)$ (line

Algorithm 6.2 CEPS [84]

Input:

B	Base solver B with configuration space C
K_i	Number of initial random component solvers
K	Number of component solvers
T	Training instance set
n	Number of temporary PAPs
t_T	Number of temporary PAPs
t_M	Number of iterations corresponding to coevolutionary cycles

Output:

Θ	Final configuration population (PAP)

procedure **CEPS**(B, K_i, K, T, n, t_M)
 $S_{\text{config}} :=$ **rconfigS**(B) ▷ Random configuration set $\{\theta_1, \ldots, \theta_{K_i}\}$, $K_i > K$
 $Q :=$ VALIDATET(S_{config}, T) ▷ $Q = \{F_{\theta_1}, \ldots, F_{\theta_{K_i}}\}$ where ▷ $F_{\theta_k} = \{f(s_1, \theta_k), \ldots, f(s_L, \theta_k)\}$ and $T = \{s_1, \ldots, s_L\}$
 initPAP(Θ) ▷ Initialize $\Theta = \emptyset$
 for $k := 1 : K$ **do**
  $\theta_k :=$ FINDC($\Theta, S_{\text{config}}, Q$) ▷ Find $\theta_k \in S_{\text{config}} \backslash \Theta$ with min $\frac{1}{|T|} \sum_{s \in T} f(s, \Theta \cup \theta_k)$
  $\Theta :=$ **combinePAP**($\Theta, \{\theta_k\}$) ▷ $\Theta := \Theta \cup \{\theta_k\}$
  $S_{\text{config}} :=$ **delPAP**($S_{\text{config}}, \{\theta_k\}$) ▷ $S_{\text{config}} := S_{\text{config}} \backslash \{\theta_k\}$
  $Q :=$ **delete**($Q, \{F_{\theta_k}\}$) ▷ $Q := Q \backslash \{F_{\theta_k}\}$
 end for
 for $i := 1 : t_M$ **do**
  $Q :=$ VALIDATET(Θ, T) ▷ Θ **Evolutionary Phase** Start
  for $j := 1 : n$ **do**
   $\theta :=$ **randSel**(Θ)
   $\bar{\Theta}_j :=$ **delranPAP**(Θ) ▷ $\bar{\Theta}_j$ is a copy of Θ with a random $\theta \in \Theta$ removed
   $Q_j :=$ **delete**($Q, \{F_\theta\}$) ▷ $Q_j := Q \backslash \{F_\theta\}$
   $\theta' :=$ **SMAC**($\bar{\Theta}_j, Q_j$) ▷ **SMAC** finds θ' so that min $\frac{1}{|T|} \sum_{s \in T} f(s, \bar{\Theta}_j \cup \theta')$
   $Q[j] :=$ **add**($Q_j, \{F_{\theta'}\}$) ▷ $Q[j] = Q_j$ where $\{F_\theta\}$ replaced with $\{F_{\theta'}\}$
   $\Theta'[j] :=$ **combinePAP**($\bar{\Theta}_j, \{\theta'\}$) ▷ $\Theta'[j] = \bar{\Theta}_j$ where $\{\theta\}$ replaced with $\{\theta'\}$
  end for
  $\Theta :=$ **bestPAP**($\Theta'[\,], Q[\,]$) ▷ Θ is best-performing PAP in $\{\Theta'_1, \ldots, \Theta'_n\}$ ▷ Θ **Evolutionary Phase** End
  if $i = t_M$ **then** ▷ T **Evolutionary Phase** Start
   break ▷ Skip instance generation as PAP Θ constructed
  end if
  $T' := T$ ▷ Backup T
  $L :=$ **sizeOf**(T) ▷ $L := |T|$
  initFit($q[\,], L, T', Q$) ▷ $q[\,] = (q_{s_1}, \ldots, q_{s_L})$, each $s_l \in T'$ assigned ▷ fitness value $q_{s_l} = f(s_l, \Theta)$
  while within time budget **do**
   $s' :=$ **mutate**(T') ▷ Mutate a randomly chosen $s \in T'$ to produce s'
   $q_{s'} :=$ **f**(s', Θ) ▷ Assign s' with a fitness value $f(s', \Theta)$
   $T' :=$ **replace**($T', q[\,], s', q_{s'}$) ▷ Select a random $s^* \in T'$ with $q[s^*] < q_{s'}$,
  end while ▷ then $T' := T' \backslash \{s^*\} \cup \{s'\}$
  $T :=$ **combine**(T', T) ▷ $T := T' \cup T$. ▷ T **Evolutionary Phase** End
 end for
 return Θ
end procedure

31). The replacement of instances in T' with new ones having better fitness ensures that with each generation (iteration) of the phase, the average fitness of instances in T' will gradually increase and that they pose a harder challenge to the configuration population representing PAP Θ. Upon completion of the phase, the final T' will be merged into the training set T (line 33), which will be used for construction of the new Θ in the next cycle of coevolution. Note in the last t_Mth iteration of coevolution, the evolution of the instance population is skipped (lines 22–23), since there is no need to evolve the population of instances further and **CEPS** can proceed to output the final PAP version.

6.3.3 Instantiations for TSP and VRPSPDTW

We emphasize that Algorithm 6.2 is a generic procedural implementation of the CEPS framework since the representations of both populations depend on the target parameterized algorithm and the target problem class, respectively. For example, the mutation operator and the fitness function for the instances depend on target problem class. Here, we introduce two instantiations of CEPS, CEPS-TSP, and CEPS-VRPSPDTW, which we have developed for the TSP and VRPSPDTW problems, respectively. We have chosen TSP because as a classic NP-hard problem it is widely investigated with readily available benchmarking problems to allow substantial computational studies. We have chosen VRPSPDTW as a much more complex routing problem that more closely takes real-world requirements into account. The significant difference between these two problems would provide a good context for assessing CEPS.

We start with a detailed presentation of CEPS-TSP. As before where we have studied GAST-TSP, we consider symmetric TSP with distances in a two-dimensional Euclidean space. A TSP instance is represented by a list of $\{(x_1, y_1), \ldots, (x_N, y_N)\}$ coordinates, with each (x_i, y_i)-coordinate representing a city. We employ an instance mutation operator based on an operator that is widely used for generating TSP instances [85] for CEPS-TSP. The procedural implementation of **MUTATETSP** is described in Algorithm 6.3 [84]. The procedure **MUTATETSP** directly mutates the instance s by introducing perturbations on all the (x_i, y_i)-coordinates. There are two types of perturbations used. The main perturbation is based on additive perturbation with a Gaussian source, i.e. a coordinate is perturbed with a random value drawn from the normal distribution $N(\mu, \sigma^2)$ with zero mean $\mu = 0$ and variance scaled to the range of the coordinate in the list, e.g. $\sigma^2 = 0.025 \cdot (x_{\max} - x_{\min})^2$. There is a probability of 0.9 that this additive perturbation is applied (lines 11–14). The other perturbation involves directly replacing the original coordinate with a new value sampled from a uniform distribution over the range of the coordinate (lines 16–17), e.g. $U(x_{\min}, x_{\max})$. The adopted parameterized algorithm is the LKH version 2.0.7 (with 23 parameters) [33]. We use PAR-10 as the performance indicator [41], which we have introduced

Algorithm 6.3 The instance mutation operator in CEPS-TSP [84]

Input:

s Instance $s = \{(x_1, y_1), \ldots, (x_N, y_N)\}$ with N number of cities

Output:

s Mutated Instance

procedure MUTATE**TSP**(s)
 $x_{\min} :=$ **minX**$(s_{1:N})$ ▷ $x_{\min} = \min\{x_1, \ldots, x_N\}$
 $x_{\max} :=$ **maxX**$(s_{1:N})$ ▷ $x_{\max} = \max\{x_1, \ldots, x_N\}$
 $y_{\min} :=$ **minY**$(s_{1:N})$ ▷ $y_{\min} = \min\{y_1, \ldots, y_N\}$
 $y_{\max} :=$ **maxY**$(s_{1:N})$ ▷ $y_{\max} = \max\{y_1, \ldots, y_N\}$
 $x^2_{\text{srange}} := 0.025 \times (x_{\max} - x_{\min})^2$
 $y^2_{\text{srange}} := 0.025 \times (y_{\max} - y_{\min})^2$
 for $i := 1 : N$ **do**
  $r :=$ **uniRand**$(0, 1)$ ▷ $r \sim U(0, 1)$ with uniform distribution
  if $r \leq 0.9$ **then**
   $\delta :=$ **normRand**$(0, x^2_{\text{srange}})$ ▷ $\delta \sim N(0, 0.025 \cdot (x_{\max} - x_{\min})^2)$
   $x_i := x_i + \delta$
   $\delta :=$ **normRand**$(0, y^2_{\text{srange}})$ ▷ $\delta \sim N(0, 0.025 \cdot (y_{\max} - y_{\min})^2)$
   $y_i := y_i + \delta$
  else
   $x_i :=$ **uniRand**$(x_{\min}, x_{\max})$ ▷ $x_i := x,\ x \sim U(x_{\min}, x_{\max})$
   $y_i :=$ **uniRand**$(y_{\min}, y_{\max})$ ▷ $y_i := y,\ y \sim U(y_{\min}, y_{\max})$
  end if
 end for
 return s
end procedure

earlier in Sect. 6.2.4 is considered as the performance indicator. The fitness function of an instance is simply $f(s, \Theta) = \min_{\theta \in \Theta} f(s, \theta)$.

For the CEPS-VRPSPDTW, we start with an introduction to the target problem. In VRPSPDTW, a number of customers require both pickup service and delivery service within a certain time window. The target of VRPSPDTW [86] is to send out a fleet of capacitated vehicles, which are stationed at a depot, to meet the customer demands at minimum total cost. More specifically, VRPSPDTW is defined on a complete graph $G = (V, E)$ with $V = \{0, 1, 2, \ldots, N\}$ as the vertex (node) set and E as the edge set $E = \{\{i, j\} : i, j \in V \wedge i \neq j\}$ defined between each pair of nodes with $\{i, j\}$ representing the edge. The depot is denoted as 0, and the customers are denoted as $1, \ldots, N$. Each node $i \in V$ has (x_i, y_i)-coordinates. The distance $\text{dist}(i, j)$ between i and j is the Euclidean distance. Each customer node is further associated with five attributes: a delivery demand d_i, a pickup demand p_i, a time window $[a_i, b_i]$, and a service time s_i. Here, d_i represents the amount of goods to be delivered from the depot to customer i and p_i represents the amount of goods to be picked up from customer i and then be delivered to the depot. a_i and b_i define the start and the end of the time window in which the customer receives service. Note that the time windows are treated as hard constraints. Arrival of a vehicle at the customer i before a_i results in a wait before service can begin, while arrival after b_i is infeasible. Finally, t_i is the time spent by the vehicle to load/unload goods at customer i. A fleet of J identical vehicles, each with a capacity of O and dispatching cost c_d, is initially located at the depot. Each vehicle starts at the depot and then serves the customers, and finally returns to the depot. The depot 0 is also associated with these five attributes, where a_0 and b_0 are the earliest time the vehicles can depart from the depot and the latest time the vehicles can return to the depot, respectively, and $d_0 = p_0 = t_0 = 0$.

A solution S to VRPSPDTW is represented by a set of vehicle routes $S = \{R_1, R_2, \ldots, R_K\}$. Each route R_i consists of a sequence of nodes that the vehicle visits $R_i = (h_{i,1}, h_{i,2}, \ldots, h_{i,L_i})$, where $h_{i,j}$ is the jth node visited in R_i, and L_i is the length of R_i. Let $TD(R_i)$ be the total travel distance in R_i, and $\text{load}(R_i)$ be the largest load on the vehicle that occurs in R_i. Let $\text{arr}(h_{i,j})$ and $\text{dep}(h_{i,j})$ be the time of arrival at $h_{i,j}$ and the time of departure from $h_{i,j}$, respectively. The total cost of S consists of two parts: (1) The dispatching cost $K \cdot c_d$ of the used vehicles. (2) The transportation cost $\sum_{i=1}^{K} \big(TD(R_i) \cdot u\big)$ given by the total travel distance in S multiplied by unit transportation cost u. We can now formulate the VRPSPDTW as the following minimization problem [84]:

$$\min_{S} \; TC(S) = \sum_{i=1}^{K} \big(c_d + TD(R_i) \cdot u\big) \tag{6.9}$$

s.t. :

$$K \le J$$

$$h_{i,1} = h_{i,L_i} = 0, \; 1 \le i \le K$$

$$\sum_{i=1}^{K} \sum_{j=2}^{L_i - 1} I[h_{i,j} = e] = 1, \; 1 \le e \le N$$

$$\text{load}(R_i) \le O, \; 1 \le i \le K$$

$$\text{dep}(h_{i,1}) \ge a_0, \; 1 \le i \le K$$

$$\text{arr}(h_{i,j}) \le b_{h_{i,j}}, \; 1 \le i \le K, \; 2 \le j \le L_i$$

where the constraints are: (1) The number of used vehicles must be smaller than the number of available ones. (2) Each customer must be served exactly once. (3) The vehicle cannot be overloaded during transportation. (4) All vehicles can only serve after the start of the time window of the depot and must return to the depot before the end of the time window of the depot. (5) The service of the vehicle to each customer must be performed within that customer's time window.

We consider a practical application scenario from the JD logistics company. In particular, a VRPSPDTW solver is required to solve a VRPSPDTW instance every day. The company has about 3000 customers in total in the city, but only 400 customers ($\approx$13%) require service every day. As such, the solver will face different VRPSPDTW instances having some underlying relationships: (1) The location and the time window of the depot, and the vehicle fleet are unchanged. (2) The locations of the customers will change. (3) The time windows of the customers will change. (4) The delivery and pickup demands of the customers will change.

Algorithm 6.4 The instance mutation operator in CEPS-VRPSPDTW [84]

Input:
s Instance $s = \{(x_0, y_0, d_0, p_0, a_0, b_0, t_0), \dots, (x_N, y_N, d_N, p_N, a_N, b_N, t_N)\}$ where the index 0 is for the depot
Output:
s Mutated Instance

procedure MUTATEVRPSPDTW(s)
  $x_{\min} :=$ **minX**$(s_{1:N})$ ▷ $x_{\min} = \min\{x_1, \dots, x_N\}$
  $x_{\max} :=$ **maxX**$(s_{1:N})$ ▷ $x_{\max} = \max\{x_1, \dots, x_N\}$
  $y_{\min} :=$ **minY**$(s_{1:N})$ ▷ $y_{\min} = \min\{y_1, \dots, y_N\}$
  $y_{\max} :=$ **maxY**$(s_{1:N})$ ▷ $y_{\max} = \max\{y_1, \dots, y_N\}$
  $x^2_{\text{srange}} := 0.025 \times (x_{\max} - x_{\min})^2$
  $y^2_{\text{srange}} := 0.025 \times (y_{\max} - y_{\min})^2$
  $w^2_{\text{srange}} := 0.025 \times (b_0 - a_0)^2$
  $d_{\min} :=$ **minD**$(s_{1:N})$ ▷ $u_{\min} = \min\{u_1, \dots, u_N\}$
  $d_{\max} :=$ **maxD**$(s_{1:N})$ ▷ $u_{\max} = \max\{u_1, \dots, u_N\}$
  $p_{\min} :=$ **minP**$(s_{1:N})$ ▷ $p_{\min} = \min\{p_1, \dots, p_N\}$
  $p_{\max} :=$ **maxP**$(s_{1:N})$ ▷ $p_{\max} = \max\{p_1, \dots, p_N\}$
  for $i := 1 : N$ **do**
    ▷ **Coordinate Mutation**
    $r :=$ **uniRand**$(0, 1)$ ▷ $r \sim U(0, 1)$ with uniform distribution
    if $r \leq 0.9$ **then**
      $\delta :=$ **normRand**$(0, x^2_{\text{srange}})$ ▷ $\delta \sim N(0, 0.025 \cdot (x_{\max} - x_{\min})^2)$
      $x_i := x_i + \delta$
      $\delta :=$ **normRand**$(0, y^2_{\text{srange}})$ ▷ $\delta \sim N(0, 0.025 \cdot (y_{\max} - y_{\min})^2)$
      $y_i := y_i + \delta$
    else
      $x_i :=$ **uniRand**$(x_{\min}, x_{\max})$ ▷ $x_i := x,\ x \sim U(x_{\min}, x_{\max})$
      $y_i :=$ **uniRand**$(y_{\min}, y_{\max})$ ▷ $y_i := y,\ y \sim U(y_{\min}, y_{\max})$
    end if
    ▷ **Demand Mutation**
    $p_i :=$ **uniRand**$(p_{\min}, p_{\max})$ ▷ $p_i := p,\ p \sim U(p_{\min}, p_{\max})$
    $d_i :=$ **uniRand**$(d_{\min}, d_{\max})$ ▷ $d_i := d,\ d \sim U(d_{\min}, d_{\max})$
    ▷ **Time-window Mutation**
    $\delta_a :=$ **normRand**$(0, w^2_{\text{srange}})$ ▷ $\delta_a \sim N(0, 0.025 \cdot (b_0 - a_0)^2)$
    $a_i := a_i + \delta_a$
    $\delta_b :=$ **normRand**$(0, w^2_{\text{srange}})$ ▷ $\delta_b \sim N(0, 0.025 \cdot (b_0 - a_0)^2)$
    $b_i := b_i + \delta_b$
  end for
  return s
end procedure

We design a specialized mutation operator for VRPSPDTW, which is procedurally described in Algorithm 6.4, based on the above observation about the nature of the target problem. In particular, MUTATE**VRPSPDTW** can be viewed as an extension of MUTATE**TSP** given that it needs to be able to operate on VRPSPDTW instances that have additional attributes [84]. MUTATE**VRPSPDTW** employs the coordinate mutation used in MUTATE**TSP**. In addition, the pickup demand p_i and the delivery demand d_i of each customer are replaced by new values sampled from $U(p_{\min}, p_{\max})$ and $U(d_{\min}, d_{\max})$, respectively. For the time window $[a_i, b_i]$ of each customer, a_i and b_i are perturbed with values drawn from $N(0, (0.025 \cdot (b_0 - a_0)^2)$. The adopted parameterized algorithm for VRPSPDTW is a powerful coevolutionary Genetic Algorithm (Co-GA) proposed by Wang and Chen [86] (with 12 parameters). The ***Penalized Average Normalized Cost*** (***PANC***) is considered as the performance indicator. Since VRPSPDTW is formulated in a minimization setting, smaller PANC values imply better performances for the solver. More specifically, the performance $f(s, \theta)$ of a configuration θ on an instance s is the PANC of the solution found by θ. The run of θ on s would be terminated after a cutoff time of 150 s. In case that θ fails to find a feasible solution to s within the cutoff time, the corresponding run is considered timeout and $f(s, \theta)$ will be set to a large penalty value of 2000. Assume θ successfully finds a feasible solution of cost c to s. However, the scales of the solution costs might vary substantially for different VRPSPDTW instances. As such, we employ the normalized cost of c given as follows [84]:

$$f(s, \theta) = \frac{c}{\mu_{\text{dist}}(s)}, \tag{6.10}$$

where $\mu_{\text{dist}}(s)$ is the average distance between all pairs of customers in instance s.

We design our computational studies to determine whether CEPS could better tackle the few-shots challenge and that a coevolutionary process is effective at improving the generalization performance of constructed PAPs. This requires that we generate benchmark instance sets for TSP and VRPSPDTW to assess the generalization of coevolved PAPs. The TSP instance set consists of 500 instances, while the VRPSPDTW instance set consists of 233 instances. We have ensured that both instance sets are generated in a different manner compared to CEPS to avoid bias. Then, we split each benchmark instance set to a training set (around 6%) and a testing set (around 94%). The splitting process is repeated 3 times to produce 3 pairs of training and testing sets, TSP-1/2/3 and VRPSPDTW-1/2/3, to reduce the effect of the random splitting on the experimental results. Note that testing instances are only used to approximate the generalization performance of the PAPs obtained by CEPS and other methods for comparisons, while training instances are reserved for PAP construction. The TSP/VRPSPDTW instance set, the compared methods, and experimental protocol are summarized in Table 6.7.

For TSP, we have collected 10 different instance generators from the literature: ***portgen***, ***ClusteredNetwork***, ***explosion***, ***implosion***, ***cluster***, ***rotation***, ***linearprojection***, ***expansion***, ***compression***, and ***gridmutation***. In particular, ***portgen*** is from the 8th DIMACS Implementation Challenge [44] (it generates a TSP ***rue*** instance by

Table 6.7 Summary of the experimental settings [84]

	Instance sets	Performance indicator	Parameterized algorithm	K (#θ)
TSP	500 instances generated by 10 different generators. Training/Testing split: 30/470	Runtime needed to find the optima of the instances. In particular, PAR-10 with cutoff time 10 s was used (see Sect. 6.3.3)	LKH version 2.0.7 [33] with 23 parameters	4
VRPSPDTW	233 instances obtained from real-world application. Training/Testing split: 14/219	Cost of the found solutions. In particular, PANC with cutoff time 150 s was used (see Sect. 6.3.3)	Co-GA [86] (with 12 parameters)	4

uniformly randomly placing the points on a Euclidean plane), ***ClusteredNetwork*** is from the R-package ***netgen*** [16] (it generates an instance by placing points around different central points), and the remaining eight generators are proposed in [17] (in general, they generate a TSP instance mainly by simulating a phenomenon in the point clouds of a ***rue*** instance). They are expected to generate highly diversed TSP instances due to different instance-generation mechanisms underlying them. Each generator is used to generate 50 instances, which are combined to provide us with a set of 500 TSP instances. The problem sizes (i.e. the number of cities) in all these instances are 800 [84].

For VRPSPDTW, we have obtained data from a real-world application of the JD logistics company [84]. Specifically, the data contain customer requests that occurred during a period of time in a city. The total number of customers is 3000, of which 400 customers require service per day. To generate a VRPSPDTW instance, 400 customers are randomly selected from the 3000 customers with the pickup and delivery demands of each customer being randomly selected from all the demands that the customer has made during this period of time. This generation process is repeated 500 times to produce a set of 500 VRPSPDTW instances. A VRPSPDTW solver [86] is employed to determine which of the generated instances have feasible solutions and discard those without feasible solutions. As a result, we have obtained a set of 233 VRPSPDTW instances.

6.3.4 Computational Study on CEPS

We compare CEPS with the state-of-the-art PAP construction methods namely GLOBAL [53], PARHYDRA [53, 92], and PCIT [54], as in the case with GAST in Sect. 6.2.4. We also compare with a baseline method ***Evolution of Parameterized Search*** (***EPS***) as the means to assess the role of coevolution for achieving (if any) improved generalization performance. EPS differs from CEPS in that it conducts instance mutation and PAP construction in two isolated phases. Starting with the same training instance set as provided to CEPS, EPS applies the instance mutation operator to generate an augmented set of instances. The size of this augmented set is kept the same as the number of instances generated during the entire procedure of CEPS. Next, a PAP is evolved with the same approach as in CEPS using the union of the initial and the augmented training instance sets as the input. We also compare the initial PAPs obtained during the initialization phase and the final PAPs after CEPS runs complete as an additional mean to verify the effectiveness of coevolution in CEPS.

The above-described experimental settings as well as the used parameterized algorithms and performance indicators have been summarized earlier in Table 6.7. We set the number of component solvers K in PAP to 4. For CEPS, the number of iterations of the coevolution and the number of temporary PAPs generated are set to $t_M = 4$ and $n = 10$, respectively. To allow for fair comparisons, we have attempted to keep the total CPU time consumed by each compared method in the experiments to be nearly similar. The detailed time settings of each compared method are presented in Table 6.8. This is achieved by setting appropriate time budgets in the major procedural calls in the methods that involve repetitions (iterations). The most time-consuming parts of PAP construction methods are the runs of the configurations on the problem instances, and the incurred computational costs account for the vast majority of the total costs. In the case of CEPS implemented via Algorithm 6.2, the configurations would be run in the initialization phase (line 6), Θ evolutionary phase (lines 17 and 21), and T evolutionary phase (line 30). For each of these four procedural calls that we have identified in Algorithm 6.2, we set the corresponding wall-clock time budget t_{init}, t_C, t_V, and t_T to control the overall computational costs of CEPS. Then, the total CPU time consumed by CEPS could be estimated by $t_{\text{init}}+t_M \cdot K \cdot \big(n \cdot (t_C+t_V)+t_T\big)$ [84]. The total CPU time consumed by GLOBAL and PCIT could be estimated by $K \cdot n \cdot (t_C + t_V)$, while for PARHYDRA it could be estimated by $\Sigma_{k=1}^{K} k \cdot n \cdot (t_C + t_V)$ [53, 54, 92]. Note that for different

Table 6.8 Detailed time settings (in hours) of each PAP construction method [84]

	TSP				
	t_C	t_V	t_T	t_{ini}	CPU time
CEPS	1.5 h	0.5 h	1.5 h	8 h	320 h
GLOBAL	7.5 h	1 h	–	–	340 h
PCIT	7.5 h	1 h	–	–	340 h
PARHYDRA	2 h	1 h	–	–	300 h
	VRPSPDTW				
	t_C	t_V	t_T	t_{ini}	CPU time
CEPS	6 h	2 h	6 h	32 h	1312 h
GLOBAL	30 h	4 h	–	–	1360 h
PCIT	30 h	4 h	–	–	1360 h
PARHYDRA	8 h	4 h	–	–	1200 h

methods, t_C, t_V and t_T could be set to different values so as to ensure that their total CPU times are nearly the same.

For each pair of training and testing sets (TSP-1/2/3 and VRPSPDTW-1/2/3), we apply each considered method to construct a PAP on the training set and then tested the resulting PAP on the testing set. For each testing instance, the PAP is run 3 times, and the median of those three runs is recorded as the performance of the solver on the instance to take into account effects of random initializations. The performances of a PAP on different testing instances are then aggregated to obtain the number of total timeouts (#TOs), PAR-10 (for TSP solver) and PANC (for VRPSPDTW solver) on the testing set. All the experiments are conducted on a cluster of three Intel Xeon machines with 128 GB RAM and 24 cores each (2.20 GHz, 30 MB Cache), running Centos 7.5.

Results of #TOs, PAR-10, and PANC achieved by the PAPs on the testing set are summarized in Table 6.9. The spread of those results for PAPs across all the testing instances is visually summarized as boxplots with the corresponding medians and variances in Fig. 6.3. Note that the mean value is also plotted in Fig. 6.3 (indicated by ▲) to show the effects of outliers (due to timeout cases) on a PAP that otherwise might be masked by the boxplot construction. There are three important observations we can make and they are presented next.

First, PAPs obtained by CEPS have the smallest number of timeouts, and hence the highest success rate for solving the testing instances among all tested PAPs in all the experiments. This is further verified in Fig. 6.3 where CEPS results indicate the least gap between the mean value (that takes timeouts into account) and median value (that naturally filters out the timeouts). In particular, CEPS's PAPs have achieved the best PAR-10 and PANC scores in 5/6 experiments due to having the least number of timeouts. For example in TSP-1 and TSP-3 of Fig. 6.3, PARHDYRA and EPS have better performances than CEPS on the normal instances but finally achieved worse PAR-10 due to having more timeouts. Another aspect is that CEPS has more stable performances over all six experiments, whereby different experiments involve different training/testing splits, compared to other PAP construction methods. For example, the number of timeouts of PCIT and PARHYDRA fluctuates over different training/testing sets on VRPSPDTW problems. These results indicate the effectiveness of CEPS that actively searches in the instance space to identify the hard-to-solve instances that are used to improve the generalization of the PAPs. Such a mechanism makes CEPS's PAPs robust being the least affected by the hard testing instances that significantly differ from the given training instances. In summary, CEPS is not only the best-performing method, but also is less sensitive to the training data, and thereby could better tackle the few-shots challenge [84].

The remaining two observations have to do with the mechanism for instance generation. We note that EPS employs instance generation but is outperformed by methods that do not generate synthetic instances in several cases, e.g. when compared to PARHYDRA on TSP-2. This indicates that isolating instance generation away from PAP construction might have negative effects. We also note that the final PAPs obtained by CEPS have performed better than initial PAPs on all the

Table 6.9 The testing results of the PAPs constructed by each method. #TOs refers to the number of the total timeouts. PAR-10 and PANC are penalized average runtime-10 and penalized average normalized cost, respectively. Performance of a PAP is highlighted in **bold** if it achieved the best testing performance [84]

	TSP-1		TSP-2		TSP-3		VRPSPDTW-1		VRPSPDTW-2		VRPSPDTW-3	
	#TOs	PAR-10	#TOs	PAR-10	#TOs	PAR-10	#TOs	PANC	#TOs	PANC	#TOs	PANC
GLOBAL	10	3.85	15	5.18	10	3.67	4	253	3	248	4	258
PCIT	**6**	**2.51**	**4**	2.44	9	3.46	4	258	2	240	6	274
PARHYDRA	9	3.55	**4**	2.19	5	2.36	2	237	5	265	3	249
EPS	7	2.93	6	2.81	8	2.81	2	238	2	236	**1**	**229**
CEPS.initial	14	4.50	14	4.63	6	3.12	3	245	2	237	2	237
CEPS	**6**	2.74	**4**	**2.15**	**2**	**1.94**	**0**	**221**	**1**	**229**	**1**	**229**

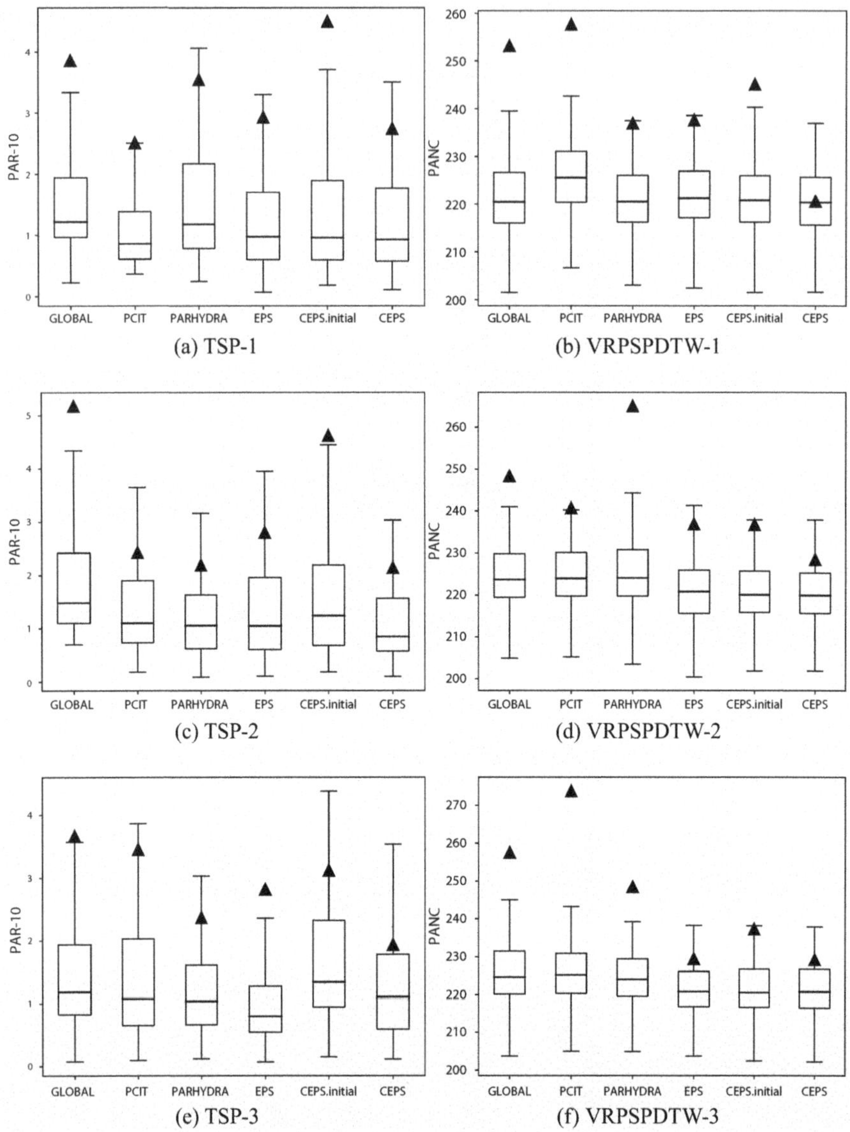

Fig. 6.3 Visual comparison in boxplots of the medians and variance of the test performance of each PAP across the testing instances. Note the mean value is also plotted, indicated by ▲ [84]

experiments. On average, the performance improvement rate (in terms of PAR-10 and PANC) is 21.78%. All these suggest that integrating instance generation with the construction of PAPs in a coevolutionary framework is effective towards enhancing the generalization of the PAP solvers [84].

We have performed additional experiments to further assess CEPS-TSP by way of comparisons against state-of-the-art TSP solvers [84]. The latter include: (1) The default configuration of the LKH algorithm (LKH-default). (2) The default configuration of EAX algorithm [65] (EAX-default), which has outperformed LKH on a broad range of TSP instances. (3) The tuned versions of LKH and EAX (LKH-tuned and EAX-tuned, respectively), which are obtained by running SMAC on their configuration spaces and the training sets for the same time budget of the PAP construction of CEPS. In addition, we consider two state-of-the-art portfolio-based Algorithm Selection (AS) methods for TSP [48, 94]. Both methods have adopted the same TSP algorithm portfolio that contains LKH, EAX, MAOS [90], and their variants. Instead of comparing both methods with CEPS, we employ a Virtual Best Solver (VBS) of their algorithm portfolio. In particular, VBS acts as an ***oracle*** that perfectly selects or chooses the best algorithm for each instance without any selection cost. As such, VBS is an ideal procedure that provides an upper bound for the performance of any algorithm selector, since in practice actual AS would incur selection costs [84].

As before, each solver is run 3 times for each testing instance, and the median of those three runs is recorded as the performance of the solver on that instance. Results for the number of timeouts and PAR-10 of these solvers on the testing set are summarized in Table 6.10, and their spreads visualized via boxplots in Fig. 6.4. The results indicate that both LKH-default and EAX-default performed badly on the testing set, with considerable timeouts. We have omitted results for LKH-default in Fig. 6.4 due to its large number of timeouts. One explanation is that the default configurations of LKH and EAX are mainly designed to handle much larger-scale TSP instances (e.g. 10,000). After tuning, LKH-tuned and EAX-tuned have achieved significant performance improvement. However, their performances are still below that of CEPS. VBS is the only solver that could match the performance level of CEPS, but on TSP-2 and TSP-3, the performance advantage of CEPS over VBS is still significant [84].

We have carried out one final study to investigate the generalization ability of CEPS [84], whereby the PAP constructed by CEPS for the VRPSPDTW-1

Table 6.10 Comparison of the state-of-the-art TSP solvers with the PAPs obtained by CEPS, on the testing set. #TOs refers to the number of total timeouts. PAR-10 is the penalized average runtime-10. Performance of a solver is highlighted in **bold** if it achieved the best testing performance [84]

	TSP-1		TSP-2		TSP-3	
	#TOs	PAR-10	#TOs	PAR-10	#TOs	PAR-10
LKH-default	131	30.84	137	31.98	150	34.73
LKH-tuned	29	8.23	34	9.40	27	7.67
EAX-default	69	17.98	73	18.91	69	17.95
EAX-tuned	33	10.97	30	10.38	29	10.12
VBS	**6**	3.82	7	4.26	6	4.13
CEPS	**6**	**2.74**	**4**	**2.15**	**2**	**1.94**

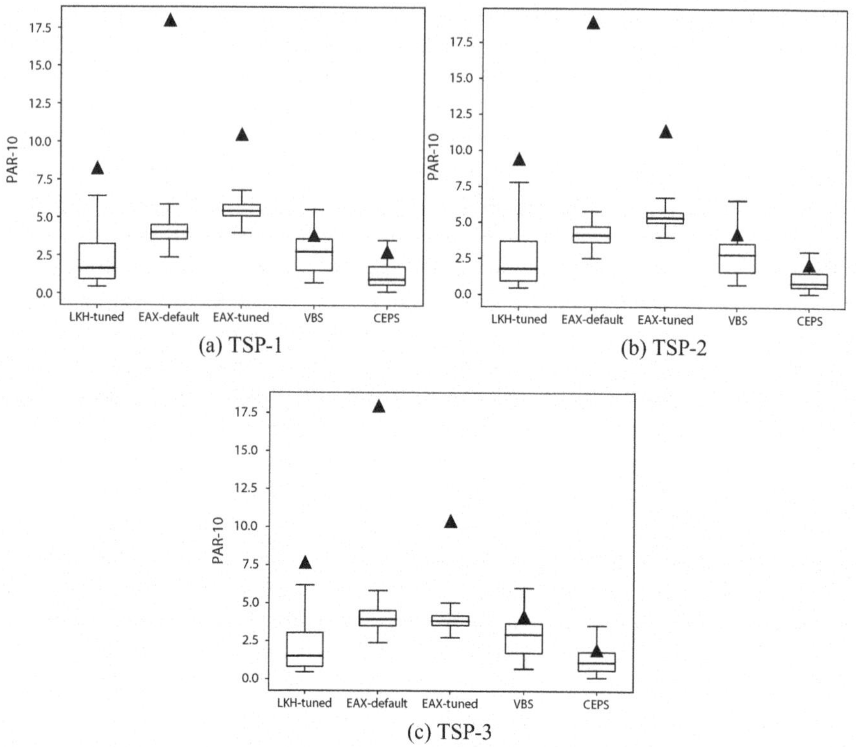

Fig. 6.4 Visual comparison in boxplots of the medians and variance of the test performance of each TSP solver across the testing instances. Note the mean value is also plotted, indicated by ▲ [84]

experiment has been tested on existing VRPSPDTW benchmarks [86] that are widely used. Note that the training instances in VRPSPDTW-1 are obtained from different sources (i.e. real-world application) and might have different problem characteristics (e.g. customer number and node distribution) compared to the benchmarks. As such, the PAP constructed by CEPS could be said to generalize well to totally unseen data if while constructed using the VRPSPDTW-1 training set it could perform well on the VRPSPDTW benchmarks. Table 6.11 compares solutions found by the PAP and the best-known solutions reported in the literature [38, 71, 77, 86, 87] (up to May 2019). Our results indicate that overall the PAP could generalize well to the VRPSPDTW benchmarks. Solutions for around 66% (43/65) instances found by the PAP are not worse than the currently best-known solutions. We note that on 10 instances, the PAP found new best solutions. However, we observe that the PAP does not perform very well on the *cdp* type instances in which the locations of customers are clustered. One possible explanation is that the parameterized algorithm (SMAC) used for tuning PAPs could not handle this type of instances. Despite this, CEPS is best viewed as a higher-level problem-solving

Table 6.11 Comparison between the solutions found by PAP and the best-known solutions (BKS) found before (as reported in the literature) on existing VRPSPDTW benchmark instances. #better, #not-worse, and #worse refer to the number of the instances on which PAP found better, not worse (i.e. either with better or the same quality), and worse solutions compared to the BKS [84]

Instance type	#instances	#better	#not-worse	#worse
RCdp (small)	9	0 (0%)	9 (100%)	0 (0%)
Rdp	23	4 (17%)	17 (57%)	6 (26%)
Cdp	17	0 (0%)	9 (53%)	8 (47%)
RCdp	16	6 (13%)	8 (38%)	8 (50%)
Count	65	10 (15%)	43 (66%)	22 (34%)

algorithm. Other parameterized algorithms for tuning PAPs that could deal with specific properties in target problem instances can be used as required [84].

6.4 Discussion and Conclusion

In this chapter, we have demonstrated how effective methods for automatic construction of PAPs with high generalization performance can be realized. First, we have proposed a novel approach GAST that involves both instance generation and portfolio construction, with feedbacks between both processes achieved using an adversarial mechanism. For two target problems (TSP and SAT), our computational studies have demonstrated that the adversarial mechanism in GAST is crucial towards obtaining PAPs that could achieve better generalization performances than other existing portfolio construction approaches and reach the performance level of the state-of-the-art parallel solvers designed by human experts. Our subsequent theoretical analysis to determine under what circumstances constructed PAPs could improve their generalization performances has informed us of a general two-step strategy on iterative procedures whereby such PAPs can be constructed. In general, this two-step strategy involves alternating between adaptation phases on problem instances and PAPs in a competitive manner, whereby within each phase the generated instances or PAPs must increase in their performance level. Guarantees for such performance increase for PAPs can be achieved via an augmentation process that increases the size of the PAP (i.e. at the worst case the new PAP Θ' has a similar generalization performance with the old PAP Θ given that $\Theta \subseteq \Theta'$). Similar arguments can be made for training instance set and its augmentation to make it more representative (i.e. provides better coverage of the entire instance space).

GAST represents one procedural implementation of this two-step strategy. We have proposed the CEPS framework to represent a family of approaches whereby the evolution of instance set approximates a greedy mechanism for instance augmentation that is exploited to improve the generalization performance of the resulting tuned or evolved PAP as much as possible. Our computational study on CEPS has demonstrated that it could better tackle the few-shots challenge than existing PAP construction methods in two target problems of TSP and VRPSPDTW. Crucially, our results strongly indicate that if CEPS is appropriately formulated for the target problems (e.g. CEPS-TSP and CEPS-VRPSPDTW), the arms-race dynamics in the coevolutionary process can be realized. This means that even when CEPS starts with only a limited number of problem instances, it is able to construct PAPs with increasingly higher generalization performances by augmenting the set of training problem instances in a competitive manner with respect to the PAPs. Note that we have applied CEPS in scenarios whereby initial training instances are available. CEPS could be used in a fully cold-start setting (zero-shot) where there are no real-world instances available for the target problem class but instead requires the means with which some instance generators can be developed. Furthermore, CEPS could either be run offline or online by incorporating new real instances whenever available.

It is important to note that the CEPS framework represents a higher problem-solving methodology. The instantiation of CEPS to a specific domain and problem class involves the specification and then design of the instance mutation operator, the parameterized algorithm, and the automatic algorithm configuration method. Among these three modules within the CEPS framework, the instance generators could be adapted from one problem to another more easily if those problems have some commonalities or the latter is an extension of the former. The automatic algorithm configuration methods are usually generic. This leaves the designers and practitioners with the final module involving the parameterized search methods. The availability of these methods might be problem dependent, e.g. for some well-studied problems such as TSP they are easily available. But for other problems, they might be difficult to obtain and as such require more development work. Nevertheless, the PAP as a problem-solving strategy still represents an improvement over base component or individual solvers in current setting whereby computational cost is decreasing.

We close this chapter with brief remarks of further research and development works that could be carried out to realize more effective automatic constructions of PAPs within the framework of coevolution. The PAP can be viewed as an ensemble of solvers that employ a perfect gating mechanism because its performance is the performance of the best solver with respect to a problem instance. Here, generalization performance of PAP can be viewed as its coverage of the problem instance space. Such a coverage can be improved and maximized by ensuring that each component solver has maximum coverage and that it has disjoint coverage with respect to all other component solvers in the PAP. The former is dependent on the baseline parameterized solver used, while the latter is dependent on the problem-solving behavioural diversity of the component solvers. The issue of improving the baseline component solvers in the PAP is to incorporate an additional mechanism to perform online tuning of the parameters associated with the solvers. One such approach could involve learning from the historical data (i.e. sampled configurations and instances) generated by automatic algorithm construction of PAPs such as CEPS and GAST to build a mapping from problem instances to their optimal parameter values. Such a learning mechanism could be viewed as an extension of the common algorithm selection systems [51]. The issue of ensuring disjoint problem instance coverage of component solvers could be addressed with mechanisms of diversity maintenance such as speciation [23] or negatively correlated search [82]. Lastly, for problem instance generation, mechanisms that measure similarities between problem instances and exploit such information to generate more dissimilar instances could be developed to more effectively generate instances that are more representative of the target problem.

References

1. Amaya, I., Ortiz-Bayliss, J.C., Conant-Pablos, S.E., Terashima-Marín, H., Coello, C.A.C.: Tailoring instances of the 1d bin packing problem for assessing strengths and weaknesses of its solvers. In: Prodeedings of the 15th International Conference on Parallel Problem Solving from Nature, PPSN'2018, pp. 373–384. Springer, Coimbra (2018)
2. Ansótegui, C., Sellmann, M., Tierney, K.: A gender-based genetic algorithm for the automatic configuration of algorithms. In: Proceedings of the 15th International Conference on Principles and Practice of Constraint Programming, CP'2009, pp. 142–157. Springer, Lisbon (2009)
3. Ansótegui, C., Malitsky, Y., Samulowitz, H., Sellmann, M., Tierney, K.: Model-based genetic algorithms for algorithm configuration. In: Proceedings of the 24th International Joint Conference on Artificial Intelligence, IJCAI'2015, pp. 733–739. AAAI Press, Buenos Aires (2015)
4. Applegate, D., Bixby, R., Chvatal, V., Cook, W.: Concorde TSP solver (2006). http://www.math.uwaterloo.ca/tsp/concorde.html
5. Asafuddoula, M., Singh, H.K., Ray, T.: An enhanced decomposition-based evolutionary algorithm with adaptive reference vectors. IEEE Trans. Cybernet. **48**(8), 2321–2334 (2018)
6. Asanovic, K., Bodík, R., Demmel, J., Keaveny, T., Keutzer, K., Kubiatowicz, J., Morgan, N., Patterson, D.A., Sen, K., Wawrzynek, J., Wessel, D., Yelick, K.A.: A view of the parallel computing landscape. Commun. ACM **52**(10), 56–67 (2009)
7. Bäck, T.: Evolutionary Algorithms in Theory and Practice: Evolution Strategies, Evolutionary Programming, Genetic Algorithms. Oxford University Press, New York (1996)
8. Back, T., Fogel, D.B., Michalewicz, Z. (eds.): Handbook of Evolutionary Computation. IOP Publishing, Bristol (1997)
9. Balint, A., Belov, A., Diepold, D., Gerber, S., Järvisalo, M., Sinz, C. (eds.): Proceedings of SAT Challenge 2012: Solver and Benchmark Descriptions, vol. B-2012-2. Department of Computer Science Series of Publications B. University of Helsinki (2012)
10. Battiti, R., Brunato, M., Mascia, F. (eds.): Reactive Search and Intelligent Optimization. Springer, Berlin (2008)
11. Bezerra, L.C.T., López-Ibáñez, M., Stützle, T.: Automatic design of evolutionary algorithms for multi-objective combinatorial optimization. In: Prodeedings of the 13th International Conference on Parallel Problem Solving from Nature, PPSN'2014, pp. 508–517. Springer, Ljubljana (2014)
12. Biere, A.: Lingeling and friends entering the SAT challenge 2012. In: Balint et al. [9], pp. 33–34 (2012)
13. Biere, A.: Splatz, Lingeling, Plingeling, Treengeling, YalSAT entering the SAT competition 2016. In: Balyo, T., Heule, M.J.H., Järvisalo, M. (eds.) Prodeedings of SAT Competition 2016: Solver and Benchmark Descriptions, vol. B-2016-1. Department of Computer Science Series of Publications B, pp. 44–45. University of Helsinki (2016)
14. Birattari, M.: On the estimation of the expected performance of a metaheuristic on a class of instances. Technical report, Technical Report TR/IRIDIA/2004-01, IRIDIA, Université Libre de Bruxelles, Brussels (2004)

15. Blot, A., Hoos, H.H., Jourdan, L., Kessaci-Marmion, M., Trautmann, H.: MO-ParamILS: a multi-objective automatic algorithm configuration framework. In: Proceedings of the 10th International Conference on Learning and Intelligent Optimization, LION'2016, pp. 32–47. Springer, Ischia (2016)
16. Bossek, J.: netgen: Network generator for combinatorial graph problems. https://github.com/jakobbossek/netgen (2015)
17. Bossek, J., Kerschke, P., Neumann, A., Wagner, M., Neumann, F., Trautmann, H.: Evolving diverse TSP instances by means of novel and creative mutation operators. In: Proceedings of the 15th ACM/SIGEVO Conference on Foundations of Genetic Algorithms, FOGA'2019, pp. 58–71. ACM, Potsdam (2019)
18. Branke, J., Pickardt, C.W.: Evolutionary search for difficult problem instances to support the design of job shop dispatching rules. Euro. J. Oper. Res. **212**(1), 22–32 (2011)
19. Burke, E.K., Gendreau, M., Hyde, M., Kendall, G., Ochoa, G., Özcan, E., Qu, R.: Hyper-heuristics: a survey of the state of the art. J. Oper. Res. Soc. **64**(12), 1695–1724 (2013)
20. Chang, C., Huang, H.: Automatic tuning of the RBF kernel parameter for batch-mode active learning algorithms: a scalable framework. IEEE Trans. Cybernet. **49**(12), 4460–4472 (2019)
21. Chen, X., Tian, Y.: Learning to perform local rewriting for combinatorial optimization. In: Proceedings of the 32ed Annual Conference on Neural Information Processing Systems, NeurIPS'2019, pp. 6278–6289. Curran Associates Inc., Vancouver (2019)
22. Chen, X., Duan, Y., Houthooft, R., Schulman, J., Sutskever, I., Abbeel, P.: InfoGAN: interpretable representation learning by information maximizing generative adversarial nets. In: Proceedings of the 29th Annual Conference on Advances in Neural Information Processing Systems, NIPS'2016, Barcelona, pp. 2172–2180 (2016)
23. Crepinsek, M., Liu, S., Mernik, M.: Exploration and exploitation in evolutionary algorithms: a survey. ACM Comput. Surv. **45**(3), 1–33 (2013)
24. Doerr, B., Neumann, F. (eds.): Theory of Evolutionary Computation: Recent Developments in Discrete Optimization. Springer, Switzerland (2020)
25. Dymond, A.S.D., Engelbrecht, A.P., Kok, S., Heyns, P.S.: Tuning optimization algorithms under multiple objective function evaluation budgets. IEEE Trans. Evol. Comput. **19**(3), 341–358 (2015)
26. Eiben, A.E., Smit, S.K.: Parameter tuning for configuring and analyzing evolutionary algorithms. Swarm Evol. Comput. **1**(1), 19–31 (2011)
27. Fogel, D.B.: Evolutionary Computation: Toward a New Philosophy of Machine Intelligence. IEEE Press, Piscataway (1995)
28. Gao, W., Nallaperuma, S., Neumann, F.: Feature-based diversity optimization for problem instance classification. In: Prodeedings of the 14th International Conference on Parallel Problem Solving from Nature, PPSN'2016, pp. 869–879. Springer, Edinburgh (2016)
29. Gomes, C.P., Selman, B.: Algorithm portfolios. Artif. Intell. **126**(1–2), 43–62 (2001)
30. Goodfellow, I.J., Pouget-Abadie, J., Mirza, M., Xu, B., Warde-Farley, D., Ozair, S., Courville, A.C., Bengio, Y.: Generative adversarial nets. In: Proceedings of the 27th Annual Conference on Advances in Neural Information Processing Systemsm, NIPS'2014, Quebec, pp. 2672–2680 (2014)
31. Han, H., Lu, W., Zhang, L., Qiao, J.: Adaptive gradient multiobjective particle swarm optimization. IEEE Trans. Cybernet. **48**(11), 3067–3079 (2018)
32. Helsgaun, K.: An effective implementation of the Lin-Kernighan traveling salesman heuristic. Euro. J. Oper. Res. **126**(1), 106–130 (2000)
33. Helsgaun, K.: General *k*-opt submoves for the Lin-Kernighan TSP heuristic. Math. Program. Comput. **1**(2–3), 119–163 (2009)
34. Hooker, J.N.: Testing heuristics: we have it all wrong. J. Heuristics **1**(1), 33–42 (1995)
35. Hoos, H.H.: Automated algorithm configuration and parameter tuning. In: Hamadi, Y., Monfroy, E., Saubion, F. (eds.) Autonomous Search, pp. 37–71. Springer, Berlin (2012)
36. Hoos, H.H.: Programming by optimization. Commun. ACM **55**(2), 70–80 (2012)
37. Hua, Y., Jin, Y., Hao, K.: A clustering-based adaptive evolutionary algorithm for multiobjective optimization with irregular Pareto fronts. IEEE Trans. Cybernet. **49**(7), 2758–2770 (2019)

38. Huang, W., Zhang, T.: Vehicle routing problem with simultaneous pick-up and delivery and time-windows based on improved global artificial fish swarm algorithm. Comput. Eng. Appl. **52**(21), 21–29 (2016)
39. Huang, C., Li, Y., Yao, X.: A survey of automatic parameter tuning methods for metaheuristics. IEEE Trans. Evol. Comput. **24**(2), 201–216 (2020)
40. Huberman, B.A., Lukose, R.M., Hogg, T.: An economics approach to hard computational problems. Science **275**(5296), 51–54 (1997)
41. Hutter, F., Hoos, H.H., Leyton-Brown, K., Stützle, T.: ParamILS: an automatic algorithm configuration framework. J. Artif. Intell. Res. **36**(1), 267–306 (2009)
42. Hutter, F., Hoos, H.H., Leyton-Brown, K.: Sequential model-based optimization for general algorithm configuration. In: Proceedings of the 5th International Conference on Learning and Intelligent Optimization, LION'2011, pp. 507–523. Springer, Rome (2011)
43. Hutter, F., López-Ibáñez, M., Fawcett, C., Lindauer, M.T., Hoos, H.H., Leyton-Brown, K., Stützle, T.: AClib: a benchmark library for algorithm configuration. In: Proceedings of the 8th International Conference on Learning and Intelligent Optimization, LION'2014, pp. 36–40. Springer, Gainesville (2014)
44. Johnson, D.S., McGeoch, L.A.: Experimental analysis of heuristics for the STSP. In: Gutin, G., Punnen, A.P. (eds.) The Traveling Salesman Problem and Its Variations, pp. 369–443. Springer, Berlin (2007)
45. Kadioglu, S., Malitsky, Y., Sellmann, M., Tierney, K.: ISAC - Instance-specific algorithm configuration. In: Proceedings of the 19th European Conference on Artificial Intelligence, ECAI'2010, pp. 751–756. IOS Press, Lisbon (2010)
46. Karafotias, G., Hoogendoorn, M., Eiben, Á.E.: Parameter control in evolutionary algorithms: trends and challenges. IEEE Trans. Evol. Comput. **19**(2), 167–187 (2015)
47. Ke, L., Zhang, Q., Battiti, R.: MOEA/D-ACO: a multiobjective evolutionary algorithm using decomposition and ant colony. IEEE Trans. Cybernet. **43**(6), 1845–1859 (2013)
48. Kerschke, P., Kotthoff, L., Bossek, J., Hoos, H.H., Trautmann, H.: Leveraging TSP solver complementarity through machine learning. Evol. Comput. **26**(4), 597–620 (2018)
49. KhudaBukhsh, A.R., Xu, L., Hoos, H.H., Leyton-Brown, K.: SATenstein: automatically building local search SAT solvers from components. Artif. Intell. **232**, 20–42 (2016)
50. Kool, W., van Hoof, H., Welling, M.: Attention, Learn to solve routing problems! In: Proceedings of the 7th International Conference on Learning Representations, ICLR'2019. OpenReview.net, New Orleans, LA (2019)
51. Kotthoff, L.: Algorithm selection for combinatorial search problems: a survey. AI Mag. **35**(3), 48–60 (2014)
52. Lin, S., Kernighan, B.W.: An effective heuristic algorithm for the traveling-salesman problem. Oper. Res. **21**(2), 498–516 (1973)
53. Lindauer, M., Hoos, H.H., Leyton-Brown, K., Schaub, T.: Automatic construction of parallel portfolios via algorithm configuration. Artif. Intell. **244**, 272–290 (2017)
54. Liu, S., Tang, K., Yao, X.: Automatic construction of parallel portfolios via explicit instance grouping. In: Proceedings of the 33rd AAAI Conference on Artificial Intelligence, AAAI' 2019, pp. 1560–1567. AAAI Press, Honolulu (2019)
55. Liu, Z., Wang, Y., Yang, S., Tang, K.: An adaptive framework to tune the coordinate systems in nature-inspired optimization algorithms. IEEE Trans. Cybernet. **49**(4), 1403–1416 (2019)
56. Liu, S., Tang, K., Lei, Y., Yao, X.: On performance estimation in automatic algorithm configuration. Proc. AAAI Conf. Artif. Intell. **34**(03), 2384–2391 (2020)
57. Liu, S., Tang, K., Yao, X.: Generative adversarial construction of parallel portfolios. IEEE Trans. Cybernet. **52**(2), 784–795 (2022)
58. López-Ibáñez, M., Dubois-Lacoste, J., Pérez Cáceres, L., Stützle, T., Birattari, M.: The irace package: iterated racing for automatic algorithm configuration. Oper. Res. Perspect. **3**, 43–58 (2016)
59. Malitsky, Y., Merschformann, M., O'Sullivan, B., Tierney, K.: Structure-preserving instance generation. In: Prodeedings of the 10th International Conference on Learning and Intelligent Optimization, LION'2016, pp. 123–140. Springer, Ischia (2016)

60. Martinez-Cantin, R.: Funneled Bayesian optimization for design, tuning and control of autonomous systems. IEEE Trans. Cybernet. **49**(4), 1489–1500 (2019)
61. Mersmann, O., Bischl, B., Trautmann, H., Wagner, M., Bossek, J., Neumann, F.: A novel feature-based approach to characterize algorithm performance for the traveling salesperson problem. Ann. Math. Artif. Intell. **69**(2), 151–182 (2013)
62. Miyato, T., Kataoka, T., Koyama, M., Yoshida, Y.: Spectral normalization for generative adversarial networks. In: Proceedings of the 6th International Conference on Learning Representations, ICLR'2018. OpenReview.net, Vancouver (2018)
63. Mohammadi, A., Asadi, H., Mohamed, S.M.K., Nelson, K., Nahavandi, S.: Multiobjective and interactive genetic algorithms for weight tuning of a model predictive control-based motion cueing algorithm. IEEE Trans. Cybernet. **49**(9), 3471–3481 (2019)
64. Moreno-Scott, J.H., Ortiz-Bayliss, J.C., Terashima-Marín, H., Conant-Pablos, S.E.: Challenging heuristics: evolving binary constraint satisfaction problems. In: Proceedings of the 12th Genetic and Evolutionary Computation Conference, GECCO'2012, pp. 409–416. ACM, Philadelphia (2012)
65. Nagata, Y., Kobayashi, S.: A powerful genetic algorithm using edge assembly crossover for the traveling salesman problem. INFORMS J. Comput. **25**(2), 346–363 (2013)
66. Nallaperuma, S., Wagner, M., Neumann, F.: Parameter prediction based on features of evolved instances for ant colony optimization and the traveling salesperson problem. In: Proceedings of the 13th International Conference on Parallel Problem Solving from Nature, PPSN'2014, pp. 100–109. Springer, Ljubljana (2014)
67. Nazari, M.R., Oroojlooy, A., Snyder, L.V., Takác, M.: Reinforcement learning for solving the vehicle routing problem. In: Proceedings of the 31st Annual Conference on Neural Information Processing Systems, NeurIPS'2018, pp. 9861–9871. Curran Associates Inc., Quebec (2018)
68. Nemhauser, G.L., Wolsey, L.A., Fisher, M.L.: An analysis of approximations for maximizing submodular set functions - I. Math. Program. **14**(1), 265–294 (1978)
69. Nguyen, S., Zhang, M., Tan, K.C.: Surrogate-assisted genetic programming with simplified models for automated design of dispatching rules. IEEE Trans. Cybernet. **47**(9), 2951–2965 (2017)
70. Peng, F., Tang, K., Chen, G., Yao, X.: Population-based algorithm portfolios for numerical optimization. IEEE Trans. Evol. Comput. **14**(5), 782–800 (2010)
71. Pu, X., Wang, K.: An evolutionary ant colony algorithm for a vehicle routing problem with simultaneous pick-up and delivery and hard time windows. In: Proceedings of the 30th Chinese Control and Decision Conference, CCDC'2018, pp. 6499–6503. IEEE, Shenyang (2018)
72. Reilly, C.H.: Synthetic optimization problem generation: show us the correlations! INFORMS J. Comput. **21**(3), 458–467 (2009)
73. Reinelt, G.: TSPLIB - A traveling salesman problem library. INFORMS J. Comput. **3**(4), 376–384 (1991)
74. Ren, Z., Jiang, H., Xuan, J., Hu, Y., Luo, Z.: New insights into diversification of hyper-heuristics. IEEE Trans. Cybernet. **44**(10), 1747–1761 (2014)
75. Rice, J.R.: The algorithm selection problem. Adv. Comput. **15**, 65–118 (1976)
76. Seipp, J., Sievers, S., Helmert, M., Hutter, F.: Automatic configuration of sequential planning portfolios. In: Bonet, B., Koenig, S. (eds.) Proceedings of the 29th AAAI Conference on Artificial Intelligence, AAAI'2015, pp. 3364–3370. AAAI Press, Austin (2015)
77. Shi, Y., Boudouh, T., Grunder, O.: An efficient tabu search based procedure for simultaneous delivery and pick-up problem with time window. IFAC-PapersOnLine **51**(11), 241–246 (2018)
78. Smith-Miles, K.: Cross-disciplinary perspectives on meta-learning for algorithm selection. ACM Comput. Surv. **41**(1), 1–25 (2008)
79. Smith-Miles, K., Bowly, S.: Generating new test instances by evolving in instance space. Comput. Oper. Res. **63**, 102–113 (2015)
80. Smith-Miles, K., van Hemert, J.I.: Discovering the suitability of optimisation algorithms by learning from evolved instances. Ann. Math. Artif. Intell. **61**(2), 87–104 (2011)

81. Smith-Miles, K., van Hemert, J.I., Lim, X.Y.: Understanding TSP difficulty by learning from evolved instances. In: Blum, C., Battiti, R. (eds.) Proceedings of the 4th International Conference on Learning and Intelligent Optimization, LION'2010, pp. 266–280. Springer, Venice (2010)
82. Tang, K., Yang, P., Yao, X.: Negatively correlated search. IEEE J. Sel. Areas Commun. **34**(3), 542–550 (2016)
83. Tang, K., Wang, J., Li, X., Yao, X.: A scalable approach to capacitated arc routing problems based on hierarchical decomposition. IEEE Trans. Cybernet. **47**(11), 3928–3940 (2017)
84. Tang, K., Liu, S., Yang, P., Yao, X.: Few-shots parallel algorithm portfolio construction via co-evolution. IEEE Trans. Evol. Comput. **25**(3), 595–607 (2021)
85. van Hemert, J.I.: Evolving combinatorial problem instances that are difficult to solve. Evol. Comput. **14**(4), 433–462 (2006)
86. Wang, H., Chen, Y.: A genetic algorithm for the simultaneous delivery and pickup problems with time window. Comput. Ind. Eng. **62**(1), 84–95 (2012)
87. Wang, C., Mu, D., Zhao, F., Sutherland, J.W.: A parallel simulated annealing method for the vehicle routing problem with simultaneous pickup-delivery and time windows. Comput. Ind. Eng. **83**, 111–122 (2015)
88. Wøhlk, S., Laporte, G.: A fast heuristic for large-scale capacitated arc routing problems. J. Oper. Res. Soc. **69**(12), 1877–1887 (2018)
89. Wotzlaw, A., van der Grinten, A., Speckenmeyer, E., Porschen, S.: pfolioUZK: solver description. In: Balint et al. [9], p. 45
90. Xie, X., Liu, J.: Multiagent optimization system for solving the traveling salesman problem (TSP). IEEE Trans. Syst. Man, Cybernet., Part B **39**(2), 489–502 (2009)
91. Xu, L., Hutter, F., Hoos, H.H., Leyton-Brown, K.: SATzilla: portfolio-based algorithm selection for SAT. J. Artif. Intell. Res. **32**, 565–606 (2008)
92. Xu, L., Hoos, H., Leyton-Brown, K.: Hydra: automatically configuring algorithms for portfolio-based selection. In: Proceedings of the 24th AAAI Conference on Artificial Intelligence, AAAI'2010, pp. 210–216. AAAI Press, Atlanta (2010)
93. Yao, X.: Evolutionary Computation: Theory and Applications. World Scientific Publication, Singapore (1999)
94. Zhao, K., Liu, S., Rong, Y., Yu, J.X.: Leveraging TSP solver complementarity via deep learning. Preprint. arXiv:2006.00715 (2020)
95. Zhong, Z., Li, J., Clausi, D.A., Wong, A.: Generative adversarial networks and conditional random fields for hyperspectral image classification. IEEE Trans. Cybernet. **50**(7), 3318–3329 (2020)

Chapter 7
Cooperative Coevolution for Large-Scale Optimization

7.1 Main Issues in Solving Large-Scale Optimization via Cooperative Coevolution

Most real-world problems such as those in engineering and science can be formulated in optimization settings. However, these optimization problems are typically complex and challenging to be solved directly. One major contributing factor to the complexity of an optimization problem is the number of decision variables [84]. Collectively, these problems can be classified as ***large-scale global optimization*** problems, and research has been actively carried out to develop methods to solve them [38, 43, 78]. Examples in Machine Learning and Data Sciences include having to optimize Deep Neural Networks that could involve over a billion variables [27, 49] and deal with learning procedures that involve very expensive optimization processes as a result of large and complex datasets [96]. Other examples include target shape design optimization for aircraft wings and turbine blades [86], satellite layout design [74], parameter estimation in large-scale systems biology models [33], seismic waveform inversion [80], and parameter calibration of water distribution system [81].

One main characteristic that makes these large-scale optimization problems exceedingly difficult to solve via iterative search procedures is in the ***curse of dimensionality*** [2], i.e. the exponential growth in the size of the search space with respect to the number of decision variables. Developments of algorithms that could scale with high-dimensional optimization problems include both approaches in classic mathematical programming [3, 23] and metaheuristics [35, 45]. The latter includes Evolutionary Algorithms (EAs) [15, 16] and other stochastic-based methods such as Differential Evolution [9, 93], Swarm Intelligence [12, 39, 95], Memetic Algorithms [34, 47, 77], and Estimation of Distribution Algorithms [17, 30], among others. These methods have gained wide interests because they solve large-scale optimization problems in a Blackbox manner (i.e. knowledge about problem structure is limited and derived only from queries to evaluate

X. Yao, S. Y. Chong, *Coevolutionary Computation and Its Applications*,
Natural Computing Series, https://doi.org/10.1007/978-981-96-2841-4_7

candidate fitness). Whether an optimization problem is considered large-scale or not depends on which family of problems it belongs to and the algorithmic procedure used to solve it. In a broad sense, a problem is large-scale if it causes scalability issues to current state-of-the-art algorithms. This chapter focuses on ***real-parameter optimization*** problems where existing algorithms exhibit scalability issues on problems with more than about a hundred decision variables. Although EAs have initially shown good promise solving high-dimensional, real-parameter optimization problems, these are made with the assumption of problem separability. EAs still suffer from the curse of dimensionality for more general large-scale global optimization problems [4, 41].

Various approaches have been developed that can be incorporated into iterative search algorithms to increase their performance to solve these challenging optimization problems. They include dimensionality reduction [7], surrogate modelling [62], local search [34, 47], and problem decomposition [88]. In particular, problem decomposition methods have been employed in classical optimization [6, 14, 22] and evolutionary optimization [42, 57]. In general, they break a large-scale optimization problem into a set of smaller subproblems, each of which is sufficiently simple so that standard iterative search procedures can optimize it. In evolutionary optimization, the framework of cooperative coevolution [58] has been proposed. It incorporates a Divide-and-Conquer problem-solving strategy with two main features: (1) subcomponent optimization—the original complete problem is decomposed into subcomponents, each of which constitutes a separate subproblem that is optimized with conventional EAs, and (2) subcomponent coadaptation—interdependencies that might exist between subcomponents are handled via a specific coevolutionary process to enable the coadaptation of candidate solutions with respect to those subcomponents. Substantial research has been devoted to develop cooperative coevolutionary algorithms (CEAs) to solve large and complex optimization problems [1, 10, 18, 31, 39, 44, 46, 51, 55, 56, 61, 64, 71, 82, 83, 88, 92, 94].

Despite early success, there remain significant technical challenges to be overcome for cooperative CEAs to be a robust problem-solving methodology applicable to a large number of high-dimensional global optimization problems. This is because cooperative coevolution relies on the right choice of problem decomposition, which could change between families of problems with different underlying problem structures. Early cooperative CEAs [42, 58, 65, 66] represent idealized implementations whereby interactions between subcomponents after problem decomposition are assumed to be minimal. The one-dimension-based strategy or approach decomposes a D-dimensional problem into D one-dimensional subcomponents. The splitting-in-half strategy decomposes a high-dimensional problem into two equal halves. Both cases involve solving highly separable problems. However, several issues related to problem decomposition have been identified at this stage of research and development. One issue is that most cooperative CEAs are applied to real-parameter optimization problems with only up to 100 dimensions, which are relatively small-scale compared to many real-world problems. Another issue includes the use of non-state-of-the-art EAs as the subcomponent optimizers despite availability of more effective and advanced EAs.

Crucially, interdependencies between different subcomponents in nonseparable problems affect the performance of cooperative CEAs. The exact nature of these interdependencies is unknown, thus making it difficult to develop mechanisms to automatically capture them and address potential performance degradation in cooperative CEAs. In [63], interdependencies between variables can greatly affect the performance of optimizers operating in continuous domains. Other relevant results could be drawn from studies in Genetic Algorithms (GAs) applied to similar problem settings. Variable interdependencies are similar to the issue of ***linkage*** (***epistasis***) on gene interaction [20, 84], the latter being investigated in the context of binary GAs where real-valued solutions are encoded in the form of binary strings (chromosomes) [11]. A similar performance-related issue is observed between early cooperative CEAs with respect to standard decomposition strategy and GAs with respect to the ordering of genes in a chromosome. In the latter, the ordering of genes could significantly impact the performance of simple GAs as a result of gene interaction [21].

The rest of this chapter is organized around a series of studies we have carried out to address main challenges applying cooperative CEAs to complex and large-scale optimization problems. In Sect. 7.2, we present our first study on cooperative CEAs to solve nonseparable functions. This is because separable functions are easily decomposed into small subcomponents and as such are not high-dimensional problems as they might initially appear. More importantly, we consider a Blackbox optimization setting whereby the variable interaction information is not available in place of the assumption of very limited knowledge about problem structure with which one could manually devise a suitable problem decomposition strategy. This means that specific mechanisms are required to identify the hidden or underlying interaction structure of the decision variables. We address this issue with a new framework of cooperative CEA based on the ***random grouping*** strategy for problem decomposition to capture the unknown variable interdependencies in nonseparable problems [87]. Furthermore, an adaptive weighting strategy is introduced to more effectively mediate coadaptation among decomposed subcomponents when they are interdependent. Our grouping and adaptive weighting strategies are attempts to optimize a group of interacting variables together as a subcomponent, which represent the ideal situation whereby each subcomponent consists of tightly interacting variables and interactions between subcomponents are minimized. We employ an advanced suboptimizer for the subcomponent based on the ***Differential Evolution with Self-adaptive Neighbourhood Search*** (***SaNSDE***). Theoretical analysis will be provided to give motivation as to why the random grouping strategy could address the issue of variable interdependency in nonseparable problems. Systematic computational studies on a large number of benchmark functions that include nonseparable problems with up to 1000 dimensions are conducted to demonstrate our framework.

The longer Sect. 7.3 presents subsequent studies [51, 52] we have made to further develop grouping strategy-based mechanisms that can automatically identify nonseparable components in high-dimensional continuous objective functions. The first part introduces the ***differential grouping*** (***DG***) strategy. We start with the theoretical foundation for identifying interacting variables that we exploit to develop a reliable mechanism that groups interacting variables and is independent of the optimizer. The latter property ensures it can be incorporated to both classical and evolutionary optimization algorithms. Systematic computational studies demonstrate that DG can provide near-optimal automatic problem decomposition. It is used with the contribution-based cooperative CEA [55] that quantifies the contribution of a subcomponent to the global fitness and exploits this information for computational budget allocation. Together, they more effectively and efficiently solve high-dimensional optimization problems up to 1000 dimensions on the CEC'2010 [73] large-scale benchmark suite [51].

The second part presents the ***improved differential grouping*** (***DG2***) that addresses four issues found in the original mechanism. They include high computational cost on fully separable functions, being unable to detect overlapping components that share decision variables (e.g. such as those encountered when solving CEC'2013 large-scale benchmark functions [53]), sensitivity to computational roundoff errors [46], and having to specify a threshold parameter ϵ. DG2 incorporates two features that address those four issues and lead to a more efficient and effective grouping strategy for automatic problem decomposition. The first feature involves a procedure for systematic generation of sample points to maximize point reuse in the identification process of interactions between decision variables. We perform a mathematical analysis to demonstrate that this procedure achieves the lower bound when the differential grouping theorem is applied to detect the interactions. As a result, the total number of objective function evaluations used to check for all pairs of variables for interactions required to identify a function with overlapping components is reduced. For example, this reduction amounts to half the total number for fully separable functions that require the most function evaluations.

The second feature involves a procedure that dynamically calculates the threshold value ϵ to detect the interaction for each pair of variables while taking into account the magnitude of roundoff errors. This is crucial for addressing sensitivity to weak interactions and thus improves on the accuracy over the mechanism in the original DG that employs fixed ϵ in solving optimization problems in the continuous domain where finite computation arithmetic becomes an issue. This is also crucial for dealing with imbalanced functions whereby the magnitude of roundoff error might be different from one component to another. The second advantage of such a feature is that it makes DG2 parameter-free with respect to the setting of ϵ. We perform error analysis that directs the implementation of this procedure. After that, systematic computational studies are carried out on the more challenging CEC'2013 benchmark suites to validate the performance improvements of DG2 to cooperative CEAs.

7.2 Random Grouping in Cooperative Coevolution

We present our first study [88] to develop a new framework for automatic problem decomposition in large-scale optimization problems and solving them through cooperative CEAs. At the core of this framework is a mechanism based on the random grouping strategy to capture the underlying variable interdependencies that could be present in problems with the specific property of nonseparability. The main working principle is that once the mechanism has correctly identified and group interdependent variables together, there would be minimal dependencies between the resulting subcomponents. Then, the subcomponents can be coadaptively optimized via cooperative CEAs. Section 7.2.1 will theoretically motivate the random grouping strategy with respect to problem decomposition of nonseparable optimization problems. Following that, Sect. 7.2.2 will present this new framework of cooperative coevolution with random grouping. Section 7.2.3 describes a particular implementation of this framework that incorporates two other features, i.e. an adaptive weighting strategy to mediate coadaptation of subcomponents with some dependencies and an advanced suboptimizer based on SaNSDE.

7.2.1 Theoretical Motivation for Random Grouping

In [57, 58], arguments have been made that problem solving in complex settings via evolutionary processes needs to explicitly introduce notions of modularity in the form of interacting coadapted subcomponents so that reasonable opportunities are provided for solutions to be effectively evolved. Other examples where modularity is taken into consideration in problem solving are rule hierarchies in classifier systems and subroutines in Genetic Programming. For evolutionary optimization, cooperative coevolution is proposed and developed as a general framework for solving large-scale problems through a coevolutionary Divide-and-Conquer approach [42, 58, 65, 66]. Its basic procedural steps are:

1. Decompose the decision vector $\boldsymbol{x}$ into m low-dimensional subcomponents $\boldsymbol{x}_i$, $i = 1, \ldots, m$.
2. Start a new coevolutionary cycle, and set $i = 1$.
3. Optimize the ith subcomponent with a suboptimizer for a predefined number of fitness evaluations (FEs).
4. If $i < m$, then $i = i + 1$, and go to Step 3.
5. Stop if termination criteria are satisfied; otherwise go to Step 2 for the next cycle.

A ***coevolutionary cycle*** consists of complete evolution of each subcomponent, which is typically performed in turn via a sequential implementation. A detailed general procedural implementation of cooperative coevolution via Algorithm 3.2 has been presented earlier in Sect. 3.2.2.

The main idea in the Divide-and-Conquer approach of cooperative coevolution is to decompose a high-dimensional optimization problem into some low-dimensional subcomponents and then evolve these subcomponents cooperatively over a predefined number of cycles. Solution representation typically takes the form of a vector of decision (design) variables $\boldsymbol{x} = (x_1, \ldots, x_D) \in \mathbb{R}^D$, which is decomposed into smaller, disjoint subcomponents $\boldsymbol{x}_i \in \mathbb{R}^{m(i)}$, $i = 1, \ldots, m$, where $2 \leq m \leq D$ and positive integers $m(i) < D$. Each subcomponent $\boldsymbol{x}_i$ is considered a species and assigned to a genetically isolated population of candidate subcomponent solutions. These populations are evolved separately since genetic isolation imposes the constraint that the evolutionary process through variation and selection operators is applied to the individual population of a subcomponent. However, the candidate solutions of a subcomponent are subjected to a coevolutionary process since their fitness evaluations are typically performed with respect to some representative candidate solutions from other subcomponents. In a typical Blackbox setting, the objective function $f(\boldsymbol{x})$ for the complete solution is available, but those for the individual subcomponents $f_i(\boldsymbol{x}_i)$ are not available.

Problem decomposition is integral to effective application of cooperative coevolution for optimization. An effective problem decomposition is one that produces subcomponents with minimal dependencies between them. This can be characterized by the separability structure of the objective function. In this chapter, we consider a minimization setting. Then, the separability structure of an objective function f is defined as follows.

Definition 7.1 ([53]) A function $f(\boldsymbol{x})$ is ***partially separable*** with m independent components if and only if

$$\operatorname*{argmin}_{\boldsymbol{x}} f(\boldsymbol{x}) = \left(\operatorname*{argmin}_{\boldsymbol{x}_1} f(\boldsymbol{x}_1, \ldots), \ldots, \operatorname*{argmin}_{\boldsymbol{x}_m} f(\ldots, \boldsymbol{x}_m) \right),$$

where $\boldsymbol{x} = (x_1, \ldots, x_D)$ is a decision vector of n dimensions, and $\boldsymbol{x}_1, \ldots, \boldsymbol{x}_m$ are disjoint subcomponents of $\boldsymbol{x}$ where $2 \leq m \leq D$.

Additive separability is a type of partial separability, which is defined as follows.

Definition 7.2 ([53]) A function is ***partially additively separable*** if it has the following general form:

$$f(\boldsymbol{x}) = \sum_{i=1}^{m} f_i(\boldsymbol{x}_i), \; m > 1,$$

where $f_i(\cdot)$ is a nonseparable function and m is the number of nonseparable subcomponents.

Early developments of cooperative CEAs have led to mixed results. For example, the original cooperative CEA [58] decomposes a D-dimensional problem into D one-dimensional subcomponents. GA-based suboptimizers are applied to populations of subcomponents in turn. Despite early success, it is shown later that it does not perform well on nonseparable benchmark problems such as Griewank and Rosenbrock [73]. Cooperative CEAs with different problem decomposition approaches based on static grouping strategies have been developed. In [5], the implementation employs Particle Swarm Optimization (PSO) and a static grouping strategy to decompose a D-dimensional problem into m s-dimensional problems where $D = ms$. In [65], the implementation employs Differential Evolution (DE) and a static grouping strategy that decomposes the decision vector into two equal-sized subcomponents. However, these implementations might not scale well with increase in problem dimensionality. Later studies have attempted to solve large-scale optimization problems up to 1000 dimensionality. In [41], Fast Evolutionary Programming is incorporated into cooperative coevolution, and computational results have shown that the coevolutionary Divide-and-Conquer approach scales better with increasing problem dimensionality on separable functions.

However, it is not known during the historical development of cooperative CEAs that employ static grouping strategies if they are effective in a Blackbox optimization setting involving nonseparable problems. The performance of these cooperative CEAs would be sensitive to the outcome of problem decomposition, i.e. whether interdependent variables are grouped together into individual subcomponents. Static grouping strategies do not provide guarantees that the right problem decomposition would be made. The resulting subcomponents with some dependencies might present search challenges to cooperative CEAs. Furthermore, outcomes of static grouping strategies are fixed, i.e. the procedure for problem decomposition is executed once and the resulting subcomponents are used throughout the cooperative coevolutionary search. The main idea of the random grouping strategy we propose in [88] is to address this issue. The general mechanism of the random grouping strategy shares some commonality with that of [5] whereby a D-dimensional problem is decomposed into m s-dimensional problems where $D = ms$. Instead of a single procedural execution of problem decomposition, the mechanism of random grouping is incorporated into the cooperative coevolutionary process whereby problem decomposition is performed at the start of each coevolutionary cycle. Furthermore, decision variables are randomly assigned to the subcomponents each time the procedure is executed.

Here, we provide a theoretical motivation why the random grouping strategy could enable cooperative coevolutionary search of nonseparable problems. First, we make one general observation about separable problems f. Let $\boldsymbol{x}, \boldsymbol{x}' \in \boldsymbol{S} \subseteq \mathbb{R}^D$, where $\boldsymbol{x} = (x_1, \ldots, x_j, \ldots, x_D), \boldsymbol{x}' = (x_1, \ldots, x_j', \ldots, x_D)$ and $f(\boldsymbol{x}) < f(\boldsymbol{x}')$ for all $j \in \{1, \ldots, D\}$. This implies that for all pairs $\boldsymbol{y}, \boldsymbol{y}' \in \boldsymbol{S}$, where $\boldsymbol{y} = (y_1, \ldots, x_j, \ldots, y_D), \boldsymbol{y}' = (y_1, \ldots, x_j', \ldots, y_D)$ with respect to the fixed values of x_j, x_j', one has $f(\boldsymbol{y}) < f(\boldsymbol{y}')$. Otherwise, the function f is nonseparable. In other words, separability of $f : \mathbb{R}^D \rightarrow \mathbb{R}$ means that the influence of any variable x_j to the fitness value $f(\boldsymbol{x})$ is independent of all other variables $x_i,\ i \neq j$. Nonseparability of f means that there are dependencies between some variables. Second, we note that no cooperative CEA could effectively solve the worst-case scenario involving a nonseparable f where all variables are highly dependent of each other.

We will show that there is a high chance a random grouping mechanism could group two interacting variables together over several coevolutionary cycles. This theoretical result provides the basis for the random grouping of interacting variables into a subcomponent. Let this search procedure be called ***cooperative CEA with random grouping*** (***CCEA-RG***). We have the following theoretical result.

Theorem 7.1 ([88]) *The probability of CCEA-RG to assign two interacting variables x_i and x_j into a single subcomponent for at least k cycles is*

$$\mathbb{P}(\#c \geq k) = \sum_{r=k}^{C} \binom{C}{r} \left(\frac{1}{m}\right)^r \left(1 - \frac{1}{m}\right)^{C-r}, \tag{7.1}$$

where C is the total number of cycles and m is the number of subcomponents.

Combinatorial arguments are employed to prove this result. For each separate cycle, the probability p of CCEA-RG to assign two interacting variables x_i and x_j into the same subcomponent is

$$p = \frac{1}{m^2}\binom{m}{1} = \frac{1}{m}. \tag{7.2}$$

Note that CCEA-RG will decompose the decision vector $\boldsymbol{x}$ for C number of times (once for each of C cycles) in a random and identical manner, whereby the outcomes of the operations are independent of each other. Let $\mathbb{P}(\#c = r)$ be the probability that CCEA-RG assigns x_i and x_j into the same subcomponent for exactly r cycles. In particular, $\mathbb{P}(\#c = r)$ satisfies the binomial distribution

$$\mathbb{P}(\#c = r) = \binom{C}{r} p^r (1-p)^{C-r} = \binom{C}{r}\left(\frac{1}{m}\right)^r \left(1 - \frac{1}{m}\right)^{C-r}.$$

Then, the probability that such an outcome occurs for at least k cycles can be calculated as $\mathbb{P}(\#c \geq k) = \sum_{r=k}^{C} \mathbb{P}(\#c = r)$.

We provide one illustrative example of Theorem 7.1. Consider an optimization problem of $D = 1000$ dimensions that is decomposed via the random grouping mechanism into $m = 10$ subcomponents, each being of $s = 100$ dimensionality. If CCEA-RG is run for $C = 50$ cycles, then we can compute the following:

$$\begin{aligned}\mathbb{P}(\#c \geq 1) &= \sum_{r=1}^{50} \mathbb{P}(\#c = r) = 1 - \mathbb{P}(\#c = 0) \\ &= 1 - \left(1 - \frac{1}{10}\right)^{50} \approx 0.9948,\end{aligned}$$

$$\mathbb{P}(\#c \geq 2) = \sum_{r=2}^{50} \mathbb{P}(\#c = r) = 1 - \mathbb{P}(\#c = 0) - \mathbb{P}(\#c = 1) \approx 0.9662.$$

Note that instead of direct calculation using the summation term, we can calculate the values of $\mathbb{P}(\#c \geq 0), \mathbb{P}(\#c \geq 1), \mathbb{P}(\#c \geq 2), \ldots$ inductively, starting with the base case of $\mathbb{P}(\#c \geq 0)$ as we have done above. We exploit the fact that the sum of probabilities of a subset of events and its complement is 1. Here, one can observe from the values of $\mathbb{P}(\#c \geq 1)$ and $\mathbb{P}(\#c \geq 2)$ that CCEA-RG has high probabilities of grouping interacting variables in the same subcomponent and then as a result optimizing the subcomponent for at least one and two cycles, respectively. In other words, the random grouping mechanism can be effective in capturing interacting variables with little use of domain knowledge [88].

7.2.2 *Cooperation Coevolution Framework with Random Grouping*

The random grouping strategy does not guarantee decomposing the problem such that there are no interdependencies between the subcomponents. Furthermore, the grouping structure changes in every coevolutionary cycle because the mechanism is implemented at the start of each cycle. As such, we introduce another mechanism of adaptive weighting strategy that is designed to provide effective coadaptation over all subcomponents when they are interdependent [88]. Its motivation is described as follows. First, we note that weights are assigned to the subcomponents that build up to form a weight vector after each coevolutionary cycle. Arguments can be formulated that for any candidate solution to the complete problem, there is an associated optimal weight vector. Finding out what this optimal weight vector is for a candidate solution is an optimization problem. Under some circumstances, this optimization problem could be of a lower dimensionality than the original high-dimensional optimization problem.

Let $\boldsymbol{w} = (w_1, \ldots, w_D) \in \mathbb{R}^D$ be the weight vector. For any $\boldsymbol{x} = (x_1, \ldots, x_D) \in \mathbb{R}^D$, we define a weighting operation $\mathrm{w}(\boldsymbol{w}, \boldsymbol{x}) = (w_1 x_1, \ldots, w_D x_D)$ written in the prefix form that performs a component-wise multiplication of its two arguments $\boldsymbol{w}$ and $\boldsymbol{x}$. We note for any $\boldsymbol{x}$ in the population, one has

$$f(\boldsymbol{x}) = f\big(\mathrm{w}(\boldsymbol{w}_c, \boldsymbol{x})\big),$$

where $\boldsymbol{w}_c = (1, \ldots, 1)$. Furthermore, better fitness values can be obtained by means of optimizing the values of the weights w_i, $i = 1, \ldots, D$, in $\boldsymbol{w}$ with respect to $\boldsymbol{x}$,

$$\min_{\boldsymbol{w}} f\big(\mathrm{w}(\boldsymbol{w}, \boldsymbol{x})\big) \leq f\big(\mathrm{w}(\boldsymbol{w}_c, \boldsymbol{x})\big) = f(\boldsymbol{x}).$$

At this point, the inequality above reexpresses the original high-dimensional optimization problem as optimizing the weight vector $\boldsymbol{w}$ with respect to $\boldsymbol{x}$, both of which have the same problem dimensionality [88]. However, one could instead take into account the problem decomposition aspect in CCEA-RG and then consider a particular subset of weight vectors that are expressed in the form of $\tilde{\boldsymbol{w}}$ with

$$\tilde{\boldsymbol{w}} = (\underbrace{\tilde{w}_1, \ldots, \tilde{w}_1}_{s}, \underbrace{\tilde{w}_2, \ldots, \tilde{w}_2}_{s}, \ldots, \underbrace{\tilde{w}_m, \ldots, \tilde{w}_m}_{s}).$$

As before, the problem decomposition involves dividing the D-dimensional problem into m subcomponents, each being an s-dimensional problem. Note that the weights associated with all the s number of elements in a subcomponent are of the same value, i.e. $\tilde{w}_i, \ldots, \tilde{w}_i,\ i = 1, \ldots, m$. Now, the problem of optimizing $\tilde{\boldsymbol{w}}$ that actually consists of only m different weight values has reduced in dimensionality. However, one now has the following inequalities:

$$\min_{\boldsymbol{w}} f\big(\mathrm{w}(\boldsymbol{w}, \boldsymbol{x})\big) \leq \min_{\tilde{\boldsymbol{w}}} f\big(\mathrm{w}(\tilde{\boldsymbol{w}}, \boldsymbol{x})\big) \leq f(\boldsymbol{x}).$$

In this manner, the adaptive weighting strategy operates as a form of trade-off between optimizing a high-dimensional weight vector $\boldsymbol{w}$ and having no weighting at all. In other words, one can view the adaptive weighting strategy as a form of coarse coadaptation over all the m subcomponents [88].

We now present our new framework of cooperative coevolution that incorporates both random grouping and adaptive weighting strategies. A general implementation of this framework is described by Algorithm 7.1, with the main procedure **CoCEAwRGRW**. Note that Algorithm 7.1 is procedurally an extended variant of the general cooperative CEA as described in Algorithm 3.2 earlier in Sect. 3.2.2. The procedural description of Algorithm 7.1 is involved but follows the general steps of cooperative coevolution we have outlined earlier in Sect. 7.2.1 to implement coevolutionary Divide-and-Conquer. The additional details include the implementation of the context vector $\boldsymbol{B}$ that represents a candidate solution of the original, full optimization problem and is crucial for the evaluation of candidate solutions with respect to the subcomponents. The two main extensions are procedures to implement the mechanisms of random grouping and adaptive weighting.

The main procedure **CoCEAwRGRW** in Algorithm 7.1 begins with the usual initializations that include those for the context vector $\boldsymbol{B}$ (lines 4–10). The specific coevolutionary process (lines 11–27) consists of three procedures that are implemented in turn. This starts with the problem decomposition and the random grouping of decision variables into their respective, disjoint subcomponents. The mechanism is implemented via the procedure **permuteRand** (line 12) that randomly permutes the indices used to address the decision variables through index[] and the associated upper and lower indices for each subcomponent calculated using s that specifies the subcomponent dimensionality (lines 14–15). Note that s is one of the input arguments of the main procedure **CoCEAwRGRW**. Then, the

Algorithm 7.1 Cooperative CEA with random grouping and random weighting [88]

Input:

D	Dimensionality of the problem, $D = ms$
s	Dimensionality of each subcomponent
m	Number of subcomponents
N	Size of population
s_{term}	Termination criterion selection
csol	Termination criterion selection for evolutionary cycle of an individual population

Output:

$\boldsymbol{B}$	Coevolved full candidate solution

```
1:  procedure CoCEAwRGRW(D, s, m, N, s_term)
2:      t := 0                                          ▷ Initialize cycle-stamp
3:      b_term := true                                  ▷ Initialize loop
4:      for i := 1 to N do
5:          initialize(X_i, D)                          ▷ Initialize X_i.x_1, ..., X_i.x_D subcomponents
6:      end for
7:      w_c := (1, ..., 1)
8:      evaluate(X_{1:N}, w_c)                          ▷ X_i.f = f(X_i.x_{1:D}), i = 1, ..., N and w_c = (1, ..., 1)
9:      b := findBest(X_{1:N})                          ▷ b = argmin_{i∈{1:N}}(X_i.f)
10:     B := X_b                                        ▷ Note B.f = min_{i∈{1:N}}(X_i.f)
11:     while b_term do
12:         permuteRand(index[], D)                     ▷ First initialize index[1] = 1, ..., index[D] = D
                                                        ▷ Then permute index[1 : D] in some random manner
13:         for j := 1 : m do
14:             l := (j − 1)s + 1
15:             u := (j)s                               ▷ l = (j − 1)s + 1 < (j − 1)s + 2 < ··· < (j − 1)s + s = u
16:             EASub(X_i.x_{index[l]:index[u]}, B, w_c)
17:             evaluate(X_{1:N}, w_c)                  ▷ X_i.f = f(X_i.x_{1:D}), i = 1, ..., N and w_c = (1, ..., 1)
18:             b := findBest(X_{1:N})                  ▷ b = argmin_{i∈{1:N}}(X_i.f)
19:             B := X_b
20:             weights(w_{index[l]:index[u]})          ▷ w_k = w, k = index[l], ..., index[u], w ~ U[0, 1]
21:         end for
22:         EAFull(X, w)                                ▷ Standard EA on complete solution X_i.x_{1:D}, i = 1, ..., N
                                                        ▷ using random weighted fitness evaluation f(X_i.x, w)
23:         b := findBest(X_{1:N})
24:         B := X_b
25:         t := t + 1                                  ▷ Increment coevolutionary cycle count
26:         b_term := termination(t, s_term)            ▷ Check for termination
27:     end while
28:     return B
29: end procedure
```

procedure for optimizing the subcomponent is implemented via **EASub**, with the context vector $\boldsymbol{B}$ being updated in turn. The coevolutionary cycle completes after all m subcomponents have been optimized (line 21). This is then followed by the procedural implementations of adaptive weighting and coadaptation of subcomponents. In particular, the weight for a subcomponent is randomly assigned via the procedure **weights** (line 20), noting that all the elements associated with a subcomponent are associated with the same weight value. Then, the procedure **EAFULL** (line 22) that performs the coadaptive optimization of the subcomponents with the generated weight vector $\boldsymbol{w}$ is called [88].

7.2.3 Differential Evolution Cooperative Coevolution with Random Grouping (DECC-G)

We will present a specific implementation of our new framework of cooperative coevolution that incorporates both mechanisms of random grouping and adaptive weighting for coadaptation of subcomponents. Furthermore, an advanced DE-based suboptimizer is employed in place of a general EA suboptimizer in Algorithm 7.1. We call our DE-based implementation as ***Differential Evolution Cooperative Coevolution with Random Grouping*** (***DECC-G***) [88]. DE [59, 69] is a simple yet effective global optimizer that has demonstrated superior performance on benchmark functions [79] and real-world problems [68, 76]. It is straightforward to develop a procedural implementation of DECC-G from the general Algorithm 7.1. As such, the focus of our presentation here is to describe specific detail implementations of the DE variant we have employed.

First, we note that the primary variation operator of DE is its specific mutation operation that adds a weighted difference vector between two individuals to a third individual in the population. Let $\boldsymbol{x}_n \in \mathbb{R}^D$, $n = 1, \ldots, N$, be the parent population, where N is the parent population size. Let $\{\boldsymbol{I}_n\}$, $n = 1, \ldots, N$, be a random sample of N selections of individuals $\boldsymbol{I}_n = (\boldsymbol{x}_i, \boldsymbol{x}_j, \boldsymbol{x}_k)_n$. For each nth selection, three distinct individuals are randomly drawn from the population with the same discrete uniform distribution, i.e. $\boldsymbol{x}_i, \boldsymbol{x}_j, \boldsymbol{x}_k \sim U\{\boldsymbol{x}_1, \ldots, \boldsymbol{x}_N\}$ and satisfying $\boldsymbol{x}_i \neq \boldsymbol{x}_j \neq \boldsymbol{x}_k \neq \boldsymbol{x}_n$. Let $\boldsymbol{y}_n \in \mathbb{R}^D$, $n = 1, \ldots, N$, be a population of individuals generated from the mutation process. The standard DE mutation operator that can be applied once to generate the nth individual is described as follows:

$$\boldsymbol{y}_n = \boldsymbol{I}_n.\boldsymbol{x}_i + F \cdot (\boldsymbol{I}_n.\boldsymbol{x}_j - \boldsymbol{I}_n.\boldsymbol{x}_k), \; n = 1, \ldots, N, \tag{7.3}$$

where $F > 0$ is a constant coefficient used to control the differential variation $\boldsymbol{d}_n = (\boldsymbol{I}_n.\boldsymbol{x}_j - \boldsymbol{I}_n.\boldsymbol{x}_k)$ [88]. Note our notation use of the period operator for direct access of members in the data structure, e.g. $\boldsymbol{I}_n.\boldsymbol{x}_i$ will access $\boldsymbol{x}_i$ in $(\boldsymbol{x}_i, \boldsymbol{x}_j, \boldsymbol{x}_k)_n$.

Next, a discrete crossover operator can be applied to increase the diversity of the values in the decision variables. Let $z_n \in \mathbb{R}^D$, $n = 1, \ldots, N$, be the child population after the mutation and crossover processes. In the following, we use the particular indexing notation $\boldsymbol{x}_n = (x_{n,1}, \ldots, x_{n,D})$ to refer to the nth individual in the population $\boldsymbol{x}_n$, $n = 1, \ldots, N$, with its associated components listed in the vector $(x_{n,l} : l = 1, \ldots, D)$. The discrete crossover operator acts on the parent individual $\boldsymbol{x}_n$ by swapping its components with those from the mutated parent individual $\boldsymbol{y}_n$ to produce a child individual z_n. The operation of discrete crossover is described as follows, whereby for all $l = 1, \ldots, D$:

$$z_{n,l} = \begin{cases} y_{n,l} & \text{if } u_l < p_r, \\ x_{n,l} & \text{otherwise,} \end{cases} \tag{7.4}$$

where $u_l \sim U(0, 1)$ is the random variable whose value is independently drawn from a uniform distribution over $[0, 1]$ for every l and $p_r \in [0, 1]$ is the crossover rate [88]. The discrete crossover operator is applied on each pair of nth individuals from the parent and mutated populations $\boldsymbol{x}_n, \boldsymbol{y}_n$, $n = 1, \ldots, N$, to produce a child population z_n, $n = 1, \ldots, N$.

The new population $\boldsymbol{x}'_n$, $n = 1, \ldots, N$, can be formed via the application of a greedy selection operator, whereby for each n:

$$\boldsymbol{x}'_n = \begin{cases} z_n & \text{if } f(z_n) < f(\boldsymbol{x}_n), \\ \boldsymbol{x}_n & \text{otherwise.} \end{cases} \tag{7.5}$$

The standard DE variant that employs a weighted difference vector for mutation, a discrete crossover operator, and greedy selection as described by (7.3)–(7.5) is denoted ***DE/rand/1/bin***. Our DE implementation [88] is based on specific extensions that have been incorporated into the standard DE. They include both neighbourhood search and self-adaptation on DE's control parameters (e.g. population size, crossover rate, and scaling factor) [8, 60].

Neighbourhood search operators have been studied and demonstrated to improve the performance of Evolutionary Programming (EP) [36, 91]. In [85], the mechanism of a generalized neighbourhood search is incorporated into DE to produce ***Neighbourhood Search DE*** (***NSDE***). In particular, NSDE extends the standard DE by replacing the mutation operator (7.3) with the following operator:

$$\begin{aligned} \boldsymbol{y}_n &= \boldsymbol{I}_n.\boldsymbol{x}_i + \boldsymbol{d}^\star_n, \ n = 1, \ldots, N, \\ \boldsymbol{d}^\star_n &= \begin{cases} \boldsymbol{d}_n \cdot r_G & \text{if } u < p_m, \\ \boldsymbol{d}_n \cdot r_C & \text{otherwise,} \end{cases} \\ \boldsymbol{d}_n &= (\boldsymbol{I}_n.\boldsymbol{x}_j - \boldsymbol{I}_n.\boldsymbol{x}_k), \end{aligned} \tag{7.6}$$

where $r_G \sim N(0.5, 0.5)$ is a random variable whose value is drawn from the Gaussian distribution with mean 0.5 and variance 0.5, $r_C \sim C(0, 1)$ is a random variable whose value is drawn from the Cauchy distribution centered at 0 with scale parameter 1, $u \sim U(0, 1)$ is a random variable whose value is drawn from a uniform distribution over $[0, 1]$, p_m is the selection probability, and $\boldsymbol{d}_n$ is the usual differential variation [88]. In general, the mutation operator in NSDE is based on the mechanism first introduced in Fast EP (FEP) [91] to replace a fixed scaling factor F with a selection of two random values drawn from different probability distributions. Computational studies over a broad range of different benchmark functions have demonstrated improvements of NSDE over the standard DE [85]. The extended mutation operator can significantly improve the diversity of NSDE's search step size and population without relying on any prior knowledge about the search space.

The second extension incorporates a self-adaptive mechanism into DE, given that the settings of optimal control parameters in DE are highly problem-dependent [19, 60], and it can be time-consuming to manually tune parameters for different problems. In [60], the ***Self-adaptive DE*** (***SaDE***) is proposed where both the crossover rate and scaling factor are adaptively controlled during the evolutionary process. The performance of SaDE has been shown to be very good on the CEC'2005 Special Session benchmark functions [70]. Subsequently, we have combined both NSDE and SaDE to develop the ***Self-adaptive NSDE*** (***SaNSDE***) [90] that improves on its two predecessors. SaNSDE employs the same neighbourhood search described in the mutation operator (7.6). However, the self-adaptive mechanism in SaDE is employed to dynamically vary the crossover rate p_r in (7.4) and the selection probability of neighbourhood search p_m in (7.6) after every fixed number of generations.

The following describes the mechanism of self-adaptation for p_m. Let ns_G and ns_C be variables that store counts of successful child individuals (i.e. those that outperform their parents) generated by a Gaussian-scaled differential variation and a Cauchy-scaled differential variation, respectively. During the selection process, nf_G and nf_C count the numbers of discarded individuals generated from Gaussian-scaled and Cauchy-scaled differential variations, respectively. Then, the setting for the selection probability p_m is calculated as follows [88]:

$$p_m = \frac{\mathrm{ns}_G \cdot (\mathrm{ns}_C + \mathrm{nf}_C)}{\mathrm{ns}_G \cdot (\mathrm{ns}_C + \mathrm{nf}_C) + \mathrm{ns}_C \cdot (\mathrm{ns}_G + \mathrm{nf}_G)}. \tag{7.7}$$

DECC-G can be procedurally developed from Algorithm 7.1. The suboptimizer we employ in DECC-G is based on SaNSDE, i.e. the DE variant with neighbourhood search and self-adaptation of control parameters (p_r, p_m). This involves replacing the call for the procedure **EASUB** with **SANSDESUB** that procedurally implements SaNSDE. At the end of each coevolutionary cycle when the process of adaptive weighted coadaptation of subcomponents is carried out, the original procedure **EAFULL** that implements a standard EA is replaced with **DEPARTIAL** that implements a standard DE. However, **DEPARTIAL** only performs adaptive weighted coadaptation of subcomponents on three distinct individuals from the child population z_n, $n = 1, \ldots, N$, that is, one with the best fitness, one with the worst fitness, and another random individual.

7.2.4 Computational Study of DECC-G

We present the results of our systematic computational studies [88] to compare and evaluate the performance of DECC-G against other cooperative CEAs such as FEPCC [42] and DECC-O [65] with SaNSDE as the baseline optimizer. We employ both the standard benchmark functions [91] and those in the CEC'2005 Special Session [70] to represent a standardized suite of large-scale optimization problems. The same number of FEs will be used on all algorithms under study as the stopping criterion to ensure fair comparison. We use the following set of problem dimensionality D and number of FEs #FEs, $\{(D, \#\text{FEs})\} = \{(100, 5 \times 10^5), (250, 1.25 \times 10^6), (500, 2.5 \times 10^6), (750, 3.75 \times 10^6), (1000, 5 \times 10^6)\}$ with a linear growth rate [42]. Other settings include SaNSDE's population size $N = 100$, subcomponent dimension $s = 100$, and the number of coevolutionary cycles in DECC-G to be $C = 50$. Each optimizer is repeatedly and independently run for 25 times with respect to a particular problem.

First, we evaluate the performance of DECC-G against the standard benchmark functions for numerical optimization [91]. In particular, the functions f_1–f_{13} in the set are scalable and applicable for use to evaluate performance of optimizers for high-dimensional problems. The first seven functions f_1–f_7 are unimodal functions, and the remaining f_8–f_{13} are multimodal functions. In particular, the number of local minima increases exponentially for the multimodal function as the problem dimensionality increases [36, 91]. Only functions f_4 and f_5 are nonseparable, while others are separable. Tables 7.1 and 7.2 summarize the average results over 25 independent runs of the optimizers we investigate. Note that the results of FEPCC are obtained from [42]. Symbol "–" means that the result is not provided, and "INF" means that no reasonable result could be found. The best mean values that are obtained among all algorithms under comparison with respect to specific test functions are marked in bold. The t-test values between DECC-G and DECC-O are also given in the last column of these tables. Figures 7.1 and 7.2 show the results of

Table 7.1 Comparison between DECC-G, SaNSDE, FEPCC, and DECC-O on functions f_1–f_7, with dimension $D \in \{500, 1000\}$. All results have been averaged over 25 independent runs. The t-value indicated by † with 24 degrees of freedom is significant at $\alpha = 0.05$ by a two-tailed t-test [88]

Test function	D	SaNSDE	FEPCC	DECC-O	DECC-G	DECC-G/O t-test
f_1	500	2.41e-11	4.90e-08	2.28e-21	**6.33e-27**	−2.06e+01†
	1000	6.97e+00	5.40e-08	1.77e-20	**2.17e-25**	−3.56e+01†
f_2	500	5.27e-02	1.30e-03	3.77e-10	**5.95e-15**	−4.73e+01†
	1000	1.24e+00	2.60e-03	INF	**5.37e-14**	–
f_3	500	2.03e-08	–	2.93e-19	**6.17e-25**	−2.15e+01†
	1000	6.43e+01	–	8.69e-18	**3.71e-23**	−2.87e+01†
f_4	500	4.07e+01	9.00e-05	60.1	**4.58e-05**	−1.60e+02†
	1000	4.99e+01	**8.50e-05**	79.2	1.01e-01	−4.85e+02†
f_5	500	1.33e+03	–	6.64e+02	**4.92e+02**	−5.23e+00†
	1000	3.31e+03	–	1.48e+03	**9.87e+02**	−9.51e+00†
f_6	500	3.12e+02	0.00e+00	0.00e+00	0.00e+00	0.00e+00
	1000	3.93e+03	0.00e+00	0.00e+00	0.00e+00	0.00e+00
f_7	500	1.28e+00	–	1.04e+01	**1.50e-03**	−6.88e+01†
	1000	1.18e+01	–	2.21e+01	**8.40e-03**	−1.39e+02†

Table 7.2 Comparison between DECC-G, SaNSDE, FEPCC, and DECC-O on functions f_8–f_{13}, with dimension $D \in \{500, 1000\}$. All results have been averaged over 25 independent runs. The t-value indicated by † with 24 degrees of freedom is significant at $\alpha = 0.05$ by a two-tailed t-test [88]

Test function	D	SaNSDE	FEPCC	DECC-O	DECC-G	DECC-G/O t-test
f_8	500	−201796.5	−209316.4	**−209491**	**−209491**	0.00e+00
	1000	−372991	−418622.6	**−418983**	**−418983**	0.00e+00
f_9	500	2.84e+02	1.43e-01	1.76e+01	**0.00e+00**	−2.30e+01†
	1000	8.69e+02	3.13e-01	3.12e+01	**3.55e-16**	−7.13e+01a
f_{10}	500	7.88e+00	5.70e-04	1.86e-11	**9.13e-14**	−3.88e+01†
	1000	1.12e+01	9.50e-04	4.39e-11	**2.22e-13**	−4.14e+01a
f_{11}	500	1.82e-01	2.90e-02	5.02e-16	**4.40e-16**	−3.93e+00†
	1000	4.80e-01	2.50e-02	2.04e-15	**1.01e-15**	−4.37e+01†
f_{12}	500	2.96e+00	–	**2.17e-25**	4.29e-21	1.75e+01†
	1000	8.97e+00	–	1.08e-24	**6.89e-25**	−8.45e+00a
f_{13}	500	1.89e+02	–	**5.04e-23**	5.34e-18	1.40e+01†
	1000	7.41e+02	–	**4.82e-22**	2.55e-21	1.05e+01†

evolutionary search by SaNSDE, DECC-O, and DECC-G. Our results indicate that DECC-G has significantly outperformed the baseline SaNSDE. The performance of DECC-G is not substantially impacted by the increase in problem dimensionality, while SaNSDE's performance has rapidly deteriorated rapidly. This confirms the advantages of cooperative CEAs over EAs for large-scale optimization problems [88].

Second, we compare the search performances of DECC-G with other cooperative CEAs. We start the comparison with FEPCC. DECC-G significantly outperformed FEPCC on most benchmark functions. Our results indicate that DECC-G is more effective in discovering and exploiting problem structures inherent in these functions that are mostly separable even though such knowledge in problem structures is not made available a priori [88]. One exception is the evolutionary search results on the nonseparable function f_4 with 1000 dimension whereby FEPCC has significantly outperformed DECC-G. We note that f_4 is a step function that is characterized by plateaus and discontinuity. The main reason why FEPCC performs better is due to its greedy fitness evaluation method that enables it to search through quickly from one plateau to a lower one [42].

Next, we compare DECC-G with DECC-O, which is a DE variant of the one-dimensional cooperative CEA that decomposes D-dimensional problems into D one-dimensional subcomponents and then optimizes these subcomponents individually. In particular, we note from our results that DECC-G has significantly better evolutionary search performance than DECC-O on nonseparable functions f_4 and f_5. Our results are consistent with earlier theoretical analysis in Sect. 7.2.1, whereby the mechanism of random grouping is able to discover and group interdependent variables in a subcomponent often enough that they can be optimized together. This does not come at significant expense to search performance on separable functions.

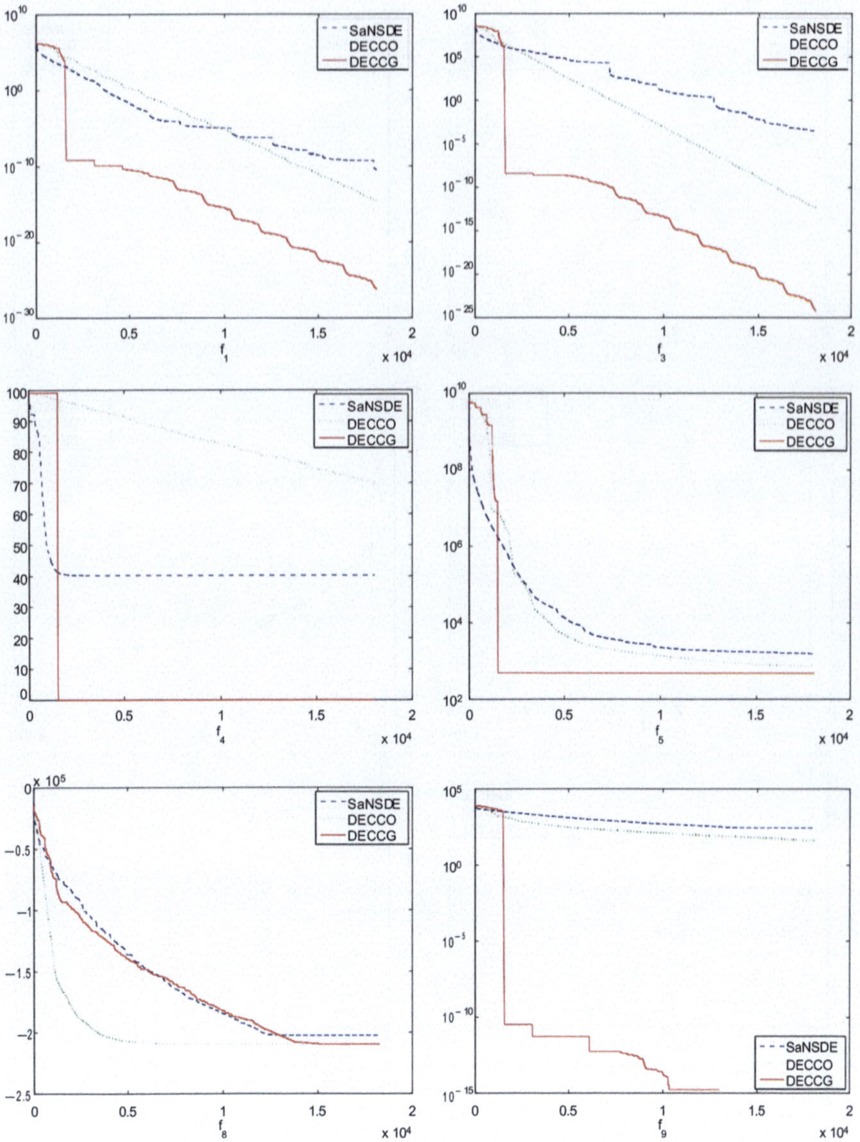

Fig. 7.1 The evolutionary search results of the best mean values over 25 runs obtained for f_1, f_3, f_4, f_5, f_8, and f_9 with dimension $D = 500$. The vertical axis is the function value, and the horizontal axis is the number of generations [88]

Our results indicate that DECC-G performs better than DECC-O for all but two functions f_{12} and f_{13}. Closer inspection of search on these two functions reveals that both DECC-G and DECC-O are able to find solutions very close to the optima [88]. One explanation is the implementation of the adaptive weighting strategy in

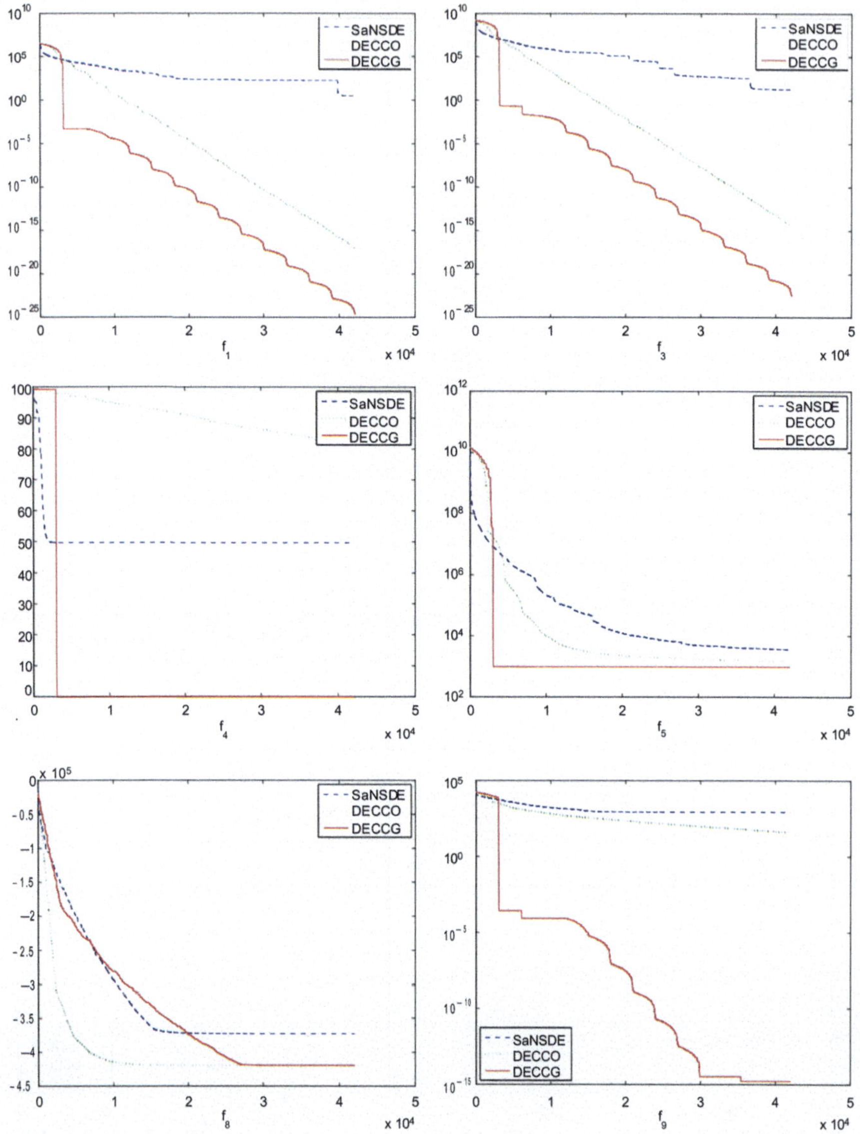

Fig. 7.2 The evolutionary search results of the best mean values over 25 runs obtained for f_1, f_3, f_4, f_5, f_8, and f_9 with dimension $D = 1000$. The vertical axis is the function value, and the horizontal axis is the number of generations [88]

DECC-G that acts as a coarse coadaptation of subcomponents, which is effective only when the global optimum is sufficiently far away from the current search point [91].

Table 7.3 Comparison between DECC-G and DECC-G-NW on functions f_1, f_4, f_5, f_8, and f_9 with dimension $D \in \{500, 1000\}$. All results have been averaged over 25 independent runs. The t-value indicated by † with 24 degrees of freedom is significant at $\alpha = 0.05$ by a two-tailed t-test [88]

Test function	D	DECC-O	DECC-G-NW	DECC-G	DECC-G/NW t-test
f_1	500	2.28e-21	1.48e-26	**6.33e-27**	−8.19e+00†
	1000	1.77e-20	**2.55e-26**	2.17e-25	1.65e+01†
f_4	500	6.01e+01	5.28e+01	**4.58e-05**	−6.78e+01†
	1000	7.92e+01	8.30e+01	**1.01e-01**	−1.69e+02†
f_5	500	6.64e+02	1.27e+03	**4.92e+02**	−1.94e+01†
	1000	1.48e+03	2.63e+03	**9.87e+02**	−3.05e+01†
f_8	500	−209491	−209491	−209491	0.00e+00
	1000	−418983	−418983	−418983	0.00e+00
f_9	500	1.76e+01	2.21e+02	**0.00e+00**	−1.17e+02†
	1000	3.12e+01	4.56e+02	**3.55e-16**	−1.17e+02†

We have conducted additional computational experiments to assess the effectiveness of the proposed adaptive weighting strategy. In particular, we compare DECC-G against its variant without the adaptive weighting strategy (DECC-G-NW). Table 7.3 summarizes the results that are representative of the following discussion. The performance of DECC-G is either significantly better than or comparable to the performance of DECC-G-NW. The single exception involves optimizing the function f_1 at $D = 1000$. Despite statistically significant difference in the result that favours DECC-G-NW, the performances of both cooperative CEAs are close to zero that it would have been negligible in practice. More importantly, results on the nonseparable functions f_4 and f_5 suggest that the incorporation of the adaptive weighting strategy in DECC-G has led to improved search performance over DECC-G-NW. The mechanism of adaptive weighting could effectively coadapt subcomponents with some interdependencies between them [88].

Given that the standard benchmark functions for numerical optimization [91] are predominantly separable, we have employed the CEC'2005 Special Session set of benchmark functions [70] for further computational evaluations of DECC-G. This latter set includes 25 functions of varying complexity, of which the first five are unimodal while the remaining 20 functions are multimodal. Many of them are shifted, rotated, expanded, and/or combined variants of the classical functions. Rotation transforms separable functions into nonseparable ones, while other changes made them more challenging for search. Furthermore, all these functions are scalable. As such, the CEC'2005 Special Session benchmark set is suitable for standardized search performance evaluation of cooperative CEAs for more realistic and complex, large-scale optimization problems. We use only eight representative functions due to very high computational cost requirements. Six of these functions are nonseparable, while two functions $f_{\text{cec}1}$ and $f_{\text{cec}9}$ are separable.

Results summarized in Table 7.4 indicate that DECC-G performs better than the baseline SaNSDE on all functions except $f_{\text{cec}8}$ where the search performances of

Table 7.4 Comparison between DECC-G, SaNSDE, and DECC-O on eight representative functions in CEC'2005 Special Session set with dimension $D \in \{500, 1000\}$. All results have been averaged over 25 independent runs. The t-value indicated by † with 24 degrees of freedom is significant at $\alpha = 0.05$ by a two-tailed t-test [88]

CEC'05 function	D	SaNSDE	DECC-O	DECC-G	DECC-G/O t-test
f_{cec1}	500	2.61e-11	1.04e-12	**3.71e-13**	−2.87e+01†
	1000	1.17e+00	3.66e-08	**6.84e-13**	−2.96e+01†
f_{cec3}	500	6.88e+08	4.78e+08	**3.06e+08**	−1.13e+01†
	1000	2.34e+09	1.08e+09	**8.11e+08**	−1.45e+01†
f_{cec5}	500	4.96e+05	2.40e+05	**1.15e+05**	−7.72e+01†
	1000	5.03e+05	3.73e+05	**2.20e+05**	−4.12e+01†
f_{cec6}	500	2.71e+03	1.71e+03	**1.56e+03**	−2.74e-01
	1000	1.35e+04	3.13e+03	**2.22e+03**	−1.10e+01†
f_{cec8}	500	2.15e+01	**2.14e+01**	2.16e+01	2.72e+01†
	1000	2.16e+01	**2.14e+01**	2.16e+01	6.11e+01†
f_{cec9}	500	6.60e+02	**8.66e+00**	4.50e+02	9.09e+01†
	1000	3.20e+03	**8.96e+01**	6.32e+02	1.39e+02†
f_{cec10}	500	6.97e+03	1.50e+04	**5.33e+03**	−5.95e+01†
	1000	1.27e+04	3.18e+04	**9.73e+03**	−2.95e+01†
f_{cec13}	500	2.53e+02	**2.81e+01**	2.09e+02	7.72e+01†
	1000	6.61e+02	**7.52e+01**	3.56e+02	5.00e+01†

both optimizers are similar. On functions where there is a performance difference in favour of DECC-G compared to SaNSDE, it becomes larger when the problem dimensionality is increased from 500 to 1000. This suggests better scalability of DECC-G compared to SaNSDE. We next compare the search performances of DECC-G and DECC-O. In particular, DECC-G significantly outperforms DECC-O on five out of eight functions. Figures 7.3 and 7.4 indicate that the evolutionary search process in DECC-G converges to the best average results faster than that of the other two optimizers on the majority of the functions (i.e. f_{cec1}, f_{cec3}, f_{cec5}, and f_{cec6}). In general, these results suggest that DECC-G is effective and efficient in solving large-scale optimization problems, including those that are nonseparable. We have performed close inspection on results where DECC-G is numerically outperformed. For f_{cec8}, our analysis reveals that all three optimizers obtain similar results corresponding to one of the local optima that is far away from the global optimum. As such, the function f_{cec8} represents a deceptive problem whereby all the optimizers are unable to solve. The function f_{cec9} is separable, and our results corroborate with the expectation that DECC-O would perform better than DECC-G. The odd exception is the result on the nonseparable function f_{cec13}, whereby DECC-G is outperformed by DECC-O. However, f_{cec13} is a composite of two different functions, making it difficult to establish the precise reason why the mechanisms of random grouping and adaptive weighting are ineffective [88].

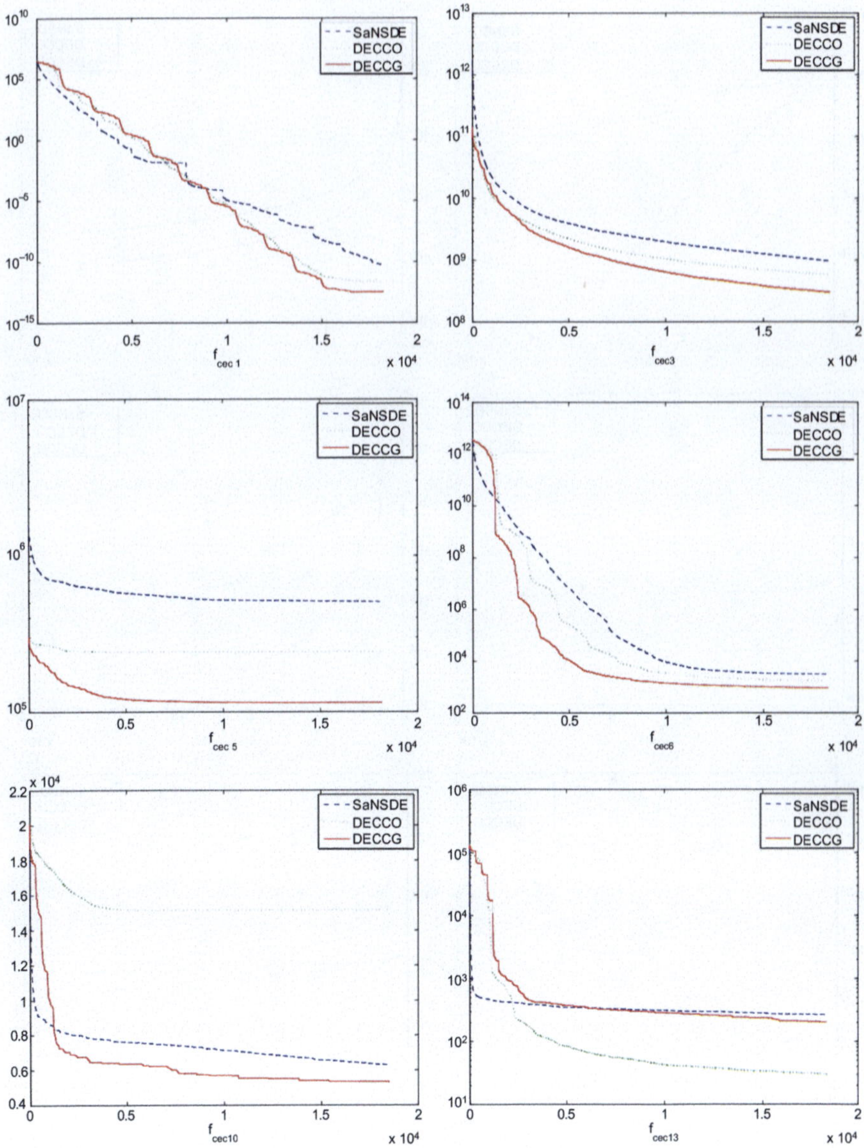

Fig. 7.3 The evolutionary search results of the best mean values over 25 runs obtained for f_{cec1}, f_{cec3}, f_{cec5}, f_{cec6}, f_{cec10}, and f_{cec13} with dimension $D = 500$. The vertical axis is the function value, and the horizontal axis is the number of generations [88]

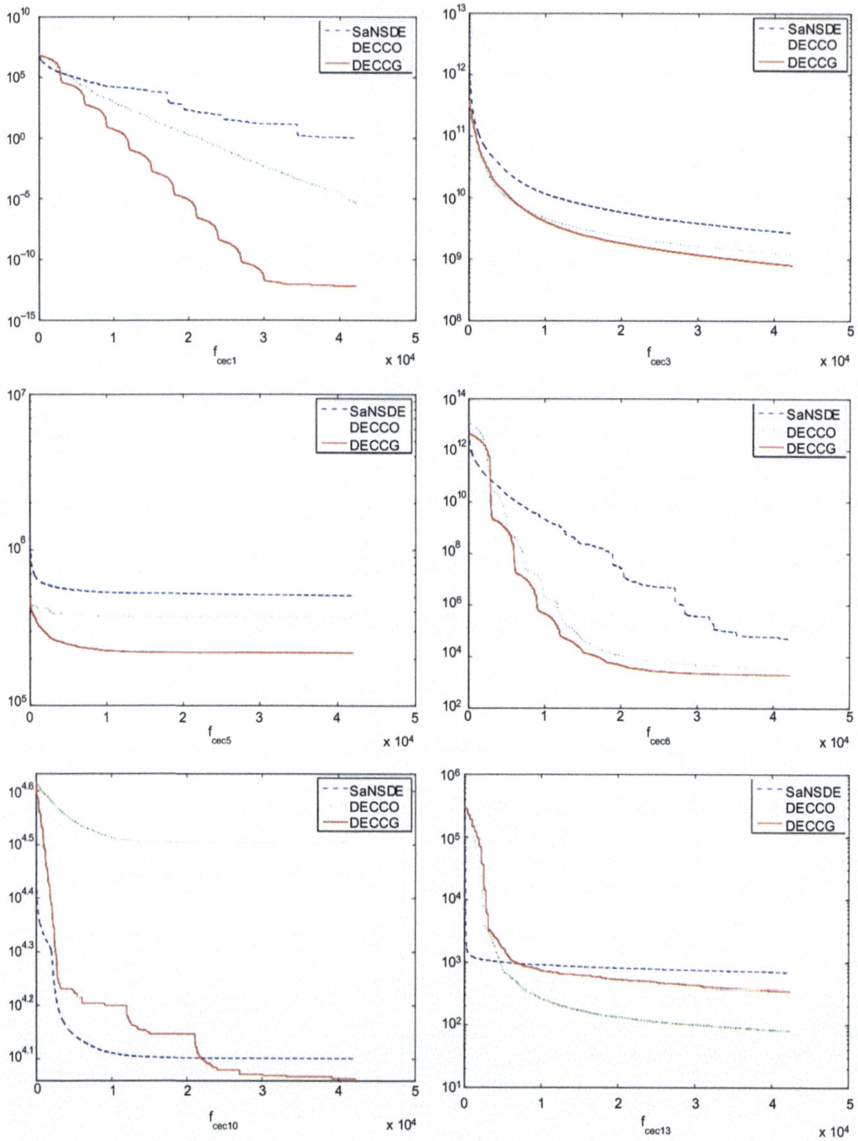

Fig. 7.4 The evolutionary search results of the best mean values over 25 runs obtained for f_{cec1}, f_{cec3}, f_{cec5}, f_{cec6}, f_{cec10}, and f_{cec13} with dimension $D = 1000$. The vertical axis is the function value, and the horizontal axis is the number of generations [88]

7.3 Differential Grouping in Cooperative Coevolution

We present our two subsequent studies [51, 52] to improve the framework of cooperative coevolution with automatic problem decomposition in large-scale optimization problems. Various decomposition mechanisms have been proposed for cooperative CEAs. The simplest approach extends the original framework of cooperative coevolution [58] and employs a static grouping strategy to decompose D-dimensional problems into m s-dimensional subcomponents by arbitrary grouping of s decision variables [5, 65]. However, this approach would not capture possible underlying variable interactions and significantly improve search performances of iterative optimizers [88]. Other approaches with more procedural complexity involve those that employ heuristics for problem decomposition. Section 7.2 has introduced the mechanism of random grouping [88]. Other studies have proposed procedures based on variable interaction characteristics of the objective function that include adaptive variable partitioning [61], delta grouping [54], and min/max variance decomposition [40]. However, these heuristics-based methods could suffer from low grouping accuracy due to arbitrary setting of the fixed number m and size s of subcomponents. Improved variants for random and delta grouping employ multiple fixed decompositions throughout the evolutionary search process [54, 89].

The more sophisticated decomposition approach involves methods that detect interactions between variables such as variable interaction learning [10], meta-modelling decomposition [44], and statistical learning decomposition [71]. In this section, we present our DG method for automatic detection of interacting variables via differential grouping [51] to address the main issue of arbitrary setting of fixed number and size of subcomponents. Section 7.3.1 presents our theoretical studies on additively separable functions that we subsequently exploit to construct a procedural implementation of the DG mechanism that automatically identifies and groups interacting variables and is independent of the optimizer. Section 7.3.2 presents systematic computational studies to evaluate the impact of DG on the search performance of cooperative CEAs with newer CEC'2010 [73] large-scale benchmark suite [51] up to 1000 dimensions. Despite good overall search performance, there are four issues with the DG mechanism that include high computational cost on fully separable functions, unable to detect overlapping components that share decision variables, sensitivity to computational roundoff errors, and having to specify a threshold parameter ϵ. We next present a subsequent study [52] to develop an improved differential grouping mechanism (DG2) that addresses all these four issues via two new features: (1) a procedure that systematically generates sample points maximizing point reuse in the identification process of interactions between decision variable, and (2) a procedure that dynamically calculates the threshold value ϵ to detect the interaction for each pair of variables while taking into account the magnitude of roundoff errors. Section 7.3.3 starts with theoretical studies we have carried out that provide results we can specifically exploit to design these two procedures. Section 7.3.4 presents systematic computational studies on the more challenging CEC'2013 benchmark suites to validate improvements DG2 offers to cooperative CEAs.

7.3.1 Theoretical Studies and Procedural Construction of Differential Grouping

For the rest of this chapter, we use the term additively separable to refer to partially additively separable property of a function. The following theoretical result gives a statement on the sufficient condition for a function $f : \mathbb{R}^D \rightarrow \mathbb{R}$ to be additively separable.

Theorem 7.2 ([51]) *Let $f(\boldsymbol{x})$ be an additively separable function. For all $a, b_1, b_2, \delta \in \mathbb{R}$ satisfying $b_1 \neq b_2$ and $\delta \neq 0$, if the following condition holds:*

$$\Delta_{\delta,x_p}[f](\boldsymbol{x})|_{x_p=a,x_q=b_1} \neq \Delta_{\delta,x_p}[f](\boldsymbol{x})|_{x_p=a,x_q=b_2}, \tag{7.8}$$

where

$$\Delta_{\delta,x_p}[f](\boldsymbol{x}) = f(\ldots, x_p + \delta, \ldots) - f(\ldots, x_p, \ldots) \tag{7.9}$$

is the forward difference of f with respect to variable x_p with interval δ, then x_p and x_q are said to be nonseparable.

In other words, Theorem 7.2 states that for an additively separable function $f(\boldsymbol{x})$, two decision variables x_p and x_q interact whenever the forward difference of f with respect to x_p given by (7.9) yields two different results for any two different values $b_1 \neq b_2$ of x_q. In summary, the inequality of Δ values calculated at $x_p = a, x_q = b_1$ and $x_p = a, x_q = b_2$ for different b_1, b_2 values implies nonseparability of x_p and x_q. The proof argument for Theorem 7.2 relies on proving the contrapositive of its main statement, i.e. if two decision variables x_p and x_q are separable, then Eq. (7.9) that is calculated for any two different values of x_q will yield the same Δ values [51]. The proof argument is contained in the following theoretical result:

Lemma 7.1 ([51]) *If $f(\boldsymbol{x})$ is additively separable according to Definition 7.2 with m disjoint subcomponents represented by subcomponents $\boldsymbol{x}_i$, $i = 1, \ldots, m$, then for all x_p in any $\boldsymbol{x}_i$, we have*

$$\frac{\partial f(\boldsymbol{x})}{\partial x_p} = \frac{\partial f_i(\boldsymbol{x}_i)}{\partial x_p}. \tag{7.10}$$

We sketch the proof arguments presented in [51]. For an additively separable $f(\boldsymbol{x})$ with m disjoint subcomponents given by $\boldsymbol{x}_i$, $i = 1, \ldots, m$, we have for all x_p

in any $\boldsymbol{x}_i$

$$\begin{aligned}\frac{\partial f(\boldsymbol{x})}{\partial x_p} &= \frac{\partial \sum_{i=1}^{m} f_i(\boldsymbol{x}_i)}{\partial x_p} \\ &= \frac{\partial f_1(\boldsymbol{x}_1)}{\partial x_p} + \cdots + \frac{\partial f_m(\boldsymbol{x}_m)}{\partial x_p} \\ &= \frac{\partial f_i(\boldsymbol{x}_i)}{\partial x_p},\end{aligned}$$

given that $\frac{\partial f(\boldsymbol{x}_j)}{\partial x_p} = 0$ for any $j \neq i$. From Lemma 7.1, one can obtain the following equalities by direct calculation. For all x_q not in $\boldsymbol{x}_i$ we have

$$\begin{aligned}\left.\frac{\partial f(\boldsymbol{x})}{\partial x_p}\right|_{x_q=b_1} &= \left.\frac{\partial f(\boldsymbol{x})}{\partial x_p}\right|_{x_q=b_2} = \frac{\partial f_i(\boldsymbol{x}_i)}{\partial x_p}, \ \forall\, b_1 \neq b_2, \\ \left.\int_a^{a+\delta} \frac{\partial f(\boldsymbol{x})}{\partial x_p} dx_p\right|_{x_q=b_1} &= \left.\int_a^{a+\delta} \frac{\partial f(\boldsymbol{x})}{\partial x_p} dx_p\right|_{x_q=b_2}, \\ \Delta_{\delta,x_p}[f](\boldsymbol{x})|_{x_p=a,x_q=b_1} &= \Delta_{\delta,x_p}[f](\boldsymbol{x})|_{x_p=a,x_q=b_2}, \\ &\forall\, a, b_1 \neq b_2, \delta \in \mathbb{R},\ \delta \neq 0.\end{aligned}$$

In the following example, we illustrate how our theoretical results can be applied to directly assess separability property of an objective function $f : \mathbb{R}^D \rightarrow \mathbb{R}$ representing an optimization problem. Consider a nonseparable objective function $f(x_1, x_2) = x_1^2 + \lambda x_1 x_2 + x_2^2$, $\lambda \neq 0$. In this case, we can apply Eq. (7.10) and calculate the following:

$$\frac{\partial f(x_1, x_2)}{\partial x_1} = 2x_1 + \lambda x_2.$$

This analytical result indicates that the change in the global objective function $f(x_1, x_2)$ with respect to x_1 is a function of x_1 and x_2. Since the function f represents an optimization problem with two decision variables $\boldsymbol{x} = (x_1, x_2)$, the result is sufficient for one to claim that both decision variables associated with f are nonseparable. In practice and for large-scale optimization problems where it might not be straightforward to perform such direct calculations, one would instead make use of Theorem 7.2. For this example, we are able to perform direct calculation via Eq. (7.9) to obtain the following analytical result:

$$\begin{aligned}\Delta_{\delta,x_1}[f] &= \left[(x_1+\delta)^2 + \lambda(x_1+\delta)x_2 + x_2^2\right] - \left[x_1^2 + \lambda x_1 x_2 + x_2^2\right] \\ &= \delta^2 + 2\delta x_1 + \lambda x_2 \delta.\end{aligned}$$

Here, the forward difference $\Delta_{\delta,x_1}[f]$ is given by a function of both x_1 and x_2. In practice, one would perform direct computations of $\Delta_{\delta,x_1}[f]$ for two different values of x_2 and obtain different answers. Then, one would apply Theorem 7.2 and claim that x_1 and x_2 interact, i.e. they are nonseparable. Note that $\lambda > 0$ reflects the strength of nonseparability. Setting $\lambda = 0$ makes the function fully separable [51].

Algorithm 7.2 Differential grouping [51]

Input:

D	Dimensionality of the problem
f	Optimization problem
lbound	Lower bound
ubound	Upper bound
ϵ	Interaction sensitivity parameter

Output:

G	Final set of all identified groups

```
procedure DGROUPING(D, f, lbound, ubound, ϵ)
    initDim(d, D)            ▷ Initialize decision variable indexing d := {1, ..., D}
    initSep(s)               ▷ Initialize variable-length s := {} of separable variable indices
    initGroup(G[])           ▷ Initialize variable-length structure G[] := {∅} of a set of all groups
    for i ∈ d do
        g := {i}                                  ▷ Initialize group as g = {i}
        for j ∈ d ∧ j ≠ i do
            initComp(y1, lbound)                  ▷ Initialize y1 := {lbound, ..., lbound}
            y2 := y1
            y2[i] := ubound
            Δ1 := evalF(f, y1) − evalF(f, y2)     ▷ evalF(f, y) returns f(y) value
            y1[j] := 0
            y2[j] := 0
            Δ2 := evalF(f, y1) − evalF(f, y2)
            if |Δ1 − Δ2| > ϵ then
                updateComp(g, j)                  ▷ g = g ∪ j
            end if
        end for
        removeVar(d, g)                           ▷ d = d\g
        if length(g) = 1 then                     ▷ length(g) returns array length of g
            updateSep(s, g)                       ▷ s = s ∪ g
        else
            updateGroup(G, g)                     ▷ G = G ∪ {g}
        end if
    end for
    updateGroup(G, s)                             ▷ G = G ∪ {s}
    return G
end procedure
```

As such, Theorem 7.2 can be directly applied to construct the DG mechanism to identify and group the interacting variables into common subcomponents. Algorithm 7.2 presents a procedural implementation of the DG mechanism. After initializations (e.g. a variable-length s that is used to store indices of identified separable variables), the main procedure **DGROUPING** begins examining the interaction of the first decision variable with all other decision variables in a pairwise manner by applying Theorem 7.2 (lines 5–14). This requires that all the variables in the decision vector $\boldsymbol{y}_1$ are initialized to their lower bounds (line 8). In order to check for interaction between the ith and jth decision variables, the vector $\boldsymbol{y}_2$ is set to be equal to $\boldsymbol{y}_1$ except for the ith dimension (lines 9–10). Then, the ith element of $\boldsymbol{y}_2$ is set to the upper bound with respect to the dimension of the domain (line 10). Now, the DG procedure can compute the value of Δ_1 (line 11). Next, the jth element of $\boldsymbol{y}_2$ is set to the center of the search space for that particular problem dimension (lines 12–13), and Δ_2 is computed (line 14). If the quantity $|\Delta_1 - \Delta_2|$ is greater than a small number ϵ, then Theorem 7.2 is applied to conclude that the ith and jth variables interact with each other. In other words, an interaction between the current variable i and another variable $j \neq i$ is detected (line 15). Then, the jth variable is grouped together with the ith variable and placed in a common subcomponent $\boldsymbol{g}$ (line 16) and removed from the set of all decision variables $\boldsymbol{d}$ (line 19) [51].

This process is repeated until all the variables that interact with the current ith variable are detected, and the subcomponent is formed and extracted (lines 22–23), e.g. the first subcomponent. If no interaction is detected, then the variable i under examination is considered to be a separable variable (lines 20–21). Note that separable decision variables are placed into a single separable group. The other alternative is to place separable decision variables in individual and separate subcomponents. We have implemented the former design choice because studies such as [5, 65] have demonstrated that the latter design choice might not necessarily constitute an optimal grouping arrangement in terms of both efficiency and effectiveness of cooperative CEAs to solve a large-scale fully separable problem. The DG algorithm then repeats the process to identify all variables interacting with the $(i + 1)$th dimension until no more dimensions can be pairwise examined (line 25). It should be noted that the choices of upper bound, lower bound, and the center of the search space to construct $\boldsymbol{y}_1$ and $\boldsymbol{y}_2$ are arbitrary. Other points can be used so long as they do not coincide with each other to give a difference value of zero. As such, the parameter ϵ in Algorithm 7.2 affects the sensitivity of the DG mechanism in detecting interactions between the variables. A smaller ϵ setting makes DG more sensitive to very weak interactions between the decision variables. We note the generality of Theorem 7.2. Not only it is applied to construct the DG mechanism as procedurally implemented in Algorithm 7.2, but it can also be applied to provide theoretical basis of other perturbation methods such as LINC-R [75]. Details of such an analysis are provided in [51].

The rest of this subsection will be devoted to providing an analysis of the DG algorithm's computational cost. In particular, we calculate an upper bound for the total number of FEs required by the DG algorithm under the assumption that there are $\frac{D}{s}$ nonseparable subcomponents, each with s variables. In this particular setting, s variables are removed from the set of remaining decision variables after each successful DG application by Algorithm 7.2. Here, one can calculate an upper bound S of the number of times the inner loop of Algorithm 7.2 is executed using the sum of an arithmetic progression [51]:

$$\begin{aligned} S &= (D-1) + (D-s-1) + \cdots + \left(D - \left(\frac{D}{s} - 1\right)s - 1\right) \\ &= (D-1) + (D-s-1) + \cdots + (s-1) \\ &= \frac{D}{2s}(D+s-2). \end{aligned} \tag{7.11}$$

Note that a perfect grouping will require a total of $4S$ fitness evaluations given that there are four fitness evaluations in the inner loop of Algorithm 7.2 (lines 11 and 14). However, Algorithm 7.2 can be optimized by exploiting the fact that Δ_1 is unchanged during the execution of the algorithm's inner loop, and thus computation for Δ_1 can be moved outside of this inner loop. Therefore, the total number of required fitness evaluations reduces to $2\left(S + \frac{D}{s}\right)$ [51]. In the following example, we illustrate the reduction in fitness evaluations from the case of a fully separable function to that of a partially separable function using (7.11) in applying this optimized Algorithm 7.2. In particular, at $D = 1000, s = 1$ a fully separable function requires 1001000 FEs, while a partially separable function at $D = 1000, s = 50$ requires 21000. It is straightforward to calculate the time complexity of the DG algorithm with respect to the maximum number of fitness evaluations as follows [51]:

$$O(\text{FE}) = O\left(2\left(S + \frac{D}{s}\right)\right) = O\left(\frac{D^2}{s}\right). \tag{7.12}$$

7.3.2 *Computational Studies on Cooperative Coevolution with Differential Grouping*

We present our systematic computational studies [51] to evaluate the performance of cooperative CEA with DG via standardized benchmark functions. In particular, Algorithm 7.1 can be modified by replacing the random grouping mechanism **permuteRand** with DG mechanism **DGROUPING**. Note that the mechanism of adaptive weighting and coadaptation of subcomponents is not implemented since **DGROUPING** explicitly detects interacting variables. The cooperative CEA we employ with DG is a variant of DECC-G whereby the suboptimizer is SaNSDE [90]. We will refer to our specific implementation as ***Differential Evolution Cooperative Coevolution with Differential Grouping*** (***DECC-DG***). We set the population size $N = 50$ and the maximum number of fitness evaluations #FE $= 3 \times 10^6$.

The experiment involving a particular setting of an algorithm to solve a specific problem is repeated 25 independent runs for statistical analysis. The default setting of $\epsilon = 10^{-3}$ is used. Other values $\{10^{-1}, 10^{-6}\}$ are investigated to examine the impact of sensitivity of DG with respect to ϵ for search performance [51].

The first half of this subsection involves a systematic analysis on the effectiveness and efficiency of the DG mechanism in DECC-DG. This involves comparison with a baseline cooperative CEA that also performs automatic identification and grouping of interacting variables. In particular, our choice of comparison involves the ***Cooperative Coevolution with Variable Interaction Learning*** (***CCVIL***) [10]. In CCVIL, two variables x_i and x_j of a decision vector $\boldsymbol{x}$ are said to interact if the following condition holds:

$$\begin{aligned} &\exists\, \boldsymbol{x}, x_i', x_j' : \\ &f(x_1, \ldots, x_i, \ldots, x_j, \ldots, x_n) < f(x_1, \ldots, x_i', \ldots, x_j, \ldots, x_n) \wedge \\ &f(x_1, \ldots, x_i, \ldots, x_j', \ldots, x_n) > f(x_1, \ldots, x_i', \ldots, x_j', \ldots, x_n), \end{aligned} \tag{7.13}$$

where x_i', x_j' represent alternative values to x_i, x_j with respect to ith and jth decision variables, respectively. The grouping mechanism in CCVIL initially places each variable in a separate subcomponent. An iterative process is applied whereby subcomponents with interacting variables x_i and x_j are repeatedly merged based on the condition (7.13) until one of the termination criteria is met (note: there are three termination conditions, one of which involves setting a fixed number of cycles for learning or discovery of variable interactions) [10]. The main difference in detection of pairwise interaction of variables of CCVIL compared to DG is that the former takes advantage of the one-dimensional cooperative coevolution process [57] that optimizes one decision variable individually as the procedural implementation for interaction detection of variables based on (7.13).

We use the specific set from the IEEE CEC'2010 special session on large-scale global optimization [73] to compare DECC-DG and CCVIL. Functions from this benchmark set can be classified into five groups:

1. Separable functions (f_1–f_3)
2. Single-group s-nonseparable functions (f_4–f_8)
3. $\frac{D}{2s}$-group s-nonseparable functions (f_9–f_{13})
4. $\frac{D}{s}$-group s-nonseparable functions(f_{14}–f_{18})
5. Nonseparable functions (f_{19}–f_{20})

We consider benchmark problems with setting of $D = 1000$ problem dimensionality and $s = 50$ variables in each nonseparable subcomponent for the following computational studies.

First, we present experimental results to assess the grouping performances of DECC-DG and CCVIL. Entries in Table 7.5 associated with the two cooperative CEAs are separated by the symbol "/". The last column gives grouping accuracies of nonseparable variables for both algorithms. Double horizontal lines separate

Table 7.5 Performance of DECC-DG and CCVIL on CEC'2010 benchmark functions (separated by "/") [51]

Function	Sep. Vars	Nonsep. Vars	Nonsep. Groups	DECC-DG ($\epsilon = 10^{-3}$) / CCVIL # Captured Sep Vars	# Captured Nonsep Vars	# Formed Nonsep Groups	# Misplaced Vars	# FE	Grouping accuracy
f_1	1000	0	0	1000 / 1000	0 / 0	0 / 0	0 / 0	1001000 / 69990	100% / 100%
f_2	1000	0	0	1000 / 1000	0 / 0	0 / 0	0 / 0	1001000 / 69990	100% / 100%
f_3	1000	0	0	1000 / 938	0 / 0	0 / 0	0 / 31	1001000 / 1798666	100% / 93.8%
f_4	950	50	1	33 / 957	50 / 43	10 / 1	0 / 7	14564 / 1797614	100% / 86%
f_5	950	50	1	950 / 950	50 / 50	1 / 1	0 / 0	905450 / 1795705	100% / 100%
f_6	950	50	1	950 / 910	50 / 47	1 / 22	0 / 3	906332 / 1796370	100% / 94%
f_7	950	50	1	247 / 951	34 / 49	4 / 1	16 / 1	67250 / 1796475	69% / 98%
f_8	950	50	1	135 / 1000	46 / 0	5 / 0	4 / 50	23608 / 69842	92% / 0%
f_9	500	500	10	500 / 583	500 / 337	10 / 33	0 / 0	270802 / 1792212	100% / 67.4%
f_{10}	500	500	10	500 / 508	500 / 492	10 / 10	0 / 8	272958 / 1774642	100% / 98.4%
f_{11}	500	500	10	501 / 476	499 / 491	10 / 26	1 / 9	270640 / 1774565	99.2% / 98.2%
f_{12}	500	500	10	500 / 516	500 / 435	10 / 11	0 / 65	271390 / 1777344	100% / 87%
f_{13}	500	500	10	131 / 1000	126 / 0	40 / 0	374 / 500	49470 / 69990	25.2% / 0%
f_{14}	0	1000	20	0 / 150	1000 / 719	20 / 63	0 / 281	21000 / 1785975	100% / 71.9%
f_{15}	0	1000	20	0 / 18	1000 / 982	20 / 20	0 / 18	21000 / 1751241	100% / 98.2%
f_{16}	0	1000	20	4 / 11	996 / 989	20 / 20	4 / 11	21128 / 1751647	99.6% / 98.9%
f_{17}	0	1000	20	0 / 25	1000 / 975	20 / 20	0 / 25	21000 / 1752340	100% / 97.5%
f_{18}	0	1000	20	85 / 1000	173 / 0	49 / 0	827 / 1000	34230 / 69990	17.3% / 0%
f_{19}	0	1000	1	0 / 0	1000 / 1000	1 / 1	0 / 0	2000 / 48212	100% / 100%
f_{20}	0	1000	1	42 / 972	82 / 20	16 / 14	918 / 980	22206 / 1798708	8.2% / 2%

results for the five different classes of functions we have described above. We start our discussion with experimental results of DECC-DG. Table 7.5 shows that its grouping accuracy for 13 out of 20 benchmark functions is 100%. In particular, DECC-DG correctly identifies all variables to be separable for fully separable functions f_1–f_3 and places all of them in one separable group. DECC-DG's grouping accuracy for the second class of benchmark functions (f_4–f_8) having one nonseparable subcomponent with 50 variables and another separable group with 950 variables is 100% on three out of five functions. Note that our computation of grouping accuracy is with respect to actual nonseparable variables of the benchmark functions. For example, the grouping accuracy on f_4 is reported as 100% since all 50 nonseparable variables are correctly grouped into one subcomponent even though the numbers of groups of separable and nonseparable variables are incorrect. Closer inspection reveals that DECC-DG divides some of the separable variables and places them into nine other nonseparable groups. Similar overall grouping accuracy is obtained by DECC-DG for the third (f_9–f_{13}) and fourth (f_{14}–f_{18}) classes of benchmark functions with 100% accuracy for three out of five functions. A function drawn from the third class has one separable subcomponent with 500 variables and ten nonseparable subcomponents of 50 variables. All functions in the fourth class have 20 nonseparable subcomponents. A small number of misplaced nonseparable variables on f_{11} and f_{16} cause a reduction to at least 99% accuracy. The grouping accuracy of DECC-DG on the last class of functions where all variables interact with one another is conflicting since it achieves 100% on f_{19} but has poor accuracy on f_{20} [51].

Second, we compare the performance of DECC-DG with CCVIL. We start with the comparison with the grouping performance of both cooperative CEAs so as to evaluate their effectiveness in solving large-scale optimization problems in a Divide-and-Conquer manner. Table 7.5 indicates that the overall grouping performance of DECC-DG is better than or comparable with that of CCVIL on the CEC'2010 benchmark functions. The sole exception involves the function f_7. We observe low overall grouping performances of both cooperative CEAs on problem instances constructed with Rosenbrock function (f_8, f_{13}, f_{18}, f_{20}), with the exception of DECC-DG on f_8 at 92%. However, closer inspection on both cooperative CEAs reveals different grouping outcomes. On the one hand, DECC-DG incorrectly groups the decision variables on f_{13} and f_{18}, having form 40 and 49 nonseparable groups where there are only 10 and 20 of such subcomponents, respectively. On the other hand, CCVIL behaves as though all variants of the Rosenbrock-type function are fully separable functions [51].

Next, we compare the efficiency of both cooperative CEAs. In general, DECC-DG achieves more accurate grouping results with considerably fewer fitness evaluations on most of the functions in the benchmark set. However, it has poor efficiency on fully separable functions (f_1–f_3) requiring around one million fitness evaluations to discover the underlying grouping structure. CCVIL is able to quickly detect fully separable variables by employing significantly fewer fitness evaluations at around 0.07% #FE used by DECC-DG for f_1–f_2. The main reason is that DECC-DG always employs pairwise comparison to detect possible interactions between two variables as described procedurally in Algorithm 7.2 and qualitatively by (7.12). This means that on fully separable functions, the DG mechanism will have to perform around D^2 comparisons. In particular, only one i decision variable is excluded from the list of all decision variables after performing D pairwise comparisons for i with all decision variables in the inner loop of Algorithm 7.2 during scan on a fully separable function since no pairwise interaction of variables will be detected. For other variable interaction structures such as those present in the second class (f_4–f_8), fewer fitness evaluations are used when a single scan leads to extraction of 50 nonseparable variables for accurate grouping. Such efficient grouping is also observed for the fourth class where there are 20 nonseparable groups. DECC-DG employs the least number of fitness evaluations on f_{19} where all variables are excluded after the first scan [51].

Figure 7.5 illustrates a situation why the mechanism for detecting interacting variables in CCVIL is slower than the DG mechanism. The figure consists of a greyscale contour plot of a two-dimensional Schwefel's Problem 1.2, with areas between the contour lines of lighter shades indicating ranges of smaller function values. For the function representing this problem, the two variables interact over the entire search space. Three regions A, B, and C are marked in Fig. 7.5. In CCVIL, the condition (7.13) is only satisfied by points in region A. For points in other regions such as in B or C, the condition (7.13) is not satisfied, and CCVIL will need to search further away in other regions such as those in A to establish that for this problem the two decision variables are nonseparable. This search issue arises due to the existential quantification in (7.13). In practice, CCVIL employs a stochastic approach to mediate this search process by way of employing a small number of

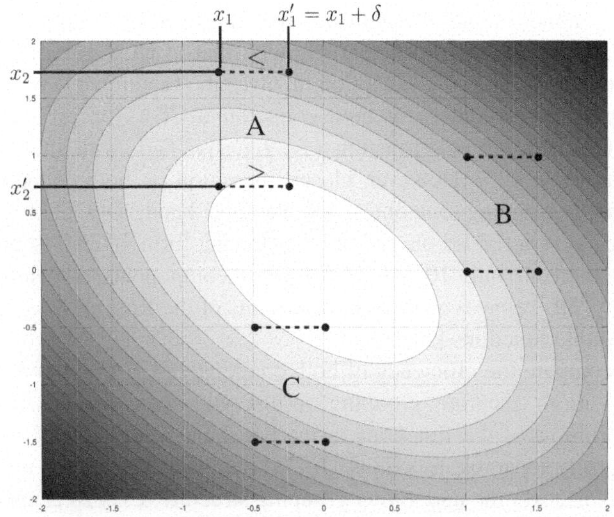

Fig. 7.5 Detection of interacting variables using DG and CCVIL on different regions of a 2D Schwefel Problem 1.2 [51]

checks to determine if condition (7.13) is satisfied for detection of interaction of variables.

The DG mechanism is more efficient in detecting interacting variables because it uses higher-order structures underlying continuous functions by way of approximating the gradient instead of low-order structures in CCVIL that performs direct comparison of pointwise function values. Figure 7.5 illustrates how the DG mechanism compares differences between elevations associated with two pairs of points marked by the pairs of horizontal dashed lines. One example is the pairs given by $|f(x_1, x_2) - f(x_1 + \delta, x_2)|$ and $|f(x_1, x_2') - f(x_1 + \delta, x_2')|$ in Fig. 7.5. If these two values are different, then one can infer that the corresponding dimensions are nonseparable. In other words, this is akin to forming a difference equation based on (7.9), evaluating $\Delta_{x_1}[f]$ for two different values of x_2, and comparing the results. Crucially in this representative example, DG can detect an interaction in its first attempt regardless of the chosen region (e.g. A, B, or C). However, there are situations that DG like most other sampling-based methods could fail to detect variable interaction. If a problem consists of a fully separable region and another region that is fully nonseparable, DG will not detect variable interaction if all two pairs of sampled points are inside the separable region. Detection requires at least one point to be sampled inside the nonseparable region. However, the mechanism in CCVIL might not detect variable interaction even if one of the four points is inside the nonseparable region because condition (7.13) might not be satisfied. Most of the nonseparable CEC'2010 functions have variable interactions occurring throughout the search space. Our results in Table 7.5 indicate that DECC-DG is able to effectively and efficiently detect interactions for nonseparable problems. In Blackbox optimization, efficient grouping mechanism that uses fewer fitness evaluation to identify variables that are interacting and those non-interacting will provide a performance advantage to cooperative CEAs given that savings in computational resource could be used instead for the search process [51].

Third, we perform a computational study to investigate the sensitivity of DG with respect to the parameter ϵ. Table 7.6 presents results of DECC-DG with the setting of $\epsilon \in \{10^{-6}, 10^{-3}, 10^{-1}\}$. Our arbitrary choice of the default setting $\epsilon = 10^{-3}$ means that we test two other values, one being three orders lower and the other being two orders higher in value. Combined results from Tables 7.5 and 7.6 indicate that the grouping accuracy of DECC-DG is not very sensitive to the setting of this parameter in general. However, we have made some specific observations upon closer inspection. In particular, more separable variables are incorrectly classified when a smaller ϵ setting of 10^{-6} is employed on separable functions f_1–f_3 compared to larger ϵ settings $\{10^{-3}, 10^{-1}\}$) that have resulted in 100% grouping accuracy. For example, DECC-DG ($\epsilon = 10^{-6}$) has poor grouping accuracy on f_9, which as a function belonging to the third class has half its problem dimensionality (i.e. 500) constitutes separable variables. Here, more separable variables are misidentified as nonseparable (i.e. 26 out of 500 are correctly classified) and as a result reduce grouping accuracy from 100% to 25.6%. Other examples of reduction in grouping accuracy include f_7, f_{12}, f_{14}, and f_{17} in Table 7.6. For a very small $\epsilon = 10^{-6}$ setting, there could be precision error in

Table 7.6 Results of DECC-DG with parameter ϵ set to 10^{-1} and 10^{-6} on CEC'2010 benchmark functions (separated by "/") [51]

Function	Sep. Vars	Nonsep. Vars	Nonsep. Groups	DECC-DG $(\epsilon = 10^{-1})$ / DECC-DG $(\epsilon = 10^{-6})$					
				# Captured Sep Vars	# Captured Nonsep Vars	# Formed Nonsep Groups	# Misplaced Vars	# FE	Grouping accuracy
f_1	1000	0	0	1000 / 96	0 / 904	0 / 20	0 / 904	1001000 / 25036	100% / 9.6%
f_2	1000	0	0	1000 / 1000	0 / 0	0 / 0	0 / 0	1001000 / 1001000	100% / 100%
f_3	1000	0	0	1000 / 862	0 / 138	0 / 4	0 / 138	1001000 / 757476	100% / 86.2%
f_4	950	50	1	34 / 34	50 / 50	9 / 9	0 / 0	14546 / 14546	100% / 100%
f_5	950	50	1	950 / 950	50 / 50	1 / 1	0 / 0	905450 / 905450	100% / 100%
f_6	950	50	1	950 / 732	50 / 50	1 / 7	0 / 0	906332 / 562748	100% / 100%
f_7	950	50	1	950 / 247	50 / 34	1 / 4	0 / 16	906822 / 67250	100% / 68%
f_8	950	50	1	135 / 135	46 / 46	5 / 5	4 / 4	23608 / 23608	92% / 92%
f_9	500	500	10	500 / 26	500 / 128	10 / 15	0 / 327	270802 / 9350	100% / 25.6%
f_{10}	500	500	10	500 / 500	500 / 500	10 / 10	0 / 0	272958 / 272958	100% / 100%
f_{11}	500	500	10	512 / 158	291 / 500	36 / 14	209 / 0	329938 / 42186	58.2% / 100%
f_{12}	500	500	10	500 / 500	500 / 492	10 / 10	0 / 8	271390 / 271182	100% / 98.4%
f_{13}	500	500	10	500 / 131	27 / 126	173 / 40	473 / 374	636686 / 49470	5.4% / 25.2%
f_{14}	0	1000	20	0 / 1	1000 / 407	20 / 19	0 / 593	21000 / 10574	100% / 40.7%
f_{15}	0	1000	20	1 / 0	999 / 1000	20 / 20	1 / 0	21012 / 21000	99.9% / 100%
f_{16}	0	1000	20	20 / 0	640 / 1000	72 / 20	360 / 640	46476 / 21000	64% / 100%
f_{17}	0	1000	20	0 / 0	1000 / 506	20 / 20	0 / 494	21000 / 11490	100% / 50.6%
f_{18}	0	1000	20	79 / 85	60 / 173	359 / 49	940 / 827	383540 / 34230	6% / 17.3%
f_{19}	0	1000	1	0 / 0	1000 / 1000	1 / 1	0 / 0	2000 / 2000	100% / 100%
f_{20}	0	1000	1	0 / 42	40 / 82	500 / 16	960 / 918	501000 / 22206	4% / 8.2%

computing the difference of the delta values that is required for **DGROUPING**. Here, all other variables that are found to interact with the variable under examination (i.e. each inner loop scan) are extracted from the set of decision variables and grouped in a common subcomponent. As such, incorrect detection of interaction between a separable variable and a set of nonseparable variables would break the nonseparable subcomponent into a set of smaller groups that reduces the overall grouping accuracy [51].

We now present the second part of this subsection on search performance results for DECC-DG against other cooperative CEAs employing other decomposition strategies, which include random grouping [88] and delta grouping [54]. These cooperative CEAs are listed in Table 7.7. The experimental results are summarized in Table 7.8. We have included DECC-I that employs an ideal decomposition based on manual construction via knowledge of the benchmark functions. DECC-I would also have the added advantage whereby all fitness evaluations are used for the optimization process, while all other cooperative CEAs would consume part of the budget of fitness evaluations to discover the grouping structure. DECC-I is used to provide a search performance reference for the family of cooperative CEAS with SaNSDE-based suboptimizer and a cooperative Divide-and-Conquer search process that exploits additive problem separability. The best average search performance with respect to a particular benchmark function associated with a cooperative CEA is determined by way of statistical comparisons with average search performances of other algorithms over 25 independent runs via two-sided Wilcoxon test at a confidence interval of 95%. These are indicated by bold font in Table 7.8. In general, results of our computational study indicate that DECC-DG has good overall performance compared to other cooperative CEAs, taking best average search performance on 11 out of the 20 benchmark functions. The overall search performance of DECC-DML is similar to that of DECC-DG, taking best

Table 7.7 Description of various algorithms that are used in the computational studies [51]

Algorithms	Description	Decomposition Strategy
DECC-G [88]	Differential Evolution Cooperative Coevolution with Random Grouping.	Random Grouping
MLCC [89]	DECC-G but uses a set of potential group sizes	Random Grouping
DECC-D [54]	Differential Evolution Cooperative Coevolution with Delta Grouping	Delta Grouping
DECC-DML [54]	DECC-D but uses a set of potential group sizes	Delta Grouping
DECC-I	DECC with ideal grouping using benchmark function knowledge	Ideal Grouping
DECC-DG	DECC with Differential Grouping	Differential Grouping
CBCC-DG [55]	Contribution Based Cooperative Coevolution with Differential Grouping	Differential Grouping
MA-SW-Chains [47]	The top rank algorithm in the IEEE CEC'210 competition on large-scale global optimization	N/A

Table 7.8 Comparison of DG against other grouping mechanisms on the CEC'2010 benchmark functions, each repeated for 25 independent runs. The **bold** entries are statistically significantly better (Wilcoxon test, $\alpha = 0.05$) [51]

Functions		DECC-DG	DECC-G	MLCC	DECC-D	DECC-DML	DECC-I
f_1	Mean	5.47e+03	2.93e-07	1.53e-27	1.01e-24	1.93e-25	1.73e+00
	Std.	2.02e+04	8.62e-08	7.66e-27	1.40e-25	1.86e-25	2.55e+00
f_2	Mean	4.39e+03	1.31e+03	5.57e-01	2.99e+02	**2.17e+02**	4.40e+03
	Std.	1.97e+02	3.26e+01	2.21e+00	1.92e+01	2.98e+01	1.90e+02
f_3	Mean	1.67e+01	1.39e+00	9.88e-13	**1.81e-13**	**1.18e-13**	1.67e+01
	Std.	3.34e-01	9.73e-02	3.70e-12	6.68e-15	8.22e-15	3.75e-01
f_4	Mean	4.79e+12	1.70e+13	9.61e+12	**3.99e+12**	**3.58e+12**	6.13e+11
	Std.	1.44e+12	5.37e+12	3.43e+12	1.30e+12	1.54e+12	2.08e+11
f_5	Mean	**1.55e+08**	2.63e+08	3.84e+08	4.16e+08	2.98e+08	1.34e+08
	Std.	2.17e+07	8.44e+07	6.93e+07	1.01e+08	9.31e+07	2.31e+07
f_6	Mean	**1.64e+01**	4.96e+06	1.62e+07	1.36e+07	7.93e+05	1.64e+01
	Std.	2.71e-01	8.02e+05	4.97e+06	9.20e+06	3.97e+06	2.66e-01
f_7	Mean	**1.16e+04**	1.63e+08	6.89e+05	6.58e+07	1.39e+08	2.97e+01
	Std.	7.41e+03	1.37e+08	7.37e+05	4.06e+07	7.72e+07	8.59e+01
f_8	Mean	**3.04e+07**	6.44e+07	4.38e+07	5.39e+07	**3.46e+07**	3.19e+05
	Std.	2.11e+07	2.89e+07	3.45e+07	2.93e+07	3.56e+07	1.10e+06
f_9	Mean	**5.96e+07**	3.21e+08	1.23e+08	**6.19.+07**	**5.92e+07**	4.84e+07
	Std.	8.18e+06	3.38e+07	1.33e+07	6.43e+06	4.71e+06	6.56e+06
f_{10}	Mean	**4.52e+03**	1.06e+04	**3.43e+03**	1.16e+04	1.25e+04	4.34e+03
	Std.	1.41e+02	2.95e+02	8.72e+02	2.68e+03	2.66e+02	1.46e+02
f_{11}	Mean	1.03e+01	2.34e+01	1.98e+02	4.76e+01	**1.80e-13**	1.02e+01
	Std.	1.01e+00	1.78e+00	6.98e-01	9.53e+01	9.88e-15	1.13e+00
f_{12}	Mean	**2.52e+03**	8.93e+04	3.49e+04	1.53e+05	3.79e+06	1.47e+03
	Std.	4.86e+02	6.87e+03	4.92e+03	1.23e+04	1.50e+05	4.28e+02
f_{13}	Mean	4.54e+06	5.12e+03	2.08e+03	**9.87e+02**	**1.14e+03**	7.51e+02
	Std.	2.13e+06	3.95e+03	7.27e+02	2.41e+02	4.31e+02	3.70e+02
f_{14}	Mean	3.41e+08	8.08e+08	3.16e+08	1.98e+08	**1.89e+08**	3.38e+08
	Std.	2.41e+07	6.07e+07	2.77e+07	1.45e+07	1.49e+07	2.40e+07
f_{15}	Mean	**5.88e+03**	1.22e+04	7.11e+03	1.53e+04	1.54e+04	5.87e+03
	Std.	1.03e+02	8.97e+02	1.34e+03	3.92e+02	3.59e+02	9.89e+01
f_{16}	Mean	**7.39e-13**	7.66e+01	3.76e+02	1.88e+02	5.08e-02	2.47e-13
	Std.	5.70e-14	8.14e+00	4.71e+01	2.16e+02	2.54e-01	9.17e-15
f_{17}	Mean	**4.01e+04**	2.87e+05	1.59e+05	9.03e+05	6.54e+06	3.91e+04
	Std.	2.85e+03	1.98e+04	1.43e+04	5.28e+04	4.63e+05	2.75e+03
f_{18}	Mean	1.11e+10	2.46e+04	7.09e+03	**2.12e+03**	**2.47e+03**	1.17e+03
	Std.	2.04e+09	1.05e+04	4.77e+03	5.18e+02	1.18e+03	9.66e+01
f_{19}	Mean	**1.74e+06**	1.11e+06	1.36e+06	1.33e+07	1.59e+07	1.74e+06
	Std.	9.54e+04	5.15e+04	7.35e+04	1.05e+06	1.72e+06	9.54e+04
f_{20}	Mean	4.87e+07	4.06e+03	2.05e+03	**9.91e+02**	**9.91e+02**	4.14e+03
	Std.	2.27e+07	3.66e+02	1.80e+02	2.61e+01	3.51e+01	8.14e+02

average search performance on 10 out of the 20 benchmark functions. However, closer inspections reveal that DECC-DG outperforms DECC-DML on nonseparable functions where it achieves high grouping accuracy (Table 7.5). DECC-DG's search performance on separable functions (f_1–f_3) is poor in comparison to that of DECC-DML. Similar observation can be made comparing DECC-DG against DECC-D. The two cooperative CEAs that employ the simple Random Grouping mechanism (DECC-G and MLCC) achieve low overall average search performance in comparison to other cooperative CEAs under investigation by our computational study [51].

Next, we investigate the convergence behaviour of the cooperative CEAs. For discussion purposes, we produce Fig. 7.6 that provides a representative qualitative illustration of the points we summarize here. The individual plot associated with each cooperative CEA in the figure is produced by taking the average over 25 independent runs. In general, we observe that the increased number of fitness evaluations used to discover the grouping structure for some functions can be compensated for during the optimization stage by the cooperative CEAs. For example, cooperative CEAs that employ DG such as DECC-DG do not register significant early improvement for functions f_6, f_9, and f_{12} while consuming computational resources for operations. However, significant improvements can be observed after the grouping structure is identified. In summary, our computational study and its results in Table 7.8 and Fig. 7.6 strongly suggest that automatic grouping mechanisms that correctly identify the underlying structure of the problems (i.e. separability of the decision variables associated with the benchmark functions) are crucial towards cooperative coevolutionary Divide-and-Conquer problem solving of continuous optimization problems. There is an advantage to be gained for consuming a fraction of the computational budget to discover such a structure before running the evolutionary optimizers [51].

The remaining second half of this subsection is focused on our study [51] to improve the search performance of cooperative CEAs via a ***contribution-based mechanism*** that allocates more computational budget to optimize subcomponents with higher contributions towards the global fitness of the complete solution [55]. This differs with classical cooperative CEAs where the subcomponents are typically optimized in a round-robin fashion with an equal amount of computational budget for Blackbox problems. In [55], manual grouping of decision variables is used to show that the proposed contribution-based mechanism is beneficial to cooperative CEAs. Here, we incorporate the contribution-based mechanism into DECC-DG that we reference as ***Contribution-based Cooperative Coevolution with Differential Grouping*** (***CBCC-DG***). The following computational study evaluates CBCC-DG against DECC-DG to assess the impact of the contribution-based mechanism towards the search performance of cooperative CEAs. Note that all cooperative CEAs that are studied here employ DG mechanism for automatic identification and grouping of interacting decision variables. Our study in [55] has developed two policies for dividing the computational budget between the subcomponents, which we reference here as CBCC1 and CBCC2.

Fig. 7.6 Convergence plots of various cooperative CEAs on selected CEC'2010 benchmark functions. The plots are generated from averages of 25 independent runs [51]

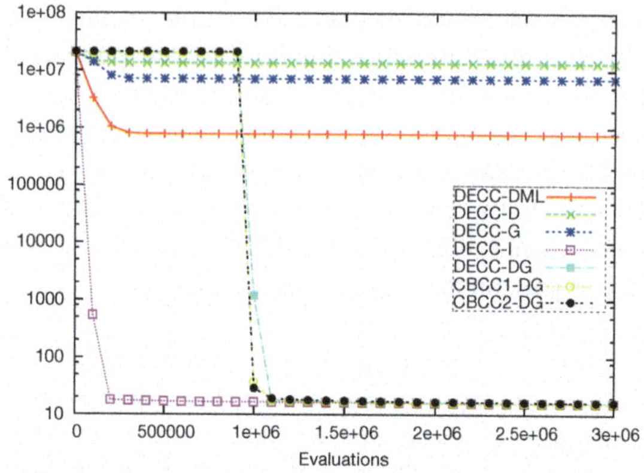

(a) f_6

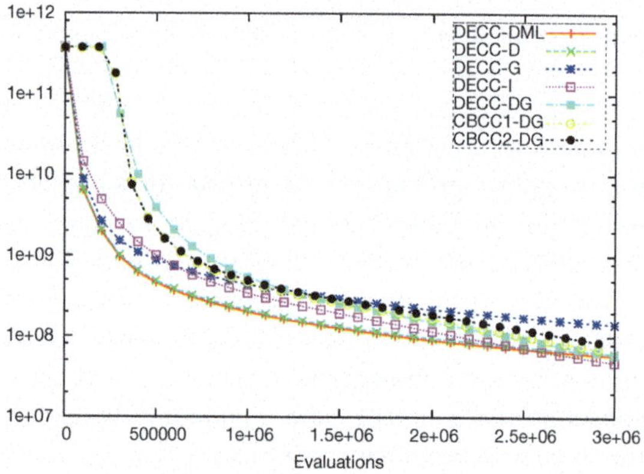

(b) f_9

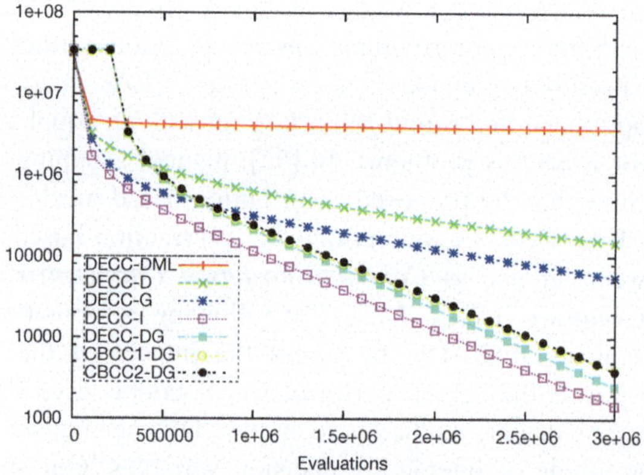

(c) f_{12}

Table 7.9 summarizes results of the computational study [51]. Our results indicate that both instances of CBCC-DGs have outperformed DECC-DG with the former achieving the best average search performance for 8 out of 20 benchmark functions in comparison to the latter achieving the best average search performance for six benchmark functions. Closer inspection reveals CBCC-DGs outperform DECC-DG for the majority of the second class of functions (f_4–f_8). For the other classes of functions such as f_9–f_{18}, CBCC-DGs and DECC-DG have similar overall average search performance. This is not unexpected given that we have found out earlier in [55] that on these benchmark functions there are equal contributions from all subcomponents. In addition, we have compared the search performance of CBCC-DGs against MA-SW-Chains that represents the top rank algorithm in the IEEE CEC'210 competition on large-scale global optimization [47]. Our results show that MA-SW-Chains has outperformed CBCC-DGs on 15 out of 20 as indicated by entries marked by the symbol "†" in Table 7.9.

We highlight two important points about historical practices of large-scale global optimization. First, the optimizers under comparison are different in nature. MA-SW-Chains is a top performing optimizer on the IEEE CEC'2010 benchmark functions, while the SaNSDE suboptimizer used in CBCC-DGs has been developed around the previous IEEE CEC'2005 benchmark functions. As such, there is scope to improve the search performance of cooperative CEAs with DG by employing more modern suboptimizers. Second, there is no imbalance of contributions between the subcomponents in most of the functions such as f_9–f_{18}. However, for functions such as f_4–f_8 that do demonstrate imbalance, results indicate marginal performance in favour of CBCC-DGs over MA-SW-Chains. This is an important consideration given that real-world problems might have some imbalance between various subcomponents. In such a problem setting, equal contribution of subcomponent to the global fitness would be unlikely [51].

As such, we have developed a set of benchmark problems to assess the performance of EAs on imbalanced problems based on modified functions of f_9–f_{18}. These functions incorporate some artificial imbalance between the subcomponents. In particular, functions in the third F_{class3} and fourth F_{class4} classes of the CEC'2010 benchmark can be formulated with imbalance in the following manner [51]:

$$F_{\text{class3}} = \sum_{i=0}^{\frac{D}{2s}-1} 10^{2(i-9)} \times F_{\text{nonsep}} + F_{\text{sep}}$$

$$F_{\text{class4}} = \sum_{i=0}^{\frac{D}{s}-1} 10^{(i-9)} \times F_{\text{nonsep}} + F_{\text{sep}} \,.$$

The overall structure of the functions remains unchanged. However, the contribution of nonseparable component is multiplied by a coefficient to introduce an imbalance effect.

Table 7.9 Comparison of classical cooperative CEAs and CBCCs with DG on the CEC'2010 benchmark functions, each repeated for 25 independent runs. The **bold** entries are statistically significantly better (Wilcoxon test, $\alpha = 0.05$) results with respect to the cooperative CEAs. The entries marked with "†" indicate significantly better results for comparisons of CBCCs with MA-SW-Chains [51]

Functions		DECC-DG	CBCC1-DG	CBCC2-DG	MA-SW-Chains [47]
f_1	Mean	5.47e+03	1.32e+04	8.34e+03	2.10e-14†
	Std.	2.02e+04	6.25e+04	3.41e+04	1.99e-14
f_2	Mean	**4.39e+03**	4.44e+03	4.44e+03	8.10e+02†
	Std.	1.92e+01	1.60e+02	1.80e+02	5.88e+01
f_3	Mean	1.67e+01	1.66e+01	1.67e+01	7.28e-13†
	Std.	3.34e-01	3.79e-01	3.28e-01	3.40e-13
f_4	Mean	4.79e+12	**2.31e+12**	**2.36e+12**	3.53e+11†
	Std.	1.44e+12	7.43e+11	7.92e+11	3.12e+10
f_5	Mean	1.55e+08	**1.35e+08**†	**1.36e+08**†	1.68e+08
	Std.	2.17e+07	2.18e+07	2.46e+07	1.04e+08
f_6	Mean	1.64e+01	1.65e+01†	1.64e+01†	8.14e+04
	Std.	2.71e-01	3.99e-01	3.46e-01	2.84e+05
f_7	Mean	**1.16e+04**	1.81e+04	1.35e+04	1.03e+02†
	Std.	7.41e+03	4.59e+04	3.92e+04	8.70e+01
f_8	Mean	3.04e+07	**3.34e+06**†	**8.70e+05**†	1.41e+07
	Std.	2.11e+07	2.29e+06	1.71e+06	3.68e+07
f_9	Mean	**5.96e+07**	6.79e+07	7.97e+07	1.41e+07†
	Std.	8.18e+06	6.92e+06	1.08e+07	1.15e+06
f_{10}	Mean	4.52e+03	**4.01e+03**	**4.04e+03**	2.07e+03†
	Std.	1.41e+02	1.37e+02	1.21e+02	1.44e+02
f_{11}	Mean	1.03e+01	1.05e+01†	1.03e+01†	3.80e+01
	Std.	1.01e+00	9.31e-01	8.47e-01	7.35e+00
f_{12}	Mean	**2.52e+03**	4.19e+03	4.00e+03	3.62e-06†
	Std.	4.86e+02	1.25e+03	8.63e+02	5.92e-07
f_{13}	Mean	4.54e+06	**9.10e+03**	**4.54e+03**	1.25e+03†
	Std.	2.13c+06	3.75e+03	1.91e+03	5.72e+02
f_{14}	Mean	**3.41e+08**	3.64e+08	3.69e+08	3.11e+07†
	Std.	2.41e+07	2.61e+07	2.42e+07	1.93e+06
f_{15}	Mean	5.88e+03	5.89e+03	5.88e+03	2.74e+03†
	Std.	1.03e+02	9.10e+01	8.81e+01	1.22e+02
f_{16}	Mean	7.39e-13	**3.08e-12**†	**4.44e-12**†	9.98e+01
	Std.	5.70e-14	3.19e-13	4.22e-13	1.40e+01
f_{17}	Mean	**4.01e+04**	4.50e+04	4.73e+04	1.24e+00†
	Std.	2.85e+03	3.18e+03	2.77e+03	1.25e-01
f_{18}	Mean	1.11e+10	**1.34e+09**	**3.47e+08**	1.30e+03†
	Std.	2.04e+09	4.94e+08	1.39e+08	4.36e+02
f_{19}	Mean	1.74e+06	1.74e+06	1.74e+06	2.85e+05†
	Std.	9.54e+04	8.46e+04	8.46e+04	1.78e+04
f_{20}	Mean	4.87e+07	**9.53e+04**	**8.42e+03**	1.07e+03†
	Std.	2.27e+07	1.02e+05	2.36e+03	7.29e+01

Table 7.10 summarizes the results of our computational study involving the new benchmark (f_4–f_8, f_9'–f_{18}') that include the new imbalanced functions, which are represented by the addition of a prime symbol (e.g. f_9'). We refer to this new set of problems as the imbalanced benchmark. Note that functions f_4–f_8 in the original benchmark are imbalanced. To distinguish between the two benchmarks, we will refer to the original benchmark as balanced. Our results indicate that with more problems having some imbalance, the gap in the overall search performance between CBCC-DG and DECC-DG widens. This suggests that the contribution-based mechanism allows cooperative CEA to better address imbalanced problems. More importantly, CBCC-DG performs at least as well as, if not marginally better than DECC-DG on the original, balanced benchmark. In other words, there is only advantage to be had implementing the contribution-based mechanism in a cooperative CEA. We have also compared the search performance of CBCC-DG and MA-SW-Chains. Table 7.10 indicates that with more imbalanced problems in the new benchmark, MA-SW-Chains only slightly outperforms CBCC-DG. Crucially, closer inspection of our results reveals that the search performance gap between CBCC-DG and MA-SW-Chains has significantly reduced on the new problems that incorporate more imbalance. For example, the gap in average search performance of CBCC-DG against MA-SW-Chains on individual functions is reduced from the setting of balanced f_{12} and f_{17} (see Table 7.9) to that of imbalanced f_{12}' and f_{17}' (see Table 7.11) [51]. In particular, the performance gap is reduced to only one order of magnitude in difference.

Next, we perform analysis on the search behaviour of algorithms under study and provide representative convergence plots in Fig. 7.7. We observe different search behaviours between MA-SW-Chains and the cooperative CEAs in the figure. In particular, the search behaviour of MA-SW-Chains is typically characterized by an initial, drastic improvement in the fitness value that then stagnates as the search process continues. This search behaviour can be attributed to local search during the evolutionary search process of MA-SW-Chains as a Memetic Algorithm [48]. In contrast, convergence plots associated with both cooperative CEAs (CBCC-DG and DECC-DG) typically show a steady improvement in the fitness value throughout the cooperative coevolutionary search process. As one example on f_{11}' and f_{16}', the search trajectories of both cooperative CEAs overtake that of MA-SW-Chains at some point during the evolutionary process in Fig. 7.7. On functions such as f_{12}', CBCC-DG and DECC-DG demonstrate steady improvement over their coevolutionary cycles and would overtake MA-SW-Chains in performance if more

Table 7.10 CBCC-DG's number of wins, losses, and ties against DECC-DG and MA-SW-Chains before and after inclusion of imbalance in benchmark problems (based on functions f_4–f_8, f_9'–f_{18}') [51]

	Balanced			Imbalanced		
Algorithm	Wins	Loses	Ties	Wins	Loses	Ties
DECC-DG	7	5	3	9	2	4
MA-SW-Chains	5	10	0	6	7	2

Table 7.11 Experimental results for imbalanced functions. The experiments are based on modified benchmark functions; each set involves 25 repeated and independent runs. The **bold** entries are statistically significantly better (Wilcoxon test, $\alpha = 0.05$) results with respect to the cooperative CEAs. The entries marked with "†" indicate significantly better results for comparisons of CBCCs with MA-SW-Chains [51]

Functions		DECC-DG	CBCC1-DG	CBCC2-DG	MA-SW-Chains [47]
f'_9	Mean	3.81e+11	**2.18e+11**	**2.04e+11**	8.43e+10†
	Std.	1.15e+11	5.51e+10	5.16e+10	3.21e+10
f'_{10}	Mean	2.89e+07	**2.36e+07†**	**2.43e+07†**	2.24e+07†
	Std.	1.51e+06	1.79e+06	1.89e+06	1.34e+07
f'_{11}	Mean	**9.98e+00**	1.03e+01†	1.05e+01†	2.72e+04
	Std.	1.16e+00	1.11e+00	8.93e-01	4.74e+04
f'_{12}	Mean	1.62e+07	**6.27e+06**	**5.02e+06**	6.93e+05†
	Std.	7.53e+06	7.27e+06	1.73e+06	4.24e+05
f'_{13}	Mean	**1.07e+07**	**1.06e+07**	**1.04e+07**	4.31e+06†
	Std.	4.77e+06	4.69e+06	4.08e+06	3.31e+06
f'_{14}	Mean	1.11e+13	**6.62e+12**	**6.50e+12**	4.69e+11†
	Std.	2.64e+12	4.20e+12	3.40e+12	5.38e+10
f'_{15}	Mean	1.98e+08	**1.73e+08†**	**1.73e+08†**	1.30e+08†
	Std.	6.84e+06	1.35e+07	1.19e+07	8.40e+07
f'_{16}	Mean	**2.78e-08**	**2.98e-08†**	4.22e-08†	1.71e+05
	Std.	1.55e-08	1.33e-08	2.87e-08	1.32e+05
f'_{17}	Mean	1.36e+09	**7.34e+08**	**6.86e+08**	3.13e+07†
	Std.	2.88e+08	5.01e+08	4.23e+08	1.56e+07
f'_{18}	Mean	**4.25e+10**	**4.33e+10†**	5.64e+10†	5.39e+12
	Std.	9.76e+09	1.23e+10	3.13e+10	9.59e+12

computational budget is allocated. Our computational study suggests that a local search approach such as MA-SW-Chains could find better solutions in the short term with a limited number of fitness evaluations for Blackbox optimization. However, cooperative CEAs with effective grouping and contribution-based mechanisms have the potential for further improvements. Indeed, our results indicate that CBCC-DG outperforms MA-SW-Chains on almost all instances of multimodal functions in the benchmark such as f_5–f_8 and f'_9-f'_{18} with 6 wins, 1 lose, and 2 draws in pairwise comparisons of their average search performances [51].

The general conclusion we can make from our systematic computational studies is that the DG mechanism is an effective and efficient strategy for automatic identification and grouping of interacting variables present in large-scale global optimization problems that are additively separable. It can be procedurally incorporated into the cooperative coevolution framework so as to construct cooperative CEAs having near-optimal grouping accuracy in a large number of functions when evaluated against the CEC'2010 benchmark, which then leads to improved average search performance. More importantly from an algorithmic design perspective, the DG mechanism constitutes a reliable performance enhancing component in

Fig. 7.7 Convergence of classical and contribution-based cooperative CEAs using DG against MA-SW-Chains on some representative imbalanced problems [51]

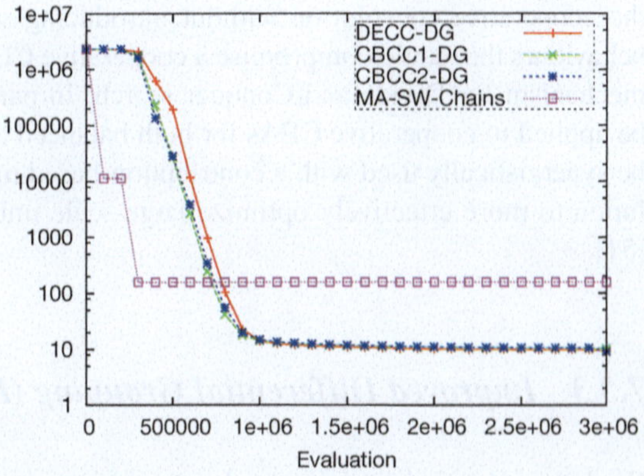

(a) f'_{11}

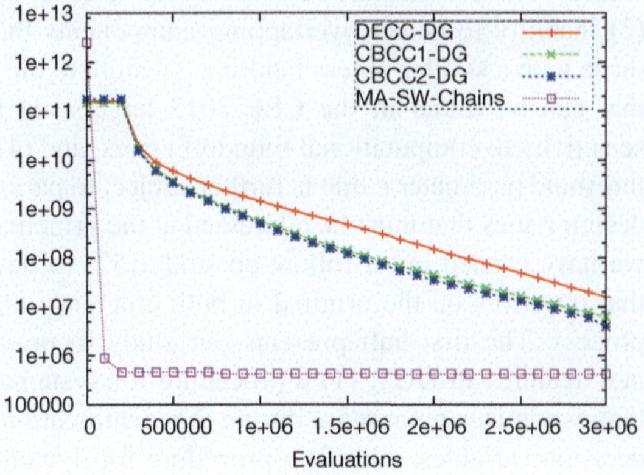

(b) f'_{12}

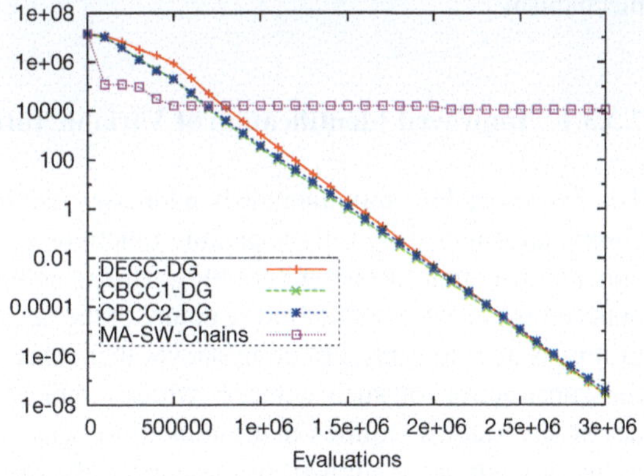

(c) f'_{16}

the cooperative coevolution without introducing significant and unwanted search behaviours that could compromise a cooperative CEA that incorporates its grouping mechanism for Divide-and-Conquer search. In particular, the DG mechanism can be applied to cooperative CEAs for both balanced and imbalanced problems. It can be synergistically used with a contribution-based mechanism in cooperative coevolution to more effectively optimize large-scale imbalanced optimization problems [51].

7.3.3 Improved Differential Grouping (DG2)

We have identified four issues that could impact the effectiveness and efficiency of the DG mechanism despite its success as demonstrated by our computational studies. These issues are (1) high computational cost on fully separable functions, (2) inability to detect overlapping components that share decision variables (e.g. those with a similar fitness landscape feature to the Rosenbrock function and others that can be found in the CEC'2013 large-scale benchmark functions [53]), (3) sensitivity to computational roundoff errors, and (4) the requirement of specifying a threshold parameter ϵ that is further subject to parameter tuning. However, these are design issues that must be addressed at the principal mechanism level. In response, we have carried out a follow-up study [52] to develop the new DG2 mechanism that improves on the original in both efficiency and effectiveness of the grouping process. The first half presents our study to provide theoretical support for two new features in DG2: (1) a procedure for systematic generation of sample points that maximizes point reuse during the identification process of interactions between decision variables, and (2) a procedure for dynamic computation of the threshold value ϵ to detect the interaction for each pair of variables that consider the magnitude of roundoff errors. After that, we present our procedural implementation of the DG2 mechanism.

7.3.3.1 Improved Identification of Variable Interactions

The first procedure simultaneously addresses two limitations of DG involving high computational cost on fully separable functions and detection of overlapping components that share decision variables. This is achieved by capturing the complete, underlying pairwise interaction structure of the decision variables via a graph, which in turn is able to represent components that share decision variables. In principle, the construction of such a graph whose connectivities represent interactions of decision variables requires that interactions between all pairs of decision variables in the problem be examined. In this aspect, the only means with which an efficient procedural implementation can be developed is to reuse information required during the identification process of interactions between decision variables.

We perform a formal analysis demonstrating that a procedure can be developed that achieves the lower bound when Theorem 7.2 is applied to detect interactions. The main outcome is that the total number of objective function evaluations used to check interactions in all pairs of variables is reduced (e.g. half the total number for fully separable functions that require the most function evaluations). We will present the formal arguments underlying our theoretical results with the aid of simple examples and illustration for general readability. For a D-dimension function, the total number interactions is $\binom{D}{2}$. Since each pairwise interaction requires four fitness evaluations for detection following Theorem 7.2, then a total of $4 \cdot \binom{D}{2} = 2D(D-1)$ evaluations is required for complete detections of all pairwise interactions in a D-dimension function [52].

It is possible to significantly reduce the total number of fitness evaluations via systematic selection of sample points used for the calculation of the difference equation (7.9). We consider a simple function with only three decision variables $f(x_1, x_2, x_3)$. The total number of function evaluations according to Theorem 7.2 is given by

$$
\begin{aligned}
x_1 \leftrightarrow x_2 :& \Delta^{(1)} = f(a',b,c) - f(a,b,c), \Delta^{(2)} = f(a',b',c) - f(a,b',c) \\
x_1 \leftrightarrow x_3 :& \Delta^{(1)} = f(a',b,c) - f(a,b,c), \Delta^{(2)} = f(a',b,c') - f(a,b,c') \\
x_2 \leftrightarrow x_3 :& \Delta^{(1)} = f(a,b',c) - f(a,b,c), \Delta^{(2)} = f(a,b',c') - f(a,b,c'),
\end{aligned}
\tag{7.14}
$$

where a, b, and c are values taken by x_1, x_2, and x_3 respectively, and $a' = x_1 + \delta$, $b' = x_2 + \delta$, and $c' = x_3 + \delta$ [52]. This is illustrated in Fig. 7.8 geometrically whereby the points that are evaluated with function f are similarly colour-coded as with the expressions above. For this function of $D = 3$ dimensions, the total number of function evaluations required for detection of all interactions is $2 \cdot 3(3-1) = 12$. Figure 7.8 clearly shows that only seven unique points are involved. This graphical observation coincides with what one can obtain by way of manual identification of unique points (x_1, x_2, x_3) in the expressions above that indicate where computational savings can be made.

We expand our observation further for clarity. Note that four points are required to calculate $\Delta^{(1)}$ and $\Delta^{(2)}$, and they are chosen from the search space so as to form a rectangle following Theorem 7.2. Figure 7.8 illustrates the base point (a, b, c) required to calculate $\Delta^{(1)}$. At the start, one could determine self-interaction with respect to x_1 by simply varying the first variable to calculate $\Delta^{(1)}$, which is achieved by selecting the second point (a', b, c), where $a' = a + \delta$. Therefore, the pairwise interaction $x_1 \leftrightarrow x_2$ can now be established with the second difference equation $\Delta^{(2)}$ that is evaluated with a different value of x_2, i.e. via the two points (a', b', c) and (a, b', c). The same argument can be made for the other two pairwise interactions $x_1 \leftrightarrow x_3$ and $x_2 \leftrightarrow x_3$, and all the expressions are given by (7.14). For a D-dimension function, we have the following evaluations: (1) The base point (a, b, c) is repeated exactly three times. (2) Cases where only one dimension is varied with respect to the base point, e.g. involving (a', b, c), (a, b', c), and (a, b, c'), are each repeated two times. (3) Cases where two dimensions are varied

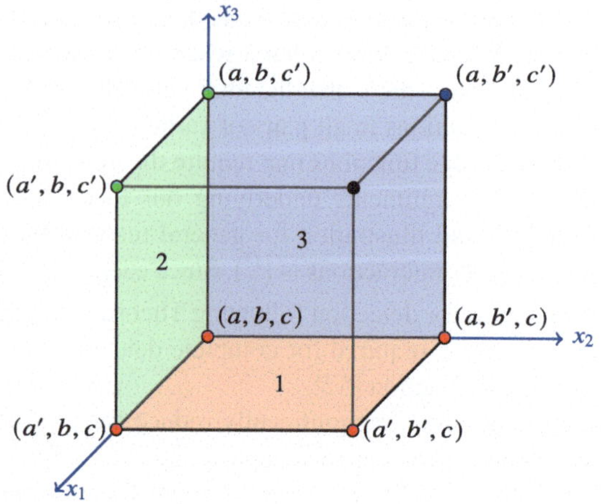

Fig. 7.8 Geometric representation of point generation in DG2 for a function with $D = 3$ [52]

with respect to the base point such as (a', b', c), (a', b, c'), and (a, b', c') are each evaluated once [52].

This argument can be generalized for an arbitrary number of decision variables $D \geq 3$. In general, the following evaluations are required to detect x_i–x_j pairwise interaction [52]:

$$x_i \leftrightarrow x_j = \begin{cases} \Delta^{(1)} = f(\ldots, x_i', \ldots) - f(x_1, \ldots, x_D) \\ \Delta^{(2)} = f(\ldots, x_i', \ldots, x_j', \ldots) - f(\ldots, x_j', \ldots). \end{cases} \tag{7.15}$$

The formulation in (7.15) lets us calculate the total number of unique evaluations required to detect all pairwise interactions. First, the base point $(x_1, \ldots, x_n)$ is evaluated exactly $\binom{D}{2} = D(D-1)/2$ times, which is the same as the total number of interactions that needs to be detected. Second, the point $(\ldots, x_i', \ldots)$ is evaluated exactly $D - 1$ times. The calculation is as follows. The point $(\ldots, x_i', \ldots)$ is evaluated $D - i$ times to detect the interactions between the ith dimension and all those *after* it via $\Delta^{(1)}$s. Then, the same point $(\ldots, x_i', \ldots)$ is evaluated $i - 1$ times to detect the interactions between the ith dimension and all those *before* it via $\Delta^{(2)}$s. Thus, the point $(\ldots, x_i', \ldots)$ is evaluated $(D-i)+(i-1) = D-1$ times. Third, the number of times the point $(\ldots, x_i', \ldots, x_j', \ldots)$ is evaluated is equal to the number of pairwise comparisons $\binom{D}{2}$. The sum of all these evaluations is [52]

$$2 \cdot \binom{D}{2} + D(D-1) = \frac{2D(D-1)}{2} + D(D-1) = 2D(D-1) = 4 \cdot \binom{D}{2}.$$

What is important in the construction of our arguments above is that we are able to use it to establish the number of redundant evaluations as follows [52]:

$$\begin{cases} \frac{D(D-1)}{2} - 1 & : \text{redundant evaluations of } (x_1, \ldots, x_D) \\ D(D-2) & : \text{redundant evaluations of } (\ldots, x_i', \ldots). \end{cases}$$

Then, the *number of unique evaluations* can be directly calculated. This is performed by subtracting the total number of redundant evaluations from the total number of evaluations to give [52]

$$2D(D-1) - D(D-2) + \frac{D(D-1)}{2} - 1 = \frac{D(D+1)}{2} + 1.$$

These arguments are used in the proofs of our theoretical results that provide principled support for the construction of the procedure in DG2 mechanism that systematically generates sample points for identification of variable interactions that maximizes point reuse. The main result is summarized in Theorem 7.3, and its proof arguments make use of Lemma 7.2. They are presented in the following.

Lemma 7.2 ([52]) *Let $h(K)$ be the minimum number of unique function evaluations for detecting the interactions between all pairs of variables when dimension is K; then*

$$h(K+1) \geq h(K) + K + 1. \tag{7.16}$$

Theorem 7.3 ([52]) *The minimum number of unique function evaluations in order to detect the interactions between all pairs of variables is*

$$h(D) \geq \frac{D(D+1)}{2} + 1. \tag{7.17}$$

The full proof arguments are available in the supplementary document of [52].

7.3.3.2 Parameter-Free Differential Grouping

The second procedure addresses two issues of DG involving sensitivity to weak interactions due to finite computation errors in DG operations and having ϵ as a parameter setting. Finite computation errors could affect actual operations of algorithms on computing machines. In DG, this typically involves having to resolve (determine the truth-value of) the statement $\lambda = |\Delta^{(1)} - \Delta^{(2)}| > \epsilon$. Consider the $\epsilon = 0$ setting whereby a positive λ value implies the presence of variable interaction. In practice, finite computations of $\Delta^{(1)}$ and $\Delta^{(2)}$ could incur roundoff errors and lead to nonzero λ values even for separable variables. There is a need to distinguish between a nonzero λ due to variable interaction and a nonzero λ due to finite computation errors when $\epsilon = 0$ is used. However, other nonzero ϵ settings are affected by these computing errors in having to resolve the inequality statement $\lambda > \epsilon$. Crucially, our following analysis will show that the magnitude of roundoff errors is a function of the magnitude of the quantities used in a computation. As such a grouping procedure that uses a fixed ϵ on finite computing machines would face this issue when it has no control over the chosen sample points used to resolve the inequality $\lambda > \epsilon$.

What the second procedure does to more reliably decide whether variable interaction is present under the effects of finite computation errors is to employ different inequalities and resolve the one that better matches the interaction structure present in the problem. This involves the use of the greatest lower bound (infimum) e_{inf} and the least upper bound (supremum) e_{sup} of the roundoff error. These values are estimated in advance and separately for each pair of variables based on the available information such as function values and the quantity $\lambda = |\Delta^{(1)} - \Delta^{(2)}|$ to maximize the detection accuracy. This can be achieved by resolving $\lambda < e_{\text{inf}}$ to decide variables are separable and $\lambda > e_{\text{sup}}$ to decide variables are interacting. The interval formed by the two inequalities constitutes a *safe* region with which a finite computing machine could establish λ values being genuine zero (when $\lambda < e_{\text{inf}}$ is computed) and nonzero (when $\lambda > e_{\text{sup}}$ is computed) [52].

The same machine could not directly establish nonzero λ values outside of the safe region whether they are indeed genuine nonzero values. For this $\lambda \in (e_{\text{inf}}, e_{\text{sup}})$ case, a heuristic is employed that involves the use of a weighted average of the estimated bounds to set the threshold ϵ. Let $\mathbf{\Lambda}$ denote the ***interaction structure matrix*** that stores the computed quantity $|\Delta^{(1)} - \Delta^{(2)}|$ for all pairs of variables $i, j = 1, \ldots, D$. Then, the threshold value is given by

$$\epsilon = \frac{\eta_0}{\eta_0 + \eta_1} e_{\text{inf}} + \frac{\eta_1}{\eta_0 + \eta_1} e_{\text{sup}}, \tag{7.18}$$

where η_0 and η_1 are the number of entries in $\mathbf{\Lambda}$ that are less than e_{inf} and greater than e_{inf}, respectively [52]. Equation (7.18) represents a heuristic for setting a threshold ϵ for non-reliable λ values and can be interpreted as follows. The value of ϵ can range between $\epsilon = e_{\text{inf}}$ (if $\eta_1 = 0$) and $\epsilon = e_{\text{sup}}$ (if $\eta_0 = 0$). ϵ takes the arithmetic mean of e_{inf} and e_{sup} if $\eta_0 = \eta_1$. This represents the least biased choice of setting ϵ when the numbers of reliably detected separable and nonseparable variables are equal. For other cases, ϵ is biased towards either e_{inf} or e_{sup} depending on the proportions of reliably detected separable and nonseparable variables [52].

We now present our mathematical analysis in [52] for estimating e_{inf} or e_{sup} for DG2 operations. The full details are available in the supplementary document of our study. We start with key principles of error analysis on the IEEE 754 standard that is employed in hardware of modern computing machines [29] and then apply them for our main arguments to develop the second procedure in DG2. Let $\mathbb{N}$, $\mathbb{Z}$, and $\mathbb{R}$ be the sets of natural, integer, and real numbers, respectively. An element n of the set of floating-point numbers $\mathbb{F} \subset \mathbb{R}$ is machine-represented as

$$n = (-1)^s \times c \times \beta^{e-p},$$

where $s \in \{0, 1\}$ indicates the sign (positive or negative), c is the ***significand*** satisfying $0 \leq c \leq \beta^p - 1$ and whose magnitude is determined by the base $\beta \in \mathbb{N}$ (typically 2 or 10), the ***precision*** $p \in \mathbb{N}$, and the ***exponent*** $e \in \mathbb{Z}$ satisfying $e_{\min} \leq e \leq e_{\max}$ [52]. In particular, the IEEE 754 [29] standard for representing a 64-bit double precision floating-point number has the specification of $\beta = 2$, $e_{\min} = -1021$, $e_{\max} = 1024$, and $p = 53$. This gives $\log_{10}(2^{53})$ decimal digits or approximately 15.95. Figure 7.9 illustrates how floating-point numbers of a hypothetical system with specification $\beta = 2$, $e_{\min} = -1$, $e_{\max} = 3$, and $p = 3$ are distributed on a closed interval of the real line.

Such a number representation in a computing machine is described by the map $\mathrm{fl} : \mathbb{R} \to \mathbb{F}$. One immediate source of error in the floating-point number representation of fl is having to round the real number x to the nearest representable number $\mathrm{fl}(x)$. Figure 7.9 illustrates how roundoff error increases with the magnitude of x due to the nonuniform spacing of floating-point numbers. This is formalized in Theorem 7.4 [26].

Theorem 7.4 ([26]) *If $x \in \mathbb{R}$ lies in the range of $\mathbb{F}$, then*

$$\mathrm{fl}(x) = x(1 + \delta), \quad |\delta| < \mu_{\mathrm{M}},$$

where μ_{M} is called the unit roundoff, which is equal to $\frac{1}{2}\beta^{1-p}$.

Theorem 7.4 not only explains the growth in the spacing between the increasing floating-point numbers but also crucially specifies that the error term δx grows with x because $\mathrm{fl}(x) = x + \delta x$. The magnitude of the maximum representation error is $\mu_{\mathrm{M}} x$. The ***machine epsilon*** μ_{M} is specified by β^{1-p}, which gives an upper bound on the relative roundoff error in floating-point operations. A ***unit roundoff*** ϵ_{M} is half of machine epsilon, and their relationship is given by $\mu_{\mathrm{M}} = \epsilon_{\mathrm{M}}/2$. The constant ϵ_{M} provides a specified guarantee to the degree of roundoff error as a result of using a specific finite computing system. For the double precision floating-point number (64 bits), $\epsilon_{\mathrm{M}} = 2^{-52} \approx 2.22 \times 10^{-16}$.

Another source of error in operations on finite computing machines is floating-point arithmetic error. The IEEE 754 standard guarantees that the result of a floating-point summation operator of two numbers $x \oplus y$ is equal to a floating-point

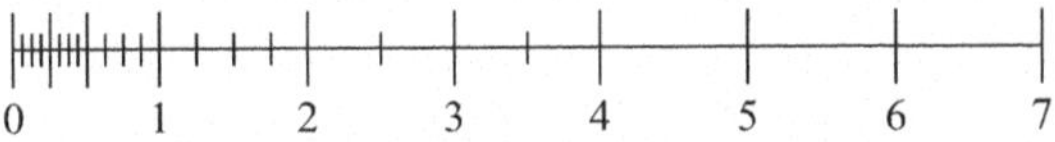

Fig. 7.9 Nonuniform distribution of floating-point numbers for a hypothetical system ($\beta = 2$, $e_{\min} = -1$, $e_{\max} = 3$, $p = 3$ with $s \in \{0, \ldots, 7\}$). The vertical bars denote all the representable numbers in this system [52]

number closest to the real sum of the two numbers $\text{fl}(x + y)$. In most models of error analysis, this is generalized to other operations such as subtraction, multiplication, and division. Note that this does not hold for a sequence of floating-point operations such as $x_1 \oplus x_2 \oplus \cdots \oplus x_n$ due to accumulation of errors. Instead, one can apply Theorem 7.5 [13] to find an upper bound for the accumulated arithmetic error in any computation.

Theorem 7.5 ([13]) *Given a floating-point number system that satisfies IEEE 754 [29] such that $|\delta_i| < \mu_M$. We have*

$$\prod_{i=1}^{k} (1 + \delta_i)^{e_i} = 1 + \theta_k, \tag{7.19}$$

where

$$|\theta_k| \leq \frac{\mu_M}{1 - n\mu_M} := \gamma_k, \; e_i = \pm 1,$$

provided that $k\mu_M < 1$.

Now, we can apply these principal theoretical results for error analysis on DG2 operations [52]. We apply Theorem 7.5 to find reasonable upper and lower bounds for the error involved in computing $\mathbf{\Lambda}$. To estimate the greatest lower bound (infimum) for the magnitude of the roundoff error, we could assume that the computation of $f(\boldsymbol{x})$ is error-free, and the only source of error is in the computation of $\lambda = |\Delta^{(1)} - \Delta^{(2)}|$. We accentuate the variables for their floating-point number representations, e.g. $\hat{\lambda}$. Then, we have the following [52]:

$$\begin{aligned}
\hat{\Delta}_1 &= f(\boldsymbol{x}) \ominus f(\boldsymbol{x}') = (f(\boldsymbol{x}) - f(\boldsymbol{x}'))(1 + \delta_1) = \Delta^{(1)}(1 + \delta_1), \\
\hat{\Delta}_2 &= f(\boldsymbol{y}) \ominus f(\boldsymbol{y}') = (f(\boldsymbol{y}) - f(\boldsymbol{y}'))(1 + \delta_2) = \Delta^{(2)}(1 + \delta_2), \\
\hat{\lambda} &= |\hat{\Delta}_1 \ominus \hat{\Delta}_2| = |\hat{\Delta}_1 - \hat{\Delta}_2|(1 + \delta_3) \\
&= |f(\boldsymbol{x})(1 + \delta_1)(1 + \delta_3) - f(\boldsymbol{x}')(1 + \delta_1)(1 + \delta_3) \\
&\quad - f(\boldsymbol{y})(1 + \delta_2)(1 + \delta_3) + f(\boldsymbol{y}')(1 + \delta_2)(1 + \delta_3)|.
\end{aligned}$$

In the expression for $\hat{\lambda}$, we have $k = 2$ for the maximum number of products of $(1 + \delta_i)$-term. Applying Theorem 7.5, we obtain the following:

$$\begin{aligned}
|\lambda - \hat{\lambda}| &\leq \gamma_2 \big| \big(f(\boldsymbol{x}) - f(\boldsymbol{x}')\big) - \big(f(\boldsymbol{y}) - f(\boldsymbol{y}')\big) \big| \\
&= \gamma_2 \big| \big(f(\boldsymbol{x}) + f(\boldsymbol{y}')\big) - \big(f(\boldsymbol{y}) + f(\boldsymbol{x}')\big) \big| \\
&\leq \gamma_2 \cdot \max \big\{ \big(f(\boldsymbol{x}) + f(\boldsymbol{y}')\big), \big(f(\boldsymbol{y}) + f(\boldsymbol{x}')\big) \big\} := e_{\text{inf}}.
\end{aligned} \tag{7.20}$$

Equation (7.20) is based on the assumption that the codomain of f is nonnegative, i.e. $f : \mathbb{R} \to \mathbb{R}_{\geq 0}$ [52].

To estimate the lowest upper bound (supremum) for the roundoff error, we could not assume that function evaluations are error-free. This means we have to address the challenge of error analysis of Blackbox functions where the exact number of error terms $(1 + \delta_i)$ is unknown a priori in the computation of $f(\star)$. Instead, we adopt a general assumption in error analysis that error grows with the square root of the number of floating-point operations ϕ involved in a computation, $k \approx \sqrt{\phi}$ [26]. For estimating the lowest upper bound of the roundoff error, we could assume that computational error in $|\lambda - \hat{\lambda}|$ is negligible with respect to that of $f(\star)$. Applying Theorem 7.5, we obtain the following [52]:

$$|f(\star) - \hat{f}(\star)| \leq \gamma_{\sqrt{\phi}} \cdot f(\star) := e_{\text{sup}}.$$

We assume a linear relationship between the problem dimensionality D and the number of floating-point operations ϕ. This lets us obtain the following [52]:

$$e_{\text{sup}} = \gamma_{\sqrt{D}} \cdot \max\{f(\boldsymbol{x}), f(\boldsymbol{x}'), f(\boldsymbol{y}), f(\boldsymbol{y}')\}. \tag{7.21}$$

We make several assumptions in formulating (7.21). First, complexity classes can be used to categorize a large body of numerical operations. The linear complexity class $O(n)$ of $f : \mathbb{R}^n \to \mathbb{R}$ includes those involving summation (e.g. $f(\boldsymbol{x}) = \sum_{i=1}^{n} x_i$ where $\phi(n) = n - 1$) and dot-product (e.g. $\boldsymbol{x} \cdot \boldsymbol{y} = \sum_{i=1}^{n} x_i y_i$ where $\phi(n) = 2n - 1$ due to n multiplications and $n - 1$ additions). This complexity class is representative of Blackbox optimization problems. Second, studies [13, 67] have shown that upper bounds derived from Theorem 7.5 are conservative and roundoff errors are smaller than those bounds in practice. In modern computers, the multiply-add (FMA) instruction involving a floating-point multiplication and then an addition commits to only one roundoff error in the worst case. Lastly, our computational studies (Sect. 7.3.4) suggest that the underestimation of e_{sup} is not detrimental to the cooperative Divide-and-Conquer search process with grouping. In general, it results in accurate detection of interacting variables at the expense of missing some separable variables. Conversely, overestimation of e_{sup} results in high detection accuracy of separable variables at the expense of missing interacting variables. The latter case is more detrimental to the coevolutionary search performance [52].

7.3.3.3 Procedural Implementation of DG2

Algorithm 7.3 describes a high-level procedural implementation of DG2 that could be achieved via appropriate modern high-level language (e.g. those that easily handle variable-length structures) [52]. The main procedure **DGROUPING2** first calls on **ISM** that implements the first new procedure for systematic generation of sample points while maximizing point reuse to enable identification of variable interactions (line 2). The procedural implementation of **ISM** is described in Algorithm 7.4 and

Algorithm 7.3 Improved differential grouping [52]

Input:

D Dimensionality of the problem
f Optimization problem
$\overline{x}$ Upper bounds of all decision variables
$\underline{x}$ Lower bounds of all decision variables

Output:

$\boldsymbol{G}$ Final problem decomposition structure $\boldsymbol{G} = (g, \boldsymbol{x}_1, \ldots, \boldsymbol{x}_g, \boldsymbol{x}_{\text{sep}}, \Gamma)$

1: **procedure** DGROUPING2($D, f, \overline{x}, \underline{x}$)
2: $(\boldsymbol{\Lambda}, \mathbf{F}, \check{f}, f_{\text{base}}, \boldsymbol{G}.\Gamma) :=$ **ISM**$(D, f, \overline{x}, \underline{x}, \boldsymbol{G}.\Gamma)$ ▷ Compute entries $|\Delta^{(1)} - \Delta^{(2)}|$ of ▷ interaction structure matrix $\boldsymbol{\Lambda}$
3: $\boldsymbol{\Theta} :=$ **DSM**$(\boldsymbol{\Lambda}, \mathbf{F}, \check{f}, f_{\text{base}}, D)$ ▷ Convert entries of interaction structure matrix $\boldsymbol{\Lambda}$ ▷ to entries of design structure matrix $\boldsymbol{\Theta}$
4: $(k, \boldsymbol{y}_1, \ldots, \boldsymbol{y}_k) :=$ **CONNCOMP**$(\boldsymbol{\Theta})$ ▷ Decompose into nonseparable groups by ▷ identification of connected components in ▷ graph structure (node adjacency matrix) via $\boldsymbol{\Theta}$
5: **initSep**$(\boldsymbol{G}.\boldsymbol{x}_{\text{sep}})$ ▷ Initialize separable group structure $\boldsymbol{G}.\boldsymbol{x}_{\text{sep}} = \{\}$
6: $\boldsymbol{G}.g := 0$
7: **for** $i := 1 : k$ **do**
8: **if** $|\boldsymbol{y}_i| = 1$ **then** ▷ Identify single-variable group
9: **updateSep**$(\boldsymbol{G}.\boldsymbol{x}_{\text{sep}}, \boldsymbol{y}_i)$ ▷ $\boldsymbol{G}.\boldsymbol{x}_{\text{sep}} = \boldsymbol{G}.\boldsymbol{x}_{\text{sep}} \cup \boldsymbol{y}_i$
10: **else**
11: $\boldsymbol{G}.\boldsymbol{x}_g := \boldsymbol{y}_i$
12: $\boldsymbol{G}.g := \boldsymbol{G}.g + 1$ ▷ Update nonseparable groups
13: **end if**
14: **end for**
15: **return** $\boldsymbol{G}$
16: **end procedure**

achieves the lower bound of required fitness evaluations according to Theorem 7.3 to detect variable interactions. The primary output of **ISM** is the interaction structure matrix $\boldsymbol{\Lambda}$ with entries being quantities of $|\Delta^{(1)} - \Delta^{(2)}|$ computations for all pairs of variables $i, j = 1, \ldots, D$. Next, a call is made on **DSM** that implements the second new procedure that dynamically computes the threshold parameter ϵ for detection of variable interactions that consider the magnitude of roundoff errors (line 3). The procedural implementation of **DSM** is described in Algorithm 7.5. In particular, **DSM** finds an appropriate ϵ we have detailed above to convert the interaction structure matrix $\boldsymbol{\Lambda}$ to the design structure matrix $\boldsymbol{\Theta}$ having binary entries $\boldsymbol{\Theta}_{ij} = 1$ if $\boldsymbol{\Lambda}_{ij} > \epsilon$, 0 otherwise. Subsequently, the main procedure **DGROUPING2** performs problem decomposition to establish the separability or nonseparability of all pairs of variables based on information in $\boldsymbol{\Theta}$ that corresponds to a node adjacency matrix (i.e. having the appropriate entries of ones and zeroes) of the graph of variable interactions (lines 4–14). In particular, the decomposition of the decision variables into nonseparable groups can be performed by identifying the connected components of the graph via $\boldsymbol{\Theta}$ with the procedure **CONNCOMP** (e.g. this can be efficiently accomplished in linear time [28]).

Algorithm 7.4 Procedure of DG2 to generate interaction structure matrix [52]

Input:

D	Dimensionality of the problem
f	Optimization problem
$\overline{x}$	Upper bounds of all decision variables
$\underline{x}$	Lower bounds of all decision variables
$G.\Gamma$	Fitness evaluation count

Output:

I	Interaction structure $\boldsymbol{I} = (g, \boldsymbol{x}_1, \ldots, \boldsymbol{x}_g, \boldsymbol{x}_{\text{sep}}, \boldsymbol{G}.\Gamma)$

procedure ISM$(D, f, \overline{x}, \underline{x}, G.\Gamma)$
    initMatrix(Λ, D) ▷ Initialize interaction structure matrix $\Lambda = \mathbf{0}_{D\times D}$
    initNMatrix$(\mathbf{F}, D)$ ▷ Initialize matrix of all NaNs $\mathbf{F}_{D\times D} = \text{NaN}_{D\times D}$
    initNVector$(\check{f}, D)$ ▷ Initialize vector of all NaNs $\check{f}_{D\times 1} = \text{NaN}_{D\times 1}$
    $\boldsymbol{x}^{(1)} := \underline{\boldsymbol{x}}$
    $f_{\text{base}} :=$ **evalF**$(f, \boldsymbol{x}^{(1)})$ ▷ $f_{\text{base}} = f(\boldsymbol{x}^{(1)})$
    $\boldsymbol{G}.\Gamma := 1$
    $\mathbf{m} = \frac{1}{2}(\overline{x} + \underline{x})$
    for $i := 1 : (D-1)$ **do**
        if $\neg$**isNaN**$(\check{f}_i)$ **then**
            $\boldsymbol{x}^{(2)} := \boldsymbol{x}^{(1)}$
            $\boldsymbol{x}_i^{(2)} := \mathbf{m}_i$
            $(\check{f}_i, \boldsymbol{G}.\Gamma) :=$ **evalFC**$(f, \boldsymbol{x}^{(2)}, \boldsymbol{G}.\Gamma)$ ▷ $\check{f}_i = f(\boldsymbol{x}^{(2)}), \boldsymbol{G}.\Gamma = \boldsymbol{G}.\Gamma + 1$
        end if
        for $j := (i+1) : D$ **do**
            if $\neg$**isNaN**$(\check{f}_j)$ **then**
                $\boldsymbol{x}^{(3)} := \boldsymbol{x}^{(1)}$
                $\boldsymbol{x}_j^{(3)} := \mathbf{m}_j$
                $(\check{f}_j, \boldsymbol{G}.\Gamma) :=$ **evalFC**$(f, \boldsymbol{x}^{(3)}, \boldsymbol{G}.\Gamma)$ ▷ $\check{f}_j = f(\boldsymbol{x}^{(3)}), \boldsymbol{G}.\Gamma = \boldsymbol{G}.\Gamma + 1$
            end if
            $\boldsymbol{x}^{(4)} := \boldsymbol{x}^{(1)}$
            $\boldsymbol{x}_i^{(4)} := \mathbf{m}_i$
            $\boldsymbol{x}_j^{(4)} := \mathbf{m}_j$
            $(\mathbf{F}_{ij}, \boldsymbol{G}.\Gamma) :=$ **evalFC**$(f, \boldsymbol{x}^{(4)}, \boldsymbol{G}.\Gamma)$ ▷ $\mathbf{F}_{ij} = f(\boldsymbol{x}^{(4)}), \boldsymbol{G}.\Gamma = \boldsymbol{G}.\Gamma + 1$
            $\Delta^{(1)} := \check{f}_i - f(\boldsymbol{x}^{(1)})$
            $\Delta^{(2)} := \mathbf{F}_{ij} - \check{f}_j$
            $\Lambda_{ij} :=$ **abs**$(\Delta^{(1)} - \Delta^{(2)})$ ▷ $\Lambda_{ij} = |\Delta^{(1)} - \Delta^{(2)}|$
        end for
    end for
    $\boldsymbol{I} := (\Lambda, \mathbf{F}, \check{f}, f_{\text{base}}, \boldsymbol{G}.\Gamma)$
    return $\boldsymbol{I}$
end procedure

Algorithm 7.5 Procedure of DG2 to generate design structure matrix [52]

Input:

$\mathbf{\Lambda}$ Interaction structure matrix
$\mathbf{F}$ $D \times D$ matrix of fitness values associated with detection sample points
$\check{f}$ $D \times 1$ vector of fitness values associated with base points
f_{base} $f(\underline{x})$
D Dimensionality of the problem

Output:

$\mathbf{\Theta}$ Design structure matrix

procedure DSM($\mathbf{\Lambda}$, $\mathbf{F}$, $\check{f}$, f_{base}, D)
  initMatrix($\mathbf{\Theta}$, D) ▷ Initialize design structure matrix $\mathbf{\Theta} = \texttt{NaN}_{D\times D}$
  $\eta_1 := 0$ $\eta_2 := 0$
  $\eta_0 := 0$
  for $i := 1 : (D-1)$ **do**
    for $j := (i+1) : D$ **do**
      $e_{\text{inf}} := \gamma_2 \times \mathbf{max}(f_{\text{base}} + \mathbf{F}_{ij}, \check{f}_i + \check{f}_j)$ ▷ $e_{\text{inf}} = \gamma_2 \times \max\{f_{\text{base}} + \mathbf{F}_{ij}, \check{f}_i + \check{f}_j\}$
      $e_{\text{sup}} := \gamma_{\sqrt{D}} \times \mathbf{max}(f_{\text{base}}, \mathbf{F}_{ij}, \check{f}_i, \check{f}_j)$ ▷ $e_{\text{sup}} = \gamma_{\sqrt{D}} \times \max\{f_{\text{base}}, \mathbf{F}_{ij}, \check{f}_i, \check{f}_j\}$
      if $\mathbf{\Lambda}_{ij} < e_{\text{inf}}$ **then**
        $\mathbf{\Theta}_{i,j} := 0$
        $\eta_0 := \eta_0 + 1$
      else if $\mathbf{\Lambda}_{ij} > e_{\text{sup}}$ **then**
        $\mathbf{\Theta}_{i,j} := 1$
        $\eta_1 := \eta_1 + 1$
      end if
    end for
  end for
  for $i := 1 : (D-1)$ **do**
    for $j := (i+1) : D$ **do**
      $e_{\text{inf}} := \gamma_2 \times \mathbf{max}(f_{\text{base}} + \mathbf{F}_{ij}, \check{f}_i + \check{f}_j)$ ▷ $e_{\text{inf}} = \gamma_2 \times \max\{f_{\text{base}} + \mathbf{F}_{ij}, \check{f}_i + \check{f}_j\}$
      $e_{\text{sup}} := \gamma_{\sqrt{D}} \times \mathbf{max}(f_{\text{base}}, \mathbf{F}_{ij}, \check{f}_i, \check{f}_j)$ ▷ $e_{\text{sup}} = \gamma_{\sqrt{D}} \times \max\{f_{\text{base}}, \mathbf{F}_{ij}, \check{f}_i, \check{f}_j\}$
      if $\mathbf{\Theta}_{i,j} \neq \text{NaN}$ **then**
        $\epsilon := \frac{\eta_0}{\eta_0+\eta_1} \times e_{\text{inf}} + \frac{\eta_1}{\eta_0+\eta_1} \times e_{\text{sup}}$
        if $\mathbf{\Lambda}_{ij} > \epsilon$ **then**
          $\mathbf{\Theta}_{i,j} := 1$
        else
          $\mathbf{\Theta}_{i,j} := 0$
        end if
      end if
    end for
  end for
  return $\mathbf{\Theta}$
end procedure

7.3.4 *Computational Studies on Cooperative Coevolution with Improved Differential Grouping*

We present our systematic and controlled computational studies in [52] to assess the use of DG2 in cooperative CEAs. The first half focuses on grouping performance, and the second half determines the improvements it can provide to cooperative coevolutionary search. In particular, this involves comparisons of DG2 against several state-of-the-art decomposition mechanisms that include DG [51], Global DG (GDG) [46], eXtended DG (XDG) [72], and CCVIL [10]. First, we are able to compare the efficiency of DG2 with other DG-based mechanisms qualitatively [52]. We note that DG2 requires the least number of function evaluations to detect *all* interactions in comparison to other DG-based mechanisms following Theorem 7.3. The total number of function evaluations is $(D^2 + D + 2)/2$ for a D-dimensional problem. GDG requires $(D^2 + 3D + 2)/2$ function evaluations [46], which is larger than DG2's. Unlike DG2 and GDG, both DG and XDG do not use information from a full interaction structure matrix and as such cannot identify overlapping components that share decision variables. Despite this, DG requires $D^2 + D$ function evaluations. XDG improves on DG with a procedure that reduces the total number of function evaluations to slightly less than $D^2 + D$ [72]. This reduction comes about by forgoing checks on x_i–x_j interaction if they both interact with a common variable x_k. Figure 7.10 illustrates the case where XDG fails to distinguish between interaction structures of those associated with overlapping components with various degrees of shared, interacting variables. This can have implications on problem decomposition of such overlapping functions.

Next, we assess the grouping accuracy of DG2 on the CEC'2010 and CEC'2013 large-scale benchmark suites with other decomposition mechanisms. DG, GDG, and XDG require a single parameter setting (note: we choose values based on the original studies). Tables 7.12 and 7.13 summarize the results of all these grouping mechanisms on the CEC'2010 and the CEC'2013 large-scale benchmark suites, respectively. To help us better assess grouping accuracy of these mechanisms, we employ three measurements on the generated interaction structure matrices that are proposed in [46]: ρ_1 (interaction), ρ_2 (independence), and ρ_3 (interaction and independence). ρ_1 measures the accuracy of detecting interactions, ρ_2 measures the

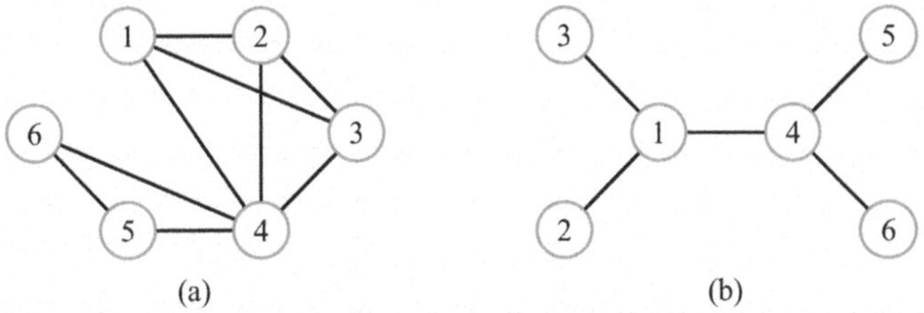

Fig. 7.10 The interaction structures represented by (**a**) and (**b**) cannot be distinguished by DG and XDG [52]

Table 7.12 Grouping accuracy of DG, GDG, and DG2 on the CEC'2010 large-scale benchmark suite. Successful ideal decompositions are shown with "✓" and "✗" otherwise. Overlapping functions are denoted by "-" [52]

	CCVIL				DG												GDG												XDG				DG2			
	–				$\epsilon = 10^{-1}$				$\epsilon = 10^{-3}$				$\epsilon = 10^{-6}$				$\alpha = 10^{-8}$				$\alpha = 10^{-9}$				$\alpha = 10^{-10}$				$\epsilon = 10^{-3}$				–			
Fun.	ρ_1	ρ_2	ρ_3		ρ_1	ρ_2	ρ_3		ρ_1	ρ_2	ρ_3		ρ_1	ρ_2	ρ_3		ρ_1	ρ_2	ρ_3		ρ_1	ρ_2	ρ_3		ρ_1	ρ_2	ρ_3		ρ_1	ρ_2	ρ_3		ρ_1	ρ_2	ρ_3	
f_1	–	100	100	✓	–	100	100	✓	–	100	100	✓	–	65.4	65.4	✗	–	100	100	✓	–	100	100	✓	–	100	100	✓	–	100	100	✓	–	100	100	✓
f_2	–	100	100	✓	–	100	100	✓	–	100	100	✓	–	100	100	✓	–	100	100	✓	–	100	100	✓	–	100	100	✓	–	100	100	✓	–	100	100	✓
f_3	–	99.9	99.9	✗	–	100	100	✓	–	100	100	✓	–	100	100	✓	–	72.6	72.6	✗	–	20.5	20.5	✗	–	4.6	4.6	✗	–	100	100	✓	–	0	0	✗
f_4	2.8	100	99.7	✗	100	76.3	76.3	✗	100	76.3	76.3	✗	100	76.3	76.3	✗	100	100	100	✓	100	100	100	✓	100	100	100	✓	100	7.1	7.3	✗	100	100	100	✓
f_5	4	100	99.7	✓	100	100	100	✓	100	100	100	✓	100	100	100	✓	100	100	100	✓	100	100	100	✓	100	100	100	✓	100	100	100	✓	100	100	100	✓
f_6	3.9	99.9	99.7	✗	100	100	100	✓	100	100	100	✗	100	99.5	99.5	✗	100	100	100	✓	100	100	100	✓	100	100	100	✓	100	100	100	✓	100	39.83	39.98	✗
f_7	3.8	100	99.7	✗	100	100	100	✓	100	69.0	69.0	✗	100	69.0	69.0	✗	100	100	100	✓	100	100	100	✓	100	100	100	✓	100	3.8	4.1	✗	100	100	100	✓
f_8	0	100	99.9	✗	100	76.4	76.4	✗	100	76.4	76.4	✗	100	76.4	76.4	✗	100	100	100	✓	100	100	100	✓	100	100	100	✓	100	11.6	11.6	✗	100	100	100	✓
f_9	3.3	100	97.6	✗	100	100	100	✓	100	100	100	✓	100	64.9	65.8	✗	100	100	100	✓	100	100	100	✓	100	100	100	✓	100	100	100	✓	100	100	100	✓
f_{10}	3.9	100	97.6	✗	99.9	100	99.9	✓	100	100	100	✓	100	100	100	✓	100	100	100	✓	100	100	100	✓	100	100	100	✓	100	100	100	✓	100	100	100	✓
f_{11}	3.9	99.9	97.6	✗	63.5	100	99.1	✓	99.6	100	99.9	✓	100	93.7	93.9	✗	100	98.5	98.6	✗	100	83.9	84.3	✗	100	76.2	76.8	✗	99.9	100	100	✓	100	74.39	75.02	✗
f_{12}	3.9	100	97.6	✗	100	100	100	✓	100	100	100	✓	100	97.1	97.1	✗	100	100	100	✓	100	100	100	✓	100	100	100	✓	100	100	100	✓	100	100	100	✓
f_{13}	0	100	99.9	✗	100	100	100	✓	100	87.0	87.0	✗	100	87.0	87.0	✗	100	100	100	✓	100	100	100	✓	100	100	100	✓	100	56.7	56.7	✗	100	100	100	✓
f_{14}	3.2	100	95.2	✗	100	100	100	✓	100	100	100	✓	100	87.7	88.3	✗	100	100	100	✓	100	100	100	✓	100	100	100	✓	100	100	100	✓	100	100	100	✓
f_{15}	3.9	100	95.2	✗	99.9	100	99.9	✓	100	100	100	✓	100	100	100	✓	100	100	100	✓	100	100	100	✓	100	100	100	✓	100	100	100	✓	100	100	100	✓
f_{16}	0.1	99.8	94.9	✗	60.5	100	98.0	✓	99.5	100	99.9	✓	100	100	100	✓	100	100	100	✓	100	100	100	✓	100	100	100	✓	99.9	100	99.9	✓	100	100	100	✓
f_{17}	3.8	100	95.2	✗	100	100	100	✓	100	100	100	✓	100	83.3	84.1	✗	100	100	100	✓	100	100	100	✓	100	100	100	✓	100	100	100	✓	100	100	100	✓
f_{18}	1.6	100	99.8	✗	100	100	100	✓	100	74.4	74.5	✗	100	74.4	74.5	✗	99.9	100	100	✗	100	100	100	✓	100	100	100	✓	100	15.0	15.2	✗	100	100	100	✓
f_{19}	0.2	–	0.20	✓	100	–	100	✓	100	–	100	✓	100	–	100	✓	99.8	–	99.8	✗	100	–	100	✓	100	–	100	✓	100	–	100	✓	100	–	100	✓
f_{20}	0	100	99.8	–	100	100	100	–	100	91.6	91.6	–	100	91.6	91.6	–	100	100	100	–	100	100	100	–	100	100	100	–	100	70.3	70.4	–	100	100	100	–
# ✓				4				17				13				7				15				17				17				14				16
Median	3.5	100	99.7		100	100	100		100	100	100		100	91.6	92.8		100	100	100		100	100	100		100	100	100		100	100	100		100	100	100	
Mean	2.6	99.9	93.4		95.5	97.5	97.5		99.9	93.4	93.7		100	87.7	88.4		100	98.5	98.5		100	95.0	95.2		100	93.7	94.1		99.9	77.0	78.2		100	90.2	90.7	
Std.	1.6	0.05	22.0		12.6	7.5	7.2		0.2	10.9	10.7		0.0	12.9	12.8		0.1	6.3	6.1		0.0	18.4	17.9		0.0	22.3	21.7		0.03	37.7	37.0		0.0	26.3	25.7	

Table 7.13 Grouping accuracy of DG, GDG, and DG2 on the CEC'2013 large-scale benchmark suite. Successful ideal decompositions are shown with "✓" and "X" otherwise. Overlapping functions are denoted by "-" [52]

	CCVIL				DG												GDG												XDG				DG2			
	–				$\epsilon = 10^{-1}$				$\epsilon = 10^{-3}$				$\epsilon = 10^{-6}$				$\alpha = 10^{-8}$				$\alpha = 10^{-9}$				$\alpha = 10^{-10}$				$\epsilon = 10^{-3}$				–			
Fun.	ρ_1	ρ_2	ρ_3		ρ_1	ρ_2	ρ_3		ρ_1	ρ_2	ρ_3		ρ_1	ρ_2	ρ_3		ρ_1	ρ_2	ρ_3		ρ_1	ρ_2	ρ_3		ρ_1	ρ_2	ρ_3		ρ_1	ρ_2	ρ_3		ρ_1	ρ_2	ρ_3	
f_1	–	100	100	✓	–	100	100	✓	–	100	100	✓	–	71.8	71.8	X	–	100	100	✓	–	100	100	✓	–	100	100	✓	–	100	100	✓	–	100	100	✓
f_2	–	100	100	✓	–	100	100	✓	–	100	100	✓	–	100	100	✓	–	100	100	✓	–	100	100	✓	–	100	100	✓	–	100	100	✓	–	100	100	✓
f_3	–	100	100	✓	–	100	100	✓	–	100	100	✓	–	74.4	74.4	X	–	34.5	34.5	X	–	7.5	7.5	X	–	1.3	1.3	X	–	100	100	✓	–	0	0	X
f_4	3.0	100	98.3	X	100	85.9	86.2	X	100	85.5	85.8	X	100	85.5	85.8	X	82.2	100	99.7	✓	98.2	100	99.9	✓	99.8	100	99.9	✓	100	4.3	6.0	X	100	100	100	✓
f_5	3.2	100	98.3	X	41.9	100	99.0	X	62.3	100	99.3	X	99.9	100	99.9	✓	38.5	100	98.9	X	40.6	100	98.9	X	42.1	100	99.0	X	98.0	100	99.9	✓	99.97	100	100	✓
f_6	3.0	100	98.3	X	63.5	100	99.3	X	90.6	100	99.8	X	99.5	76.7	77.1	X	72.7	100	99.5	X	90.4	100	99.8	X	92.7	100	99.8	X	93	100	99.8	X	99.98	50.45	51.30	X
f_7	2.7	100	98.3	X	100	76.7	77.1	X	100	76.7	77.1	X	100	76.7	77.1	X	96	100	99.9	X	99.2	100	99.9	✓	99.9	100	99.9	✓	100	2.2	3.9	X	100	100	100	✓
f_8	2.3	100	93.3	X	83.9	98.9	97.9	X	83.9	98.9	97.9	X	83.9	98.9	97.9	X	11.3	100	94.0	X	17.8	100	94.4	X	33.7	100	95.5	X	99.4	6.8	13.1	X	70.72	100	98.01	X
f_9	2.7	100	93.4	X	98.2	100	99.8	X	99.9	100	99.9	✓	100	88.4	89.2	X	60.5	100	97.3	X	66.7	100	97.7	X	85.5	100	99.0	X	100	100	100	✓	99.99	100	100	✓
f_{10}	2.7	100	93.4	X	57.5	100	97.1	X	83.2	100	98.8	X	98.6	100	99.9	✓	27.7	100	95.1	X	58.1	100	97.1	X	79.9	100	98.6	X	83.6	100	98.8	X	99.93	100	99.99	✓
f_{11}	2.5	100	93.3	X	99.9	72.6	74.5	X	99.9	72.6	74.5	X	99.9	72.6	74.5	X	82.0	100	98.8	X	87.1	100	99.1	X	90.9	100	99.3	X	100	2.0	8.7	X	99.95	100	99.99	✓
f_{12}	0	100	99.8	–	100	100	100	–	100	91.0	91.0	–	100	91.0	91.0	–	100	100	100	–	100	100	100	–	100	100	100	–	100	70.2	70.3	–	100	100	100	–
f_{13}	2.5	100	91.9	–	100	91.1	91.9	–	100	91.1	91.9	–	100	91.1	91.9	–	91.6	100	99.3	–	97.4	100	99.7	–	99.4	100	99.9	–	100	1.3	9.4	–	100	100	100	–
f_{14}	2.4	100	91.9	–	100	96.5	96.8	–	100	96.5	96.8	–	100	96.5	96.8	–	96.3	100	99.7	–	97.9	100	99.8	–	98.3	100	99.8	–	100	4.5	12.4	–	99.97	100	99.99	–
f_{15}	0	–	0	✓	100	–	100	✓	100	–	100	✓	100	–	100	✓	100	–	100	✓	100	–	100	✓	100	–	100	✓	100	–	100	✓	100	–	100	✓
#✓				4				4				5				4				4				5				5				6				9
Median	2.60	100	98.3		100	100	99.0		100	99.5	98.8		100	89.7	91.0		82.1	100	99.5		93.9	100	99.8		95.5	100	99.8		100	85.1	98.8		100	100	100	
Mean	2.25	100	90.0		87.1	94.4	94.6		93.3	93.7	94.2		98.5	87.4	88.5		71.6	95.3	94.4		79.5	93.4	92.9		85.2	93.0	92.8		97.8	56.5	61.4		97.5	89.3	90.0	
Std.	1.08	0	25.1		20.8	9.4	8.6		11.7	9.3	8.6		4.6	11.0	10.8		30.5	17.5	16.7		27.6	24.7	23.7		23.1	26.4	25.3		4.9	48.2	45.0		8.4	28.9	27.9	

accuracy of detecting separable variables, and ρ_3 measures the overall accuracy. The check marks "✓" and crosses "X" indicate whether the final decomposition of each function matches the ideal grouping of the benchmark problem. A count # of correctly grouped functions is provided for each decomposition mechanism. However, this count is not performed on functions with overlapping components because their ideal decompositions are not known. Instead, a dash symbol "–" is indicated for each of these functions. For DG2, the reported accuracy of ρ measurements could be slightly less than 100%, but the overall grouping matches the ideal grouping as indicated by ✓. This is due to the connected components algorithm **ConnComp** in DG2 that returns the correct grouping even if some of the interactions are missed. For XDG and CCVIL, full separability is assumed at the start with the generated interaction structure matrices initialized with zero entries $\boldsymbol{\Theta} = \mathbf{0}_{D\times D}$. This can have an impact on the measurements because if a pair of variables is not checked for interaction, the associated entries in $\boldsymbol{\Theta}$ will take on the default value of zero for the latter ρ computations [52].

In general, our results in Tables 7.12 and 7.13 indicate that DG2 outperforms all other decomposition mechanisms on the CEC'2010 and CEC'2013 benchmark suites. The overall performance gap is wider on the more difficult CEC'2013 benchmark suite. There are some individual exceptions such as DG2 having lower performance measurements on f_3 and f_6 that constitute instances of the Ackley function. In particular, the Ackley function is not additively separable [24] but is correctly identified by DG2. However, these functions are reported as separable in the benchmark suites according to Definition 7.1. In the following, we make several observations with respect to the other decomposition mechanisms. CCVIL has the least overall grouping performance, XDG has a similar overall grouping performance as DG with the same $\epsilon = 10^{-3}$ setting, and DG demonstrates sensitivity to its control parameter ϵ especially on the CEC'2010 benchmark suite. This is due to the use of a fixed ϵ setting, which in practice can be affected by finite computation error in $\lambda = |\Delta^{(1)} - \Delta^{(2)}|$ that grows with the magnitude of function evaluation. The ϵ setting is smaller than computational errors on some functions, in which case some separable variables will be considered as nonseparable [52].

GDG sets ϵ proportional to the magnitude of the function evaluation, with the rationale that a larger objective function value results in a larger computational error value in λ. The following is a simple heuristic implementation of setting ϵ in GDG:

$$\epsilon = \alpha \cdot \min\{f(\boldsymbol{x}_1), \ldots, f(\boldsymbol{x}_n)\}, \tag{7.22}$$

where $\boldsymbol{x}_1, \ldots, \boldsymbol{x}_n$ are n random sample points. Our results indicate that GDG is overall effective on the CEC'2010 benchmarks but less effective on the CEC'2013 benchmarks. Closer inspection reveals that GDG systematically overestimates the computational error that results in large ϵ values. Table 7.14 summarizes the ϵ values computed by GDG and DG2 for selected functions. For DG2, we provide statistical indicators via mean, median, and maximum values given that DG2 employs different ϵ settings for detection of variable interactions. Although calculations of ϵ in DG2 via (7.20) and (7.21) and GDG via (7.22) are functions of objective function

Table 7.14 ϵ values of GDG and DG2 on selected functions [52]

	GDG			DG2		
Functions	$\alpha = 10^{-8}$	$\alpha = 10^{-9}$	$\alpha = 10^{-10}$	Min	Median	Max
f_1	8.93e+03	8.93e+02	8.93e+01	4.06e-04	4.16e-04	4.18e-04
f_4	2.75e+06	2.75e+05	2.75e+04	2.34e-01	3.14e-01	3.78e-01
f_8	1.94e+11	1.94e+10	1.94e+09	1.76e+04	2.37e+04	3.48e+04
f_{13}	8.11e+12	8.11e+11	8.11e+10	1.64e+06	2.72e+06	1.20e+07
f_{15}	1.19e+04	1.19e+03	1.19e+02	1.25e-02	1.25e-02	1.27e-01

values, there are two main differences. First, the constant $\alpha = \{10^{-8}, 10^{-9}, 10^{-10}\}$ used in GDG is set to values significantly larger than both $\gamma_2 \approx 2.2204 \times 10^{-16}$ and $\gamma_{\sqrt{D}} \approx 3.5108 \times 10^{-15}$, $D = 1000$ used in DG2. Second, GDG employs a sample of n objective function values to detect underlying interactions that could be present in a problem, whereas DG2 employs quantities produced from (7.20) and (7.21) that are directly involved in computing λ for detection of interaction in a particular pair of variables. These differences have led to overestimation of ϵ by GDG, which allows for high accuracy of detecting separable variables as measured by ρ_2 at the expense of a low interaction detection accuracy as measured by ρ_1. In general, GDG with a high ϵ setting tends to classify most variables as fully separable, which as a consequence means it is less sensitive to weak interactions. Such a grouping behaviour of GDG is made more apparent on the CEC'2013 benchmarks with imbalance in the contribution of each component to the overall objective value [52].

We close the first half of the subsection with sensitivity analysis on the settings of complexity class used to estimate roundoff errors in DG2 operations and imbalance levels in contributions of subcomponents to the global fitness in the benchmark problems. We start our investigation on the assumption of linear relationship between complexity of problems (e.g. those encountered in Blackbox optimization such as the CEC'2010 and CEC'2013 benchmark suites) and the number of floating-point operations in formulating (7.21) that provides the means to calculate e_{sup}. This lets us analyse how robust DG2 is with respect to deviations from the assumed linear $O(D)$ complexity class to higher orders such as quadratic $O(D^2)$ and cubic $O(D^3)$ complexity classes. Table 7.15 summarizes the grouping accuracy of DG2 with respect to different complexity class settings. Similar overall results are obtained for linear and quadratic complexity class settings, which then decrease when a cubic complexity class setting is made. The latter results are due to the overestimation of e_{sup} and subsequently overestimation of roundoff errors that causes DG2 to treat weakly interacting variables as separable. For example, grouping accuracy of DG2 decreases when it is applied to partially separable problems of f_{10} and f_{11}. In general, our results indicate that underestimation of e_{sup} is less detrimental to DG2 grouping performance in comparison to the case of overestimation of e_{sup} [52].

Next, we analyse the robustness of DG2 on the benchmark problems with varying degrees of imbalance level among the components that contribute to the global fitness. This is because DG2 uses function evaluations to identify interaction

Table 7.15 Analysis of DG2 to various complexity class settings for calculation of e_{sup} on the CEC'2013 large-scale benchmark suite [52]

	$O(D)$				$O(D^2)$				$O(D^3)$			
Fun.	ρ_1	ρ_2	ρ_3		ρ_1	ρ_2	ρ_3		ρ_1	ρ_2	ρ_3	
f_1	–	100	100	✓	–	100	100	✓	–	100	100	✓
f_2	–	100	100	✓	–	100	100	✓	–	100	100	✓
f_3	–	0	0	×	–	0.0016	0.0016	×	–	0.08	0.08	×
f_4	100	100	100	✓	100	100	100	✓	100	100	100	✓
f_5	99.97	100	100	✓	99.93	100	99.99	✓	98.69	100	99.97	✓
f_6	99.98	50.45	51.30	×	99.97	54.56	55.35	×	99.77	68.86	69.40	×
f_7	100	100	100	✓	100	100	100	✓	100	100	100	✓
f_8	70.72	100	98.01	×	70.07	100	97.97	×	63.58	100	97.53	×
f_9	99.99	100	100	✓	99.99	100	100	✓	99.78	100	99.98	✓
f_{10}	99.93	100	99.99	✓	99.07	100	99.93	✓	85.56	100	99.02	×
f_{11}	99.95	100	99.99	✓	99.47	100	99.96	✓	97.72	100	99.84	×
f_{12}	100	100	100	–	100	100	100	–	100	100	100	–
f_{13}	100	100	100	–	100	100	100	–	99.99	100	100	–
f_{14}	99.97	100	99.99	–	99.75	100	99.98	–	99.12	100	99.92	–
f_{15}	100	–	100	✓	100	–	100	✓	100	–	100	✓
Success rate				9				9				7
Median	100	100	100		100	100	100		99.8	100	100	
Mean	97.5	89.3	90.0		97.4	89.6	90.2		95.4	90.6	91.0	
Std.	8.4	28.9	27.9		8.6	28.5	27.5		10.8	27.4	26.4	

structures in a problem. In particular, the functions f_4–f_{11} and f_{13}–f_{14} have the general form $\sum_{i=1}^{m} w_i f_i(\boldsymbol{x}_i)$, $w_i = 10^{c\mathcal{N}(0,1)}$. The parameter c determines the variance among the weights and is fixed to a value to generate a problem, e.g. $c = 3$ in the CEC'2013 large-scale benchmark suite. For our sensitivity analysis on the performance of cooperative CEA with DG2, we consider $c \in \{0, 1, 2, 3, 4\}$. Table 7.16 summarizes our results, and they indicate that the overall grouping accuracy of DG2 is stable with various imbalance levels, except at $c = 4$ problem setting. The measurements on independence for detecting separable variables (ρ_2) are very high and stable across various imbalance levels c, but those on detecting interacting variables (ρ_1) drop in values when the imbalance level increases. Closer inspections on DG2 operations of f_4–f_{11} at $c = 4$ reveal that the nonseparable components missed by DG2 are always those with the smallest weight. In summary, our analysis suggests DG2 is robust to moderate imbalance levels and that any inaccuracy when imbalance level in the problem is very high comes from identifying very weak interacting variables as fully separable. Note that very high imbalance levels as simulated with CEC'2013 problems with $c = 4$ setting rarely occur in real-world optimization. Furthermore, treating very weak interacting variables as fully separable may not be detrimental to cooperative coevolutionary search. For example, DG2's grouping error in f_8 is mostly attributed to detecting weakly interacting variables as fully separable, resulting in lower ρ_1 measurements. We

Table 7.16 Sensitivity analysis of DG2 to various imbalance levels (CEC'2013) [52]

	$w_i = 1$				$w_i = 10^{\mathcal{N}(0,1)}$				$w_i = 10^{2\mathcal{N}(0,1)}$				$w_i = 10^{3\mathcal{N}(0,1)}$				$w_i = 10^{4\mathcal{N}(0,1)}$			
Fun.	ρ_1	ρ_2	ρ_3	ρ_1	ρ_2	ρ_3	ρ_1	ρ_2	ρ_3	ρ_1	ρ_2	ρ_3	ρ_1	ρ_2	ρ_3					
f_4	100	100	100	✓	100	100	100	✓	100	100	100	✓	100	100	100	✓	100	100	100	✓
f_5	100	100	100	✓	100	100	100	✓	100	100	100	✓	99.97	100	99.99	✓	100	100	100	✓
f_6	100	50.16	51.02	×	100	50.16	51.02	×	100	50.22	51.08	×	99.98	50.45	51.30	×	99.98	50.30	51.15	×
f_7	100	100	100	✓	100	100	100	✓	100	100	100	✓	100	100	100	✓	96.64	100	99.94	×
f_8	100	100	100	✓	100	100	100	✓	99.99	100	99.99	✓	70.72	100	98.01	×	95.85	100	99.71	×
f_9	100	100	100	✓	100	100	100	✓	100	100	100	✓	99.99	100	99.99	✓	78.43	100	98.53	×
f_{10}	100	100	100	✓	100	100	100	✓	100	100	100	✓	99.93	100	99.99	✓	80.28	100	98.66	×
f_{11}	100	100	100	✓	100*	100	100*	×*	100	100	100	✓	99.95	100	99.99	✓	80.34	100	98.66	×
f_{13}	100	100	100	–	100	100	100	–	100	100	100	–	100	100	100	–	97.94	100	99.83	–
f_{14}	100	100	100	–	100	100	100	–	100	99.99	99.99	–	99.97	100	99.99	–	99.85	100	99.98	–
Success rate				7				6				7				6				2
Mean	100	95.0	95.1		100	95.0	95.1		100.0	95.0	95.1		97.1	95.0	94.9		92.9	95.0	94.6	
Median	100	100	100		100	100	100		100	100	100		100	100	100		97.3	100	99.8	
Std.	0.0	15.8	15.5		0.0	15.8	15.5		0.0	15.7	15.5		9.3	15.7	15.3		9.3	15.7	15.3	

* The values are not exact due to rounding. Two interactions are miscalculated which cause a small error. Hence, the final grouping does not match the ideal case.

will show later in our results (e.g. Table 7.17) that such inaccuracies do not affect the overall cooperative coevolutionary search performance [52].

In the second half of this subsection, we assess the impact DG2 has on the search performance of cooperative CEA when applied to solve large-scale optimization problems. We will evaluate the cooperative CEA with different grouping mechanisms on the more challenging CEC'2013 benchmark suite [37]. There are four classes of functions which we designate C1–C4. For the cooperative CEAs we compare, they include the classical implementation that optimizes subcomponents in a round-robin fashion and one that further incorporates a contribution-based mechanism that allocates more computing resources for optimization of the subcomponent with higher contribution to the overall solution quality. For the latter, we employ the CBCC3 variant with $p_t = 0$ setting [50]. As before in Sect. 7.3.2 where we have assessed the original DG, the other common settings for our computational study here include the subcomponent optimizer SaNSDE, the population size $N = 50$ [90], and the maximum number of fitness evaluations #FE $= 3 \times 10^6$ [37] that are allocated for both problem decomposition and optimization processes (with the exception of ideal grouping variant). Each experiment set is repeated for 25 independent runs to allow statistical comparison via a two-tailed Wilcoxon rank-sum test with $\alpha = 0.05$ [52].

Table 7.17 summarizes our results comparing the performance of DG2 against DG, XDG, CCVIL, and ideal grouping in both settings of classical and contribution-based cooperative coevolution. The entries in the table indicate the number of wins, ties, and losses. Our results clearly indicate that cooperative CEA with DG2 has a better overall search performance compared to those employing other decomposition mechanisms in both classical and contribution-based settings. The cooperative CEA with ideal grouping is used to provide the search performance ceiling. We observe that in a classical cooperative coevolutionary setting, DG2 is able to match the ideal grouping on nine functions and then loses out on six functions, despite having fewer computational budget for optimization with the DG2 process consuming $0.5 \times 10^6 \approx (1/6) \cdot$ #FE. In a contribution-based cooperative coevolutionary setting, DG2 is outperformed by ideal grouping on 11 out of 15 functions. The gap in overall search performance between DG2 and ideal grouping

Table 7.17 Performance comparison of DG2 against DG, XDG, CCVIL, and ideal grouping on classical and contribution-based cooperative coevolutionary settings. The numbers of wins, ties, and losses of DG2 against other decomposition mechanisms are reported [52]

	Classical CC				Contribution-based CC			
	DG w/t/l	XDG w/t/l	CCVIL w/t/l	Ideal w/t/l	DG w/t/l	XDG w/t/l	CCVIL w/t/l	Ideal w/t/l
C1: f_1–f_3	1/2/0	2/1/0	0/1/2	0/1/2	1/2/0	3/0/0	2/0/1	0/1/2
C2: f_4–f_7	3/1/0	2/1/1	3/0/1	0/3/1	3/1/0	1/2/1	4/0/0	0/3/1
C3: f_8–f_{11}	2/1/1	2/0/2	3/0/1	0/4/0	2/1/1	1/2/1	3/1/0	0/0/4
C4: f_{12}–f_{15}	2/0/2	1/0/3	2/0/2	0/1/3	2/0/2	1/0/3	2/0/2	0/0/4
Total	8/4/3	7/2/6	8/1/6	0/9/6	8/4/3	6/4/5	11/1/3	0/4/11

is due to the former having fewer computational resources for the optimization process [52].

Next, we investigate the impact of having accurate problem decomposition on the performance of contribution-based cooperative CEA against the classical cooperative CEA that employs the same decomposition mechanism. The results are summarized in Table 7.18, and they generally indicate that accurate decomposition mechanism can synergistically work with the contribution-based mechanism to further improve the search performance of cooperative CEA. DG2 and ideal grouping provided similar overall performance enhancement with 8 wins and 7 ties. This is followed by CCVIL with 7 wins and 4 ties. Both DG and XDG provided similar overall performance enhancement with 5 wins and 7 ties. Closer inspections reveal that ties for comparisons involving DG2 are mostly on fully separable functions (C1) and overlapping, nonseparable functions (C4) where no decomposition is performed. In effect, cooperative CEA with mechanisms that could enhance search performance (DG2 and CBCC3) reduces to that of the default suboptimizer (SaNSDE). One exception involves both DG and CCVIL with increased search performance in contribution-based cooperative CEA over the classical counterpart on two functions. Closer inspection reveals that DG and CCVIL decompose some C4 functions into smaller subcomponents even though they constitute those with shared variables. Despite search performance benefit in experimental settings, the ideal problem decompositions of these C4 functions are unknown. For the remaining partially separable functions (C2 and C3), the contribution-based cooperative CEA demonstrates superior performance over the classical counterpart when DG2 and the ideal grouping are used. In general, our results suggest that contribution-based cooperative CEA can be used with relatively accurate grouping mechanism. This is consistent with a detailed sensitivity analysis in [32], indicating that contribution-based cooperative CEA has some degree of sensitivity to grouping errors but is able to perform at the level of its classical counterpart. This makes contribution-based mechanism a safe choice for Blackbox optimization [52].

Lastly, we compare the performance of the best overall variant of cooperative CEA we have studied (i.e. CBCC3-DG2) against some state-of-the-art optimizers

Table 7.18 Performance comparison of contribution-based against classical cooperative coevolution using different decomposition mechanisms. The numbers of wins, ties, and losses of the contribution-based cooperative CEA against classical cooperative CEA with respect to the use of specific decomposition mechanisms are reported [52]

	CCVIL	DG	DG2	XDG	Ideal
Categories	w/t/l	w/t/l	w/t/l	w/t/l	w/t/l
C1: f_1–f_3	1/1/1	0/3/0	0/3/0	0/3/0	0/3/0
C2: f_4–f_7	1/0/3	1/1/2	4/0/0	2/0/2	4/0/0
C3: f_8–f_{11}	3/1/0	2/1/1	4/0/0	3/0/1	4/0/0
C4: f_{12}–f_{15}	2/2/0	2/2/0	0/4/0	0/4/0	0/4/0
Total	7/4/4	5/7/3	8/7/0	5/7/3	8/7/0

such as the Multiple Offspring Framework (MOS) [34], MA-SW-Chains [47], and CMA-ES [25]. The first two are, respectively, the top rank optimizer in the CEC'2013 and CEC'2010 competition on large-scale optimization. We use the recommended parameter settings of their original studies. Table 7.19 summarizes results on partially separable functions f_4–f_{11} since our focus is on the performance of cooperative CEA with automatic problem decomposition where DG2 is able to discover the underlying variable interaction structure of these functions. Our results show that no single algorithm is superior across all functions. It is crucial to note that although CBCC3-DG2 uses the outdated SaNSDE as the suboptimizer, the synergistic combination of the DG2 grouping and CBCC3 contribution-based mechanisms allows it to be competitive against the other three state-of-the-art optimizers. This is consistent with past studies [5, 40, 41, 46, 88] indicating that the cooperative coevolutionary framework allows for the construction of cooperative CEAs with different suboptimizers in which their search performance can be scaled up with appropriate mechanisms such as grouping and contribution-based [52].

Table 7.19 Results of CBCC3-DG2, MOS, and CMA-ES on the CEC'2013 large-scale benchmark functions, each repeated with 25 independent runs. The **bold** entries are statistically significantly better via Wilcoxon rank-sum test with Holm p-value correction, $\alpha = 0.05$ [52]

	Stats.	CBCC3-DG2	MOS [34]	CMA-ES [25]	MA-SW-Chains [47]
f_4	Median	**2.77e+07**	1.56e+08	4.10e+08	4.27e+09
	Mean	3.39e+07	1.74e+08	4.30e+08	4.58e+09
	StDev	1.77e+07	8.02e+07	1.17e+08	2.51e+09
f_5	Median	**2.11e+06**	6.79e+06	**2.06e+06**	**1.81e+06**
	Mean	2.14e+06	6.94e+06	2.04e+06	1.87e+06
	StDev	4.24e+05	9.03e+05	2.64e+05	3.13e+05
f_6	Median	1.05e+06	**1.39e+05**	6.09e+05	1.01e+06
	Mean	1.05e+06	1.48e+05	6.01e+05	1.01e+06
	StDev	3.37e+03	6.56e+04	1.28e+05	1.56e+04
f_7	Median	2.94e+07	1.62e+04	**6.83e+02**	3.92e+06
	Mean	2.95e+07	1.62e+04	3.00e+03	3.45e+06
	StDev	2.78e+07	9.29e+03	5.23e+03	1.29e+06
f_8	Median	**1.41e+10**	8.08e+12	1.00e+13	4.90e+13
	Mean	4.28e+10	8.00e+12	1.13e+13	4.85e+13
	StDev	8.86e+10	3.13e+12	6.08e+12	1.04e+13
f_9	Median	1.68e+08	3.87e+08	1.74e+08	**1.08e+08**
	Mean	1.70e+08	3.83e+08	1.80e+08	1.07e+08
	StDev	3.16e+07	6.42e+07	2.28e+07	1.71e+07
f_{10}	Median	9.30e+07	**1.18e+06**	1.42e+07	9.18e+07
	Mean	9.28e+07	9.01e+05	1.64e+07	9.18e+07
	StDev	7.16e+05	5.17e+05	1.44e+07	1.08e+06
f_{11}	Median	5.83e+08	4.48e+07	**7.72e+06**	2.15e+08
	Mean	6.59e+08	5.22e+07	9.37e+06	2.19e+08
	StDev	2.80e+08	2.10e+07	6.53e+06	3.04e+07

Our computational studies have verified improvements DG2 has over its predecessor DG with the introduction of two new procedures. The first procedure makes DG2 more efficient and addresses one main weakness of DG on fully separable functions by employing half the total number of function evaluations required in interaction tests over all pairs of variables. The same procedure also constructs the full interaction structure matrix that allows DG2 to detect overlapping components with shared decision variables. This increases the use of DG2 over DG on more problems, demonstrated by DG2's superior grouping performance over both CEC'2010 and CEC'2013 benchmark suites. The second procedure makes DG2 more effective and addresses another main weakness of DG that requires tuning of the threshold parameter ϵ setting crucial for high interaction detection accuracy across different problems. This procedure enables DG2 to be parameter-free through automatic and dynamic calculation of ϵ while taking computational errors into account from arithmetic operations in DG2. This has led to DG2 having higher overall grouping accuracy than DG across more problems. It is also robust to roundoff errors present in the differential grouping approach to variable interaction detection and imbalance levels in problems whereby different components have varied contributions to the fitness value [52].

7.4 Discussion and Conclusion

In this chapter, we have presented studies to develop more effective and efficient approaches of cooperative CEAs to solve large-scale optimization problems common in real-world settings in a Divide-and-Conquer manner. The cooperative CEAs we have proposed and investigated in detail come about from theoretical studies we have carried out that are directly informative towards construction of new procedures that address main issues faced in cooperative coevolutionary Divide-and-Conquer problem solving. We have focused on a new framework of cooperative coevolution to solve large-scale additively separable continuous optimization problems. This is a significant and practical improvement over the original cooperative coevolution framework that solves separable problems in which the decomposed subcomponents consist of a single decision variable. Crucially, this allows us to design new mechanisms that address those main issues and be incorporated into cooperative CEAs. We have validated these new cooperative CEAs that feature these mechanisms with systematic and controlled computational studies with benchmark suites. Our presentation of these studies follows the historical development of those mechanisms with respect to gaps in the literature. Each time the mechanism is incorporated into the cooperative CEA it is shown to be competitive with state-of-the-art optimizers with respect to the benchmark suite when they are introduced.

In Sect. 7.2, we have presented our first study [88] to develop a new cooperative coevolutionary framework for automatic problem decomposition in large-scale optimization problems. Here, the main issue we have addressed is the need for an effective approach that allows for automatic decomposition of a large class of nonseparable problems encountered in real-world settings. At this stage of development in the literature, we aim for effective procedural implementation of cooperative CEAs to solve additively separable problems without substantial access to domain knowledge. This means developing a mechanism to capture the underlying variable interdependencies in those additively separable problems. For cooperative CEAs incorporating such a mechanism, the main working principle is that there would be minimal dependencies between decomposed subcomponents. This occurs when the mechanism correctly identifies and groups interdependent variables together. Then, cooperative coevolutionary search works with subcomponents going through evolutionary optimization processes in turn. Section 7.2.1 provided theoretical motivation for the design of the random grouping strategy. Our theoretical analysis has demonstrated a high chance the random grouping mechanism groups two interacting variables together over several coevolutionary cycles. We have procedurally implemented the new framework as DECC-G. Our systematic and controlled computational studies have demonstrated the effectiveness of DECC-G against the baseline optimizer SaNSDE (which is used as the suboptimizer in DECC-G).

Next in Sect. 7.3 we have presented two other studies [51, 52] to improve the new cooperative coevolution framework with automatic problem decomposition for large-scale optimization problems [88]. This involves a new mechanism for automatic detection of interacting variables called differential grouping. The first study [51] introduces the DG mechanism to address the main issue of having to set the parameters involving a fixed number and size of subcomponents required for the random grouping mechanism [88]. Our theoretical studies on additively separable functions provided the means with which nonseparability of variables in a function can be computationally determined. This informs the design of differential grouping that carries out direct computations of the forward difference $\Delta_{\delta,x_p}[f](\boldsymbol{x})|_{x_p,x_q}$ of the function f for two different values of x_q with respect to a particular x_p value to detect interaction between the two decision variables x_p, x_q. Subsequently, we are able to construct the DG mechanism for automatic identification and grouping of interacting variables that is independent of the optimizer. We have carried out systematic and controlled computational studies to evaluate the impact of DG on the search performance of cooperative CEAs with the CEC'2010 large-scale benchmark suite up to 1000 dimensions [51]. They have validated the effectiveness and efficiency of a cooperative CEA with DG (i.e. DECC-DG) against other cooperative CEAs and evolutionary optimizers that include the top ranking optimizer in the CEC'2010 large-scale optimization competition.

Our second study [52] introduced the improved differential grouping (DG2) mechanism. It aimed to address four issues in the original DG with two new features, each enabled by the incorporation of a procedure. The first procedure involves systematic generation of sample points maximizing point reuse for the identification process of variable interactions. This procedure addresses two limitations of DG involving high computational cost on fully separable functions and detection of overlapping components that share decision variables. This is achieved by the capture of the complete, underlying pairwise interaction structure of the decision variables. Such a graph structure has to be constructed by examining interactions of all pairs of decision variables. In response, our theoretical analysis [52] reduces the total number of function evaluations that differential grouping needs to examine all pairwise interactions by rigorously examining redundant function evaluations with respect to the same base points. At best, our analysis indicates that the total number of objective function evaluations used to check for all pairs of variables for interactions is reduced by half for fully separable functions.

The second procedure dynamically calculates the threshold value ϵ to detect the interaction for each pair of variables that take into account the magnitude of round-off errors. This addresses the remaining two limitations of DG involving sensitivity to computational roundoff errors and having to specify a threshold parameter ϵ. The procedure is informed by our formal error analysis [52] demonstrating the need to set an appropriate ϵ value. Two sources contribute towards sensitivity to roundoff errors when finite computing machines are used to detect variable interactions via $\lambda = |\Delta^{(1)} - \Delta^{(2)}| > \epsilon$: floating-point representation and arithmetic. The rounding process in floating-point representation incurs an error term that grows with the magnitude of the real number to be represented, thus demonstrating that a fixed ϵ setting cannot be used because DG2 has no control of the chosen sample points used to resolve $\lambda > \epsilon$. More importantly, the parameter-free procedure operates on a safe interval where finite computing errors have been taken into account, which is formed by two truth-value statements: $\lambda < e_{\text{inf}}$ for deciding variables are separable, and $\lambda > e_{\text{sup}}$ for deciding variables are interacting. Our computational studies [52] have validated the improvements in effectiveness and efficiency of DG2 over DG with competitive performance when incorporated into a cooperative CEA against the top ranking optimizer in the CEC'2013 large-scale optimization competition.

In closing this chapter, we provide two brief remarks about some observations we have made on the search performance of cooperative CEAs with differential grouping mechanisms. We note that no problem decomposition is performed by DG2 for fully separable and nonseparable problems. The design decision is made at the time DG2 is introduced and that the optimal problem decomposition and optimization process for those problems require further investigations. There are observed instances whereby other cooperative CEAs have performed some problem decomposition on nonseparable problems for which improved search performances can be observed. Lastly, the differential grouping mechanisms are designed to be independent of the optimization process with regard to solving additively separable problems. As such, there remains the potential for the adoption of such mechanisms with other Divide-and-Conquer methods that employ the latest, state-of-the-art optimizers to construct more effective and efficient approaches for large-scale optimization problems and apply them to real-world problem-solving settings.

References

1. Antonio, L.M., Coello, C.A.C.: Use of cooperative coevolution for solving large scale multiobjective optimization problems. In: Proceedings of the IEEE Congress Evolutionary Computation, pp. 2758–2765 (2013)
2. Bellman, R.E.: Dynamic Programming. ser. Dover Books on Mathematics. Princeton University, Princeton (1957)
3. Benson, H.Y., Shanno, D.F., Vanderbei, R.J.: A comparative study of large-scale nonlinear optimization algorithms. In: High Performance Algorithms and Software for Nonlinear Optimization, pp. 95–127. Springer, Berlin (2003)
4. van den Bergh, F., Engelbrecht, A.: A cooperative approach to participle swam optimization. IEEE Trans. Evol. Comput. **8**(3), 225–239 (2004)
5. van den Bergh, F., Engelbrecht, A.P.: A cooperative approach to particle swarm optimization. IEEE Trans. Evol. Comput. **8**(3), 225–239 (2004)
6. Bertsekas, D.: Nonlinear programming. In: Optimization and Neural Computation Series. Athena Scientific, Nashua (1995)
7. Bhowmik, T., Liu, H., Ye, Z., Oraintara, S.: Dimensionality reduction based optimization algorithm for sparse 3-d image reconstruction in diffuse optical tomography. Sci. Rep. **6**, 22242 (2016)
8. Brest, J., Greiner, S., Bošković, B., Mernik, M., Zumer, V.: Self-adapting control parameters in differential evolution: A comparative study on numerical benchmark problems. IEEE Trans. Evol. Comput. **10**(6), 646–657 (2006)
9. Brest, J., Zamuda, A., Bošković, B., Maučec, M.S., Žumer, V.: High-dimensional real-parameter optimization using self-adaptive differential evolution algorithm with population size reduction. In: Proceedings of the IEEE Congress Evolutionary Computation, pp. 2032–2039 (2008)
10. Chen, W., Weise, T., Yang, Z., Tang, K.: Large-scale global optimization using cooperative coevolution with variable interaction learning. In: Proceedings of the International Conference Parallel Problem Solving from Nature, vol. 6239, pp. 300–309. Springer, Berlin (2011)
11. Chen, Y.p., Yu, T.l., Sastry, K., Goldberg, D.E.: A survey of linkage learning techniques in genetic and evolutionary algorithms. Tech. rep., Illinois Genetic Algorithms Library (2007)
12. Cheng, R., Jin, Y.: A competitive swarm optimizer for large scale optimization. IEEE Trans. Cybern. **45**(2), 191–204 (2015)
13. Corless, R.M., Fillion, N.: A Graduate Introduction to Numerical Methods. Springer, Berlin (2013)
14. Dantzig, G.B., Wolfe, P.: Decomposition principle for linear programs. Oper. Res. **8**(1), 101–111 (1960)
15. Deb, K., Myburgh, C.: Breaking the billion-variable barrier in real-world optimization using a customized evolutionary algorithm. In: Proceedings of the Genetic and Evolutionary Computation Conference, pp. 653–660 (2016)
16. Deb, K., Reddy, A.R., Singh, G.: Optimal scheduling of casting sequence using genetic algorithms. Mater. Manuf. Process. **18**(3), 409–432 (2003)
17. Dong, W., Chen, T., Tino, P., Yao, X.: Scaling up estimation of distribution algorithms for continuous optimization. IEEE Trans. Evol. Comput. **17**(6), 797–822 (2013)
18. Fan, J., Wang, J., Han, M.: Cooperative coevolution for large-scale optimization based on kernel fuzzy clustering and variable trust region methods. IEEE Trans. Fuzzy Syst. **22**(4), 829–839 (2014)
19. Gamperle, R., Muller, S.D., Koumoutsakos, P.: A parameter study for differential evolution. In: Proceedings WSEAS International Conference on Advances in Intelligent Systems, Fuzzy Systems, Evolutionary Computation, pp. 293–298 (2002)
20. Goldberg, D.E.: Genetic Algorithms in Search, Optimization, and Machine Learning. Addison-Wesley, New York (1989)

21. Goldberg, D.E., Korb, B., Deb, K.: Messy genetic algorithms: Motivation, analysis, and first results. Complex Syst. **3**(5), 493–530 (1989)
22. Griewank, A., Toint, P.L.: Local convergence analysis for partitioned quasi-newton updates. Numer. Math. **39**, 429–448 (1982). 10.1007/BF01407874
23. Hager, W.W., Hearn, D.W., Pardalos, P.M.: Large Scale Optimization: State of the Art. Springer, Berlin (2013)
24. Hansen, N., Kern, S.: Evaluating the CMA evolution strategy on multimodal test functions. In: Proceedings of the International Conference Parallel Problem Solving from Nature, pp. 282–291 (2004)
25. Hansen, N., Muller, S., Koumoutsakos, P.: Reducing the time complexity of the derandomized evolution strategy with covariance matrix adaptation (CMA-ES). Evol. Comput. **11**(1), 1–18 (2003)
26. Higham, N.J.: Accuracy and Stability of Numerical Algorithms. SIAM, New York (2002)
27. Hinton, G.E., Salakhutdinov, R.R.: Reducing the dimensionality of data with neural networks. Science **313**(5786), 504–507 (2006)
28. Hopcroft, J.E., Tarjan, R.E.: Algorithm 447: efficient algorithms for graph manipulation. Commun. ACM **16**(6), 372–378 (1973)
29. IEEE: IEEE standard for floating-point arithmetic. In: IEEE Std 754-2008 (Revision of IEEE Std 754-1985), pp. 1–82 (2008)
30. Kabán, A., Bootkrajang, J., Durrant, R.J.: Towards large scale continuous EDA: a random matrix theory perspective. Evol. Comput. **24**(2), 255–291 (2015)
31. Kazimipour, B., Omidvar, M.N., Li, X., Qin, A.: A novel hybridization of opposition-based learning and cooperative co-evolutionary for large-scale optimization. In: Proceedings of the IEEE Congress on Evolutionary Computation, pp. 2833–2840 (2014)
32. Kazimipour, B., Omidvar, M.N., Li, X., Qin, A.: A sensitivity analysis of contribution-based cooperative co-evolutionary algorithms. In: Proceedings of the IEEE Congress Evolutionary Computation, pp. 417–424 (2015)
33. Kimura, S., Ide, K., Kashihara, A., Kano, M., Hatakeyama, M., Masui, R., Nakagawa, N., Yokoyama, S., Kuramitsu, S., Konagaya, A.: Inference of s-system models of genetic networks using a cooperative coevolutionary algorithm. Bioinformatics **21**(7), 1154–1163 (2005)
34. LaTorre, A., Muelas, S., Peña, J.M.: Large scale global optimization: experimental results with MOS-based hybrid algorithms. In: Proceedings of the IEEE Congress Evolutionary Computation, pp. 2742–2749 (2013)
35. LaTorre, A., Muelas, S., Peña, J.M.: A comprehensive comparison of large scale global optimizers. Inf. Sci. **316**, 517–549 (2015)
36. Lee, C.Y., Yao, X.: Evolutionary programming using mutations based on the lévy probability distribution. IEEE Trans. Evol. Comput. **8**(1), 1–13 (2004)
37. Li, X., Tang, K., Omidvar, M.N., Yang, Z., Qin, K.: Benchmark functions for the CEC'2013 special session and competition on large-scale global optimization. Tech. rep., RMIT University, Melbourne (2013)
38. Li, X., Tang, K., Suganthan, P., Yang, Z.: Editorial for the special issue of information sciences journal (ISJ) on nature-inspired algorithms for large scale global optimization. Inf. Sci. **316**, 437–439 (2015)
39. Li, X., Yao, X.: Cooperatively coevolving particle swarms for large scale optimization. IEEE Trans. Evol. Comput. **16**(2), 210–224 (2012)
40. Liu, J., Tang, K.: Scaling up covariance matrix adaptation evolution strategy using cooperative coevolution. In: Proceedings of the International Conference Intelligent Data Engineering and Automated Learning, pp. 350–357. Springer, Berln (2013)
41. Liu, Y., Yao, X., Zhao, Q., Higuchi, T.: Scaling up fast evolutionary programming with cooperative coevolution. In: Proceedings of the IEEE Congress on Evolutionary Computation, pp. 1101–1108 (2001)
42. Liu, Y., Zhao, Q., Yao, X., Higuchi, T.: Scaling up fast evolutionary programming with cooperative coevolution. In: Proceedings of the IEEE Conference on Evolutionary Computation, ICEC, vol. 2, pp. 1101–1108 (2001)

43. Lozano, M., Molina, D., Herrera, F.: Editorial scalability of evolutionary algorithms and other metaheuristics for large-scale continuous optimization problems. Soft Comput. **15**(11), 2085–2087 (2011)
44. Mahdavi, S., Shiri, M.E., Rahnamayan, S.: Cooperative co-evolution with a new decomposition method for large-scale optimization. In: Proceedings of the IEEE Congress Evolutionary Computation, pp. 1285–1292 (2014)
45. Mahdavi, S., Shiri, M.E., Rahnamayan, S.: Metaheuristics in large-scale global continues optimization: A survey. Inf. Sci. **295**, 407–428 (2015)
46. Mei, Y., Omidvar, M.N., Li, X., Yao, X.: Competitive divide-and-conquer algorithm for unconstrained large scale black-box optimization. ACM Trans. Math. Softw. **42**(2), 13 (2015)
47. Molina, D., Lozano, M., Herrera, F.: MA-SW-Chains: Memetic algorithm based on local search chains for large scale continuous global optimization. In: Proceedings of the IEEE Congress Evolutionary Computation, pp. 3153–3160 (2010)
48. Moscato, P.: On evolution, search, optimization, genetic algorithms and martial arts: towards memetic algorithms. Tech. rep., Caltech Concurrent Computation Program (1989)
49. Ngiam, J., Coates, A., Lahiri, A., Prochnow, B., Le, Q.V., Ng, A.Y.: On optimization methods for deep learning. In: Proceedings of the 28th International Conference on Machine Learning, pp. 265–272 (2011)
50. Omidvar, M.N., Kazimipour, B., Li, X., Yao, X.: CBCC3—a contribution-based cooperative co-evolutionary algorithm with improved exploration/exploitation balance. In: Proceedings of the IEEE Congress Evolutionary Computation, pp. 3541–3548 (2016)
51. Omidvar, M.N., Li, X., Mei, Y., Yao, X.: Cooperative co-evolution with differential grouping for large scale optimization. IEEE Trans. Evol. Comput. **18**(3), 378–393 (2014)
52. Omidvar, M.N., Li, X., Mei, Y., Yao, X.: DG2: A faster and more accurate differential grouping for large-scale black-box optimization. IEEE Trans. Evol. Comput. **21**(6), 929–942 (2017)
53. Omidvar, M.N., Li, X., Tang, K.: Designing benchmark problems for large-scale continuous optimization. Inf. Sci. **316**, 419–436 (2015)
54. Omidvar, M.N., Li, X., Yao, X.: Cooperative co-evolution with delta grouping for large scale non-separable function optimization. In: Proceedings of the IEEE Congress Evolutionary Computation, pp. 1762–1769 (2010)
55. Omidvar, M.N., Li, X., Yao, X.: Smart use of computational resources based on contribution for cooperative co-evolutionary algorithms. In: Proceedings of the Genetic and Evolutionary Computation Conference, pp. 1115–1122 (2011)
56. Omidvar, M.N., Mei, Y., Li, X.: Effective decomposition of large-scale separable continuous functions for cooperative co-evolutionary algorithms. In: Proceedings of the IEEE Congress Evolutionary Computation, pp. 1305–1312 (2014)
57. Potter, M.: The design and analysis of a computational model of cooperative coevolution. Ph.D. thesis, George Mason University, Fairfax (1997)
58. Potter, M., De Jong, K.: A cooperative coevolutionary approach to function optimization. In: Lecture Notes in Computer Science (including subseries Lecture Notes in Artificial Intelligence and Lecture Notes in Bioinformatics), vol. 866 LNCS, pp. 249–257 (1994)
59. Price, K., Storn, R., Lampinen, J.: Differential Evolution: A Practical Approach to Global Optimization. Springer, Berlin (2005)
60. Qin, A., Suganthan, P.: Self-adaptive differential evolution algorithm for numerical optimization. In: 2005 IEEE Congress on Evolutionary Computation, IEEE CEC 2005. Proceedings, vol. 2, pp. 1785–1791 (2005)
61. Ray, T., Yao, X.: A cooperative coevolutionary algorithm with correlation based adaptive variable partitioning. In: Proceedings of the IEEE Congress Evolutionary Computation, pp. 983–989 (2009)
62. Regis, R.G., Shoemaker, C.A.: Local function approximation in evolutionary algorithms for the optimization of costly functions. IEEE Trans. Evol. Comput. **8**(5), 490–505 (2004)
63. Salomon, R.: Reevaluating genetic algorithm performance under coordinate rotation of benchmark functions—a survey of some theoretical and practical aspects of genetic algorithms. BioSystems **39**, 263–278 (1995)

64. Sayed, E., Essam, D., Sarker, R.: Dependency identification technique for large scale optimization problems. In: Proceedings of the IEEE Congress Evolutionary Computation, pp. 1–8 (2012)
65. Shi, Y.J., Teng, H.F., Li, Z.Q.: Cooperative co-evolutionary differential evolution for function optimization. In: Lecture Notes in Computer Science, vol. 3611, pp. 1080–1088 (2005)
66. Sofge, D., De Jong, K., Schultz, A.: A blended population approach to cooperative coevolution for decomposition of complex problems. In: Proceedings of the 2002 Congress on Evolutionary Computation, CEC 2002, vol. 1, pp. 413–418 (2002)
67. Sterbenz, P.H.: Floating-Point Computation. Prentice Hall, New York (1973)
68. Storn, R.: System design by constraint adaptation and differential evolution. IEEE Trans. Evol. Comput. **3**(1), 22–34 (1999)
69. Storn, R., Price, K.: Differential evolution—a simple and efficient heuristic for global optimization over continuous spaces. J. Glob. Optim. **11**(4), 341–359 (1997)
70. Suganthan, N., Hansen, N., Liang, J.J., Deb, K., Chen, Y.P., Auger, A., Tiwari, S.: Problem definitions and evaluation criteria for the CEC 2005 special session on real-parameter optimization. Tech. Rep. Technical Report, Nanyang Technological University, Singapore (2005)
71. Sun, L., Yoshida, S., Cheng, X., Liang, Y.: A cooperative particle swarm optimizer with statistical variable interdependence learning. Inf. Sci. **186**(1), 20–39 (2012)
72. Sun, Y., Kirley, M., Halgamuge, S.K.: Extended differential grouping for large scale global optimization with direct and indirect variable interactions. In: Proceedings of the Genetic and Evolutionary Computation Conference, pp. 313–320 (2015)
73. Tang, K., Li, X., Suganthan, P.N., Yang, Z., Weise, T.: Benchmark functions for the CEC'2010 special session and competition on large-scale global optimization. Tech. rep., Nature Inspired Computation and Applications Laboratory, USTC, China (2009)
74. Teng, H.F., Chen, Y., Zeng, W., Shi, Y.J., Hu, Q.H.: A dual-system variable-grain cooperative coevolutionary algorithm: satellite-module layout design. IEEE Trans. Evol. Comput. **14**(3), 438–455 (2010)
75. Tezuka, M., Munetomo, M., Akama, K.: Linkage identification by nonlinearity check for real-coded genetic algorithms. In: Proceedings of the Genetic and Evolutionary Computation Conference. Lecture Notes in Computer Science, vol. 3103, pp. 222–233. Springer, Berlin (2004)
76. Thomsen, R.: Flexible ligand docking using differential evolution. In: 2003 Congress on Evolutionary Computation, CEC 2003—Proceedings, vol. 4, pp. 2354–2361 (2003)
77. Tseng, L.Y., Chen, C.: Multiple trajectory search for large scale global optimization. In: Proceedings of the IEEE Congress Evolutionary Computation, pp. 3052–3059 (2008)
78. Vanderplaats, G.N.: Very large scale optimization. Tech. Rep. NASA/CR-2002-211768, National Aeronautics and Space Administration (NASA), Langley Research Center, Colorado, US (2002)
79. Vesterstrom, J., Thomsen, R.: A comparative study of differential evolution, particle swarm optimization, and evolutionary algorithms on numerical. In: Proceedings of IEEE Congress on Evolutionary Computation, pp. 1980–1987 (2004)
80. Wang, C., Gao, J.: High-dimensional waveform inversion with cooperative coevolutionary differential evolution algorithm. IEEE Geosci. Remote Sens. Lett. **9**(2), 297–301 (2012)
81. Wang, Y., Huang, J., Dong, W.S., Yan, J.C., Tian, C.H., Li, M., Mo, W.T.: Two-stage based ensemble optimization framework for large-scale global optimization. Eur. J. Oper. Res. **228**(2), 308–320 (2013)
82. Wang, Y., Li, B., Lai, X.: Variance priority based cooperative co-evolution differential evolution for large scale global optimization. In: Proceedings of the IEEE Congress Evolutionary Computation, pp. 1232–1239. IEEE, New York (2009)
83. Wei, F., Wang, Y., Zong, T.: A novel cooperative coevolution for large scale global optimization. In: Proceedings of the IEEE International Conference Systems, Man and Cybernetics, pp. 738–741 (2014)

84. Weise, T., Chiong, R., Tang, K.: Evolutionary Optimization: Pitfalls and Booby Traps. J. Comput. Sci. Technol. (JCST) **27**(5), 907–936 (2012). Special Issue on Evolutionary Computation
85. Yang, Z., He, J., Yao, X.: Making a difference to differential evolution. In: Advances in Metaheuristics for Hard Optimization, pp. 397–414 (2008)
86. Yang, Z., Sendhoff, B., Tang, K., Yao, X.: Target shape design optimization by evolving b-splines with cooperative coevolution. Appl. Soft Comput. **48**, 672–682 (2016)
87. Yang, Z., Tang, K., Yao, X.: Differential evolution for high-dimensional function optimization. In: 2007 IEEE Congress on Evolutionary Computation, CEC 2007, pp. 3523–3530 (2007)
88. Yang, Z., Tang, K., Yao, X.: Large scale evolutionary optimization using cooperative coevolution. Inf. Sci. **178**, 2985–2999 (2008)
89. Yang, Z., Tang, K., Yao, X.: Multilevel cooperative coevolution for large scale optimization. In: Proceedings of the IEEE Congress Evolutionary Computation, pp. 1663–1670 (2008)
90. Yang, Z., Tang, K., Yao, X.: Self-adaptive differential evolution with neighborhood search. In: 2008 IEEE Congress on Evolutionary Computation, CEC 2008, pp. 1110–1116 (2008)
91. Yao, X., Liu, Y., Lin, G.: Evolutionary programming made faster. IEEE Trans. Evol. Comput. **3**(2), 82–102 (1999)
92. Zamuda, A., Brest, J., Bošković, B., Žumer, V.: Large scale global optimization using differential evolution with self-adaptation and cooperative co-evolution. In: Proceedings of the IEEE Congress Evolutionary Computation, pp. 3718–3725 (2008)
93. Zhang, J., Sanderson, A.C.: JADE: adaptive differential evolution with optional external archive. IEEE Trans. Evol. Comput. **13**(5), 945–958 (2009)
94. Zhang, K., Li, B.: Cooperative coevolution with global search for large scale global optimization. In: Proceedings of the IEEE Congress Evolutionary Computation, pp. 1–7 (2012)
95. Zhao, S.Z., Liang, J.J., Suganthan, P.N., Tasgetiren, M.F.: Dynamic multi-swarm particle swarm optimizer with local search for large scale global optimization. In: Proceedings of the IEEE Congress Evolutionary Computation, pp. 3845–3852 (2008)
96. Zhou, Z.H., Chawla, N.V., Jin, Y., Williams, G.J.: Big data opportunities and challenges: discussions from data analytics perspectives. IEEE Comput. Intell. Mag. **9**(4), 62–74 (2014)

Chapter 8
High-Dimensional Optimization as Computationally Expensive Optimization

8.1 Solving High-Dimensional Optimization Problems via Cooperative Coevolution

Evolutionary Algorithms (EAs) constitute a large family of population-based, stochastic algorithms that are effective alternatives for complex optimization problems [16, 35, 39, 41]. However, the performances of existing EAs are affected when the dimensionality of the problem increases and becomes high, in which case, the effectiveness of EAs is significantly reduced [3]. This is not unexpected given that the solution space of a problem exponentially increases with the problem dimensionality so that EAs quickly become inefficient to fully explore the space within a reasonable time budget. Maintaining the problem-solving coverage of EAs for such high-dimensional optimization problems is crucial, and that in the last decade, this issue has attracted substantial research interests [22]. There are two general ideas to enhance existing EAs and improve their search abilities so that they are able to handle a high-dimensional search space.

The first idea is the more straightforward general approach to improve the search efficiency of existing EAs by either explicitly or implicitly rebalancing the exploration and exploitation [5, 25, 42] so that a successful search [7] can be made on high-dimensional optimization problems. The main challenge is that the exploration-exploitation balance might vary from one problem to another. Furthermore, mechanisms that allocate computational resource for explorative and exploitative search processes might be sensitive to dimensionality even within the same problem, e.g. the balance for a problem of a given dimensionality might not be transferable to the same problem on a higher dimensionality.

The second idea involves approaches that simplify high-dimensional problems to a low-dimensional representation for which existing evolutionary search can be applied. One approach involves ***dimensionality reduction*** techniques that seek to exploit the fact that the crucial structures underpinning problems specified in high dimensions live in some lower dimensional subspaces. Two proposed dimen-

X. Yao, S. Y. Chong, *Coevolutionary Computation and Its Applications*,
Natural Computing Series, https://doi.org/10.1007/978-981-96-2841-4_8

sionality reduction approaches include ***Random Embedding*** [44] and ***Random Projections*** [15]. In the second approach, it has been demonstrated that search based on an ensemble of random projections to low-dimensional subspaces of high-dimensional search points is representative of what the search could achieve in the original high-dimensional space. Computational cost savings come from being able to perform search operations on this ensemble of random projections. In general, these theoretically informed approaches have competitive performances on established benchmark problems against known state-of-the-art EAs.

An alternative approach makes more explicit use of the ***Divide-and-Conquer*** methodology that constitutes a generic problem-solving approach with successful application in many other fields [1, 33]. In the context of evolutionary optimization for high-dimensional problems, Divide-and-Conquer would perform problem decomposition of the original high-dimensional problems into multiple low-dimensional subproblems (subcomponents). In particular, if there are no dependencies between these subcomponents, then in effect one could solve them separately and efficiently with existing EAs. After that, the partial solutions (i.e. the solutions to the subcomponents) have to be combined to form the full solution for the original high-dimensional problem. In particular, the cooperative coevolution framework incorporates such a Divide-and-Conquer methodology into its search process [9, 28, 34]. The focus of the study presented in this chapter is on this family of cooperative Coevolutionary Algorithms (CEAs).

In the case of fully separable additive optimization problems, cooperative coevolution is particularly effective in which case the global optimum can be found by optimizing one-dimensional subcomponents separately [2, 21, 34]. If there are some dependencies or couplings between the solution variables on some dimensions, then the performance of cooperative CEAs would depend on the decomposition method [4, 19, 23, 24, 28, 49]. The general aim of a decomposition method is to divide a high-dimensional problem in manner whereby search for the global optimum can be obtained by solving the subcomponents independently. A large variety of decomposition methods have been proposed [22, 28, 52]. In the case of methods that employ variable grouping strategy to group interdependent variables together, they range from the simple random grouping methods to the more sophisticated approaches (e.g. Differential Grouping [28]) that detect variable interactions in the problem structures.

Most of these decomposition methods target high-dimensional problems that are decomposable in some sense (e.g. additive separability) for which prior knowledge of decomposable structures is not available. As such, guaranteed performance improvements are dependent on these methods correctly identifying interdependent groups for problem decomposition. In the case whereby the optimization problem is fully nonseparable, then cooperative CEAs would face the challenge of having to deal with interdependent subcomponents and then to be able to correctly evaluate and rank partial solutions in each subcomponent. Otherwise, cooperative CEAs would very likely prematurely converge to a suboptimum [30, 46, 47]. To understand this issue, note that a partial solution cannot be naturally evaluated with respect to the original objective function because of the mismatch in dimensionality between the subspace associated with the subcomponents and the full search space of the original problem.

One way to address this problem of dimensionality mismatch is to complement a partial solution to be a complete solution (with respect to the original problem) and then evaluate it with the original objective function. One would immediately note that the measured qualities of partial solutions could largely depend on the manner with which the complements are selected, thereby affecting the ranks of partial solutions and subsequently constitute a form of search biases. In the case where independent subcomponents exist and can be correctly identified via decomposition technique [19, 21, 24, 28, 49], then changes to complements will not perturb the rank of partial solutions [2, 4, 13, 21, 34]. The issues arise because existing decomposition techniques produce incorrect outputs with some interdependencies between subcomponents or that the original problems are nonseparable. For this reason, the complements should be carefully chosen to correctly rank the partial solutions. However, it has been formally proven that performing such a task will be computationally prohibitive because one has to exhaustively test all feasible samples in the whole solution space of the other subcomponents for each single partial solution to guarantee obtaining the correct rank [30]. In practice, existing cooperative CEAs usually sample one or a few complements for each partial solution to balance computational burden against effective search for higher quality solutions, which in itself is another nontrivial task when subcomponents are interdependent [4, 28, 31, 32, 38].

Despite this, there are opportunities with which under certain problem settings the issue can be adequately addressed. We have carried out a study in [48] to address this issue. In Sect. 8.2, we have performed an analysis of this issue of dimensionality mismatch in cooperative CEAs as a computationally expensive problem. This informs us of an approach to address such a problem through meta-modelling whereby a surrogate function can be identified that allows for the direct and approximate comparisons of partial solutions of subcomponents. Such a surrogate function is computationally cheaper compared to what exhaustive testing world require. As such, Sect. 8.3 presents the Self-Evaluation Evolution framework that in essence frames and solves high-dimensional optimization problems as computationally expensive optimization problems. Section 8.4 presents our computational study to demonstrate the effectiveness of SEE on all the 20 problems in the CEC'2010 large-scale optimization benchmark [37] that range from 1000 to 5000 dimensionalities.

8.2 The Main Challenge in Coevolutionary Divide-and-Conquer

In this section, our main motivation is to show that the main challenge of coevolutionary Divide-and-Conquer is with the dimensionality mismatch between partial solution evaluations and the objective function of the original problem.

Furthermore, this challenge of correctly ranking the partial solutions can be formulated as a computationally expensive problem. We have first performed this analysis and subsequent reformulation in [48]. We first introduce several notations and definitions that are used in the subsequent analysis. Let $\boldsymbol{x}_{1:P}$ be a given set of complete candidate solutions to the original objective function $\mathcal{F}$, $\boldsymbol{x}_i$ be the ith complete solution, and $\boldsymbol{x}_{1:N}$ be the first N complete solutions in the set $\boldsymbol{x}_{1:P}$ where $N \leq P$. Consider a D-dimensional problem divided into M exclusive d_j-dimensional subproblems or subcomponents whereby $\sum_{j=1}^{M} d_j = D$. Let $x_{i,j}$ be the ith partial solution to the jth subcomponent. In the same manner, $x_{1:N,j}$ is the first N partial solutions to the jth subcomponent. We can more explicitly write a complete solution $\boldsymbol{x}_i$ as $\boldsymbol{x}_{i,1:D}$, which expresses the stacked version of D partial solutions with the same i-index. As such, $x_{i,j}$ is also said to be the jth component solution that belongs to the complete solution $\boldsymbol{x}_i$.

Obviously, the original D-dimensional objective function $\mathcal{F} : \mathbb{R}^D \rightarrow \mathbb{R}$ cannot be used to directly evaluate any of the d_j-dimensional partial solutions in the jth subcomponent, given that they are of type $\mathbb{R}^{d_j} \rightarrow \mathbb{R}$ with $d_j < D$ for all $j = 1, 2, \ldots, M$. One straightforward approach to address this dimensionality mismatch is to complement a partial solution to be a complete solution before evaluating it with the objective function. Here, ***complementing*** a d_j-dimensional partial solution $x_{i,j}$ involves taking fixed values for the decision variables associated with the remaining $M - 1$ subcomponents. For each partial solution, there are $|\mathcal{X}|^{D-d_j}$ candidate complements available, where $\mathcal{X} \subseteq \mathbb{R}$ denotes a bounded range of each decision variable. Since a partial solution and its complement are passed as the input to $\mathcal{F}$ to obtain the objective function value, then for a fixed partial solution, different complements would result in different objective values. What is crucial here that in comparing multiple distinct partial solutions with respect to a subcomponent, their ordering, or ranks stays the same or differs over different complements. Given that the ranks of partial solutions are usually used to determine the search direction in the corresponding subcomponent, it is important to select a correct complement from $|\mathcal{X}|^{D-d_j}$ candidates for each partial solution. However, this is a nontrivial and computationally expensive task.

For a set of partial solutions, their correct ordering must be unique with respect to their real objective qualities in the solution space. In particular, if a partial solution belongs to a good complete solution, it is said to be with good real quality. Such an assertion comes about from the building blocks hypothesis [10, 12, 26, 27, 45]. As such, the correct rank of a set of partial solutions should be obtained by comparing the function values of the best complete solutions they belong to. This is formalized as follows. Consider a minimization setting. For the jth subcomponent, if the ath partial solution $x_{a,j}$ is said to be better than the bth partial solution $x_{b,j}$, then the following inequality is satisfied [48]:

$$\min_{\boldsymbol{x}_\mathbf{c} \in S_\mathbf{c}} \mathcal{F}(x_{a,j}, \boldsymbol{x}_\mathbf{c}) < \min_{\boldsymbol{x}_\mathbf{c} \in S_\mathbf{c}} \mathcal{F}(x_{b,j}, \boldsymbol{x}_\mathbf{c}), \tag{8.1}$$

where $\boldsymbol{x}_{\mathbf{c}}$ denotes the candidate complements with $\mathbf{c} = (1, \ldots, j-1, j+1, \ldots, M)$ denoting all the subcomponents except the jth subcomponent, and $S_{\mathbf{c}}$ denotes the corresponding complement space $\mathcal{X}^{D-d_j}$.

From the inequality given by Eq. (8.1), one can define the correct complement for each partial solution $x_{i,j}$ as follows [48]:

$$\boldsymbol{x}_{\mathbf{c}}^* = \underset{\boldsymbol{x}_{\mathbf{c}} \in S_{\mathbf{c}}}{\operatorname{argmin}}\, \mathcal{F}(x_{i,j}, \boldsymbol{x}_{\mathbf{c}}). \tag{8.2}$$

Despite specifying the existence of the correct complement $\boldsymbol{x}_{\mathbf{c}}^*$ for a partial solution $x_{i,j}$, the guarantee for finding $\boldsymbol{x}_{\mathbf{c}}^*$ requires evaluating the combinations of $x_{i,j}$ with all possible complements in $\mathcal{X}^{D-d_j}$ by $\mathcal{F}$ and identifies the best one. Unfortunately, this is a computationally overwhelming task. This implies that the task of obtaining the correct rank of a set of partial solutions is computationally expensive. In a cooperative CEA framework, the aim is to approximate the correct rank by heuristic sampling from the complement space, thereby reducing the computational cost.

Let us define a function $\mathcal{G}$ that can directly evaluate the real quality of a partial solution $x_{i,j}$ as follows [48]:

$$\mathcal{G}(x_{i,j}) = \min_{\boldsymbol{x}_{\mathbf{c}} \in S_{\mathbf{c}}} \mathcal{F}(x_{i,j}, \boldsymbol{x}_{\mathbf{c}}). \tag{8.3}$$

In this case, the correct rank of a set of partial solution can be obtained by directly comparing their function values in terms of $\mathcal{G}$. Given N partial solutions $x_{1:N,j}$ with respect to the jth subcomponent, their correct ranks $\mathcal{R}(x_{1:N,j})$ are defined as follows [48]:

$$\mathcal{R}(x_{1:N,j}) = \underset{1 \le i \le N}{\operatorname{sort}}\, \mathcal{G}(x_{i,j}), \tag{8.4}$$

where the function sort ranks the set of partial solutions $x_{i:N,j}$ in an ascending order in terms of their qualities $\mathcal{G}(x_{i,j}), i = 1, \ldots, N$. Direct calculation of $\mathcal{G}(x_{i,j})$ is computationally prohibitively expensive, thus necessitating some form of heuristic sampling of the complement space to approximate $\mathcal{G}(x_{i,j})$. One natural idea is to directly build a rule or surrogate as the approximation of each $\mathcal{G}(x_{i,j})$ or even $\mathcal{R}(x_{1:N,j})$. In the following section, we propose such an approximation-based framework based on the analysis we have performed here.

8.3 Self-Evaluation Evolution

Earlier in Sect. 8.3, we have formulated the task of correctly ranking the partial solutions as a computationally expensive problem. In studies of EAs, computationally expensive problems are solved with the aid of meta-model (surrogate model) techniques [14, 36]. Here, the underlying idea is to introduce a computationally less expensive function that approximates the original function. We propose the ***Self-Evaluation Evolution*** (***SEE***) framework to address the issue of computationally expensive formulation of ranking the partial solutions in cooperative CEAs [48]. Section 8.3.1 first introduces the general framework of SEE. Section 8.3.2 next introduces a simple and efficient meta-model that is designed to be paired with local search used as one-dimensional optimizer as an instantiation of a specific SEE implementation that is introduced in Sect. 8.3.3. After this, Sect. 8.4 presents a systematic computational study where we compared the performances of SEE against representatives of three categorical approaches to solve high-dimensional optimization problems we have introduced in the introduction and to investigate the impact of the meta-model on SEE.

8.3.1 The Framework of SEE

Algorithm 8.1 procedurally describes a generic implementation of SEE [48]. The procedure **SEE** starts with the original objective function $\mathcal{F}$ being decomposed into subcomponents $j = 1, \ldots, M$, each exclusively containing d_j decision variables, via **problemDecom**. For each $j = 1, \ldots, M$ subcomponent, (1) a meta-model $\mathcal{H}_j$ is initialized for to approximately rank $\mathcal{R}$ partial solutions via **INITMETAM**, (2) a subcomponent optimizer is constructed with associated parameters $\boldsymbol{P}_j$ initialized via **InitOp**, and (3) the parent partial solutions are initialized via **initParPop**. Note that the subcomponent optimizer could be any type of EAs. In Algorithm 8.1, we consider a $(\mu + \lambda)$ implementation whereby μ parent partial solutions $x^p_{1:\mu,j}$ are uniformly randomly initialized. Then, λ offspring partial solutions $x^o_{1:\lambda,j}$ are generated from the parents at each iteration. The offsprings $x^o_{1:\lambda,j}$ and parents $x^p_{1:\mu,j}$ are then ranked by the meta-model H_j via **sort**, where the best μ partial solutions are chosen as the new parents for the next iteration via **selection**.

In addition to the phase involving the evolutionary search phase of the partial solutions for each subcomponent, we also consider in SEE a training phase of the associated subcomponent meta-models and optimizers. The procedure **constructSol** first assembles the set of complete solutions $z[i]$, $i = i, \ldots, \mu$, for example, by combining all the partial solutions $x^p_{i,1:M}$ with respect to each ith sample. Training data based on samples $z[i]$, $i = i, \ldots, \mu$ of the original solution space are collected via **collectData** and stored in an appropriate data structure $\boldsymbol{T}_j[i] = \mathcal{F}(z_i)$, $i = i, \ldots, \mu$, containing their objective function values. These training data are then used to tune the subcomponent meta-models and optimizers via **TRAINMETAM** and **adjustOp**, respectively. These tuning processes are crucial because it is usually not possible to obtain the right meta-models that correctly rank the partial solutions during the initial stages, and for the subcomponent optimizers their associated parameters might need to be optimized at different stages of search. At the completion of **SEE** after exhausting the time budget t_e, the procedure

Algorithm 8.1 The framework of self-evaluation evolution (SEE) [48]

Input:

$\mathcal{F}$ Objective function
M Number of subcomponents
$\mathcal{H}$ Meta-model
t_e Total number of function evaluation (time budget)
λ Number of offspring solutions
μ Number of parent solutions

Output:

$\boldsymbol{x}$ Best complete solution

procedure SEE($\mathcal{F}, M, \mathcal{H}, t_e, \lambda, \mu$)
  $\boldsymbol{F}$:= **problemDecom**($\mathcal{F}, M$) ▷ Divide $\mathcal{F}$ to M exclusive subcomponents, ▷ $\boldsymbol{F}$ stores associated indices
  for $j := 1 : M$ **do**
    INITMETAM($\mathcal{H}_j, \boldsymbol{F}$) ▷ Initialize meta-model $\mathcal{H}_j$
    initOp($\boldsymbol{P}_j, \boldsymbol{F}$) ▷ Initialize jth subcomponent optimizer parameters $\boldsymbol{P}_j$
    $x^p_{1:\mu,j}$:= **initParPop**($\mu, \boldsymbol{F}$) ▷ Initialize μ parent partial solutions $x^p_{1:\mu,j}$ randomly
  end for
  while within time budget t_e **do**
    for $j := 1 : M$ **do** ▷ **Evolutionary Search Phase**
      $x^o_{1:\lambda,j}$:= **variation**($x^p_{1:\mu,j}, \boldsymbol{F}$) ▷ Generate λ offsprings $x^o_{1:\lambda,j}$ based on $x^p_{1:\mu,j}$
      $x^{\text{full}}_{1:(\mu+\lambda),j}$:= **fullPop**($x^p_{1:\mu,j}, x^o_{1:\lambda,j}$) ▷ $x^{\text{full}}_{1:(\mu+\lambda),j} := x^p_{1:\mu,j} \cup x^o_{1:\lambda,j}$
      $x^{\text{full}}_{1:(\mu+\lambda),j}$:= **sort**($x^{\text{full}}_{1:(\mu+\lambda),j}, \mathcal{H}_j$) ▷ Rank $x^{\text{full}}_{1:(\mu+\lambda),j}$ in terms of $\mathcal{H}_j$
      $x^p_{1:\mu,j}$:= **selection**($x^{\text{full}}_{1:(\mu+\lambda),j}$) ▷ Update $x^p_{1:\mu,j}$: choose best μ from $x^{\text{full}}_{1:(\mu+\lambda),j}$
      $\boldsymbol{y}[j] := x^p_{1:\mu,j}$
    end for
    ▷ **Meta-Model Training Phase**
    $z[]$:= **constructSol**($\boldsymbol{y}[], \boldsymbol{F}$) ▷ Construct complete solutions $z[i]$, $i = 1, \ldots, \mu$
    for $j := 1 : M$ **do**
      $\boldsymbol{T}_j[]$:= **collectData**($\mathcal{F}, z[]$) ▷ Sampled training data for $\mathcal{H}_j$: ▷ $\boldsymbol{T}_j[i] = \mathcal{F}(z_i)$, $i = 1, \ldots, \mu$
      TRAINMETAM($\mathcal{H}_j, \boldsymbol{T}_j[], \boldsymbol{F}$) ▷ Train $\mathcal{H}_j$
      adjustOp($\boldsymbol{P}_j, \boldsymbol{F}$) ▷ Adjust the jth subcomponent optimizer
    end for
  end while
  $\boldsymbol{x}$:= BESTSOL($z[], \boldsymbol{F}$) ▷ Find best solution $\boldsymbol{x}$
  return $\boldsymbol{x}$
end procedure

BESTSOL will select the best solution from the population $z[i]$, $i = 1, \ldots, \mu$ to find the best complete solution $\boldsymbol{x} = (x_1, \ldots, x_D)$.

8.3.2 Meta-Modelling in SEE

There are different kinds of ***meta-models*** that have been extensively studied, in which most of them do not explicitly require prior knowledge of specific problem structures [14, 36]. Some examples of commonly used approaches in ***meta-modelling*** include ***Support Vector Machines*** [6] and ***Artificial Neural Networks*** [40] that employ parametric models and the more sophisticated approach of ***Gaussian Processes*** [20] that constitute probability distributions over functions. As such, one would expect these meta-models to perform well in the framework of SEE in the context of approximating the ranks of partial solutions. However, these existing, sophisticated meta-models might not be the best choices to be used in SEE in the context of computational cost requirements.

In order to better appreciate why computational cost requirement is more important than accuracy in assessing ranks of partial solutions, consider that real-world high-dimensional optimization problem usually exceeds 1000 dimensions, and could even be in the range of billions in dimensionality [8]. The effectiveness of current EAs is mostly observed on problems whose dimensionalities do not exceed 50. One would require that the dimensionality of each subcomponent to be manageable and solvable for the subcomponent optimizers. As a result, one might have to use a large number of meta-models because after problem decomposition of a very high-dimensional optimization problem there are a large number of subcomponents. In the framework of SEE, these meta-models have to be trained in each iteration of the search and thus incur huge computational costs if meta-models have high computational requirements.

For such a consideration, we propose and specifically design a very simple and efficient ***rank-based meta-model*** for SEE so that the computational cost of the meta-model is kept very low [48]. Unlike the case of existing meta-models that aim to approximate the global landscape of the solution space, our proposed meta-model only approximates the pairwise rank within the ***local area*** of each one-dimensional subspace. We are able to build an effective meta-model that is also efficient by setting a simpler objective for the meta-model. In the following, we will present our specific meta-model to be implemented in SEE.

Consider the rank between a parent partial solution $x^p_{1,j}$ and its offspring $x^o_{1,j}$ for the jth subcomponent. In the one-dimensional case, $x^o_{1,j}$ is either smaller or larger than $x^p_{1,j}$ based on their scalar decision variable values. Assume that the local fitness landscape changes monotonically in either an increasing or a decreasing direction. Then, the rank of $x^p_{1,j}$ and $x^o_{1,j}$ is directly obtained without explicit calculations of their real qualities $\mathcal{G}(x^p_{1,j})$ and $\mathcal{G}(x^o_{1,j})$. In a minimization setting, there are two cases that $\mathcal{G}(x^o_{1,j})$ is better than (having a smaller value compared to) $\mathcal{G}(x^p_{1,j})$ if (1) $x^o_{1,j}$ is smaller than $x^p_{1,j}$ for an increasing local fitness landscape and (2) $x^o_{1,j}$ is larger than $x^p_{1,j}$ for a decreasing local fitness landscape. Otherwise, $\mathcal{G}(x^o_{1,j})$ is worse than $\mathcal{G}(x^p_{1,j})$.

We note that the global fitness landscape might be complex, e.g. in the case multi-modal problems where the associated fitness landscape will not be a nondecreasing (nonincreasing) function only. However, it is still possible to assume the local fitness landscape changes monotonically within the small interval between $x^p_{1,j}$ and $x^o_{1,j}$ if they are generated via some local search operator. In this scenario, the meta-model only needs to learn the likelihoods between the two cases we have described above. In the following, we define these two likelihoods $\mathrm{PS}_{1,j}$ and $\mathrm{PL}_{1,j}$ [48]

$$\begin{aligned}\mathrm{PS}_{1,j} &= \mathbb{P}\Big(\mathcal{G}(x^o_{1,j}) < \mathcal{G}(x^p_{1,j}) \mid x^o_{1,j} < x^p_{1,j}\Big)\\ \mathrm{PL}_{1,j} &= \mathbb{P}\Big(\mathcal{G}(x^o_{1,j}) < \mathcal{G}(x^p_{1,j}) \mid x^o_{1,j} > x^p_{1,j}\Big).\end{aligned} \tag{8.5}$$

In the highly unlikely event that $x^o_{1,j}$ equals $x^p_{1,j}$, $x^o_{1,j}$ is resampled. Equation (8.5) states that for a pair of partial solutions $x^p_{1,j}$ and $x^o_{1,j}$, one could predict that $\mathcal{G}(x^o_{1,j})$ is better than $\mathcal{G}(x^p_{1,j})$ with probability $\mathrm{PS}_{1,j}$ (or $\mathrm{PL}_{1,j}$) if $x^o_{1,j}$ is smaller (or larger) than $x^p_{1,j}$. As such, we formulate the meta-model $\mathcal{H}_j$ as follows [48]:

$$\begin{cases}\mathcal{G}\big(x^o_{1,j}\big) \le \mathcal{G}\big(x^p_{1,j}\big) \text{ if } & (x^o_{1,j} < x^p_{1,j} \wedge \mathrm{PS}_{1,j} \ge r) \;\vee \\ & (x^o_{1,j} > x^p_{1,j} \wedge \mathrm{PL}_{1,j} \ge r) \\ \mathcal{G}\big(x^o_{1,j}\big) > \mathcal{G}\big(x^p_{1,j}\big) \text{ otherwise,} & \end{cases} \tag{8.6}$$

where $r \sim U[0, 1]$ is a random variable taking on values from the closed interval of real numbers [0, 1] with uniform probability distribution.

Both $\mathrm{PS}_{1,j}$ and $\mathrm{PL}_{1,j}$ are initialized to a value of 1 in Eq. (8.6), which indicate that the initialized parent partial solution $x^p_{1,j}$ is assumed to be worse than any of its neighbours. However, $\mathrm{PS}_{1,j}$ and $\mathrm{PL}_{1,j}$ would be adjusted during the search since the parent $x^p_{1,j}$ moves during the course of the search process and that the corresponding local fitness landscape would be different. We have adapted a well-known parameters tuning technique, the 1/5 successful rule [11], to adjust $\mathrm{PS}_{1,j}$ and $\mathrm{PL}_{1,j}$ for the training process of the meta-model [48]

$$\begin{aligned}\mathrm{PS}_{1,j} &= \mathrm{PS}_{1,j} \cdot \exp\Big(1/\sqrt{2} \cdot \mathbf{1}_{x^o_{1,j} < x^p_{1,j}} \cdot \big(\mathbf{1}_{\mathcal{F}(\boldsymbol{x}^p_1) \ge \mathcal{F}(\boldsymbol{x}^o_1)} - 1/5\big)\Big),\\ \mathrm{PL}_{1,j} &= \mathrm{PL}_{1,j} \cdot \exp\Big(1/\sqrt{2} \cdot \mathbf{1}_{x^o_{1,j} > x^p_{1,j}} \cdot \big(\mathbf{1}_{\mathcal{F}(\boldsymbol{x}^p_1) \ge \mathcal{F}(\boldsymbol{x}^o_1)} - 1/5\big)\Big),\end{aligned} \tag{8.7}$$

where $\mathbf{1}_a$ is an indicator function that returns 1 if a is true and 0 otherwise, and $\boldsymbol{x}^o_1 = (x^o_{1,1}, \ldots, x^o_{1,D})$ and $\boldsymbol{x}^p_1 = (x^p_{1,1}, \ldots, x^p_{1,D})$ are the complete solutions that $x^o_{1,j}$ and $x^p_{1,j}$ are associated with, respectively. Equation (8.7) expresses the mechanism used to train the meta-model, i.e. the corresponding probability is increased for correct meta-model prediction; otherwise it is reduced. To verify whether the prediction is correctly made, we use the original objective function and as such compare the values of $\mathcal{F}(\boldsymbol{x}^o_1)$ and $\mathcal{F}(\boldsymbol{x}^p_1)$.

The assumption we have made in the design of the training process for the meta-model expressed by Eq. (8.6) is that the local fitness landscape changes smoothly. This is because the meta-model from the previous iterations could be helpful for the current pairwise comparison under such an assumption. Another requirement is that the subcomponent optimizer employs local search. As such, the tuning or adjustment of $PS_{1,j}$ and $PL_{1,j}$ could be made gradually with a constant factor.

8.3.3 A Specific Implementation of SEE

Here, we present a specific implementation of SEE whereby to better facilitate the proposed meta-model we have employed local search operators for the subcomponent optimizers [48]. This implementation is described procedurally in Algorithm 8.2. In particular, note that a simple problem decomposition of a D-dimension $\mathcal{F}$ into D one-dimensional subcomponents can be achieved via indexing alone. All the procedural steps required for the adaptations of partial solutions and meta-models are detailed out following the SEE framework (Algorithm 8.1). Our presentation here will focus on the key principles and design motivations of the specific procedures we have implemented in Algorithm 8.2.

In the general Algorithm 8.1, we have considered a (1 + 1)-EA [11] since only a pair of parent and offspring partial solutions is considered in each subcomponent. Here, we generalized this operator further to a $(1 + \lambda)$-EA-based subcomponent optimizer for SEE. For each jth subcomponent, SEE will now first randomly initialize a single parent partial solution $x^p_{1,j}$. At each iteration, λ offsprings $x^o_{1:\lambda,j}$ are generated from $x^p_{1,j}$ with different local search operators. After that, only the best of the full population $1 + \lambda$ partial solutions is selected for the next generation.

In particular, both Gaussian and Cauchy perturbation-based mutation operators are used as the local search operators. For each jth subcomponent and the parent partial solution $x^p_{1,j}$, they are both applied to generate λ offspring partial solutions. The ***Gaussian mutation*** operator generates an offspring partial solution $x^o_{i,j}$, $i = 1, \ldots, n$, as follows: [48]

$$x^o_{i,j} = x^p_{1,j} + \sigma_{i,j} \cdot N(0, 1), \tag{8.8}$$

where $N(0, 1)$ denotes a Gaussian random variable with zero mean and standard deviation of one. The ***Cauchy mutation*** operator generates an offspring partial solution $x^o_{i,j}$, $i = n + 1, \ldots, \lambda$, as follows: [48]

$$x^o_{i,j} = x^p_{1,j} + \sigma_{i,j} \cdot C(0, 1), \tag{8.9}$$

where $C(0, 1)$ denotes a random variable subject to the standard Cauchy distribution. Note that n offspring partial solutions with index $1 \leq i \leq \lambda$ are generated via Gaussian mutation, while the remaining $\lambda - n$ offspring partial solutions with index $(n + 1) \leq i \leq \lambda$ are generated via Cauchy mutation.

Both mutation operators employ search step-size parameters $\sigma_{i,j}$ that can be adapted during the search process and also vary over different subcomponents. At the start, all $\sigma_{i,j}$ are initialized to the same value of 1. In every subsequent iteration,

Algorithm 8.2 Specific SEE implementation [48]

Input:

$\mathcal{F}$ Objective function
D Problem dimensionality
$\mathcal{H}$ Meta-model
t_e Total number of function evaluation (time budget)
λ Number of offspring solutions
μ Single parent solution $\mu = 1$
n Number of offspring solutions generated via Gaussian perturbations

Output:

$\boldsymbol{x}_1^p$ Final best complete solution

procedure **SEE**($\mathcal{F}, D, \mathcal{H}, t_e, \lambda, \mu, n$)
  for $j := 1 : D$ **do** ▷ Simply divide $\mathcal{F}$ to D 1-dimension subcomponents
    for $i := 1 : \lambda$ **do** ▷ Initializations
      $\text{PS}_{i,j} := 1.0$
      $\text{PL}_{i,j} := 1.0$
      $\sigma_{i,j} := 1.0$
    end for
    $x_{1,j}^p :=$ **initParPop**(μ) ▷ Initialize parent partial solution $x_{1,j}^p$ randomly
  end for
  while within time budget t_e **do**
    for $j := 1 : D$ **do**
      for $i := 1 : n$ **do**
        $x_{i,j}^o := x_{1,j}^p + \sigma_{i,j} \times N(0, 1)$ ▷ Gaussian generated n offsprings
      end for
      for $i := (n+1) : \lambda$ **do**
        $x_{i,j}^o := x_{1,j}^p + \sigma_{i,j} \times C(0, 1)$ ▷ Cauchy generated n offsprings
      end for
      for $i := 1 : \lambda$ **do**
        if $x_{1,j}^o < x_{1,j}^p \wedge \text{PS}_{1,j} < r$ **then**
          $x_{i,j}^o := x_{1,j}^p$
        end if
        if $x_{1,j}^o > x_{1,j}^p \wedge \text{PL}_{1,j} < r$ **then**
          $x_{i,j}^o := x_{1,j}^p$
        end if
      end for
    end for
    if $\mathcal{F}(\boldsymbol{x}_1^p) >$ **minF**$(\mathcal{F}(\boldsymbol{x}_i^o), \lambda)$ **then** ▷ Check for $\mathcal{F}(\boldsymbol{x}_1^p) > \min_{1 \le i \le \lambda} \mathcal{F}(\boldsymbol{x}_i^o)$
      $\boldsymbol{x}_1^p :=$ **argminF**$(\mathcal{F}(\boldsymbol{x}_i^o), \lambda)$ ▷ $\boldsymbol{x}_1^p = \text{argmin}_{1 \le i \le \lambda} \mathcal{F}(\boldsymbol{x}_i^o)$
    end if
    for $j := 1 : D$ **do**
      $\sigma_{i,j} := \sigma_{i,j} \times \exp\left(1/\sqrt{2} \times \mathbf{1}_{x_{i,j}^o \neq x_{1,j}^p} \times \left(\mathbf{1}_{\mathcal{F}(\boldsymbol{x}_1^p) \ge \mathcal{F}(\boldsymbol{x}_i^o)} - 1/5\right)\right)$
      $\text{PS}_{1,j} := \text{PS}_{1,j} \times \exp\left(1/\sqrt{2} \times \mathbf{1}_{x_{1,j}^o < x_{1,j}^p} \times \left(\mathbf{1}_{\mathcal{F}(\boldsymbol{x}_1^p) \ge \mathcal{F}(\boldsymbol{x}_1^o)} - 1/5\right)\right)$
      $\text{PL}_{1,j} := \text{PL}_{1,j} \times \exp\left(1/\sqrt{2} \times \mathbf{1}_{x_{1,j}^o > x_{1,j}^p} \times \left(\mathbf{1}_{\mathcal{F}(\boldsymbol{x}_1^p) \ge \mathcal{F}(\boldsymbol{x}_1^o)} - 1/5\right)\right)$
    end for
  end while
  return $\boldsymbol{x}_1^p$
end procedure

each $\sigma_{i,j}$ is adapted via the 1/5 successful rule [11] that we have formulated as follows [48]:

$$\sigma_{i,j} = \sigma_{i,j} \cdot \exp\left(1/\sqrt{2} \cdot \mathbf{1}_{x^o_{i,j} \neq x^p_{1,j}} \cdot \left(\mathbf{1}_{\mathcal{F}(\boldsymbol{x}^p_1) \geq \mathcal{F}(\boldsymbol{x}^o_i)} - 1/5\right)\right). \quad (8.10)$$

Here, if the search process has successfully made a progress in which $x^o_{i,j}$ is different from $x^p_{1,j}$, then corresponding search step-size is enlarged; otherwise, it is reduced.

The probability density function describing a Gaussian distribution is such that the probability measures are concentrated around the mean as compared to the case of a standard Cauchy distribution. As such, a Gaussian-perturbed solution is more likely to remain close to the original (source) solution compared to a Cauchy-perturbed solution. In this manner, the Gaussian mutation operator might be more compatible with the proposed meta-model as a local optimizer since it is more exploitative than explorative [18, 51]. In this way, we use the parameter n setting as the means to balance the requirements of having an effective meta-model through localized search against and the actual overall search for complete solutions.

Selection of the partial solution from the full population is required to update the parent solution for the next iteration. We have employed λ meta-models $\mathcal{H}_{1:\lambda,j}$ for each jth subcomponent, given that our proposed meta-model is only a simple, local distance-based predictor [48]. Each meta-model $\mathcal{H}_{i,j}$ is used to compare the corresponding offspring $x^o_{i,j}$ and the parent $x^p_{1,j}$ partial solutions. The offspring $x^o_{i,j}$ that is predicted by $\mathcal{H}_{i,j}$ to be better than the parent $x^p_{1,j}$ is kept for the next stage of processing. Note that in Algorithm 8.2, we procedurally implement the conditional of $\mathcal{G}(x^o_{1,j}) > \mathcal{G}(x^p_{1,j})$ in Eq. (8.6) that specifies the prediction of $\mathcal{H}_{i,j}$. That is, SEE will determine if $x^o_{i,j}$ is predicted to be worse than $x^p_{1,j}$ and then discard the offspring by setting $x^o_{i,j} = x^p_{1,j}$. Procedurally, this is equivalent to removing such an offspring from memory with a pairwise comparison cost rather than having rebuilt the data structure of $x^o_{1:\lambda,j}$ that would be more complex and incur higher computational cost.

After completing the executions of meta-models $\mathcal{H}_{i,j}$ for all subcomponents, the resulting λ offspring complete solutions $\boldsymbol{x}^o_{1:\lambda}$ will be evaluated with the objective function $\mathcal{F}$. Crucially, each partial solution $x^o_{i,j}$ shares the same objective function value $f(x^o_{i,j})$ with that of its corresponding complete solution, i.e. $f(x^o_{i,j}) = \mathcal{F}(\boldsymbol{x}^o_i)$, $j = 1, \ldots, D$. After that, $x^p_{1,j}$, $j = 1, \ldots, D$, are updated with their corresponding best offspring in terms of the objective function values. Note that since the ith partial solutions for all jth subcomponents $x^o_{i,j}$, $j = 1, \ldots, D$ share the same objective function value equivalent to $\mathcal{F}(\boldsymbol{x}^o_i)$, then we can directly replace the parent complete solution with the best offspring complete solution. This is expressed as $\boldsymbol{x}^p_1 = \text{argmin}_{1 \leq i \leq \lambda} \mathcal{F}(\boldsymbol{x}^o_i)$ if $\mathcal{F}(\boldsymbol{x}^p_1) > \min_{1 \leq i \leq \lambda} \mathcal{F}(\boldsymbol{x}^o_i)$; otherwise, $\boldsymbol{x}^p_1$ is kept unchanged [48].

One can view the ensemble of meta-models $\mathcal{H}_{1:\lambda,j}$ as a predictor that take λ samples of the local fitness landscape centered at $x^p_{1,j}$ with different step-sizes or radius and then infer via pairwise comparisons of all pairs of offspring and parent partial solutions that offspring generated via local search is better than the parent. This predictor calculates out the likelihood that the locally searched offspring is better than the parent, but it cannot determine the best offspring partial solution. Thus, after the pairwise comparisons are completed, those offsprings that are predicted to be better than the parent will be verified in terms of the objective functions values of their complete solutions and compared to find the best one.

8.4 Computational Study of SEE

We start the presentation of our computational study we have carried out in [48] with brief details of the algorithms that we have used for comparison. In the introduction, we have described three approaches that have been studied to solve high-dimensional optimization problems. In particular, there are existing EAs with rebalanced exploitation and exploration, EAs equipped with dimensionality, and EAs that employ Divide-and-Conquer. For a more comprehensive comparison study, we select representative algorithms from all these three approaches. SEE belongs to the third category of approaches we have considered here. For comparisons, we have studied two other cooperative CEAs that employ Divide-and-Conquer, i.e. DECC-D [29] and DECC-DG [28]. Both of these cooperative CEAs employ a self-adaptive variant of Differential Evolution (SaNSDE [50]) to solve each subcomponent cooperatively by sharing the partial solutions with best function values. Their difference is with the specific problem decomposition strategy. DECC-DG employs Differential Grouping to decompose the original problem in an offline manner by first learning the interdependencies between variables and dividing them into multiple subcomponents. The decomposition strategy of DECC-DG relies on establishing the property that if two variables are independent, then changing the value of either variable will not perturb the rank of the corresponding complete solutions. In DECC-D, Random Grouping decomposes the original problem in an online manner, and as such the subcomponents could differ from one iteration to the next. The underlying motivation of the mechanism design is about identifying interdependencies between two variables such that the performance improvement that comes about from optimizing one of them is limited.

For the other two categorical approaches, we have chosen MA-SW-Chains [25] and RP-Ens [15] to represent the first and second categories, respectively. In particular, MA-SW-Chains that has won the IEEE WCCI'2010 large-scale global optimization competition is a Memetic Algorithm that allocates a local search intensity for each candidate solution in terms of its historical search behaviour. The local search intensity is a parameter that is used to balance the exploration and exploitation during search. RP-Ens is a random projection-based Estimation of Distribution Algorithm (EDA) [17, 43, 53]. Informally, RP-Ens employs a problem decomposition technique of projecting the original problem to a number of lower dimensional random subspaces (i.e. random projections). After that, Gaussian-distribution-based EDAs are employed to solve the lower dimensional subproblems. RP-Ens has a single parameter, which is the number of lower dimensional random projections. In general, a larger number of random projections enhance the performance of RP-Ens at the expense of higher computational cost.

All the algorithms are configured based on suggestions of the original papers that have introduced them with the exception of RP-Ens. The very high working memory requirements for stable operations necessitate that we adjust the number of random projections at the default suggestion of 1000 for problems involving $D = 1000$ and then reduce to 200 random projections when the problem dimensionality is increased. DECC-DG has high function evaluation requirements to complete the learning stage, and as a result those function evaluations are omitted. As such, the DECC-DG implementation we have here represents the ideal case. For SEE, the population size is set to $\lambda = 10$, and then $n = 5$ so that both Gaussian and Cauchy mutation operators are allocated the same search resources.

Our computational study compares SEE against four state-of-the-art algorithms on the test suite of 20 problems at five different dimensionality settings $D = \{1000, 2000, 3000, 4000, 5000\}$ and also investigates the impact of the proposed meta-model on the performance of SEE. We have employed the widely used CEC'2010 large-scale optimization benchmark [37] to assess the performance of high-dimensional optimization algorithms [4, 15, 24, 25, 28, 29, 49]. This test suite consists of four types of high-dimensional problems in terms of different degrees of separability. For each type of problems, both unimodal and multi-modal problems are included. This test suite is flexible with two parameters to change the underlying properties of the problem that as a result influence its complexity. The first parameter is on the dimensionality D of each problem. The problem dimensionality can be adjusted from the original 1000 to a higher number of decision variables. The second parameter is the size M of each group of interdependent variables for each problem.

There is an interplay between problem complexity and the effectiveness of a particular search heuristic. We have prepared our computational study and use five groups of problems with different configuration settings of (D, M, t_e) that we list in Table 8.1. The third setting is the time budget t_e that refers to the total number of function evaluations used in all the algorithms before termination. In general, we linearly increase the time budget. As such, our controlled computational study lets us investigate how SEE scales on those problems in comparison with the other four algorithms. If an algorithm obtains the final solutions with similar quality on one particular problem with different dimensionalities, then one could claim that the algorithm scales linearly with the dimensionality. The quality of the final solution is measured in terms of function error, i.e. the difference between the objective function value of the output solution and that of the optimal solution with respect to the problem. For the benchmark problems we use, the optimal objective function

Table 8.1 Five groups of problems with different configurations [48]

	Group 1	Group 2	Group 3	Group 4	Group 5
D	1000	2000	3000	4000	5000
M	50	100	150	200	250
t_e	6e5	1.2e6	1.8e6	2.4e6	3e6

value is 0.0. For a minimization setting that is considered here, the solution with a smaller function error means it is a better solution. Note that the output solution with function error smaller than 1e-13 will be regarded as the global optimum.

Given that all the algorithms we study are randomized algorithms, we repeat the procedural run of each algorithm on a problem 20 times. The average and standard deviation over these 20 runs of all five algorithms applied to solve each of the 20 benchmark problems in the test suite for the five problem configuration settings (D, M, t_e) are listed in Tables 8.2, 8.3, 8.4, 8.5, and 8.6. The best result on each problem is marked in **bold**. We have performed the two-sided Wilcoxon rank-sum test at a 0.05 level of significance level to see whether the performances of two algorithms are statistically significantly different. Pairwise comparison outcomes of SEE against the other four algorithms are summarized at the end of each table as the number of wins, draws, and loses (**w-d-l**). In addition to the quantitative results that we present in a tabular format, we provide qualitative results that are summarized in Figs. 8.1 and 8.2. Each consists of individual plot of the averaged results of function errors (in orders of magnitude at $\log_{10}$ scale) at the end of the runs from each algorithm applied to a particular problem over dimensionality setting $D = \{1000, 2000, 3000, 4000, 5000\}$. Essentially, each plot indicates the scalability of the algorithm with respect to a problem. If the plotted curve slowly increases in function errors as D increases, then the algorithm has good scalability with respect to that problem [48].

In general, the quantitative results summarized in Tables 8.2, 8.3, 8.4, 8.5, and 8.6 indicate that SEE performs better compared to the other four algorithms on the configured benchmark problems. Furthermore, the pairwise comparison outcomes (**w-d-l**) that increasingly favour SEE as the problem dimensionality is increased indicate that SEE has better scalability compared to the other algorithms on these configured benchmark problems. This is also verified from Figs. 8.1 and 8.2 whereby for the majority of the 20 problems, the plots of SEE indicate a slower increase in function errors compared to the other four algorithms with increase in problem dimensionality [48].

Next, we will analyse results from our computational study in more detail with respect to individual problems that have specific problem characteristics. We start with the shifted Elliptic function f_1, which is unimodal and fully separable. This problem can be easily solved and allows for a solver to separately and independently optimize each decision variable. The proposed SEE always outputs good final solutions with function errors around 1e-10 that is very close to the global optimum. However, the other compared algorithms perform very poorly on this simple problem whereby the function errors of their final solutions are at least 12 orders of magnitude away from that of SEE. The performance advantage SEE compared to the other algorithms is maintained when they are all applied to solve partially separable and nonseparable problems. For example, f_4, f_9, and f_{14} are rotated variants of f_1, in which case they are no longer fully separable. On these problems, SEE maintains significant performance advantage over the other algorithms. Another example involves f_7, f_{12}, f_{17}, and f_{19} problems, which are constructed from a fully nonseparable Schwefel problem 1.2 combined with the fully separable

Table 8.2 The average and standard deviation of function errors on the 1000-dimensional CEC'2010 large-scale global optimization benchmark. The best result is marked in **bold** [48]

Algorithm		MA-SW-Chain	RP-Ens	DECC-D	DECC-DG	SEE
f_1	Mean	3.09e+04	1.68e+07	3.33e+02	7.97e+06	**6.99e-11**
	Std	1.03e+04	1.77e+06	2.41e+01	4.93e+06	1.98e-11
f_2	Mean	2.74e+03	**7.25e+02**	3.13e+03	4.62e+03	8.77e+03
	Std	1.11e+02	1.94e+01	6.99e+01	1.67e+02	2.99e+02
f_3	Mean	3.82e+00	**7.77e-05**	6.54e-03	1.70e+01	1.99e+01
	Std	2.32e-02	3.67e-06	2.57e-04	4.14e-01	1.61e-02
f_4	Mean	6.08e+11	2.64e+12	3.02e+13	6.86e+13	**2.58e+11**
	Std	4.48e+10	5.90e+11	7.64e+12	1.21e+13	6.41e+10
f_5	Mean	**8.50e+07**	3.05e+08	2.68e+08	2.39e+08	5.85e+08
	Std	2.02e+07	1.17e+07	7.57e+07	2.42e+07	1.37e+08
f_6	Mean	2.13e+01	9.09e+01	1.00e+06	**1.69e+01**	1.99e+07
	Std	5.88e-02	2.91e+00	4.49e+06	5.23e-01	6.01e+04
f_7	Mean	5.68e+06	1.49e+09	1.76e+09	1.29e+09	**3.14e-02**
	Std	3.49e+05	4.38e+08	7.41e+08	4.50e+08	1.56e-02
f_8	Mean	1.33e+07	4.59e+07	1.30e+08	6.21e+07	**1.82e+06**
	Std	5.87e+06	3.71e+05	5.54e+07	2.72e+07	2.11e+06
f_9	Mean	7.14e+07	4.56e+07	4.52e+08	8.27e+08	**2.67e+07**
	Std	4.82e+06	1.74e+06	4.71e+07	6.95e+07	2.13e+06
f_{10}	Mean	3.00e+03	**7.24e+02**	1.33e+04	8.60e+03	1.27e+04
	Std	1.57e+02	1.08e+01	3.02e+02	1.13e+02	5.06e+02
f_{11}	Mean	4.97e+01	**1.57e-04**	3.74e+00	1.67e+01	2.19e+02
	Std	1.91e+01	7.21e-06	1.39e+01	3.78e-01	2.44e-01
f_{12}	Mean	4.75e+02	6.12e+02	4.82e+06	3.29e+05	**2.60e+02**
	Std	2.86e+01	2.37e+01	2.30e+05	1.57e+04	4.81e+01
f_{13}	Mean	9.01e+03	3.26e+04	5.75e+03	2.60e+10	**7.12e+02**
	Std	6.68e+03	8.50e+03	5.14e+03	4.32e+09	2.61e+02
f_{14}	Mean	1.44e+08	1.65e+08	1.52e+09	2.34e+09	**9.88e+07**
	Std	5.01e+06	2.72e+06	1.22e+08	1.10e+08	7.61e+06
f_{15}	Mean	5.43e+03	**7.36e+02**	1.63e+04	7.85e+03	1.50e+04
	Std	3.57e+03	1.64e+01	3.62e+02	8.90e+01	3.10e+02
f_{16}	Mean	1.13e+02	**1.64e-04**	3.37e+01	2.38e+01	3.97e+02
	Std	2.85e+01	8.40e-06	8.91e+01	1.05e+00	2.92e-01
f_{17}	Mean	2.73e+04	1.79e+04	8.39e+06	7.70e+05	**7.40e+03**
	Std	9.27e+02	4.35e+03	5.39e+05	2.85e+04	9.59e+02
f_{18}	Mean	1.32e+04	2.48e+04	3.02e+04	3.98e+11	**3.14e+03**
	Std	8.37e+03	3.63e+03	7.08e+03	2.79e+10	1.00e+03
f_{19}	Mean	1.38e+06	3.26e+06	2.22e+07	3.66e+06	**7.13e+05**
	Std	4.00e+04	6.80e+03	1.68e+06	2.16e+05	4.24e+04
f_{20}	Mean	**1.12e+03**	1.64e+03	5.75e+03	6.48e+10	1.43e+03
	Std	7.22e+01	1.42e+03	4.65e+02	8.50e+09	1.85e+02
w-d-l		11-1-8	11-0-9	14-0-6	12-0-8	–

Table 8.3 The average and standard deviation of function errors on the 2000-dimensional CEC'2010 large-scale global optimization benchmark. The best result is marked in **bold** [48]

Algorithm		MA-SW-Chain	RP-Ens	DECC-D	DECC-DG	SEE
f_1	Mean	7.39e+05	4.82e+09	1.34e+03	2.28e+08	**2.19e-10**
	Std	9.50e+04	2.04e+08	1.05e+02	1.07e+08	3.51e-11
f_2	Mean	6.66e+03	**4.76e+03**	6.29e+03	1.27e+04	1.81e+04
	Std	2.62e+02	2.09e+02	1.15e+02	2.50e+02	4.93e+02
f_3	Mean	7.57e+00	1.03e+01	**9.74e-03**	1.92e+01	1.99e+01
	Std	3.13e-01	3.49e-01	3.56e-04	1.23e-01	1.02e-02
f_4	Mean	8.07e+13	1.07e+15	7.08e+15	1.07e+16	**4.89e+11**
	Std	5.11e+13	2.28e+14	1.82e+15	1.90e+15	1.10e+11
f_5	Mean	4.34e+08	**8.79e+07**	1.20e+09	9.81e+08	1.23e+09
	Std	1.01e+08	1.67e+07	1.49e+08	9.79e+07	2.16e+08
f_6	Mean	2.01e+07	2.00e+07	2.09e+07	2.13e+07	**1.99e+07**
	Std	9.86e+04	2.58e+04	7.57e+04	2.06e+04	6.40e+04
f_7	Mean	1.17e+07	9.90e+08	1.94e+10	3.46e+10	**7.33e+00**
	Std	3.51e+05	2.67e+08	2.94e+09	3.78e+09	2.97e+00
f_8	Mean	7.47e+07	1.18e+08	1.77e+08	3.20e+08	**9.99e+05**
	Std	6.83e+07	5.73e+07	5.61e+07	5.54e+08	1.77e+06
f_9	Mean	1.10e+10	2.24e+11	1.25e+11	2.61e+11	**6.95e+07**
	Std	9.91e+08	7.90e+09	7.85e+09	1.51e+10	4.97e+06
f_{10}	Mean	2.54e+04	2.85e+04	2.74e+04	**2.36e+04**	2.49e+04
	Std	3.14e+03	9.19e+02	1.07e+03	3.54e+02	4.33e+02
f_{11}	Mean	2.23e+02	2.24e+02	2.37e+02	2.33e+02	**2.19e+02**
	Std	5.13e-01	2.04e-01	2.79e-01	3.09e-01	1.22e-01
f_{12}	Mean	2.26e+04	4.11e+06	1.24e+07	1.28e+06	**1.68e+04**
	Std	8.79e+02	9.62e+04	4.79e+05	4.39e+04	2.49e+03
f_{13}	Mean	1.50e+04	6.38e+06	1.08e+04	2.66e+10	**1.43e+03**
	Std	4.04e+03	8.30e+05	5.21e+03	3.98e+09	3.27e+02
f_{14}	Mean	1.95e+10	3.89e+11	2.38e+11	8.59e+11	**2.61e+08**
	Std	1.39e+09	1.85e+10	1.40e+10	5.40e+10	1.16e+07
f_{15}	Mean	3.70e+04	7.28e+04	3.46e+04	**2.86e+04**	3.08e+04
	Std	3.07e+03	3.14e+03	7.69e+02	5.71e+02	8.11e+02
f_{16}	Mean	4.04e+02	4.09e+02	4.32e+02	4.27e+02	**3.98e+02**
	Std	1.00e+00	4.28e-01	1.32e-01	1.70e-01	2.56e-01
f_{17}	Mean	3.15e+05	4.78e+06	2.55e+07	2.85e+06	**1.59e+05**
	Std	1.52e+04	1.17e+05	1.25e+06	6.44e+04	1.15e+04
f_{18}	Mean	**3.97e+03**	2.99e+09	2.38e+04	1.06e+12	4.65e+03
	Std	1.36e+03	6.69e+08	5.40e+03	4.64e+10	1.09e+03
f_{19}	Mean	4.55e+06	1.32e+07	6.15e+07	1.06e+07	**2.61e+06**
	Std	1.31e+05	6.99e+05	4.27e+06	5.01e+05	1.15e+05
f_{20}	Mean	**2.33e+03**	3.18e+09	7.20e+03	2.98e+12	3.19e+03
	Std	1.56e+02	6.65e+08	3.84e+02	1.74e+11	2.25e+02
w-d-l		14-1-5	17-0-3	17-1-2	15-0-5	–

Table 8.4 The average and standard deviation of function errors on the 3000-dimensional CEC'2010 large-scale global optimization benchmark. The best result is marked in **bold** [48]

Algorithm		MA-SW-Chain	RP-Ens	DECC-D	DECC-DG	SEE
f_1	Mean	3.17e+06	1.07e+10	1.92e+03	1.15e+09	**6.37e-10**
	Std	3.41e+05	3.73e+08	1.37e+02	4.92e+08	9.64e-11
f_2	Mean	1.10e+04	1.10e+04	**9.39e+03**	2.23e+04	2.74e+04
	Std	3.81e+02	3.39e+02	7.61e+01	5.18e+02	6.19e+02
f_3	Mean	1.05e+01	1.42e+01	**9.65e-03**	1.97e+01	1.99e+01
	Std	3.95e-01	2.18e-01	3.32e-04	8.48e-02	1.22e-02
f_4	Mean	1.36e+14	3.24e+15	1.58e+16	1.75e+16	**6.25e+11**
	Std	5.20e+13	3.70e+14	2.86e+15	3.23e+15	1.56e+11
f_5	Mean	9.73e+08	**2.57e+08**	2.48e+09	2.43e+09	2.03e+09
	Std	1.90e+08	2.98e+07	3.01e+08	2.08e+08	2.78e+08
f_6	Mean	2.01e+07	2.01e+07	2.10e+07	2.14e+07	**1.99e+07**
	Std	9.25e+04	3.16e+04	3.18e+04	1.72e+04	3.96e+04
f_7	Mean	1.74e+07	1.79e+10	4.01e+10	8.33e+10	**4.21e+02**
	Std	4.10e+05	2.80e+09	7.75e+09	7.64e+09	2.01e+02
f_8	Mean	1.17e+08	1.96e+09	1.85e+08	2.70e+08	**8.63e+05**
	Std	2.83e+07	1.80e+09	5.04e+07	6.69e+07	1.63e+06
f_9	Mean	2.38e+10	6.02e+11	2.71e+11	6.08e+11	**1.33e+08**
	Std	1.85e+09	1.99e+10	1.60e+10	5.09e+10	1.02e+07
f_{10}	Mean	5.02e+04	7.17e+04	4.65e+04	4.22e+04	**3.78e+04**
	Std	4.96e+03	2.81e+03	2.14e+03	8.07e+02	7.06e+02
f_{11}	Mean	2.23e+02	2.24e+02	2.37e+02	2.34e+02	**2.19e+02**
	Std	6.23e-01	2.01e-01	2.07e-01	1.46e-01	2.04e-01
f_{12}	Mean	1.18e+05	8.50e+06	2.30e+07	2.64e+06	**9.34e+04**
	Std	5.06e+03	1.79e+05	9.56e+05	6.71e+04	9.15e+03
f_{13}	Mean	4.62e+04	2.30e+08	5.91e+03	1.70e+11	**1.97e+03**
	Std	7.46e+03	7.29e+07	2.87e+03	1.66e+10	3.98e+02
f_{14}	Mean	5.17e+10	1.48e+12	5.99e+11	4.06e+12	**4.91e+08**
	Std	3.38e+09	5.27e+10	2.64e+10	3.28e+11	6.25e+07
f_{15}	Mean	1.14e+05	2.24e+05	6.01e+04	5.77e+04	**4.64e+04**
	Std	2.99e+04	5.80e+03	1.59e+03	1.07e+03	1.14e+03
f_{16}	Mean	4.03e+02	4.11e+02	4.33e+02	4.28e+02	**3.98e+02**
	Std	3.17e+00	2.15e-01	3.66e-01	1.49e-01	2.43e-01
f_{17}	Mean	9.65e+05	1.05e+07	4.74e+07	5.49e+06	**5.74e+05**
	Std	8.76e+04	3.63e+05	2.44e+06	1.13e+05	3.21e+04
f_{18}	Mean	**3.63e+03**	3.98e+10	2.64e+04	2.51e+12	6.57e+03
	Std	1.87e+02	6.49e+09	7.07e+03	1.07e+11	6.88e+02
f_{19}	Mean	8.91e+06	3.04e+07	1.15e+08	1.97e+07	**5.16e+06**
	Std	3.82e+05	1.61e+06	9.68e+06	5.06e+05	1.51e+05
f_{20}	Mean	**3.62e+03**	4.09e+10	1.11e+04	5.21e+12	5.10e+03
	Std	1.66e+02	2.82e+09	1.26e+03	5.94e+10	3.07e+02
w-d-l		15-0-5	17-0-3	18-0-2	18-0-2	–

Table 8.5 The average and standard deviation of function errors on the 4000-dimensional CEC'2010 large-scale global optimization benchmark. The best result is marked in **bold** [48]

Algorithm		MA-SW-Chain	RP-Ens	DECC-D	DECC-DG	SEE
f_1	Mean	8.14e+06	2.05e+10	2.45e+03	5.60e+10	**1.55e-09**
	Std	5.34e+05	8.68e+08	1.15e+02	3.70e+09	2.66e-10
f_2	Mean	1.53e+04	1.91e+04	**1.25e+04**	3.20e+04	3.72e+04
	Std	3.59e+02	4.10e+02	1.64e+02	3.82e+02	7.59e+02
f_3	Mean	1.32e+01	1.61e+01	**9.33e-03**	1.97e+01	1.99e+01
	Std	8.13e-01	1.98e-01	2.51e-04	4.03e-02	9.59e-03
f_4	Mean	2.25e+14	7.11e+15	2.73e+16	1.21e+16	**1.16e+12**
	Std	5.67e+13	9.59e+14	5.29e+15	1.97e+15	2.74e+11
f_5	Mean	1.61e+09	**5.12e+08**	4.15e+09	4.11e+09	2.85e+09
	Std	2.70e+08	4.14e+07	5.12e+08	4.00e+08	2.70e+08
f_6	Mean	2.01e+07	2.01e+07	2.10e+07	2.14e+07	**1.99e+07**
	Std	6.83e+04	2.35e+04	5.88e+04	1.42e+04	4.90e+04
f_7	Mean	2.34e+07	6.94e+10	8.38e+10	2.86e+11	**1.02e+04**
	Std	4.48e+05	8.66e+09	1.14e+10	2.69e+10	3.86e+03
f_8	Mean	1.56e+08	2.09e+08	2.31e+08	4.91e+08	**1.20e+06**
	Std	1.30e+07	5.34e+05	4.21e+07	7.19e+07	1.84e+06
f_9	Mean	3.86e+10	1.21e+12	4.74e+11	9.56e+11	**2.01e+08**
	Std	3.89e+09	4.24e+10	1.86e+10	7.16e+10	1.12e+07
f_{10}	Mean	8.04e+04	1.50e+05	6.69e+04	6.17e+04	**5.03e+04**
	Std	4.92e+03	5.41e+03	3.44e+03	1.65e+03	1.11e+03
f_{11}	Mean	2.23e+02	2.25e+02	2.38e+02	2.35e+02	**2.19e+02**
	Std	5.83e-01	1.73e-01	1.58e-01	9.15e-02	1.06e-01
f_{12}	Mean	3.28e+05	1.40e+07	3.47e+07	3.96e+06	**2.49e+05**
	Std	5.24e+03	1.86e+05	1.37e+06	9.62e+04	1.63e+04
f_{13}	Mean	6.06e+04	2.27e+09	8.36e+03	2.00e+11	**2.90e+03**
	Std	6.53e+03	2.89e+08	3.46e+03	1.44e+10	5.17e+02
f_{14}	Mean	9.40e+10	3.76e+12	1.03e+12	2.54e+12	**7.67e+08**
	Std	7.19e+09	8.49e+10	3.70e+10	1.30e+11	4.31e+07
f_{15}	Mean	2.98e+05	1.26e+06	8.40e+04	9.47e+04	**6.14e+04**
	Std	3.44e+04	1.59e+05	1.63e+03	1.51e+03	1.18e+03
f_{16}	Mean	4.03e+02	4.17e+02	4.33e+02	4.28e+02	**3.98e+02**
	Std	1.75e+00	5.83e+00	5.29e-01	1.29e-01	2.51e-01
f_{17}	Mean	1.85e+06	2.74e+07	7.50e+07	8.63e+06	**1.22e+06**
	Std	2.88e+05	3.64e+05	3.66e+06	1.84e+05	4.80e+04
f_{18}	Mean	**6.05e+03**	9.29e+11	2.54e+04	1.19e+12	9.03e+03
	Std	2.80e+02	8.46e+10	5.44e+03	6.71e+10	9.15e+02
f_{19}	Mean	1.40e+07	7.64e+07	1.58e+08	2.85e+07	**8.28e+06**
	Std	8.05e+05	3.81e+06	6.25e+06	5.33e+05	3.32e+05
f_{20}	Mean	**4.53e+03**	4.65e+11	1.65e+04	5.57e+12	7.17e+03
	Std	1.60e+02	9.91e+09	2.79e+03	2.48e+11	3.65e+02
w-d-l		15-0-5	17-0-3	18-0-2	18-0-2	–

Table 8.6 The average and standard deviation of function errors on the 5000-dimensional CEC'2010 large-scale global optimization benchmark. The best result is marked in **bold** [48]

Algorithm		MA-SW-Chain	RP-Ens	DECC-D	DECC-DG	SEE
f_1	Mean	1.60e+07	6.72e+10	3.09e+03	6.72e+10	**3.44e-09**
	Std	1.47e+06	2.77e+09	1.64e+02	2.68e+09	3.95e-10
f_2	Mean	1.99e+04	4.39e+04	**1.57e+04**	4.39e+04	4.64e+04
	Std	5.65e+02	6.26e+02	1.64e+02	6.26e+02	9.76e+02
f_3	Mean	1.42e+01	1.93e+01	**9.58e-03**	1.93e+01	1.99e+01
	Std	4.27e-02	7.05e-02	2.73e-04	7.05e-02	5.34e-03
f_4	Mean	2.16e+14	3.11e+16	4.29e+16	3.11e+16	**1.29e+12**
	Std	2.10e+13	2.56e+15	5.36e+15	2.56e+15	2.15e+11
f_5	Mean	2.73e+09	**2.27e+09**	5.76e+09	**2.27e+09**	3.27e+09
	Std	1.13e+09	2.16e+08	7.89e+08	2.16e+08	1.92e+08
f_6	Mean	2.02e+07	2.02e+07	2.11e+07	2.02e+07	**1.99e+07**
	Std	4.43e+04	3.42e+04	3.91e+04	3.42e+04	2.54e+04
f_7	Mean	2.83e+07	2.34e+11	1.33e+11	2.34e+11	**1.14e+05**
	Std	5.21e+05	1.67e+10	1.25e+10	6.75e+10	4.49e+04
f_8	Mean	1.93e+08	2.62e+08	3.74e+08	2.62e+08	**6.26e+05**
	Std	1.39e+07	7.35e+05	7.06e+07	7.35e+05	1.45e+06
f_9	Mean	5.36e+10	4.38e+12	7.86e+11	4.38e+12	**3.07e+08**
	Std	1.45e+10	1.84e+11	3.86e+10	1.84e+11	1.72e+07
f_{10}	Mean	1.18e+05	5.90e+05	9.17e+04	5.90e+05	**6.34e+04**
	Std	7.85e+03	1.83e+04	5.46e+03	1.83e+04	8.89e+02
f_{11}	Mean	2.23e+02	2.28e+02	2.38e+02	2.28e+02	**2.19e+02**
	Std	2.57e-01	1.73e-01	2.94e-01	1.73e-01	1.26e-01
f_{12}	Mean	6.72e+05	2.35e+07	4.70e+07	2.35e+07	**4.78e+05**
	Std	5.63e+03	4.51e+05	2.03e+06	4.51e+05	2.54e+04
f_{13}	Mean	1.34e+05	2.32e+11	1.55e+04	2.32e+11	**3.47e+03**
	Std	2.01e+04	9.56e+09	5.95e+03	9.56e+09	4.11e+02
f_{14}	Mean	1.37e+11	1.59e+13	1.59e+12	1.59e+13	**1.08e+09**
	Std	6.68e+09	4.01e+11	4.65e+10	4.01e+11	3.70e+07
f_{15}	Mean	5.62e+05	2.34e+06	1.15e+05	2.34e+06	**7.67e+04**
	Std	4.70e+04	1.32e+04	1.66e+03	1.32e+04	1.56e+03
f_{16}	Mean	4.04e+02	4.19e+02	4.32e+02	4.19e+02	**3.98e+02**
	Std	5.05e-01	3.49e-01	6.16e-01	3.49e-01	1.89e-01
f_{17}	Mean	3.03e+06	4.06e+07	1.07e+08	4.06e+07	**2.11e+06**
	Std	5.54e+04	1.10e+06	4.03e+06	1.10e+06	5.75e+04
f_{18}	Mean	**6.82e+03**	1.85e+12	4.13e+04	1.85e+12	1.15e+04
	Std	8.40e+02	5.48e+10	8.15e+03	5.48e+10	1.07e+03
f_{19}	Mean	1.94e+07	1.10e+08	2.58e+08	1.10e+08	**1.15e+07**
	Std	5.61e+05	3.24e+06	7.44e+06	3.24e+06	3.78e+05
f_{20}	Mean	**5.60e+03**	1.85e+12	1.72e+04	1.85e+12	9.40e+03
	Std	9.62e+01	1.51e+11	4.38e+02	1.51e+11	3.59e+02
w-d-l		15-1-4	17-0-3	18-0-2	18-0-2	–

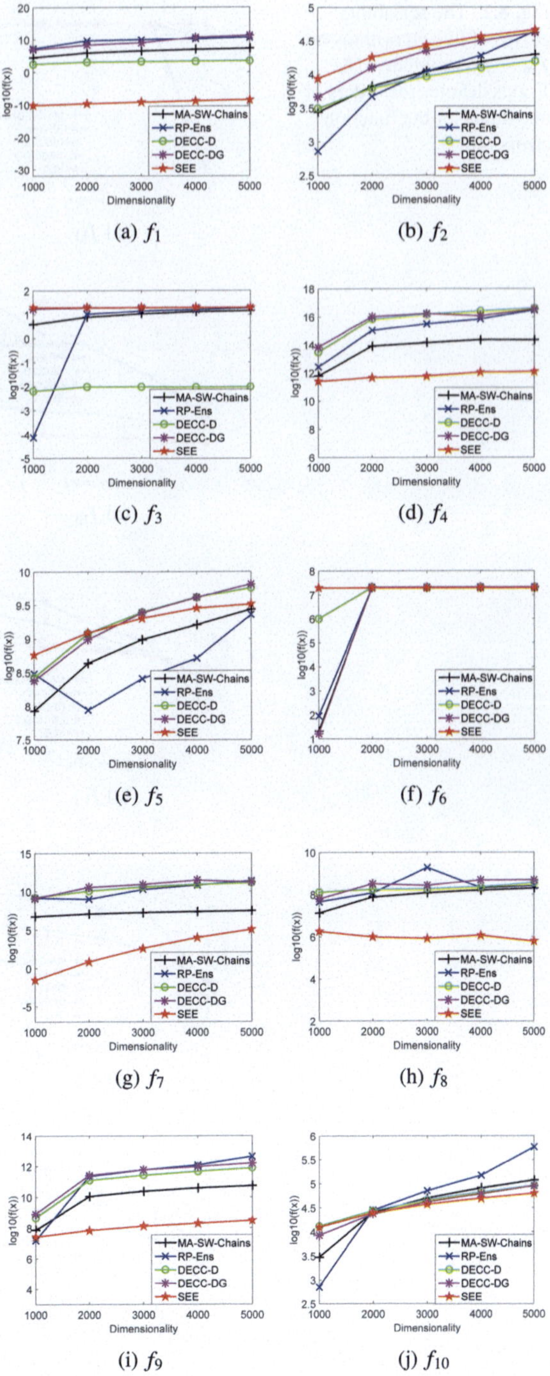

Fig. 8.1 The scalability curves of five algorithms on $f_1 - f_{15}$ problems. The Y-axis denotes the orders of magnitude of the function errors [48]

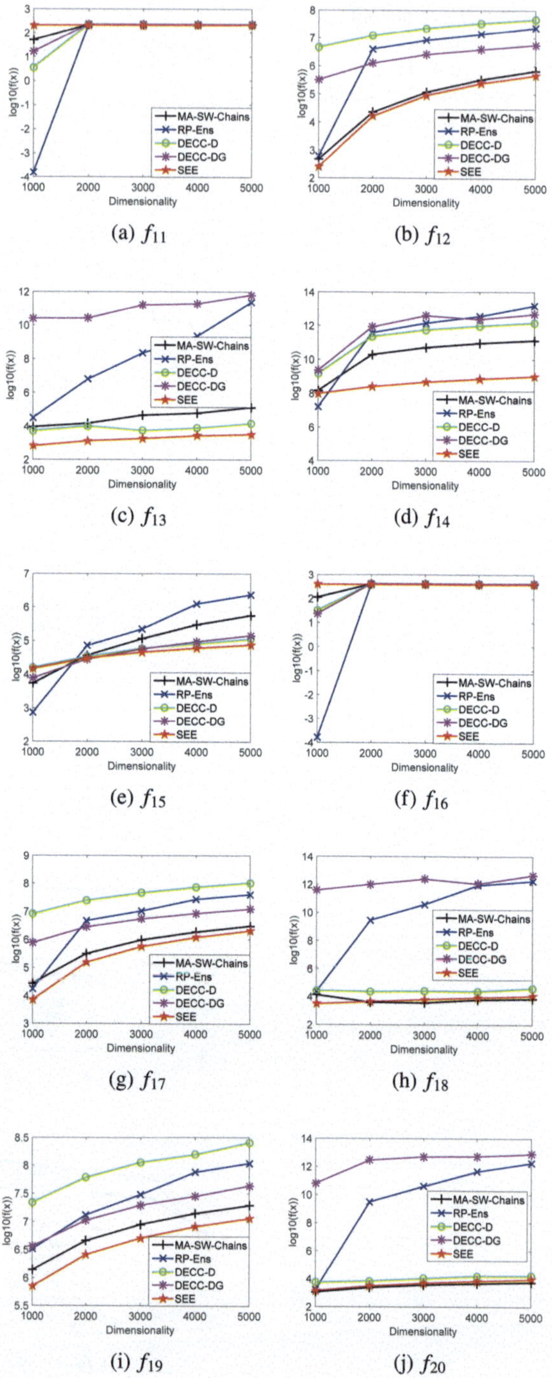

(a) f_{11} (b) f_{12}

(c) f_{13} (d) f_{14}

(e) f_{15} (f) f_{16}

(g) f_{17} (h) f_{18}

(i) f_{19} (j) f_{20}

Fig. 8.2 The scalability curves of five algorithms on $f_{16} - f_{20}$ problems. The Y-axis denotes the orders of magnitude of the function errors [48]

sphere function of different degrees. In particular, f_7 has one single group of m interdependent variables. f_{12} has $D/2m$ groups of m interdependent variables, f_{17} has D/m groups of m interdependent variables, and f_{19} is fully nonseparable with one group of D interdependent variables. As before, the scalability plots in Figs. 8.1 and 8.2 indicate that SEE has better overall performances compared to the other algorithms as the dimensionality is increased across these four problems [48].

We now turn our attention for other problems, for example, f_8, f_{13}, f_{18}, and f_{20} that constitute the well-known multi-modal problems that are based on the Rosenbrock function. Similar observations on the overall higher performances of SEE in comparison to the other algorithms can be made. In the case of f_{18} and f_{20}, MA-WS-Chains has a slight edge in performances compared to SEE. However, both exhibit significantly superior search performances compared to the other algorithms as the problem dimensionality is increased. For the remaining group of problems (f_2, f_3, f_5, f_6, f_{10}, f_{11}, f_{15}, and f_{16}), SEE performs poorer than the compared algorithms for problems at 1000-dimension. However, as the problem dimensionality is increased, the overall performances of the four other algorithms are reduced in most cases. In some cases such as on f_{10} and f_{15} at higher problem dimensions, SEE's better scalability means a slower drop in performances so that it would end up having better performances compared to the other algorithms. One possible reason for SEE to have such performances is that these problems are based on either Rastrigin's or Ackley's functions that possess rugged fitness landscapes. As a result, there could be rapid changes to the local fitness landscapes that the proposed meta-model would become less effective and unable to keep up with the changes [48].

In general, the two cooperative CEAs (DECC-D and DECC-DG) have lower overall performances than SEE and in most problem cases several orders of magnitude below that of SEE when problem dimensionality increases. Such a result indicates that the effectiveness of meta-models to assist with the local search of decomposed partial solutions of subcomponents that respect the directionality of search that operates on full complete solutions. Furthermore, our proposed approach to a Divide-and-Conquer cooperative search could be a useful alternative to existing cooperative CEAs whose search processes are based on the use of complemented partial solutions. Both MA-SW-Chains and RP-Ens have obtained good performances on 1000-dimensional benchmark problems, but they would deteriorate as the problem dimensionality is increased. For MA-SW-Chains, the local search intensity setting might require additional tuning so that the algorithm could scale with respect to higher dimensional problems. For RP-Ens, the reduced performances for problems with larger dimensionality are likely the result of having to set the number of random projections to 200 to trade off against computational cost and operational stability. The small number of random projections might not jointly cover the original solution space [48].

Lastly, we have carried out an additional investigation to assess the impact of the meta-model on the performance of SEE. A more direct analysis would require finding out the ground truth of the rank of any pair of partial solutions, which is computationally prohibitive. As such, it might not be feasible to directly calculate the accuracy of the meta-model as the means to assess its impact on SEE. An alternative would be to employ an indirect analysis approach whereby the baseline meta-model is constructed by fixing $\mathrm{PS}_{i,j}$ and $\mathrm{PL}_{i,j}$ to 0.5. Here, the effect would be as though the meta-model performs a random guess in having to rank any pair of partial solutions. We denote the algorithm that uses this specific meta-model configuration as SEE-baseline. After that, we compare SEE against SEE-baseline with the same suite of benchmark problems we have employed earlier [48].

Here, we focus on six selected 1000-dimensional problems where we have plotted convergence curves of both SEE and SEE-baseline on different problems in Fig. 8.3. Each plot consists of average function errors against the iterations of the corresponding search process. These problems are selected in terms of their multi-modality, separability, and landscape smoothness. They consist of f_1, f_2, f_7, f_{11}, f_{18}, and f_{20} with their problem features listed in Table 8.7. Our selection choice is to show that the proposed meta-model is less sensitive to multi-modality and separability but relies on the smoothness of the problem's fitness landscape to work well. Indeed, qualitative results that are summarized and illustrated in Fig. 8.3 indicate that the proposed meta-model has a positive impact on the performance of SEE. For f_1, f_7, f_{18}, and f_{20} problems that feature smooth fitness landscape but could have different modality and separability, the convergence curves of SEE diverge away from SEE-baseline with increasing iterations of the search processes. However, on f_2 and f_{11} problems that feature rough fitness landscapes, the plots do not show a divergence in performances between SEE and SEE-baseline [48].

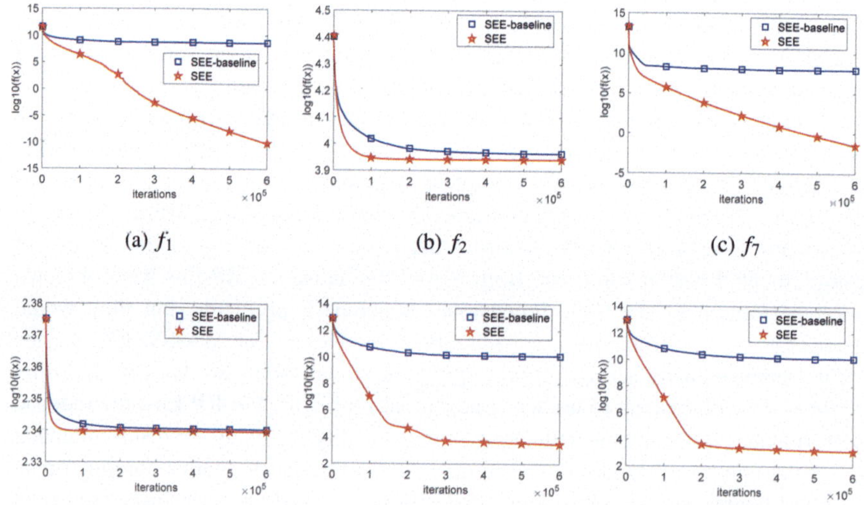

(a) f_1 (b) f_2 (c) f_7

(d) f_{11} (e) f_{18} (f) f_{20}

Fig. 8.3 The convergence curves of SEE and SEE-baseline on six problems. The Y-axis denotes the orders of magnitude of the function errors [48]

Table 8.7 The features of six selected 1000-dimensional problems [48]

	f_1	f_2	f_7
Modality	Unimodal	Multi-modal	Unimodal
Separability	Fully separable	Fully separable	Partially separable
Smoothness	Smooth	Rugged	Smooth
	f_{11}	f_{18}	f_{20}
Modality	Multi-modal	Multi-modal	Multi-modal
Separability	Partially separable	Partially separable	Fully nonseparable
Smoothness	Rugged	Smooth	Smooth

8.5 Discussion and Conclusion

This chapter has investigated an alternative approach to coevolutionary Divide-and-Conquer to solve high-dimensional optimization problems. The development of this approach is premised on a particular analysis we have performed. In particular, this analysis has identified that the main challenge of coevolutionary Divide-and-Conquer problem-solving of high-dimensional optimization has to do with the dimensionality mismatch in which partial solutions to each decomposed subcomponents have to be evaluated and correctly ranked. In formulating this dimensionality mismatch as a computationally expensive problem, the analysis has opened up the possibility of employing meta-models having significantly lower computational costs to provide approximate ranking of the partial solutions of the subcomponents. As a result, the SEE framework is formulated to provide one such procedural implementation. In particular, we have proposed a specific SEE whereby the original objective function $\mathcal{F}$ is decomposed to D one-dimensional subcomponents. For each subcomponent, a $(1 + \lambda)$-EA is employed as the local search operator. A simple but effective meta-model that performs probabilistic inference on the pairwise performance rank between the parent and offspring partial solutions has been introduced. A systematic computational study with which we have compared SEE against representatives of three categorical approaches to solving high-dimensional optimization problems has demonstrated the effectiveness of SEE and detailed how the meta-model could have a positive impact of the search performance of SEE.

In the context of a coevolutionary Divide-and-Conquer problem-solving framework, the difference between SEE and a standard cooperative CEA that employs a grouping strategy for problem decomposition might on first sight appear to predominantly focus on the mechanism of selection of partial solutions. Informally, this difference might be minimal framework-wise. However, procedurally and conceptually the difference between the two could be substantial. For a procedural or algorithmic perspective, SEE performs evaluation of complete solutions as the means with which the current search state of the whole algorithm is queried for verification of predictions by and for training of the meta-models. However, a cooperative CEA involves a complex procedure of sequential search of partial solutions requiring their complementation (to form the complete solutions) so that selection (ranking) of generated offspring partial solutions can be performed.

For a conceptual perspective one could view the difference between SEE and cooperative CEA in the context of their iterative applications on a transformed search space. In particular, a search operator that is part of a coevolutionary Divide-and-Conquer problem-solving algorithm is applied on a new population $\mathbf{X}' = \mathcal{T}(\mathbf{X})$ in the transformed low-dimensional subspace, where $\mathbf{X}$ is the $\mu \times D$ parent population matrix and $\mathcal{T}()$ is a transformation function. In standard cooperative CEAs, $\mathcal{T}(\mathbf{X})$ is associated with the element-wise product of $\mathbf{X}$ and M separate $\mu \times D$ matrices $\mathbf{M}_i$, $i = 1, \ldots, M$. Each matrix $\mathbf{M}_i$ exclusively contains d_i columns of 1 with the rest of the columns set to 0. This is because a cooperative CEA sequentially varies the variables in each subcomponent i with all other variables in other subcomponents fixed. In SEE, the $\mathcal{T}(\mathbf{X})$ is the element-wise product of $\mathbf{X}$ and a single $\mu \times D$ dimension matrix, where any of its elements can be either 0 or 1, the probability of which is determined by the meta-models. Together, the two perspectives could offer an explanation why SEE is less sensitive to nonseparable problems compared to standard cooperative CEAs. Indeed, one could view SEE as represented by the framework in Algorithm 8.1 as a coevolution of local solvers and meta-models that work together to determine the direction of search, with the meta-models learning the problem structures so as to provide more accurate ranking of partial solutions of each subcomponent.

In the context of computational complexity, the proposed SEE with a $(1+\lambda)$-EA would incur a cost in each iteration amounting to $D \cdot \lambda$ times of Eqs. (8.6)–(8.7) of associated costs for the meta-model and λ objective function evaluations $\mathcal{F}$. Note that the costs associated with applying both equations involve scalar calculations, which is lower compared to that of performing objective function evaluations. In the second equation, the reduction comes from the offspring's partial solutions of decomposed subcomponents sharing the same objective function value as their fitness values. However, in a standard cooperative CEA that decomposes a problem into M subcomponents, the computational cost would amount to at least $\lambda \cdot M$ objective function evaluations at each iteration. Although the proposed meta-model cannot guarantee the correct rank of partial solutions, it provides an efficient way to produce an approximated rank and would be at most M times faster than that of a standard cooperative CEA.

In closing this chapter, we examine again the implicit requirement that the proposed meta-model in SEE works for problems having smooth local fitness landscapes. This has been verified by our computational study, whereby SEE has good performances on problems with smooth landscapes but with other properties such as multi-modality and nonseparability, which then deteriorate when SEE is applied to problems with rugged fitness landscapes. It should be emphasized that the framework of SEE is independent of such a pairing of simple meta-models with local search as one-dimension subcomponent optimizer. If there is prior information that the problems one would encounter possess rugged fitness landscapes, more appropriate and sophisticated meta-models could be paired with global optimizers for a specific instantiation of SEE to tackle those problems.

References

1. Bentley, J.L.: Multidimensional divide-and-conquer. Commun. ACM **23**(4), 214–229 (1980)
2. Van den Bergh, F., Engelbrecht, A.P.: A cooperative approach to particle swarm optimization. IEEE Trans. Evol. Comput. **8**(3), 225–239 (2004)
3. Cabrera, D.M.: Evolutionary algorithms for large-scale global optimisation: a snapshot, trends and challenges. Prog. Artif. Intell. **5**(2), 85–89 (2016)
4. Chen, W., Weise, T., Yang, Z., Tang, K.: Large-scale global optimization using cooperative coevolution with variable interaction learning. In: International Conference on Parallel Problem Solving from Nature, pp. 300–309. Springer, Krakow (2010)
5. Cheng, R., Jin, Y.: A competitive swarm optimizer for large scale optimization. IEEE Trans. Cybern. **45**(2), 191–204 (2015)
6. Ciccazzo, A., Pillo, G.D., Latorre, V.: Support vector machines for surrogate modeling of electronic circuits. Neural Comput. Appl. **24**(1), 69–76 (2014)
7. Črepinšek, M., Liu, S.H., Mernik, M.: Exploration and exploitation in evolutionary algorithms: a survey. ACM Comput. Surv. (CSUR) **45**(3), 35 (2013)
8. Deb, K., Myburgh, C.: Breaking the billion-variable barrier in real-world optimization using a customized evolutionary algorithm. In: Proceedings of the 2016 on Genetic and Evolutionary Computation Conference, pp. 653–660. ACM, New York (2016)
9. Dong, W., Chen, T., Tiňo, P., Yao, X.: Scaling up estimation of distribution algorithms for continuous optimization. IEEE Trans. Evol. Comput. **17**(6), 797–822 (2013)
10. Goldberg, D.E., Deb, K., Kargupta, H., Harik, G.R.: Rapid accurate optimization of difficult problems using fast messy genetic algorithms. In: Proceedings of the 5th International Conference on Genetic Algorithms, vol. 5, pp. 56–64. Urbana-Champaign, USA (1993)
11. Hansen, N., Arnold, D.V., Auger, A.: Evolution strategies. In: Springer Handbook of Computational Intelligence, pp. 871–898. Springer, Berlin (2015)
12. Iqbal, M., Browne, W.N., Zhang, M.: Reusing building blocks of extracted knowledge to solve complex, large-scale Boolean problems. IEEE Trans. Evol. Comput. **18**(4), 465–480 (2014)
13. Jansen, T., Wiegand, R.P.: The cooperative coevolutionary (1+1) ea. Evol. Comput. **12**(4), 405–434 (2004)
14. Jin, Y.: Surrogate-assisted evolutionary computation: recent advances and future challenges. Swarm Evol. Comput. **1**(2), 61–70 (2011)
15. Kabán, A., Bootkrajang, J., Durrant, R.J.: Toward large-scale continuous EDA: a random matrix theory perspective. Evol. Comput. **24**(2), 255–291 (2016)
16. Langdon, W.B., Harman, M.: Optimizing existing software with genetic programming. IEEE Trans. Evol. Comput. **19**(1), 118–135 (2015)
17. Larranaga, P., Lozano, J.A.: Estimation of Distribution Algorithms: A New Tool for Evolutionary Computation, vol. 2. Springer Science & Business Media, Berlin (2002)
18. Lee, C.Y., Yao, X.: Evolutionary programming using mutations based on the levy probability distribution. IEEE Trans. Evol. Comput. **8**(1), 1–13 (2004)
19. Li, X., Yao, X.: Cooperatively coevolving particle swarms for large scale optimization. IEEE Trans. Evol. Comput. **16**(2), 210–224 (2012)
20. Liu, B., Zhang, Q., Gielen, G.G.E.: A Gaussian process surrogate model assisted evolutionary algorithm for medium scale expensive optimization problems. IEEE Trans. Evol. Comput. **18**(2), 180–192 (2014)
21. Liu, Y., Yao, X., Zhao, Q., Higuchi, T.: Scaling up fast evolutionary programming with cooperative coevolution. In: 2001 IEEE Congress on Evolutionary Computation, vol. 2, pp. 1101–1108. IEEE, Seoul (2001)
22. Mahdavi, S., Shiri, M.E., Rahnamayan, S.: Metaheuristics in large-scale global continues optimization: a survey. Inf. Sci. **295**, 407–428 (2015)
23. Mei, Y., Li, X., Yao, X.: Cooperative coevolution with route distance grouping for large-scale capacitated arc routing problems. IEEE Trans. Evol. Comput. **18**(3), 435–449 (2014)

24. Mei, Y., Omidvar, M.N., Li, X., Yao, X.: A competitive divide-and-conquer algorithm for unconstrained large-scale black-box optimization. ACM Trans. Math. Softw. (TOMS) **42**(2), 13 (2016)
25. Molina, D., Lozano, M., Herrera, F.: MA-SW-Chains: memetic algorithm based on local search chains for large scale continuous global optimization. In: IEEE Congress on Evolutionary Computation, pp. 1–8. IEEE, Barcelona (2010)
26. Munetomo, M., Goldberg, D.E.: A genetic algorithm using linkage identification by nonlinearity check. In: IEEE International Conference on Systems, Man, and Cybernetics, vol. 1, pp. 595–600. IEEE, Tokyo (1999)
27. Munetomo, M., Goldberg, D.E.: Linkage identification by non-monotonicity detection for overlapping functions. Evol. Comput. **7**(4), 377–398 (1999)
28. Omidvar, M.N., Li, X., Mei, Y., Yao, X.: Cooperative co-evolution with differential grouping for large scale optimization. IEEE Trans. Evol. Comput. **18**(3), 378–393 (2014)
29. Omidvar, M.N., Li, X., Yao, X.: Cooperative co-evolution with delta grouping for large scale non-separable function optimization. In: 2010 IEEE Congress on Evolutionary Computation, pp. 1–8. IEEE, Barcelona (2010)
30. Panait, L.: Theoretical convergence guarantees for cooperative coevolutionary algorithms. Evol. Comput. **18**(4), 581–615 (2010)
31. Panait, L., Luke, S.: Time-dependent collaboration schemes for cooperative coevolutionary algorithms. In: Proceedings of the 2005 AAAI Fall Symposium on Coevolutionary and Coadaptive Systems. Virginia, USA (2005)
32. Panait, L., Luke, S., Harrison, J.F.: Archive-based cooperative coevolutionary algorithms. In: Proceedings of the 8th Annual Conference on Genetic and Evolutionary Computation, pp. 345–352. ACM, Seattle (2006)
33. Posner, E.A., Spier, K.E., Vermeule, A.: Divide and conquer. J. Legal Analysis **2**(2), 417–471 (2010)
34. Potter, M.A., De Jong, K.A.: A cooperative coevolutionary approach to function optimization. In: International Conference on Parallel Problem Solving from Nature, pp. 249–257. Springer, Jerusalem (1994)
35. Preen, R.J., Bull, L.: Toward the coevolution of novel vertical-axis wind turbines. IEEE Trans. Evol. Comput. **19**(2), 284–294 (2015)
36. Tabatabaei, M., Hakanen, J., Hartikainen, M., Miettinen, K., Sindhya, K.: A survey on handling computationally expensive multiobjective optimization problems using surrogates: non-nature inspired methods. Struct. Multidiscip. Optim. **52**(1), 1–25 (2015)
37. Tang, K., Li, X., Suganthan, P.N., Yang, Z., Weise, T.: Benchmark functions for the cec'2010 special session and competition on large scale global optimization. Tech. rep., Nature Inspired Computation and Applications Laboratory, University of Science and Technology of China, China (2009)
38. Tang, R.L., Wu, Z., Fang, Y.J.: Adaptive multi-context cooperatively coevolving particle swarm optimization for large-scale problems. Soft Comput. **21**, 4735–4754 (2017)
39. Tayarani-N., M.H., Yao, X., Xu, H.: Meta-heuristic algorithms in car engine design: a literature survey. IEEE Trans. Evol. Comput. **19**(5), 609–629 (2015)
40. Tenne, Y., Armfield, S.W.: A framework for memetic optimization using variable global and local surrogate models. Soft Comput. **13**(8), 781 (2008)
41. Vasicek, Z., Sekanina, L.: Evolutionary approach to approximate digital circuits design. IEEE Trans. Evol. Comput. **19**(3), 432–444 (2015)
42. Wang, H., Wu, Z., Rahnamayan, S.: Enhanced opposition-based differential evolution for solving high-dimensional continuous optimization problems. Soft Comput. **15**(11), 2127–2140 (2011)
43. Wang, J., Tang, K., Lozano, J.A., Yao, X.: Estimation of the distribution algorithm with a stochastic local search for uncertain capacitated arc routing problems. IEEE Trans. Evol. Comput. **20**(1), 96–109 (2016)

44. Wang, Z., Zoghi, M., Hutter, F., Matheson, D., Freitas, N., et al.: Bayesian optimization in high dimensions via random embeddings. In: International Joint Conferences on Artificial Intelligence (IJCAI). Beijing, China (2013)
45. Watson, R.A.: Analysis of recombinative algorithms on a non-separable building-block problem. Found. Genet. Algorithms **6**, 69–89 (2001)
46. Wiegand, R.P.: An analysis of cooperative coevolutionary algorithms. Ph.D. thesis, George Mason University, Fairfax, VA, USA (2004). AAI3108645
47. Wiegand, R.P., Sarma, J.: Spatial embedding and loss of gradient in cooperative coevolutionary algorithms. In: International Conference on Parallel Problem Solving from Nature, pp. 912–921. Springer, Birmingham (2004)
48. Yang, P., Tang, K., Yao, X.: Turning high-dimensional optimization into computationally expensive optimization. IEEE Trans. Evol. Comput. **22**(1), 143–156 (2017)
49. Yang, Z., Tang, K., Yao, X.: Large scale evolutionary optimization using cooperative coevolution. Inf. Sci. **178**(15), 2985–2999 (2008)
50. Yang, Z., Tang, K., Yao, X.: Self-adaptive differential evolution with neighborhood search. In: 2008 IEEE Congress on Evolutionary Computation, pp. 1110–1116. IEEE, Hong Kong (2008)
51. Yao, X., Liu, Y., Lin, G.: Evolutionary programming made faster. IEEE Trans. Evol. Comput. **3**(2), 82–102 (1999)
52. Yu, T.L., Goldberg, D.E., Sastry, K., Lima, C.F., Pelikan, M.: Dependency structure matrix, genetic algorithms, and effective recombination. Evol. Comput. **17**(4), 595–626 (2009)
53. Zhou, A., Sun, J., Zhang, Q.: An estimation of distribution algorithm with cheap and expensive local search methods. IEEE Trans. Evol. Comput. **19**(6), 807–822 (2015)

Chapter 9
Design Space Exploration for Multimode Dataflow Mapping

9.1 Introduction to Heterogeneous Multiprocessor Mapping

Some signal processing and multimedia applications can be specified by ***Synchronous Dataflow*** (***SDF***) models [22]. An SDF graph consists of nodes (called actors) that are connected via edges. An actor represents a task. An edge represents communication between a pair of actors whereby one actor sends data samples (called tokens) to another actor. A fundamental feature of SDF graphs is that when an actor executes the task it is associated with, it consumes the same amount of tokens from its input edges and produces the same amount of tokens on its output edges. This amount is called the sample rate. An actor can only execute the task if there are enough tokens from its input edge. In this manner, an edge represents the data or control dependency between two connected tasks and tokens capture dependencies between task executions. The frequencies of actors are determined by the sample rates, so that their consumptions and productions can be balanced. These sample rates are made constant so that an SDF graph executes in a repetitive pattern called iteration.

Modern devices are built over heterogeneous multiprocessor architectures. These platforms provide better performance and energy tradeoffs than their homogeneous counterparts. As such, recent studies have targeted heterogeneous platforms for analysis [19]. Given an SDF model, the assignment of nodes is decided in the task mapping step, while the execution order of mapped nodes is decided in the task scheduling step during the ***Design Space Exploration*** (***DSE***) process. The decisions of both mapping and scheduling are static because the sample rates are made constant. This allows the evaluation of the performance and resource requirement of the platform according to these static decisions. However, the problem of SDF mapping to a given set of heterogeneous processors has been known to be NP-hard [23], as it is an instance of a constrained quadratic assignment problem (QAP) that is NP-hard [20]. It has been widely studied in the design automation field. In [34], a

X. Yao, S. Y. Chong, *Coevolutionary Computation and Its Applications*,
Natural Computing Series, https://doi.org/10.1007/978-981-96-2841-4_9

survey has been carried out on emerging trends for node mapping on multiprocessor platforms, while in [18], reviews have been made on fault-tolerant node mapping.

Various heuristic search approaches such as those based on Evolutionary Algorithms (EAs) have been developed to find good and efficient mapping solutions. One example includes the development of ***Multi-Objective EAs*** (***MOEAs***) [10] (e.g. NSGA-II [9] and SPEA2 [51]) for balancing processing time and power consumption and architecture cost and the symmetry-eliminating MOEAs [33] that employ clustering of nodes and their mappings to core types. Other examples include Tabu search (TS) [27], the nested Genetic Algorithms (GAs) [2, 4], cooperative coevolution [46] with the exploration of both node mappings and fault-tolerance techniques, the multistage discrete Particle Swarm Optimization (PSO) [25, 31] for minimizing communication cost and peak temperature in both 2D and 3D architectures, and the fruit fly optimization algorithm [12] for preventing fault injection attacks of the security protection on processors. In addition to heuristic search approaches such as EAs, machine learning algorithms that can learn from data and make decisions based on new data [5, 6] have been employed in DSE node mapping. They include single-objective and multi-objective STAGE [3, 8, 15] that construct predictive evaluation functions based on historical search trajectories to guide search in future iterates.

Despite this, the SDF model cannot fully express modern applications that are increasingly more complex with dynamic behaviours that changes over time. For example, advanced video CODEC algorithms [39] consist of several function modules that are conditionally invoked according to the input frames. Furthermore, there are various implementations of the same algorithm in a video application to meet different quality of service (QoS) requirements. Several extensions to the SDF model have been proposed to express such dynamic behaviours of applications while keeping its static estimation capability. For example, scenario-based DSE approaches [41] capture dynamic behaviour changes whereby each scenario corresponds to a behaviour (mode). In [41], coevolutionary principle is used whereby two GAs operate in parallel, with one to search for node mapping solutions and another for scenario selection. Scenario clustering is proposed in [29, 32], and a MOEA-based DSE is performed at design time to find a mapping with the best average performance for that particular cluster. These studies are extended in [30] to include support for heterogeneous architectures and new applications. NSGA-II-based DSE is performed at design time to find two mappings: one that maximizes the throughput and another that maximizes the throughput under a predefined energy budget.

However, in most of these studies tasks are not allowed to migrate from one mode to another, and the static mapping and scheduling are independently determined for each mode. Studies have been carried out to consider applications with a finite number of behaviours (modes). Approaches to solve this extended mapping problem include [16, 17, 21, 35, 38]. In [16], a GA-based hybrid task mapping approach is developed to minimize the total energy consumption with constraints on throughput. Task migration is allowed, and migration overhead is taken into account for evaluation of the energy consumption. However, the approach analyzes each

SDF graph independently regardless of how the mode transition delay affects the throughput performance of the application. A failure-aware task mapping approach is proposed in [21] for throughput maximization according to a varying number of allocated processors. Tasks can be migrated to another allocated processor if the current allocated processor fails, with changes in the corresponding mapping. The migration overhead between two SDF mappings is minimized, but the approach performs a separate GA-based DSE for each failure scenario.

The study [17] in particular has proposed a heterogeneous multiprocessor mapping approach for the computational model represented by ***Multi-Mode Dataflow*** (***MMDF***) graphs. Mode transition is made at the iteration boundary, and the mode switching decision can be made any time during the execution with the associated SDF graph statically mapped and scheduled for each mode. Mode switching incurs extra time overhead for the mode transition, which then affects system performance. In such a setting, the optimization objective is to minimize the maximum mode transition interval. Multiprocessor mapping of an SDF is already a well-known NP-hard problem [23]. The multiprocessor mapping of an MMDF is significantly more challenging since a MMDF graph consists of a set of modes and each mode is specified by an SDF graph. The design space associated with the MMDF mapping increases with the number of modes. The GA-based DSE method in [17] maps and schedules all SDF graphs in all modes by encoding the design space as a whole whereby a chromosome consists of a set of vectors and each vector represents the mapping in a mode. However, traditional GAs suffer from the curse of dimensionality with rapid performance deterioration for an increasing number of decision variables. One would expect a GA-based DSE method used to solve the MMDF mapping problem to be similarly affected because of the latter's growth in problem complexity.

Here, we present our study [49] that has proposed a more effective DSE method based on cooperative coevolution [28] to find the solutions of static mapping for each mode in a single optimization framework whereby a successful problem decomposition on the MMDF mapping problem could partially address the curse of dimensionality. In Sect. 9.2, we formalize the setting of the heterogeneous multiprocessor mapping within the application and architecture models defined in [17]. We note that there are several MMDF graph mapping approaches, and they all assume that each node is mapped to the same processor in all modes [37]. However, our proposed approach does not make such an assumption. After that, Sect. 9.3 presents the cooperative coevolution framework for the DSE method. In particular, the problem-specific decomposition strategy assigns each mode to an individual population. In order to further improve the search effectiveness, we introduce a problem-specific local search operator as a supplement to the global search of the ***Cooperative Coevolutionary GA*** (***CCGA***). We employ a hybrid fitness evaluation strategy to reduce the time consumption of fitness evaluations after prior analysis of the fitness function form. Section 9.4 presents our computational study. In comparison to a GA-based DSE method, the proposed CCGA-based DSE method can obtain an optimization result with better quality (2–3 times) using less optimization time (1/2–1/3). Section 9.4 concludes the chapter with a discussion on more complex settings found in real-world scenarios and potential approaches to address them.

9.2 Problem Model

In this section, we present the MMDF model defined in [17], which is a general model that encompasses other similar models such as Finite State Machine Scenario-Aware Dataflow (FSM-SADF) [38] and Mode Transition Machine Scenario-Aware Dataflow (MTM-SADF) [16]. Figure 9.1 illustrates an MMDF graph whereby the mode transition is specified by an FSM and all modes are integrated into a single graph with varying configuration parameters. We first provide related definitions of the model, which are later use to formulate the main problem of MMDF mapping in an optimization setting:

Application Model The MMDF graph is specified by a combination of a task graph TG $= (T, L, D)$ and a mode transition graph MTG $=$ (Mode, Trans) [17], where:

- For a given mode set Mode, Trans $= \{(m_p, m_n) : e(m_p, m_n) = 1,\ m_p, m_n \in$ Mode$\}$ whereby $e(m_p, m_n) = 1$ if m_p is a previous mode and m_n is a next mode; otherwise $e(m_p, m_n) = 0$ (i.e. a directed edge $m_p \rightarrow m_n$ in MTG is indicated by $e(m_p, m_n) = 1$).
- T is a finite set of computational tasks (e.g. nodes A, B, C, D in Fig. 9.1), and we use $T_m \subseteq T$ to denote the executed tasks in mode m. For each task $t \in T$, there are input ports denoted by IP_t for data reception from its predecessors and output ports denoted by OP_t for data that are sent to its successors. We assign a fixed rate $R_{p,m}$ for each port $p \in \mathrm{IP}_t \cup \mathrm{OP}_t$ in each execution mode m. Note that the rate is omitted from the figures if its value is 1 in all modes.

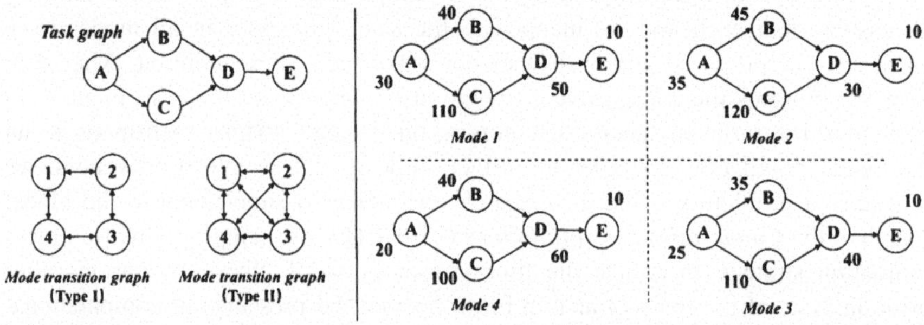

Fig. 9.1 An MMDF graph example and two general topologies of mode transition graphs (Type I and Type II) [49]

- L is a finite set of links, and a link indicates a one-to-one connection between an output port and an input port. For each link $l \in L$, src(l) and dst(l) represent the source task and destination task of link l, respectively.
- $D = \{d_{l,m} : l \in L \wedge m \in \text{Mode}\}$, where $d_{l,m}$ denotes the number of initial tokens in the link l in mode m.

Architecture Model The target architecture consists of a set of processing elements PE. It is assumed that the architecture is characterized with a full connectivity communication fabric, and there is no connectivity constraint for the processing elements.

Task Mapping For each $p \in \text{PE}$ and $m \in \text{Mode}$, $\text{map}(m, p) = \{t : t \in T$, where task t is mapped to the processor p in mode $m\}$.

Profiling Information The following profiling information is given for evaluating the task mapping solution:

- The worst-case execution time is denoted as WCET(t, m, p) for task $t \in T$ and mode $m \in$ Mode and processor $p \in$ PE. In Fig. 9.1, the WCET of a task is annotated in each mode, e.g. the WCETs of task A are 30 in mode 1, 35 in mode 2, 25 in mode 3, and 20 in mode 4.
- For each $t \in T$, we use MC(t) to denote the cost of task migration. If the platform is a distributed memory system, the cost should include the time overhead for moving the code and the context of a task between two processors. If the platform is a shared memory system, the cost should be as small as cache miss penalty for the first reference (called cold miss penalty) of the task [17].

Given the application model, architecture model, and profiling information described above, the main problem is defined as finding a task mapping solution that minimizes the maximum mode transition interval $\text{MaxInterval}_{\text{overall}}$:

$$\min \text{MaxInterval}_{\text{overall}} = \max_{(m_i, m_j) \in \text{Trans}} \text{TransInterval}(m_i, m_j),$$

$$\text{TransInterval}(m_i, m_j) = \text{Lat}(m_j) + \text{StartOffset}(m_i, m_j) - \text{Lat}(m_i), \tag{9.1}$$

where Lat(m_k) denotes the latency of mode m_k (see Fig. 9.2). The mode transition interval TransInterval(m_i, m_j) between two modes m_i, m_j is the time interval between the last output production time of the previous mode (t_1 in Fig. 9.3) and the first output production time of the next mode (t_2 in Fig. 9.3). We note that the maximum mode transition interval determines the required output buffer size and the achievable throughput of an MMDF graph [17]. As such, we consider the maximum mode transition interval as the optimization objective as given by (9.1).

The mode transition interval is a relative term as its value does not depend on the start time of the last iteration of the previous mode. However, we consider that the last iteration of the previous mode to start at $t = 0$ (see Fig. 9.3) to better illustrate the following definitions. In particular, we will formulate the start offset

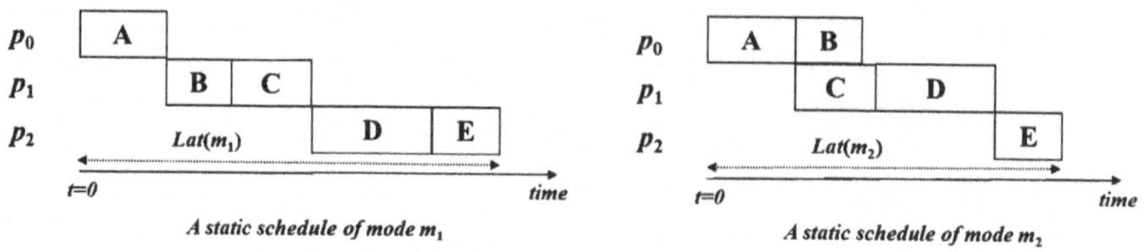

Fig. 9.2 Static schedule of mode m_1 and mode m_2 [49]

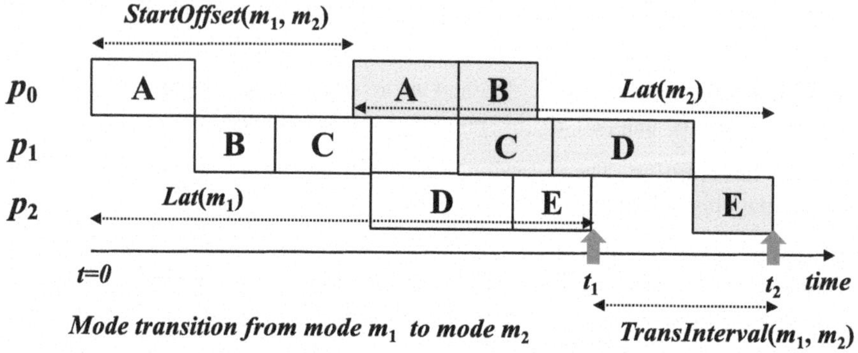

Fig. 9.3 Mode transition from mode m_1 to mode m_2 [49]

(StartOffset) of the first iteration of the next mode, which is determined by the scheduling delay (D_{sched}), the task ordering delay (D_{order}), and the task migration delay (D_{mig}):

(a) Scheduling Delay The start time of the next mode should be shifted to guarantee consistent execution of the MMDF model. Here, consistent execution indicates there exists a schedule of task executions that does not change the sample rates on all links. For example, the scheduling delay between modes m_1 and m_2 in Fig. 9.2 is illustrated by the time interval labeled as (1) in Fig. 9.4a.

Definition 9.1 (Scheduling Delay from Mode m_i to Mode m_j) The first factor that determines the scheduling delay is the time delay D_{proc} to keep the temporal property of the static schedule of the next mode. The start time of the next mode should be shifted to guarantee that for each processor, the first task in the next mode cannot be executed until the last task in the previous mode is finished. It is given as follows:

$$D_{\text{proc}}^{m_i m_j} = \max_{p \in \text{PE}} \max\left(\max_{t \in \text{Map}(m_i, p)} E_t^{m_i} - \min_{t \in \text{Map}(m_j, p)} S_t^{m_j},\ 0 \right), \tag{9.2}$$

where $E_t^{m_i}$ is the end time of task t in mode m_i and $S_t^{m_j}$ is the start time of task t in mode m_j.

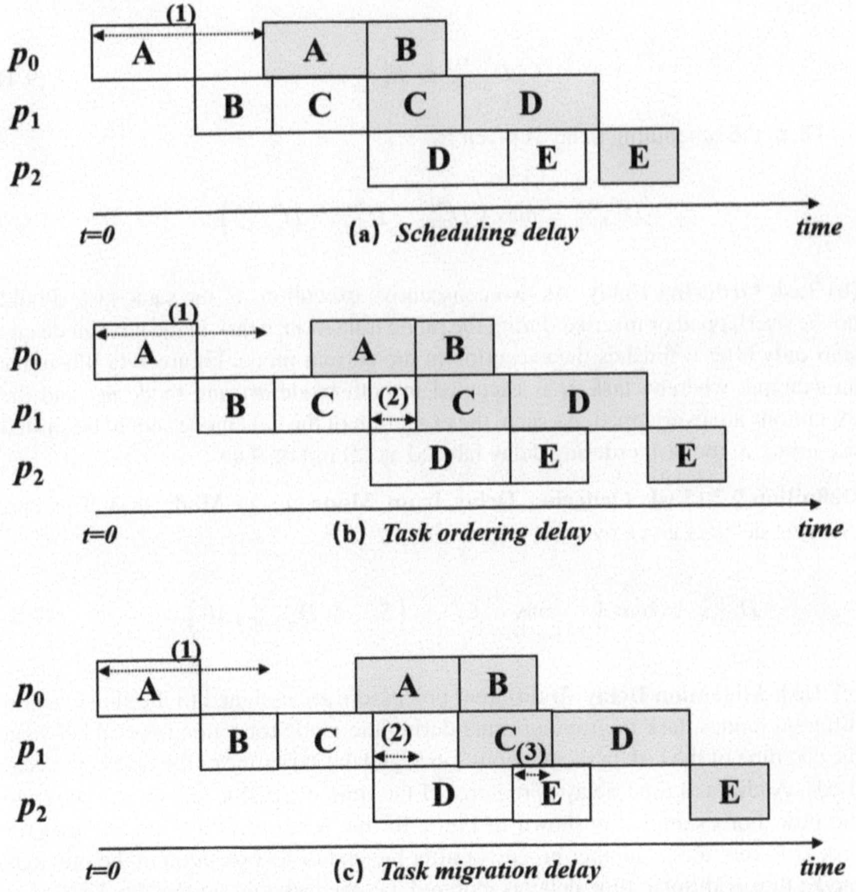

Fig. 9.4 Mode transition delay from mode m_1 to mode m_2 with (1) the scheduling delay dictated by task C, (2) the task ordering delay dictated by task D, and (3) the task migration delay dictated by task D [49]

The second factor D_{data} accounts for the data dependency between two modes. In case there is no initial token in links, the start of the next mode should be delayed until the required number of tokens is produced in the previous mode. It is given as follows:

$$D_{\text{data}}^{m_i m_j} = \max_{l \in L, d_{l,m_j} > 0} \max\left(E_{\text{src}(l)}^{m_i} - S_{\text{dst}(l)}^{m_j}, 0\right). \tag{9.3}$$

The third factor D_{mode} is the finish time of the last instance of the mode decision task t_{mode} in the previous mode since the next mode might start afterward. It is given as follows:

$$D_{\text{mode}}^{m_i m_j} = E_{t_{\text{mode}}}^{m_i}. \tag{9.4}$$

Then, the scheduling delay is given by

$$D_{\text{schedl}}^{m_i m_j} = \max\left(D_{\text{proc}}^{m_i m_j}, D_{\text{data}}^{m_i m_j}, D_{\text{mode}}^{m_i m_j}\right). \tag{9.5}$$

(b) Task Ordering Delay As two consecutive executions of the same task should not be overlapped or inverted during the mode transition, a task in the next mode can start only after it finishes the execution in the current mode. Figure 9.4a illustrates an example whereby task D is executed in both mode m_1 and mode m_2 and the executions are overlapped. As such, the execution of the next mode should be shifted according to the task ordering delay labeled as (2) in Fig. 9.4b.

Definition 9.2 (Task Ordering Delay from Mode m_i to Mode m_j) The task ordering delay is given by

$$D_{\text{order}}^{m_i m_j} = \max\left(\max_{t \in T_{m_i} \cap T_{m_j}} E_t^{m_i} - \left(S_t^{m_j} + D_{\text{schedl}}^{m_i m_j}\right), 0\right). \tag{9.6}$$

(c) Task Migration Delay If different processors are assigned to the same task in different modes, task migration occurs during the mode transition interval between the end time of the task in the previous mode and the start time of the task in the next mode. Additional time delay is required if the time interval is too short to migrate the task. For example, as shown in Fig. 9.4b, the time interval is not enough for task D to migrate to another processor after the end of its execution in the previous mode; thus additional time delay is required, i.e. the task migration delay labeled as (3) in Fig. 9.4c.

Definition 9.3 (Task Migration Delay from Mode m_i to Mode m_j) The task migration delay is given by

$$D_{\text{mig}}^{m_i m_j} = \max\left(\max_{\substack{t \in T_{m_i} \cap T_{m_j} \\ \text{proc}(t, m_i) \neq \text{proc}(t, m_j)}} \text{MC}(t) - \left(S_t^{m_j} + D_{\text{schedl}}^{m_i m_j} + D_{\text{order}}^{m_i m_j} - E_t^{m_i}\right), 0\right). \tag{9.7}$$

(d) Start Offset The start offset of the next mode is formulated by taking the sum of all delay terms defined earlier in **(a)**–**(c)**:

Definition 9.4 (Start Offset of Mode m_j in the Case of Mode Transition $m_i \rightarrow m_j$) The start offset is given by

$$\text{StartOffset}(m_i, m_j) = D_{\text{schedl}}^{m_i m_j} + D_{\text{order}}^{m_i m_j} + D_{\text{mig}}^{m_i m_j}. \quad (9.8)$$

9.3 Cooperative Coevolutionary Framework

In this section, we present the cooperative cooperation framework we have developed in [49] that is applied to the DSE method to optimize the MMDF mapping problem specified by (9.1). Figure 9.5 illustrates the whole framework whereby the mapping solutions for each mode are searched cooperatively in a single optimization framework. Our specific implementation is based on CCGA to effectively explore the design space by a problem-specific decomposition strategy that assigns each mode to an individual population (Fig. 9.5(middle)). The CCGA-based DSE method takes as inputs the application model, the architecture model, and their associated profiling information (Fig. 9.5(top)). The outputs are the decisions of the processor assignment for each task in each mode and the schedule table for each processor in each mode (Fig. 9.5(bottom)).

The main conceptual difference between the previous GA-based method [17] and our proposed CCGA-based method is illustrated using Fig. 9.6. On the one hand, Fig. 9.6a shows that the GA-based method encodes the design space as a whole whereby the chromosome consists of a set of vectors and each vector represents the mapping in a mode. On the other hand, Fig. 9.6b shows that in the CCGA-based method, the design space for static mapping in M modes is explored through M

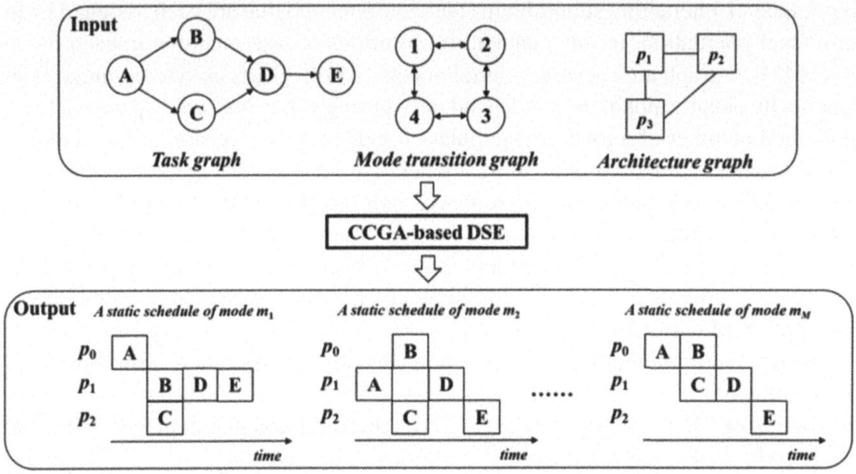

Fig. 9.5 The proposed DSE framework [49]

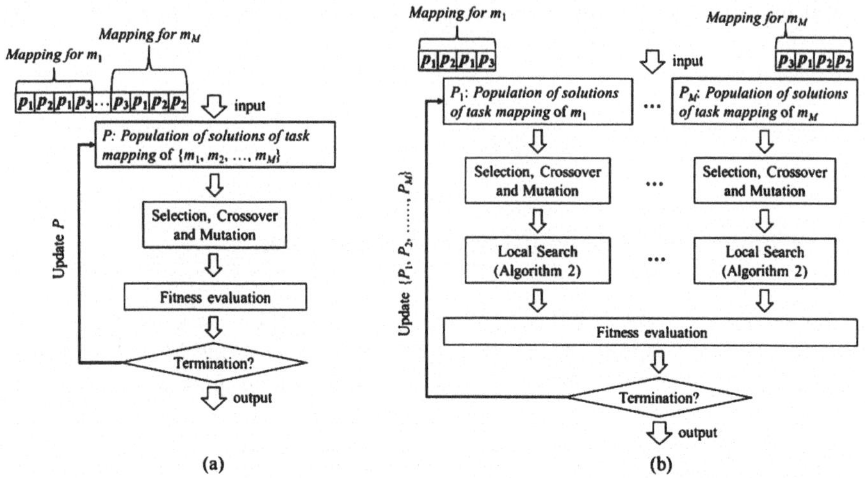

Fig. 9.6 The flowcharts of (**a**) the previous GA-based method [17] and (**b**) the proposed CCGA-based method [49]

interactive populations $\{P_1, P_2, \ldots, P_M\}$, where $M = |\text{Mode}|$ is the number of modes. A problem-specific local search operator is introduced for task mapping as a supplement to the global search of CCGA. For fitness evaluation of the individual, the solutions of task mapping in M modes are integrated as one complete solution, and its fitness value is the maximum mode transition interval $\text{MaxInterval}_{\text{overall}}$ that is minimized during the DSE process.

Cooperative Coevolutionary Algorithms (CEAs) have been used with success in many applications to speed up the original, baseline EAs [1, 11, 24, 26]. Our use of cooperative CEA exploits problem decomposition on task mappings over M modes to produce M interacting subproblems (subcomponents) that are each assigned to an individual population. In this manner, the solutions of task mapping in each mode of an MMDF graph are coevolved simultaneously [49]. GA is used to optimize each genetically isolated population independently through an evolutionary process. Each individual member of a particular population indicates the decision of the mapping in that particular mode. As in most cooperative CEAs, the complete solution is constructed by combining a representative individual from each population. This construction process is also required for the fitness value of an individual in a population with respect to its partners selected from other populations. Therefore, the search of cooperative CEA is still guided by a global evaluation function during the whole exploration process.

The details of the major steps of the proposed cooperative CEA framework [49] are as follows:

(a) Encoding: M populations $\{\boldsymbol{P}_1, \ldots, \boldsymbol{P}_M\}$ are used and maintained to represent the encoding of complete mapping solutions. The task mappings of M modes are decomposed to M independent subcomponents, and each subcomponent is

assigned to a genetically isolated population. For each particular task mapping population $\boldsymbol{P}_m = \{\text{map}_1^m, \text{map}_2^m, \ldots, \text{map}_{\text{PS}}^m\}$ of size PS, the ith individual or chromosome is represented by a $|T_m|$-dimensional integer vector $\text{map}_i^m = \{\text{map}_{i,1}^m, \text{map}_{i,2}^m, \ldots, \text{map}_{i,|T_m|}^m\} \in \{1, 2, \ldots, |\text{PE}|\}^{|T_m|}$, where $|T_m|$ is the number of tasks in mode m and $|\text{PE}|$ is the number of processors. As such, $\text{map}_{i,j}^m = p_k$ indicates that processor p_k is assigned to task t_j in mode m as given to the ith individual. Note that the processors are numbered sequentially.
(b) Crossover: The one-point crossover operator is used in the GA. The operator randomly selects a single point that partitions each of the two parent chromosomes, a left part and a right part, and then they produce two child chromosomes whose left and right parts are exchanged with respect to the parent chromosomes. The operator outputs one of the two child chromosomes.
(c) Mutation: The GA uses the uniform mutation operator whereby for a randomly selected position of the chromosome, its data value is replaced by a random value uniformly drawn from a finite set. Figure 9.6b illustrates how crossover and mutation operators are applied on each separate population P_m. In this manner, the mappings of different tasks in a particular mode are selected and exchanged via crossover and mutation processes.
(d) Selection: Selection occurs twice in each generation of the GA. Selection for reproduction is required before the application of the crossover operator. At the end of a generation in the evolutionary process, a combined pool of both parent population and child population is produced. The selection operator is applied to reduce the population to its original size through an elitist scheme that chooses the best individuals from the pool.
(e) Fitness evaluation: The modified credit assignment algorithm based on CCGA-2 [28] is employed whereby each individual in a population is evaluated by combining it with the current best individual from each of the other populations and a random selection of individuals from each of the other populations. The two resulting vectors of complete solutions are evaluated according to the fitness function given by Eq. (9.9) with the better fitness value returned as the individual's fitness value. We construct the fitness function based on the objective function (9.1) for each solution map^{m_l} of the population P_{m_l} that represents the subcomponent or mode m_l as

$$\text{Fit}(\text{map}^{m_l}) = \max_{(m_i, m_j) \in \text{Trans}} \text{Lat}(m_j) + \text{StartOffset}(m_i, m_j) - \text{Lat}(m_i). \tag{9.9}$$

Algorithm 9.1 describes a procedural implementation of the CCGA framework via **CCGA** [49]. Note that there are parameters that can be set, which includes the population size (PS) for each population representing solutions for the subcomponent m, the crossover rate p_c, and the mutation rate p_m. Algorithm 9.1 starts by initializing the M number of genetically isolated populations $\mathbf{P} = \{\boldsymbol{P}_1, \ldots, \boldsymbol{P}_M\}$ with each population $\boldsymbol{P}_m$ consisting of PS random individuals for task mapping of mode m, $\boldsymbol{P}_m = \{\text{map}_i^m\}_{i=1}^{\text{PS}}$, $\text{map}_i^m = \{\text{map}_{i,t}^m\}_{t=1}^{|T_m|}$ (line 3). The fitness value f_i^m of each initialized random individual $\text{map}_i^m \in \boldsymbol{P}_m$ is obtained via the procedure

Algorithm 9.1 CCGA-based task mapping of MMDF [49]

Input:

M	**M** = {TG, MTG, PE, WCET} domain knowledge incorporated into problem representation: task graph, mode transition graph, architecture graph (fully connected PE number of processors), and worst-case execution time for all tasks
PS	Size of individual population
p_c	Crossover rate
p_m	Mutation rate
M	Problem decomposition producing M subcomponents
τ_g	Maximum number of generations
τ_e	Maximum number of fitness evaluations

Output:

mapb[] Coevolved task mappings for each mode

procedure CCGA(**M**, PS, $p_c, p_m, M, \tau_g, \tau_e$)
    for $m := 0 : M$ **do**
        $\boldsymbol{P}_m :=$ **initSubPop**(m) ▷ Initialize populations for M subcomponents
        $\mathbf{P}[m].\boldsymbol{P} := \boldsymbol{P}_m$ ▷ $\boldsymbol{P}_m = \{\text{map}_i^m\}_{i=1}^{\text{PS}}$, $\text{map}_i^m = \{\text{map}_{i,t}^m\}_{t=1}^{|T_m|}$
    end for
    for $m := 0 : M$ **do** ▷ Evaluate fitness f_i^m of each map_i^m,
        $\boldsymbol{f}_m :=$ INITEVAL$(m, \boldsymbol{P}_m, \mathbf{P}[\,], \mathbf{M})$ ▷ $i = 1, \ldots, \text{PS}$, $m = 1, \ldots, M$
    end for ▷ $\boldsymbol{f}_m = \{f_i^m\}_{i=1}^{\text{PS}}$
    for $m := 0 : M$ **do**
        $\text{mapbest}^m :=$ **selBestMap**$(\mathbf{P}[m].\boldsymbol{P}, \boldsymbol{f}_m)$ ▷ $\text{mapbest}^m = \text{map}_b^m$,
        mapb$[m] := \text{mapbest}^m$ ▷ $b = \text{argmin}_{i \in \{1,\ldots,\text{PS}\}} f_i^m$
    end for
    while within fitness evaluation budget τ_e **do**
        for $i := 0 : M$ **do**
            for $t := 0 : \tau_g$ **do**
                $\boldsymbol{P}'_m :=$ **crossOp**$(\boldsymbol{P}_m, p_c)$ ▷ With probability p_c crossover $\otimes$ is applied ▷ $\text{map}'^m_k := \text{map}_i^m \otimes \text{map}_j^m$, $i, j \sim U[1, \text{PS}]$
                $\boldsymbol{P}''_m :=$ **mutOp**$(\boldsymbol{P}'_m, p_m)$ ▷ With probability p_m $\text{map}''^m_k :=$ **mutate**(map'^m_i)
                $\boldsymbol{f}_m :=$ MODEVAL$(m, \boldsymbol{P}_m, \mathbf{P}[\,], \mathbf{mapb}[\,], \mathbf{M})$ ▷ Evaluate fitness f_i^m of parent ▷ map_i^m, $\boldsymbol{f}_m = \{f_i^m\}_{i=1}^{\text{PS}}$
                $\boldsymbol{q}_m :=$ MODEVAL$(m, \boldsymbol{P}''_m, \mathbf{P}[\,], \mathbf{mapb}[\,], \mathbf{M})$ ▷ Evaluate fitness q_i^m of child ▷ map''^m_i, $\boldsymbol{q}_m = \{q_i^m\}_{i=1}^{\text{PS}}$
                $\boldsymbol{P}_m^{\text{comb}} :=$ **combine**$(\boldsymbol{P}_m, \boldsymbol{P}''_m)$ ▷ $\boldsymbol{P}_m^{\text{comb}} = \boldsymbol{P}_m \cup \boldsymbol{P}''_m$
                $\boldsymbol{f}_m^{\text{comb}} :=$ **combine**$(\boldsymbol{f}_m, \boldsymbol{q}_m)$ ▷ $\boldsymbol{f}_m^{\text{comb}} = \boldsymbol{f}_m \cup \boldsymbol{q}_m$
                $\boldsymbol{P}_m :=$ **selOp**$(\boldsymbol{P}_m^{\text{comb}}, \boldsymbol{f}_m^{\text{comb}})$
                $\boldsymbol{f}_m :=$ **selFitOp**$(\boldsymbol{P}_m^{\text{comb}}, \boldsymbol{f}_m^{\text{comb}})$ ▷ Select best PS maps of $\boldsymbol{P}_m \cup \boldsymbol{P}''_m$
                $\mathbf{P}[m].\boldsymbol{P} := \boldsymbol{P}_m$
                $\text{mapbest}^m :=$ **selBestMap**$(\mathbf{P}[m].\boldsymbol{P}, \boldsymbol{f}_m)$
                mapb$[m] := \text{mapbest}^m$ ▷ Update mapbest^m
            end for
        end for
    end while
    return mapb[]
end procedure

INITEVAL (line 7). This requires constructing the complete solution, which is achieved by combining $\text{map}_i^m \in \boldsymbol{P}_m$ with a random selection of individuals, each from one of the other populations $\mathbf{P}_1, \ldots, \mathbf{P}_{m-1}, \mathbf{P}_{m+1}, \ldots, \mathbf{P}_M$. Then, the resulting vector of the complete solution is passed to the fitness function. The current best individual mapbest^m of each population $\boldsymbol{P}_m$ is determined via **selBestMap** that establishes the individual $b = \text{argmin}_{i \in \{1,\ldots,\text{PS}\}} f_i^m$ whose fitness value in the population is minimum (line 10).

These M populations $\mathbf{P} = \{\boldsymbol{P}_1, \ldots, \boldsymbol{P}_M\}$ are cooperatively coevolved (lines 14–25) [49]. This involves independent application of the evolutionary process on the genetically isolated populations $\{\boldsymbol{P}_1, \ldots, \boldsymbol{P}_M\}$. For each population $\boldsymbol{P}_m$, (a) the parent population $\boldsymbol{P}_m$ of PS individuals first generates the child population $\boldsymbol{P}'_m$ of PS individuals via crossover operations in **crossOp** (line 16) and then finally $\boldsymbol{P}''_m$ after mutation operations in **mutOp** (line 17). (b) The full population consisting of $\boldsymbol{P}_m$ and $\boldsymbol{P}''_m$ is evaluated using the modified credit assignment algorithm via **MODEVAL** (lines 18–19). (c) The selection process is applied on the full population $\boldsymbol{P}_m \cup \boldsymbol{P}''_m$ of 2PS individuals (note that the $\cup$ operation is a concatenation operation rather than the usual set union because the multiset $\boldsymbol{P}_m \cup \boldsymbol{P}''_m$ can have copies of the same individuals). Then, the best PS individuals are selected to form the population for the next generation. This involves copying over these best PS individuals to $\mathbf{P}[m] = \boldsymbol{P}_m$ (lines 22–24). (d) mapbest^m is updated based on the new population (lines 25–26). This evolutionary process for the population $\boldsymbol{P}_m$ of solutions for the subcomponent m is applied for τ_g generations. The loop that represents the cooperative coevolutionary process (lines 13–29) terminates if maximum fitness evaluation budget τ_e is reached. Implicit in the implementation is a break procedure to exit the loop once fitness evaluation budget is exhausted, which could be tracked with a global counter variable.

Both **INITEVAL** and **MODEVAL** that evaluate the fitness of a population of maps $\boldsymbol{P}$ implement a list scheduling heuristic that is specified by Algorithm 9.2 [49] as part of their procedural operations. It consists of the main procedure **SCHTAB** that produces the static schedule table for each processor $p \in \text{PE}$ in mode m, which is used to calculate the fitness specified by Eq. (9.9) according to the mapping information represented by the chromosome map_i^m. The general operation of Algorithm 9.2 involves having tasks whereby all their predecessors have been already mapped and are put in the ready list L. All the tasks from the ready list L will be investigated during the loop whereby the task with the highest priority is always selected for scheduling in each iteration. First, the priority of each task $t_i \in T_m$ is computed as the maximum length of the path from task t_i to sink $s_m \in T_m$ via **priority** (line 2). For each processor p_j, the initial current idle time I_{p_j} is set to 0 via **initIdle** (line 3). After that, the procedure **sourceNode** moves the source node(s) in T_m into the ready list L (line 3).

Then, the following process is repeated until the ready list is empty (lines 5–11) [49]: (a) Task t_i with the highest priority is selected from the ready list L via **selPriTask** (line 6). (b) Task t_i has been mapped to processor p_j by the solution $\text{map}_{i,t_i}^m = p_j$. Then, task t_i is scheduled in the schedule table S via the procedure

Algorithm 9.2 List scheduling heuristic [49]

Input:

M — $\mathbf{M} = \{\text{TG, MTG, PE, WCET}\}$ domain knowledge incorporated into problem representation: task graph, mode transition graph, architecture graph (fully connected PE number of processors), and worst-case execution time for all tasks

m — The specific mode

map_i^m — Solution of the task mapping represented by individual i for task set T_m

Output:

S — Schedule table for each processor $p \in$ PE in mode m

1: **procedure** SCHTAB($\mathbf{M}, m, \text{map}_i^m$)
2: **priority** := **compPriority**(m, **M**) ▷ Compute priority of each task $t_i \in T_m$ ▷ **priority** = $\{\text{priority}_{t_i}\}_{t_i=1}^{T_m}$
3: **I** := **initIdle(D)** ▷ Initialize idle time for each processor $I_{p_j} = 0, \mathbf{I} = \{I_{p_j}\}_{p_j=1}^{\text{PE}}$
4: L := **sourceNode**(m, **M**) ▷ L consists of the source node from T_m
5: **while** $L \neq \emptyset$ **do**
6: t_i := **selPriTask**(L, **priority**) ▷ Select $t_i \in L$ with highest priority
7: SCHTASK(S, t_i, map_i^m) ▷ Schedule t_i that is run on p_j as given by $\text{map}_{i,t_i}^m = p_j$ in S
8: **updateIdle**(**I**, map_{i,t_i}^m) ▷ Adjust current idle time I_{p_j} of $\text{map}_{i,t_i}^m = p_j$
9: **removeList**(L, t_i) ▷ $L = L \backslash \{t_i\}$
10: UPDATELIST(L, t_i, S, **M**) ▷ Add successors of t_i to L ▷ if all their predecessors are scheduled
11: **end while**
12: **return** S
13: **end procedure**

SCHTASK (line 7). (c) The current idle time I_{p_j} of processor p_j is adjusted via **updateIdle** (line 8). (d) Finally, task t_i is removed from the ready list L via **removeList** (line 9). The procedure UPDATELIST(f)urther updates the ready list L whereby successors of t_i are added to L if all their predecessors are scheduled (line 10). If all the tasks are scheduled in which case the ready list is empty ($L \neq \emptyset$), then the loop is terminated (line 11). Then, the main procedure SCHTAB returns the schedule table of each processor S (line 12). It should be noted that the scheduling algorithm does not require the task graph has a single source node, and a virtual task can be introduced as the source for the graph with multiple source tasks.

We have incorporated other design elements to improve the effectiveness of CCGA. The first element is based on the principle of ***Memetic Computation*** [7, 36, 44, 47, 50] that combines both individual-based local search and population-based global search. Here, meme or individual learning refers to the problem-specific knowledge that is incorporated in the local search process. In general, such a problem-specific local search operator is designed as a greedy search process. As such, we have designed a problem-specific local search operator that extends the idea of the ***Greedy Reassignment Local Search*** (***GRLS***) [48] for task mapping of an MMDF graph [49]. During the process of GRLS, the reassignment of a part of data values of the parent chromosome is guided by the greedy information extracted from the problem instance. Since it is hard to directly minimize the fitness function (9.9), the proposed GRLS operator tries to reduce the latency Lat(m) of mode m for each individual map_i^m in the population $\boldsymbol{P}_m$. This is because the latency Lat(m) would be a positive term in the fitness function (9.9) if mode m is a successor mode.

Algorithm 9.3 GRLS for task mapping [49]

Input:

M	M = {TG, MTG, PE, WCET} domain knowledge incorporated into problem representation: task graph, mode transition graph, architecture graph (fully connected PE number of processors), and worst-case execution time for all tasks
m	The specific mode
map_i^m	Given solution of the task mapping represented by individual i for task set T_m
CP	Critical path in the form of an array of task

Output:

map_i^m	Hardened solution of the task mapping represented by individual i for task set T_m

1: **procedure GRLS**(**M**, m, map_i^m, CP)
2:   $p_{minWCET}$:= **fastestPro(M)** ▷ $p_{minWCET} = \text{argmin}_{p \in PE}$ WCET(t_i, m, p)
3:   **for** $i = 1$: **abs**(CP) **do** ▷ for each task on the critical path $t_i \in$ CP
4:     $j := map_{i,t_i}^m$ ▷ t_i runs on p_j as given by $map_{i,t_i}^m = p_j$
5:     **if** $j \neq p_{minWCET}$ **then**
6:       map_{i,t_i}^m := **selPro**(map_{i,t_i}^m, **M**) ▷ A faster processor p_k) is assigned to run t_i ▷ where WCET$(t_i, m, p_k) <$ WCET(t_i, m, p_j)
7:       **break**
8:     **end if**
9:   **end for**
10:   **return** map_i^m
11: **end procedure**

Algorithm 9.3 implements the problem-specific local search operator for task mapping of an MMDF graph [49]. In order to reduce latency Lat(m), the most direct way is to reassign a faster processor p_k for the given task $t_i \in T_m$ that is currently mapped to processor $map_{i,t_i}^m = p_j$. This is specified by the condition WCET$(t, m, p_k) <$ WCET(t, m, p_j). A procedure **FINDCPATH** is first applied to find the critical path CP of the dataflow graph based on the schedule table on each processor. The main procedure **GRLS** will take the critical path CP as one of the inputs and then procedurally implement the GRLS, Algorithm 9.3. In particular, only one change is allowed during the GRLS process (lines 6–7) to relieve the time overhead introduced since the chromosome will be reevaluated after it has been made. Furthermore, only the tasks on the critical path will be considered for the reassignment (line 3). As part of the local search process, the new generated chromosome map_i^m produced by **GRLS**, which represents a hardened solution of the task mapping, will first be reevaluated first, and then it will only displace the previous chromosome if it has a better fitness value.

The second element is based on the ***Surrogate-assisted Optimization*** (or ***meta-modelling***) methodology. Surrogate models [13] such as polynomial models, Artificial Neural Networks, and Support Vector Machines among others are widely used for computationally expensive optimization or high-dimensional optimization problems. However, conventional surrogate models require extra time overhead for model learning in an offline [43] or online manner [42]. Our application of surrogate is a problem-specific approximated fitness evaluation strategy that does not require model learning [49]. This approach can improve the effectiveness of the original CCGA given that the time overhead of the fitness evaluation is high. This is because the number of computation that is involved for mode transition interval equals the number of possible mode transitions |Trans|. In Fig. 9.1 as an example, |Trans| takes on a value of 8 and 12 for the Type I and Type II MTG, respectively.

The fitness function (9.9) can be reformulated as follows:

$$\mathrm{Fit}(\mathrm{map}^{m_l}) = \max\left(\max_{\substack{(m_i,m_j)\in\mathrm{Trans}\\ i=l\vee j=l}} \mathrm{TransInterval}(m_i,m_j),\ \max_{\substack{(m_i,m_j)\in\mathrm{Trans}\\ i\neq l\wedge j\neq l}} \mathrm{TransInterval}(m_i,m_j)\right). \tag{9.10}$$

Note that the second term on the right-hand side of Eq. (9.10) is not dependent on the subsolution that is being evaluated. As such, the term could be removed to obtain the following approximated fitness function [49]:

$$\mathrm{Fit}(\mathrm{map}^{m_l}) \approx \max_{\substack{(m_i,m_j)\in\mathrm{Trans}\\ i=l\vee j=l}} \mathrm{TransInterval}(m_i,m_j). \tag{9.11}$$

The original fitness function (9.9) requires computing the values of all possible mode transitions. In the example illustrated by Fig. 9.1, they consist of $|\mathrm{Trans}| = 2M$ for Type I MTG or $|\mathrm{Trans}| = M(M-1)$ for Type II MTG since the mode transition is bidirectional. For the approximated fitness function (9.11), only the mode transitions between the target mode m_l and its directly connected modes are considered. As such, the total number is reduced to 4 for Type I MTG or $2(M-1)$ for Type II MTG. In both cases, the computation complexity of the approximated fitness functions is reduced to $2/M$ compared with that of the original fitness function [49]. However, to balance evaluation accuracy and computation efficiency, both the original and approximated fitness evaluations are used on the candidate solutions. In particular, our fitness evaluation implementation employs the original evaluation for the population in every GAP generations, where GAP is the exact evaluation gap. Such generation-based evolution control [14] could mitigate the evolutionary process from converging to a false optima as a result of using the approximated fitness model [49].

9.4 Computational Study

This section presents computational studies [49] we have carried out to investigate the performance of the proposed CCGA-based DSE method. The superiority of the cooperative coevolution framework, the GRLS operator, and the approximated fitness evaluation is each verified in turn through extensive computational experiments. In particular, systematic comparisons are performed to assess that the incorporation of additional design elements (e.g. the GRLS operator) leads to improvement in performance. In the experiments, we synthesize a large set of benchmarks with different scales [49]. The number of tasks in the MMDF graph $|T|$ is set to $\{10, 20, 30, 40, 50\}$. For the number of modes in the application M, it is set to $\{5, 10, 15\}$ (Type I MTG) and $\{4, 6, 8\}$ (Type II MTG). The number of processors on the multiprocessor architecture is set as $|\mathrm{PE}| = 9$. The task migration time is set to 30 ms, and the WCET of a task on a processor is randomly drawn from the interval of $[10, 50]$ ms. We also conduct a real-world case study involving an MP3 decoder [45]. The algorithms are implemented in MATLAB, and all the experiments are performed on a 3.20 GHz Intel Core i7-8700 platform with 64 GB memory. The algorithms are implemented as monolithic processes, and no CPU core parallelism is exploited for fair comparisons in the computational studies.

We propose to optimize each mode separately using the Divide-and-Conquer cooperative coevolution framework to tackle the scalability issue that occurs in the GA-based DSE framework [49]. The cooperative coevolution framework is implemented in the form of Algorithm 9.1 and referred to as CCGA in subsequent presentations. For the CCGA that incorporates the GRLS operator specified by Algorithm 9.3 to further improve its search effectiveness, it will be referred to as MCCGA. The parameter setting of the baseline GA involves the population size $\mathrm{PS} = M \times |T|$, the crossover rate $p_c = 0.8$, and the mutation rate $p_m = 0.8$. We employ a roulette selection operator for all the algorithms, having determined through a comparative computational study in [49] that it is superior to standard elitist selection operator. The parameter settings of CCGA and MCCGA are consistent with those of the GA, except for the population size $\mathrm{PS} = |T|$ for each population $\boldsymbol{P}_m$ representing solutions for the $m \in M$ subcomponent. For these three algorithms, the maximum number of fitness evaluations is set to $\tau_e = 500 \times M \times |T|$. Given that the procedure MODEVAL in CCGA and MCCGA evaluates an individual in a population $\boldsymbol{P}_m$ twice (i.e. it requires combining it with the current best individual and a random individual from each of the other populations to form two candidate complete solutions), then the maximum number of generations of both CCGA and MCCGA is set to $\tau_g = 250$. For fair comparison with GA, $\tau_g = 500$ is set.

Figures 9.7a and 9.8a show the average evolutionary plots of fitness (over 30 independent runs) using different algorithms (i.e. GA, CCGA, and MCCGA) for Type I MTG and Type II MTG, respectively. Figures 9.7b and 9.8b show the average computer runtimes (over 30 independent runs) for different algorithms on different benchmarks. We first make the observation that the total average computer

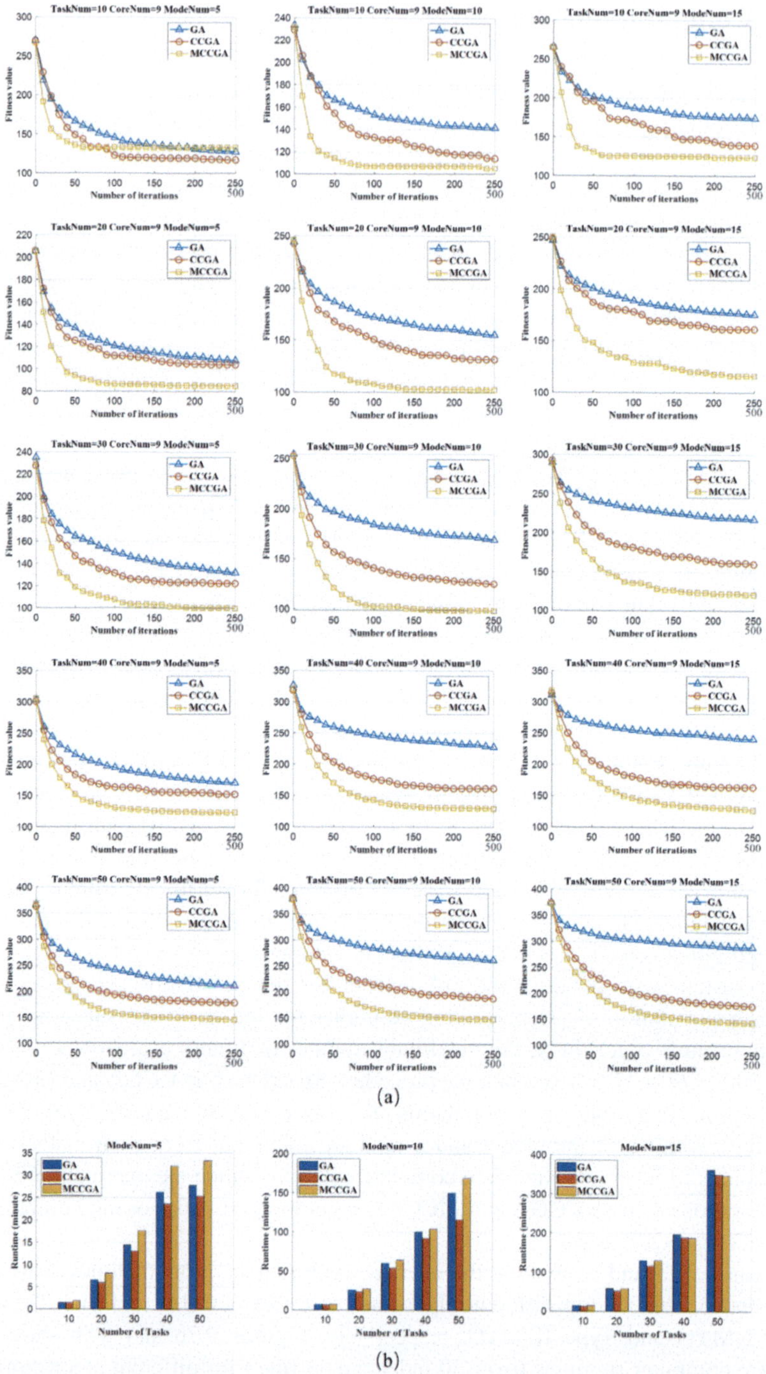

Fig. 9.7 The average (**a**) evolutionary plots and (**b**) computer runtimes of the GA, CCGA and MCCGA considering Type I MTG [49]

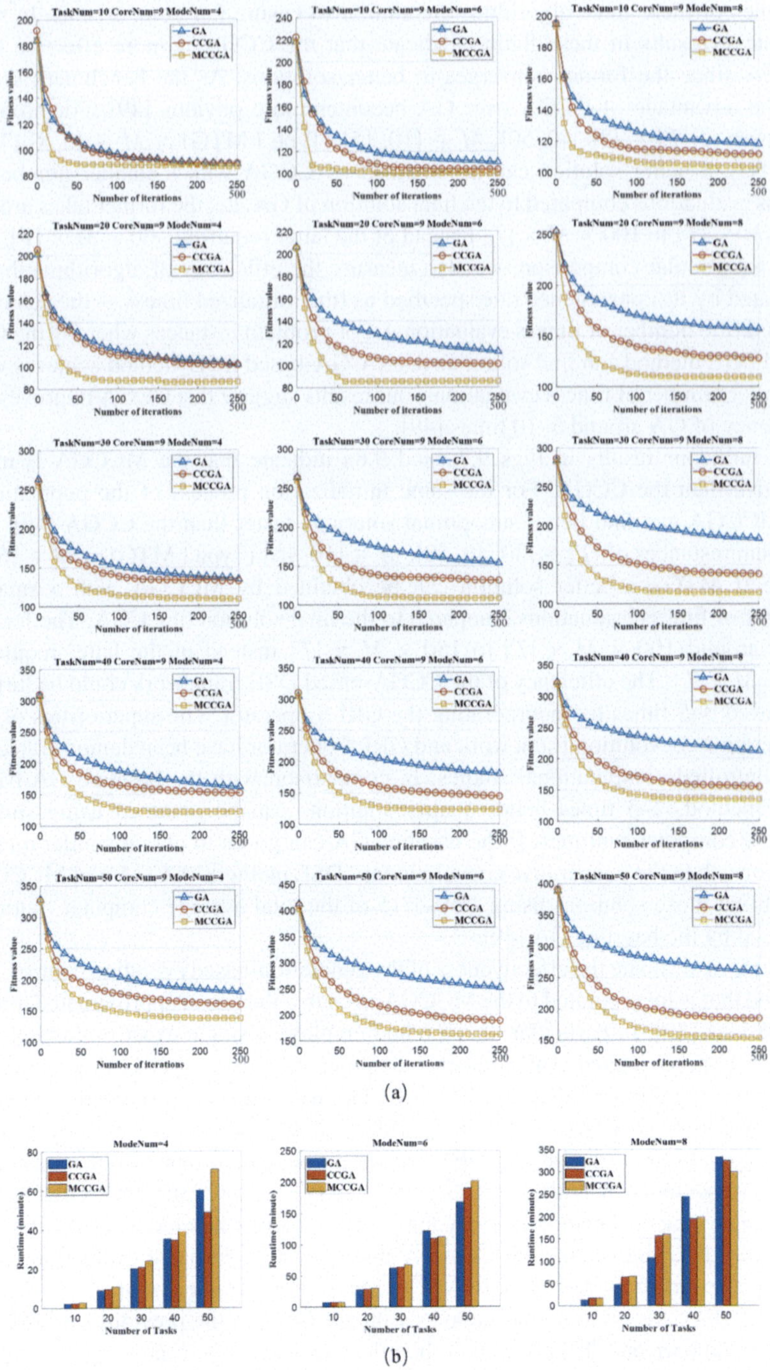

Fig. 9.8 The average (**a**) evolutionary plots and (**b**) computer runtimes of the GA, CCGA and MCCGA considering Type II MTG [49]

runtimes of these three algorithms are similar to ensure fair comparisons. In such a context, results in these figures indicate that the CCGA is more effective than the GA since the former converges to better solutions. As the benchmark scales up, the advantage of CCGA over GA becomes more obvious [49]. For problem instances of $|T| \in \{30, 40, 50\}$, $M \in \{10, 15\}$ (Type I MTG) or $M \in \{6, 8\}$ (Type II MTG), a better solution can be obtained by CCGA with a smaller number of fitness evaluations compared to the final solution of GA, i.e. the former takes around $50 \times M \times |T|$ to $100 \times M \times |T|$ instead of the latter requiring $500 \times M \times |T|$. For such a particular comparison, we also measure the efficiency of algorithms that is indicated by its convergence rate specified as (the initialized fitness − the achieved fitness)/(the number of fitness evaluations). For problem instances whereby the GA-based DSE method can find solutions that CCGA-based DSE method achieves with a smaller number of fitness evaluations, our results suggest that CCGA improves the efficiency of GA around 5–10 times [49].

In turn, our results in Figs. 9.7a and 9.8a indicate that the MCCGA is more effective than the CCGA. For the same initialization process of the populations, the MCCGA can find better suboptimal solutions faster than the CCGA [49]. For problem instances of $|T| \in \{30, 40, 50\}$, $M \in \{10, 15\}$ (Type I MTG) or $M \in \{6, 8\}$ (Type II MTG), a better solution can be obtained by MCCGA with a smaller number of fitness evaluations compared to the final solution of CCGA. The former takes around $100 \times M \times |T|$ to $150 \times M \times |T|$ instead of the latter requiring $500 \times M \times |T|$. The efficiency of the CCGA-based DSE framework could be further improved 3–5 times by incorporating the GRLS operator. The superiorities of the cooperative coevolution framework and GRLS operator have been demonstrated by the controlled computational studies. In comparison with the baseline GA-based DSE method, 2–3 times better quality solutions can be obtained using similar average computer runtimes. If the baseline GA can go on to obtain similar quality solutions, then the cooperative coevolutionary DSE methods (CCGA and MCCGA) could find those solutions using 1/50–1/15 of the total average computer runtimes required by the baseline GA [49].

Next, we evaluate the effectiveness of the generation-based evolution control [14] method that is incorporated to the MCCGA to enable the use of approximated fitness evaluations [49]. In particular, this method employs a single exact evaluation and $GAP - 1$ approximated evaluations for the population over every GAP generations during the operation of MCCGA. The MCCGA with fitness approximation strategy is referred to as MCCGA/FA. Three algorithmic implementations are considered for comparison [49]: (1) the MCCGA/FA that has the best performance as demonstrated earlier (essentially, it is the MCCGA with $GAP = 1$), (2) the MCCGA/FA that uses single exact evaluation in every ten generations that is referred to as MCCGA ($GAP = 10$), and (3) the MCCGA/FA that uses approximated evaluations only that is referred to as MCCGA ($GAP = \text{Max}$). The parameters of the two other MCCGA/FAs are set to be the same as with MCCGA in the previous experiments for fair comparisons. This is so that the effectiveness of the fitness approximation can be investigated. Results in Figs. 9.9a and 9.10a indicate that the performance of MCCGA/FAs is not degraded on most benchmarks. For large-scale instances,

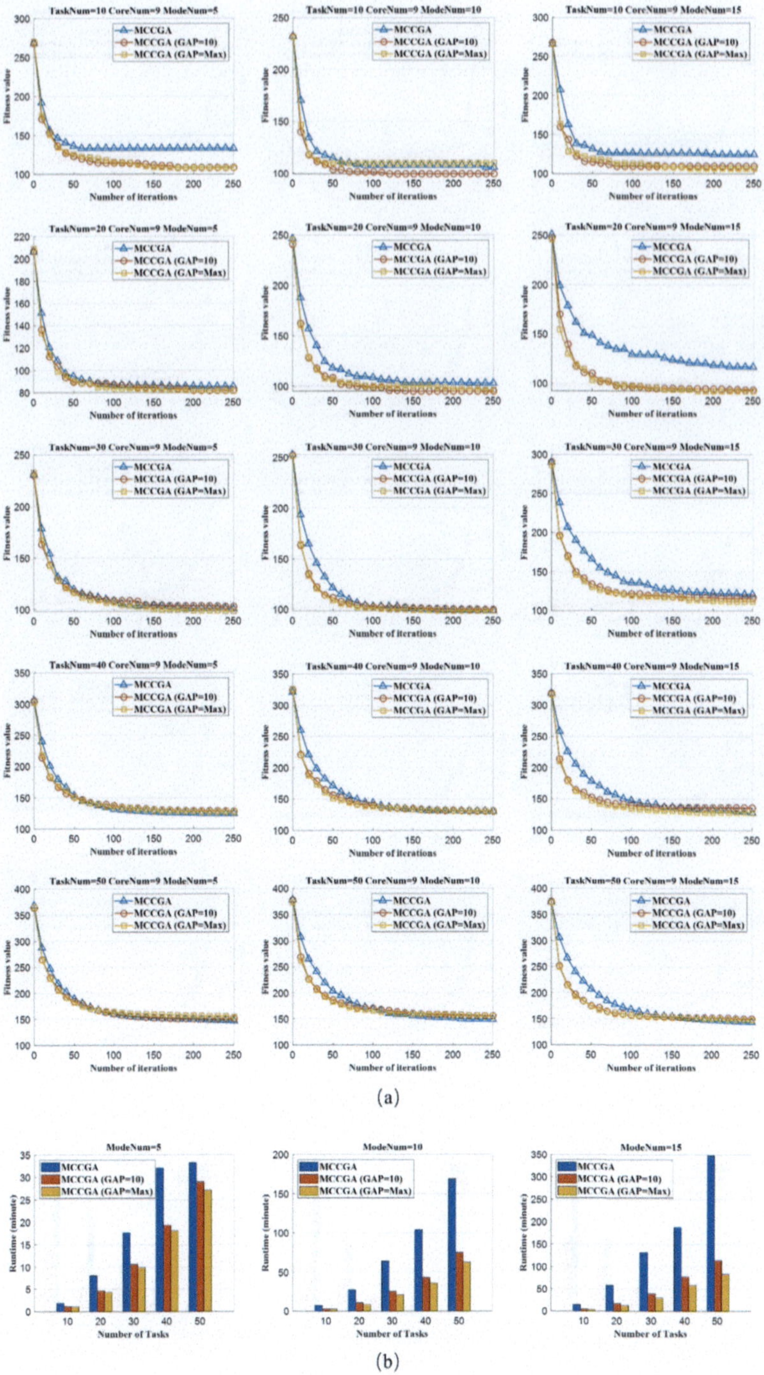

Fig. 9.9 The average (**a**) evolutionary plots and (**b**) computer runtimes of the MCCGA/FAs considering Type I MTG [49]

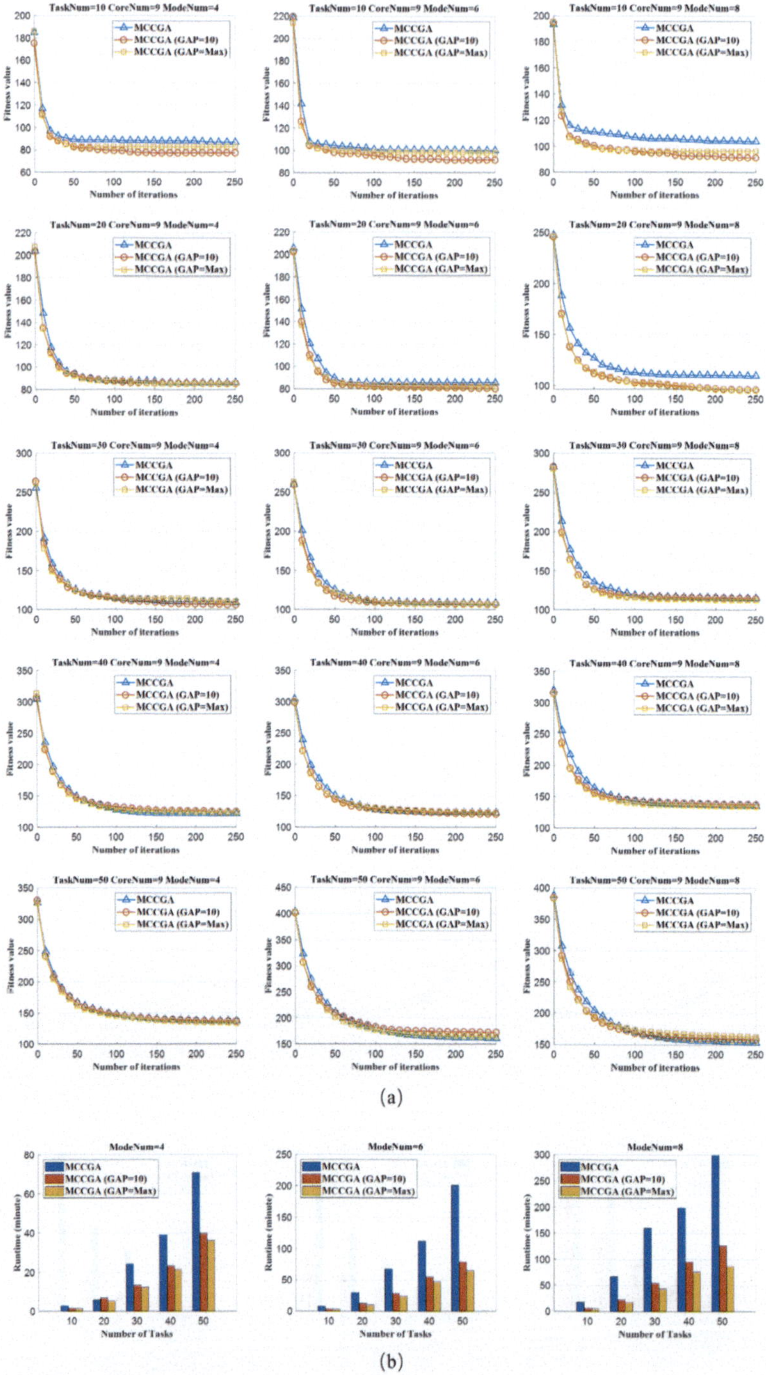

Fig. 9.10 The average (**a**) evolutionary plots and (**b**) computer runtimes of the MCCGA/FAs considering Type II MTG [49]

i.e. $|T| = 50$ and $M = 15$ (Type I MTG) or $M = 8$ (Type II MTG), the results of MCCGA/FAs are slightly worse than that of the MCCGA. Results in Figs. 9.9b and 9.10b indicate that the average computer runtimes of the MCCGA/FAs and in particular MCCGA (GAP = 10) and MCCGA (GAP = Max) are reduced to around 1/2–1/3 of the original MCCGA with the use of the fitness approximation strategy on most benchmarks, e.g. $M \in \{10, 15\}$ (Type I MTG) or $M = \{6, 8\}$ (Type II MTG) [49].

Numerical results of the computational experiments involving GA, CCGA, and MCCGAs (which include MCCGA/FAs) are summarized in Tables 9.1 and 9.2, which are obtained through 30 independent runs on every benchmark instance. They indicate superior performance of MCCGA/FAs, especially on large-scale benchmarks with 2–3 times better solution qualities obtained using 1/2–1/3 computer runtimes compared to that produced by the baseline GA [49]. We have also conducted statistical tests on these results to assess the significance of improvements on the fitness values of solutions produced by these algorithms. These are performed in a pairwise manner for every pair of algorithms, i.e. GA versus CCGA, GA versus MCCGAs, and CCGA versus MCCGAs. Each involves a two-tailed t-test with the null hypothesis that there is no significance difference between the paired algorithms in comparison. The null hypothesis is rejected if the p-value is smaller than the significance level of $\alpha = 0.05$. In summary, the statistical tests indicate that (1) the results of CCGA are significantly better than those of GA for most of the benchmarks (i.e. $|T| \in \{20, 30, 40, 50\}$ and $M = \{10, 15\}$ for Type I MTG and $|T| \in \{30, 40, 50\}$ and $M = \{6, 8\}$ for Type II MTG), (2) the results of MCCGAs are significantly better than those of GA for most of the benchmarks (i.e. $|T| \in \{20, 30, 40, 50\}$ and $M = \{10, 15\}$ for Type I MTG and $|T| \in \{20, 30, 40, 50\}$ and $M = \{6, 8\}$ for Type II MTG), and (3) the results of MCCGAs are significantly better than those of CCGA for most of the benchmarks (i.e. $|T| \in \{20, 30, 40, 50\}$ and $M = \{10, 15\}$ for Type I MTG and $|T| \in \{20, 30, 40, 50\}$ and $M = \{6, 8\}$ for Type II MTG) [49].

Finally, we evaluate the performance of the proposed methods on a real application involving an MP3 decoder. The mode transition graph and task graph of the MP3 decoder involving 15 tasks are illustrated in Fig. 9.11 (one can refer to studies in [40, 45] for specific details). Again, we consider $|\mathrm{PE}| = 9$ heterogeneous processors on the target platform. The WCET of a task on different processors is determined to be in the range of 1/2–3/2 times its standard execution cycle according to a uniform distribution. The parameters of the GA, CCGA, and MCCGA/FA are set to be the same as in the previous computational experiments. Results of this computational study on the MP3 decoder are summarized in Fig. 9.12 and Table 9.3 [49]. They indicate that the fitness values of CCGA are slightly better than those of GA. However, the fitness values of MCCGA/FA are statistically significantly better than those of GA and CCGA. Furthermore, the average computer runtimes can be reduced to 2.83 min from around 5.15 min with the use of the fitness approximation strategy. As such, our computational study demonstrates the superiority of the cooperative coevolution framework and GRLS operator in a particular real-world problem-solving setting [49].

Table 9.1 The fitness values (average $\pm$ variance) and computer runtimes (average $\pm$ variance) of various methods considering Type I MTG [49]

$\lvert T\rvert - \lvert PE\rvert - M$		GA	CCGA	MCCGA	MCCGA (GAP = 10)	MCCGA (GAP = Max)
10-9-5	Fitness (unit: ms)	127.40 ± 7.92	117.50 ± 15.26	133.30 ± 26.02	109.60 ± 17.72	109.00 ± 13.98
	Runtime (unit: min)	1.38 ± 0.28	1.36 ± 0.14	1.83 ± 0.08	1.12 ± 0.03	1.03 ± 0.02
10-9-10	Fitness	141.70 ± 3.13	116.40 ± 10.68	105.70 ± 8.25	104.90 ± 19.26	113.70 ± 20.69
	Runtime	6.95 ± 0.12	6.34 ± 0.10	7.25 ± 0.17	2.87 ± 0.09	2.36 ± 0.07
10-9-15	Fitness	174.50 ± 2.01	140.90 ± 14.02	128.20 ± 4.83	117.40 ± 17.55	107.20 ± 7.51
	Runtime	14.17 ± 0.27	12.85 ± 0.22	14.24 ± 0.30	4.30 ± 0.09	3.43 ± 0.04
20-9-5	Fitness	107.70 ± 2.63	104.30 ± 7.04	85.40 ± 4.67	83.70 ± 6.63	81.60 ± 4.12
	Runtime	6.56 ± 0.01	5.98 ± 0.02	8.03 ± 0.05	4.64 ± 0.04	4.33 ± 0.13
20-9-10	Fitness	155.20 ± 2.94	136.00 ± 15.11	103.10 ± 4.18	98.10 ± 6.23	96.40 ± 10.02
	Runtime	25.71 ± 0.04	23.21 ± 0.05	27.23 ± 0.08	10.53 ± 0.24	8.43 ± 0.08
20-9-15	Fitness	175.30 ± 2.21	167.10 ± 9.01	116.70 ± 7.87	94.20 ± 9.60	93.30 ± 7.44
	Runtime	57.49 ± 0.06	51.82 ± 0.11	58.24 ± 0.09	17.60 ± 0.08	13.10 ± 0.30
30-9-5	Fitness	132.00 ± 3.06	122.90 ± 5.86	100.10 ± 2.64	105.40 ± 15.17	98.10 ± 7.22
	Runtime	14.37 ± 0.03	13.01 ± 0.02	17.61 ± 0.07	10.62 ± 0.03	9.89 ± 0.06
30-9-10	Fitness	169.30 ± 2.16	125.40 ± 6.00	98.50 ± 1.90	99.80 ± 1.55	98.60 ± 2.76
	Runtime	60.10 ± 0.07	54.11 ± 0.10	64.14 ± 0.17	25.05 ± 0.09	20.94 ± 0.11

30-9-15	Fitness	217.20 ± 1.93	160.30 ± 7.02	121.80 ± 5.03	117.40 ± 5.38	112.80 ± 5.57
	Runtime	130.16 ± 0.21	116.70 ± 0.29	131.13 ± 0.22	39.94 ± 0.14	30.31 ± 0.12
40-9-5	Fitness	171.00 ± 4.50	153.00 ± 4.24	124.20 ± 7.24	128.00 ± 6.20	127.70 ± 8.77
	Runtime	26.19 ± 0.08	23.63 ± 0.05	32.06 ± 0.12	19.38 ± 0.06	18.10 ± 0.10
40-9-10	Fitness	228.00 ± 3.06	162.80 ± 9.40	130.20 ± 4.02	130.80 ± 6.44	131.30 ± 5.62
	Runtime	100.15 ± 3.42	92.64 ± 2.38	104.13 ± 3.50	43.15 ± 0.19	36.37 ± 0.14
40-9-15	Fitness	241.30 ± 2.71	164.40 ± 9.38	128.30 ± 4.57	136.60 ± 6.26	129.90 ± 7.55
	Runtime	197.25 ± 19.65	189.33 ± 18.86	187.66 ± 15.50	77.94 ± 0.28	58.96 ± 0.34
50-9-5	Fitness	212.00 ± 2.45	179.30 ± 8.99	146.40 ± 6.11	151.10 ± 7.08	154.40 ± 7.81
	Runtime	27.73 ± 0.05	25.37 ± 0.05	33.26 ± 0.12	28.96 ± 0.13	27.10 ± 0.03
50-9-10	Fitness	262.30 ± 4.03	189.50 ± 7.56	148.50 ± 6.50	156.60 ± 7.07	157.80 ± 9.65
	Runtime	150.07 ± 25.15	116.11 ± 0.13	168.88 ± 29.08	75.13 ± 0.13	62.92 ± 0.11
50-9-15	Fitness	289.10 ± 2.51	175.60 ± 6.67	143.90 ± 1.91	151.30 ± 8.92	149.30 ± 7.36
	Runtime	361.61 ± 46.63	349.71 ± 8.99	347.89 ± 62.82	114.50 ± 3.44	83.23 ± 3.74

Table 9.2 The fitness values (average ± variance) and computer runtimes (average ± variance) of various methods considering Type II MTG [49]

$\lvert T\rvert - \lvert PE\rvert - M$		GA	CCGA	MCCGA	MCCGA (GAP = 10)	MCCGA (GAP = Max)
10-9-4	Fitness (unit: ms)	88.70 ± 4.99	89.20 ± 8.53	88.30 ± 8.01	77.60 ± 7.29	83.20 ± 4.16
	Runtime (unit: min)	2.17 ± 0.28	2.30 ± 0.15	2.68 ± 0.06	1.50 ± 0.03	1.36 ± 0.03
10-9-6	Fitness	110.20 ± 1.69	103.60 ± 7.56	100.10 ± 6.38	91.80 ± 9.25	98.00 ± 8.45
	Runtime	7.11 ± 0.14	7.26 ± 0.15	7.61 ± 0.15	3.26 ± 0.06	2.80 ± 0.09
10-9-8	Fitness	118.20 ± 1.03	111.70 ± 5.83	103.90 ± 8.62	91.60 ± 4.20	97.10 ± 7.55
	Runtime	13.17 ± 0.03	17.07 ± 0.34	17.44 ± 0.31	5.86 ± 0.06	4.82 ± 0.14
20-9-4	Fitness	101.90 ± 2.69	101.50 ± 5.72	86.30 ± 2.54	85.90 ± 3.54	85.80 ± 3.61
	Runtime	8.84 ± 0.17	9.37 ± 0.03	10.72 ± 0.05	5.71 ± 0.05	5.24 ± 0.07
20-9-6	Fitness	112.30 ± 2.54	101.20 ± 6.71	85.70 ± 7.90	80.40 ± 1.43	82.60 ± 3.41
	Runtime	27.32 ± 0.06	28.01 ± 0.06	29.59 ± 0.08	12.25 ± 0.09	9.92 ± 0.09
20-9-8	Fitness	157.20 ± 1.55	128.50 ± 7.96	109.70 ± 3.30	96.80 ± 5.14	96.50 ± 4.33
	Runtime	47.94 ± 1.66	64.71 ± 0.08	66.24 ± 0.09	22.35 ± 0.08	17.79 ± 0.07
30-9-4	Fitness	134.70 ± 3.80	132.80 ± 7.11	109.00 ± 6.63	106.70 ± 5.36	111.20 ± 5.05
	Runtime	20.23 ± 0.06	20.79 ± 0.04	24.08 ± 0.06	13.28 ± 0.06	12.25 ± 0.09
30-9-6	Fitness	153.80 ± 3.29	131.40 ± 8.37	108.50 ± 2.12	107.50 ± 4.28	108.00 ± 4.62
	Runtime	62.24 ± 0.09	63.75 ± 0.11	67.54 ± 0.16	27.93 ± 0.11	23.97 ± 0.10
30-9-8	Fitness	183.80 ± 2.74	138.50 ± 4.90	115.00 ± 2.40	114.90 ± 3.63	113.60 ± 2.07
	Runtime	106.98 ± 0.11	156.58 ± 0.21	160.18 ± 0.82	53.97 ± 0.13	42.96 ± 0.12
40-9-4	Fitness	161.40 ± 3.10	151.20 ± 11.78	121.50 ± 6.92	125.30 ± 3.56	124.80 ± 8.42
	Runtime	35.36 ± 0.04	34.69 ± 1.35	39.15 ± 0.25	23.16 ± 0.12	21.40 ± 0.10

40-9-6	Fitness	180.00 ± 4.19	144.00 ± 4.47	123.20 ± 2.25	121.60 ± 4.53	124.30 ± 4.14
	Runtime	121.45 ± 0.21	109.54 ± 9.13	111.92 ± 10.31	54.94 ± 0.13	47.12 ± 0.11
40-9-8	Fitness	212.20 ± 3.65	155.00 ± 3.13	135.40 ± 4.74	138.00 ± 10.09	135.10 ± 5.69
	Runtime	243.34 ± 38.50	194.27 ± 8.50	198.55 ± 6.71	95.13 ± 0.27	75.58 ± 0.32
50-9-4	Fitness	179.50 ± 4.77	159.00 ± 5.21	136.80 ± 9.96	137.10 ± 7.64	134.40 ± 8.11
	Runtime	60.16 ± 0.08	48.72 ± 8.57	70.80 ± 1.67	39.75 ± 0.09	36.22 ± 0.43
50-9-6	Fitness	252.00 ± 6.41	187.70 ± 6.55	161.30 ± 3.62	173.00 ± 11.79	165.30 ± 5.33
	Runtime	167.28 ± 13.50	189.01 ± 0.62	201.26 ± 0.32	79.48 ± 2.57	65.31 ± 2.29
50-9-8	Fitness	259.30 ± 3.95	185.00 ± 6.31	153.20 ± 3.16	158.60 ± 5.91	163.10 ± 6.82
	Runtime	332.65 ± 77.19	324.37 ± 82.58	299.12 ± 78.10	126.31 ± 9.55	87.47 ± 7.71

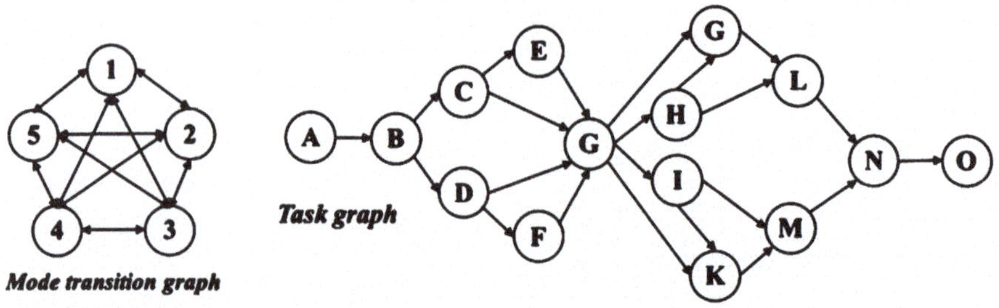

Fig. 9.11 The mode transition graph and task graph of MP3 decoder [49]

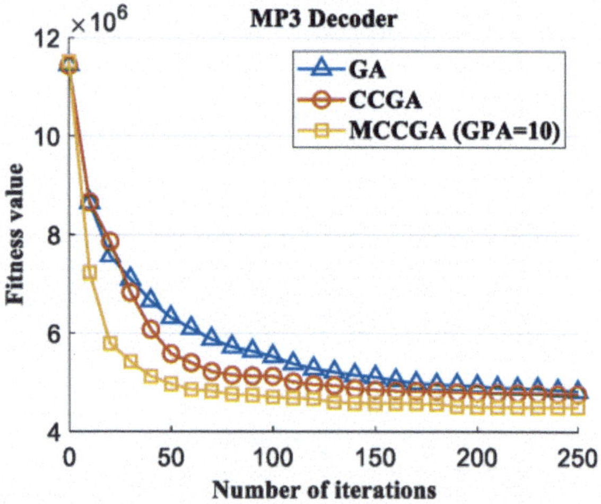

Fig. 9.12 The average evolutionary curves of the GA, CCGA, and MCCGA/FA for the MP3 decoder [49]

Table 9.3 The fitness values (average ± variance) and computer runtimes (average ± variance) of the GA, CCGA, and MCCGA/FA for the MP3 Decoder [49]

		GA	CCGA	MCCGA (GAP = 10)
MP3 Decoder	Fitness (unit: circle)	4.82e+06 ± 0.26e+06	4.77e+06 ± 0.33e+06	4.51e+06 ± 0.31e+06
	Runtime (unit: min)	5.15 ± 0.10	5.14 ± 0.11	2.83 ± 0.13

9.5 Discussion and Conclusion

In this chapter, we have presented our study on the use of the principles of cooperative coevolution to solve real-world optimization problems in some signal processing and multimedia applications [49]. In particular, we have addressed the task mapping problem of an MMDF graph on the heterogeneous multiprocessor platform with non-negligible mode transition delay. This is of crucial importance given that most modern devices running these applications are equipped with multi-core computing units. The MMDF graph consists of a finite set of modes $\{1, \ldots, M\}$, and each mode is specified by an SDF graph. Such a problem has been tackled earlier via a GA-based DSE method. However, such an approach would not scale well and suffers from the curse of dimensionality. We have proposed a cooperative coevolution framework on the DSE method that maps and schedules all SDF graphs simultaneously to minimize the maximum transition interval. A novel problem-specific decomposition strategy is developed in this cooperative CEA. In the GA-based DSE method that encodes the design space as a whole, a chromosome consists of a set of vectors, and each vector represents the mapping in a mode. In our CCGA-based DSE method, problem decomposition is achieved whereby each subcomponent represents a specific mode. As such, the cooperative coevolutionary Divide-and-Conquer problem-solving strategy will be to assign each mode to an individual genetically isolated population consisting of candidate solutions of node mapping for each individual mode. In this manner, CCGA could partially address the curse of dimensionality issue associated with the task mapping problem of an MMDF graph that is at least NP-hard.

In addition to the global search capability of the CCGA-based DSE method, the effectiveness and efficiency of the search process are enhanced with two additional elements. The first element involves a problem-specific local search (GRLS) operator that is incorporated into CCGA for an indirect optimization of the fitness function. This follows the principles of Memetic Computation on search heuristics that combined both global and local search capabilities. The second element involves the use of an approximated fitness function that reduces the computation complexity to $2/M$ of that associated with the original fitness function. This follows the principles of surrogate models but without having to employ model learning. The additional performance of the proposed method has been demonstrated by extensive and systematic computational studies in terms of both effectiveness and efficiency. Our results indicate that it is possible to achieve an optimization result with 2–3 times better quality using less optimization times (1/2–1/3) compared to the previous state-of-the-art GA-based DSE method with the full implementation specified by MCCGA/FA that incorporates the GRLS operator and approximated fitness function evaluations into the CCGA framework [49].

In closing this chapter, we discuss two situations that can arise in modern settings. First, one would note that modern devices now are equipped with more computing cores. The majority of our computational studies that we have presented have been conducted with the parameter setting of $|PE| = 9$ to represent a target architecture of a set of nine processing elements. We have a conducted an additional computational study in [49] whereby the parameter setting is at $|PE| = 36$. Results from the computational experiments have indicated that the proposed CCGA-based DSE method is effective with both parameter settings $|PE| \in \{9, 36\}$ with no significant difference in the optimization time required. For the task mapping problem of an MMDF graph that we have studied, its dimensionality is determined by the number of tasks and the number of modes.

Second, we note that in some modern applications on specific target platform the number of modes can be uncertain or bigger than what is considered in current studies. The proposed method can be used to effectively find the (sub)optimal solutions of task mapping for all modes offline when the number of modes is limited and fixed. These mapping solutions can be stored in the system memory. The tasks will be remapped according to the stored mapping solutions when a mode transition occurs. However, for situations when the number of modes is uncertain or bigger, the proposed method cannot be used as the runtime is unacceptable even in offline operations. In this situation, two-stage hybrid schemes [30, 33] are promising approaches to address the associated challenges. In the first stage, the original modes are clustered into a small number of representative modes, and then computationally complex DSE methods can be applied to find (sub)optimal mappings for these representative modes. In the second stage that involves the operation of the system, a fast greedy or heuristic method can be employed to dynamically adjust the mapping of the running tasks based on mappings established in the first stage and the current mode state.

Although the cooperative coevolution framework has been used successfully with the DSE method to solve the task mapping problem of an MMDF graph on the heterogeneous multiprocessor platform, only the maximum mode transition interval is considered as the optimization objective. It is of interest in future studies to extend the proposed method to include multiple objectives such as reliability, throughput, and resource requirement. Such extensions are more reflective of real-world problem solving given that these are valid objectives that have to be optimized. However, they also increase the problem complexity requiring additional design elements for complex DSE methods to be designed to mitigate such an increase.

References

1. Bartoli, A., De Lorenzo, A., Medvet, E., Tarlao, F.: Automatic search-and-replace from examples with coevolutionary genetic programming. IEEE Trans. Cybern. **51**(5), 2612–2624 (2021)
2. Bolchini, C., Miele, A.: Reliability-driven system-level synthesis for mixed-critical embedded systems. IEEE Trans. Comput. **62**(12), 2489–2502 (2013)
3. Boyan, J., Moore, A.: Learning evaluation functions to improve optimization by local search. J. Mach. Learn. Res. **1**, 77–112 (2000)
4. Caplan, J., Al-Bayati, Z., Zeng, H., Meyer, B.: Mapping and scheduling mixed-criticality systems with on-demand redundancy. IEEE Trans. Comput. **67**(4), 582–588 (2018)
5. Chen, H., Tino, P., Rodan, A., Yao, X.: Learning in the model space for cognitive fault diagnosis. IEEE Trans. Neural Networks Learn. Syst. **25**(1), 124–136 (2014)
6. Chen, H., Tino, P., Yao, X.: Probabilistic classification vector machines. IEEE Trans. Neural Netw. **20**(6), 901–914 (2009)
7. Chen, X., Ong, Y.S., Lim, M.H., Tan, K.: A multi-facet survey on memetic computation. IEEE Trans. Evol. Comput. **15**(5), 591–607 (2011)
8. Das, S., Doppa, J., Pande, P., Chakrabarty, K.: Design-space exploration and optimization of an energy-efficient and reliable 3-d small-world network-on-chip. IEEE IEEE Trans. Comput. Aided Des. Integr. Circuits Syst. **36**(5), 719–732 (2017)
9. Deb, K., Pratap, A., Agarwal, S., Meyarivan, T.: A fast and elitist multiobjective genetic algorithm: NSGA-II. IEEE Trans. Evol. Comput. **6**(2), 182–197 (2002)

10. Erbas, C., Cerav-Erbas, S., Pimentel, A.: Multiobjective optimization and evolutionary algorithms for the application mapping problem in multiprocessor system-on-chip design. IEEE Trans. Evol. Comput. **10**(3), 358–374 (2006)
11. Gong, M., Li, H., Luo, E., Liu, J., Liu, J.: A multiobjective cooperative coevolutionary algorithm for hyperspectral sparse unmixing. IEEE Trans. Evol. Comput. **21**(2), 234–248 (2017)
12. Jiang, W., Wen, L., Jiang, K., Zhang, X., Pan, X., Zhou, K.: System-level design to detect fault injection attacks on embedded real-time applications. ACM J. Emerg. Technol. Comput. Syst. **13**(2), 1–8 (2016)
13. Jin, Y.: A comprehensive survey of fitness approximation in evolutionary computation. Soft Comput. **9**(1), 3–12 (2005)
14. Jin, Y., Olhofer, M., Sendhoff, B.: A framework for evolutionary optimization with approximate fitness functions. IEEE Trans. Evol. Comput. **6**(5), 481–494 (2002)
15. Joardar, B., Kim, R., Doppa, J., Pande, P., Marculescu, D., Marculescu, R.: Learning-based application-agnostic 3D NOC design for heterogeneous manycore systems. IEEE Trans. Comput. **68**(6), 852–866 (2019)
16. Jung, H., Lee, C., Kang, S.H., Kim, S., Oh, H., Ha, S.: Dynamic behavior specification and dynamic mapping for real-time embedded systems: hopes approach. ACM Trans. Embed. Comput. Syst. **13**(4), 1–26 (2014)
17. Jung, H., Oh, H., Ha, S.: Multiprocessor scheduling of a multi-mode dataflow graph considering mode transition delay. ACM Trans. Des. Autom. Electron. Syst. **22**(2), 1–25 (2017)
18. Kadri, N., Koudil, M.: A survey on fault-tolerant application mapping techniques for network-on-chip. J. Syst. Archit. **92**, 39–52 (2019)
19. Kumar, R., Tullsen, D., Ranganathan, P., Jouppi, N., Farkas, K.: Single-ISA heterogeneous multi-core architectures for multithreaded workload performance. In: Proceedings of the 31st Annual International Symposium on Computer Architecture, vol. 31, pp. 64–75 (2004)
20. Lawler, E.: The quadratic assignment problem. Manag. Sci. **9**(4), 586–599 (1963)
21. Lee, C., Kim, S., Oh, H., Ha, S.: Failure-aware task scheduling of synchronous data flow graphs under real-time constraints. J. Signal Process. Syst. **73**(2), 201–212 (2013)
22. Lee, E., Messerschmitt, D.: Synchronous data flow. Proc. IEEE **75**(9), 1235–1245 (1987)
23. Lin, J., Srivatsa, A., Gerstlauer, A., Evans, B.: Heterogeneous multiprocessor mapping for real-time streaming systems. In: Proceedings of the 2011 IEEE International Conference on Acoustics, Speech and Signal Processing (ICASSP), pp. 1605–1608 (2011)
24. Lu, X., Menzel, S., Tang, K., Yao, X.: Cooperative co-evolution-based design optimization: a concurrent engineering perspective. IEEE Trans. Evol. Comput. **22**(2), 173–188 (2018)
25. Manna, K., Mukherjee, P., Chattopadhyay, S., Sengupta, I.: Thermal-aware application mapping strategy for network-on-chip based system design. IEEE Trans. Comput. **67**(4), 528–542 (2018)
26. Nguyen, S., Zhang, M., Johnston, M., Tan, K.: Automatic design of scheduling policies for dynamic multi-objective job shop scheduling via cooperative coevolution genetic programming. IEEE Trans. Evol. Comput. **18**(2), 193–208 (2014)
27. Pop, P., Izosimov, V., Eles, P., Peng, Z.: Design optimization of time- and cost-constrained fault-tolerant embedded systems with checkpointing and replication. IEEE Trans. Very Large Scale Integr. VLSI Syst. **17**(3), 389–402 (2009)
28. Potter, M.A., De Jong, K.A.: A cooperative coevolutionary approach to function optimization. In: International Conference on Parallel Problem Solving from Nature PPSN III. Lecture Notes in Computer Science, vol. 866, pp. 249–257. Springer, Berlin (1994)
29. Quan, W., Pimentel, A.: A scenario-based run-time task mapping algorithm for MPSoCs. In: 2013 50th ACM/EDAC/IEEE Design Automation Conference (DAC), pp. 1–6 (2013)
30. Quan, W., Pimentel, A.: A hybrid task mapping algorithm for heterogeneous MPSoCs. ACM Trans. Embed. Comput. Syst. **14**(1), 1–25 (2015)
31. Sahu, P., Shah, T., Manna, K., Chattopadhyay, S.: Application mapping onto mesh-based network-on-chip using discrete particle swarm optimization. IEEE Trans. Very Large Scale Integr. VLSI Syst. **22**(2), 300–312 (2014)

32. Schor, L., Bacivarov, I., Rai, D., Yang, H., Kang, S.H., Thiele, L.: Scenario-based design flow for mapping streaming applications onto on-chip many-core systems. In: Proceedings of the 2012 ACM International Conference on Compilers, Architectures and Synthesis for Embedded Systems (CASES'12), pp. 71–80 (2012)
33. Schwarzer, T., Weichslgartner, A., Glaß, M., Wildermann, S., Brand, P., Teich, J.: Symmetry-eliminating design space exploration for hybrid application mapping on many-core architectures. IEEE Trans. Comput. Aided Des. Integr. Circuits Syst. **37**(2), 297–310 (2018)
34. Singh, A., Shafique, M., Kumar, A., Henkel, J.: Mapping on multi/many-core systems: survey of current and emerging trends. In: 2013 50th ACM/EDAC/IEEE Design Automation Conference (DAC), pp. 1–10 (2013)
35. Skelin, M., Geilen, M., Catthoor, F., Hendseth, S.: Parameterized dataflow scenarios. IEEE Trans. Comput. Aided Des. Integr. Circuits Syst. **36**(4), 669–682 (2017)
36. Starke, S., Hendrich, N., Zhang, J.: Memetic evolution for generic full-body inverse kinematics in robotics and animation. IEEE Trans. Evol. Comput. **23**(3), 406–420 (2019)
37. Stuijk, S., Geilen, M., Basten, T.: A predictable multiprocessor design flow for streaming applications with dynamic behaviour. In: Proceedings of the 13th Euromicro Conference on Digital System Design: Architectures, Methods and Tools (DSD 2010), pp. 548–555 (2010)
38. Stuijk, S., Geilen, M., Theelen, B., Basten, T.: Scenario-aware dataflow: modeling, analysis and implementation of dynamic applications. In: Proceedings of the 2011 International Conference on Embedded Computer Systems: Architectures, Modeling and Simulation (IC-SAMOS 2011), pp. 404–411 (2011)
39. Theelen, B.D., Geilen, M.C.W., Basten, T., Voeten, J.P.M., Gheorghita, S.V., Stuijk, S.: A scenario-aware data flow model for combined long-run average and worst-case performance analysis. In: Proceedings of the Fourth ACM and IEEE International Conference on Formal Methods and Models for Co-Design (MEMOCODE'06), pp. 185–194 (2006)
40. Theelen, B.D., Geilen, M.C.W., Stuijk, S., Gheorghita, V., Basten, T., Voeten, J.P.M., Ghamarian, A.H.: Scenario-aware dataflow. Tech. rep., Technische Universiteit Eindhoven (2008)
41. Van Stralen, P., Pimentel, A.: Scenario-based design space exploration of MPSoCs. In: Proceedings of the IEEE International Conference on Computer Design: VLSI in Computers and Processors, pp. 305–312 (2010)
42. Wang, H., Jin, Y., Doherty, J.: Committee-based active learning for surrogate-assisted particle swarm optimization of expensive problems. IEEE Trans. Cybern. **47**(9), 2664–2677 (2017)
43. Wang, H., Jin, Y., Sun, C., Doherty, J.: Offline data-driven evolutionary optimization using selective surrogate ensembles. IEEE Trans. Evol. Comput. **23**(2), 203–216 (2019)
44. Wang, S.Y., Wang, L.: An estimation of distribution algorithm-based memetic algorithm for the distributed assembly permutation flow-shop scheduling problem. IEEE Trans Syst Man Cybern Syst **46**(1), 139–149 (2016)
45. Yang, Y.: Exploring resource/performance trade-offs for streaming applications on embedded multiprocessors. Ph.D. Dissertation, Eindhoven University of Technology, Eindhoven (2012)
46. Yuan, B., Chen, H., Yao, X.: Toward efficient design space exploration for fault-tolerant multiprocessor systems. IEEE Trans. Evol. Comput. **24**(1), 157–169 (2020)
47. Yuan, B., Li, B., Chen, H., Yao, X.: Defect-and variation-tolerant logic mapping in nanocrossbar using bipartite matching and memetic algorithm. IEEE Trans. Very Large Scale Integr. VLSI Syst. **24**(9), 2813–2826 (2016)
48. Yuan, B., Li, B., Weise, T., Yao, X.: A new memetic algorithm with fitness approximation for the defect-tolerant logic mapping in crossbar-based nanoarchitectures. IEEE Trans. Evol. Comput. **18**(6), 846–859 (2014)
49. Yuan, B., Lu, X., Tang, K., Yao, X.: Cooperative coevolution-based design space exploration for multi-mode dataflow mapping. ACM Trans. Embed. Comput. Syst. **20**(3), 1–25 (2021)
50. Zhou, Y., Hao, J.K., Glover, F.: Memetic search for identifying critical nodes in sparse graphs. IEEE Trans. Cybern. **49**(10), 3699–3712 (2019)
51. Zitzler, E., Laumanns, M., Thiele, L.: Spea2: Improving the strength pareto evolutionary algorithm. Technical Report, vol. 103 (2001)

Chapter 10
Cooperative Coevolutionary Design Optimization in Concurrent Engineering

10.1 An Introduction to Concurrent Engineering in Relation to Cooperative Coevolution

Concurrent Engineering (***CE***) is a well-known engineering practice that can reduce product development time and cost and also improve product quality [1, 56]. In CE practices, all elements that are involved in a product's life cycle are considered from the early stages of product development. These elements range from functionality, producibility, maintenance issues, and finally disposal and recycling. A CE practice advocates executing all design tasks in parallel or with some overlaps [24, 55]. This stands in contrast with a traditional sequential engineering practice that conducts different tasks separately and at different times. In CE, issues that could occur in a product life cycle are discovered earlier when the design is still flexible and as a consequence avoid laborious redesigns.

Despite this advantage, CE increases the complexity of the design problem. This is because a CE practice simultaneously considers all parts or aspects of the product design. One situation that could occur in a typical real-world design problem such as one that involves a product with a complex physical structure is that many design variables are involved. In this situation, optimizing all the design variables as a whole might not be the best approach due to the curse of dimensionality from exponential increase in the size of the search space. In another situation, different disciplinary expertise might be required to solve different parts of the product design. In ***Multidisciplinary Design Optimization*** (***MDO***) problems [34], the designer might not have access to the simulation models of each discipline, or the integration of different analysis codes might be too difficult [20, 37].

For these situations, decomposition-based optimization strategies are preferred. Studies in the literature specify three ways that decomposition can be implemented: product decomposition, problem decomposition, and process decomposition [25]. Product decomposition is based on the predefined product structure. Problem decomposition is carried out at the design parameter level whereby the optimization

X. Yao, S. Y. Chong, *Coevolutionary Computation and Its Applications*,
Natural Computing Series, https://doi.org/10.1007/978-981-96-2841-4_10

problem is decomposed based on the dependencies between design variables. However, knowledge about such dependencies might not be easily obtained. Process decomposition is often applied to MDO problems where different disciplines share design variables and are usually coupled by coupling variables. Process decomposition strategies such as Collaborative Optimization [4], Analytical Target Cascading (ATC) [50], and Bi-level Integrated System [48] break up the design problem according to engineering disciplines to ensure disciplinary autonomy.

There are many aspects with which CE could benefit from employing advanced methodologies in Evolutionary Computation. For example, CE is typically adopted in the design of complex systems. Such a problem setting often involves subproblems with nonconvex or nondifferentiable objective functions. Evolutionary Algorithms (EAs) could provide effective solvers for these subproblems in the framework of collaborative optimization [12, 57] and ATC [36, 58]. More importantly, the conceptual idea of Divide-and-Conquer in CE is mostly similar to that in cooperative Coevolutionary Algorithm (CEA). The difference between the two lies in the assumptions made on problem decomposition. On the one hand in cooperative CEAs, the main working assumption is that the problem can be divided into several disjoint parts (i.e. design parameters or variables are divided into several disjoint groups) called subcomponents or subproblems in coevolutionary literature. The problem decomposition is performed based on domain knowledge prior to [42] or automatically during [54] the cooperative coevolutionary search process. On the other hand in CE, such an assumption might not hold because problem decomposition is mainly driven by real-world applications whereby different subproblems share the same design parameters.

Although cooperative CEAs have been employed in CE problem settings [6, 7, 37, 53], the shared design variables are handled as if they are normal variables. The cooperative CEA still decomposes the problem into several fully disjoint subproblems. The shared design variables affecting several subproblems are only evolved or updated in one subproblem. The other subproblems would just take the updated variables as their inputs. Despite such a discrepancy in problem treatment, the use of such a problem decomposition strategy enables direct application of existing cooperative CEAs to CE problem settings. However, this raises the immediate issue on the choice of which subproblem a shared variable should be evolved in, as well as the larger issue of potential restriction to the inherent parallelism in the whole CE procedure if shared variables are only evolved in a chosen subproblem. In this chapter, we present our study [32] that investigates cooperative coevolution-based design optimization in the context of CE to address these two research issues. Our approach involves developing new concurrent design methods based on ***parallel cooperative CEAs*** to solve two main types of CE problems with different decomposition features.

The first problem type involves the product design problem whereby the decomposition of the product is assumed to be known in advance and that there are no shared design variables between different subproblems. Although existing cooperative CEAs can be directly used, a candidate solution for a subproblem might have to be evaluated with a large number of collaborators (candidate solutions of other subproblems) to prevent premature convergence to mediocre stable states [18, 39]. This has the implication that makes cooperative CEAs inefficient for use in CE. One approach to address this issue is to employ the novelty search approaches [16, 18] that score each individual based not only on its fitness but also on its novelty. However, the existing novelty calculation methods are still computationally inefficient. As such, we propose a new novelty-driven cooperative CEA that uses a new novelty calculation method that is computationally efficient. The cooperative CEA is further equipped with a stochastic selection process that uses an adaptive probability to adjust the trade-off between novelty and fitness.

The second problem type is the quasi-separable MDO problem in which different disciplines share part of the design variables, which is assumed to be known in advance. Such problems are frequently encountered in MDO [51], and where a general MDO problem can be easily transformed into a quasi-separable MDO problem [34]. Instead of evolving the shared variables in a single subproblem, we design a novel coevolutionary concurrent design method in which the shared variables are handled through the use of duplicates and consistency constraints. The method employs stochastic ranking [45] with an adaptive probability to deal with those constraints.

For both parallel cooperative CEA-based concurrent design methods, we have investigated the important issue of setting the appropriate communication frequency between different subtasks. In a real-world concurrent design process, communications among different subtasks will introduce a cost that differs for various problems and environments. Determining the exact time point to exchange information is a challenging task. There have been various studies in CE on this issue [23, 30, 31, 44, 56]. However, these approaches have been developed for concurrent design of dependent tasks whereby the information exchange is unidirectional. The two types of CE problems that are considered in this chapter involve interdependent subproblems for which information exchange is bidirectional. As such, we have addressed this research issue and gap in the literature through extensive computational studies to identify the best communication frequency among different populations in both cooperative CEA-based concurrent design methods. As a result, we develop a self-adaptive scheme that can automatically adjust the communication frequency among populations representing solutions for different subproblems during the optimization process.

The presentation for the remaining sections of this chapter is organized as follows. Section 10.2 will more formally present the two general types of CE problems, and then specific design problems considered for the computational experiments. We also provide a brief review of evolutionary-based problem-solving approaches that have been applied to these problems. Next, Sect. 10.3 will detail the two parallel cooperative CEA-based concurrent design methods. Section 10.4 presents computational studies to evaluate and compare the performance of the proposed methods against the baseline method that optimizes all parts (disciplines) as a whole and other state-of-the-art methods on ***Universal Electric Motor*** (***UEM***) design problems and a commonly used MDO test problem. These computational studies include experiments to determine the best communication frequencies with and without communication costs for all evaluated methods on the test problems. Section 10.5 closes the chapter with a discussion on communication frequencies of the CE problems considered in the experiments and the impact of incorporating the self-adaptive method for finding the best communication frequencies in the cooperative CEA-based methods.

10.2 Problem Formulation and Evolutionary-Based Problem-Solving Approaches

In this section, we first more formally specify the general problem formulation of two types of complex design problems in CE that are addressed by our study. After that, we will describe the specific formulation of the problems that our computational studies will address. Then, we will briefly review evolutionary-based problem-solving approaches that have been applied for these types of CE problems.

10.2.1 Complex CE Design Problem Formulation

We consider two types of complex design problems that are frequently encountered in real-world CE practices. These problems include:

1. Monodisciplinary product design with multiple parts.
2. Quasi-separable MDO problems.

The first type of problem involving ***monodisciplinary product design*** with multiple parts is considered the simpler type of CE problems that can be solved. Here, it is assumed that the product is composed of m parts and that there is a performance metric for the design of the product. Let f be the function that gives the performance metric $f(\boldsymbol{x}_1, \ldots, \boldsymbol{x}_m)$ for any valid product design specified by $\boldsymbol{x}_1, \ldots, \boldsymbol{x}_m$. Then, the problem formulation is given as follows [32]:

$$\min_{\boldsymbol{x}_1, \ldots, \boldsymbol{x}_m} f(\boldsymbol{x}_1, \ldots, \boldsymbol{x}_m) \tag{10.1}$$

where the vector $\boldsymbol{x}_i$ denotes the design variables associated with the ith part of the product $i = 1, \ldots, m$. Note that we assume throughout this paper that the decomposition of the product is known in advance and there are no shared design variables between different subproblems.

For the second type of design problem, different disciplines aim to find a complete design that improves the different physical aspects of a product. Such a design problem can be represented as the ***quasi-separable MDO*** problem [51] with m disciplines, which is formulated as follows [32]:

$$\min_{x_1,\ldots,x_m,z} \sum_{i=1}^{m} f_i(x_i, z) \tag{10.2}$$

where the tuple $(x_1, \ldots, x_m, z)$ captures all the design variables associated with the entire design. In particular, the vector x_i denotes the local design variables with respect to the ith discipline $i = 1, \ldots, m$, and f_i denotes the objective of the ith discipline. The vector z denotes the shared design variables or parameters among disciplines. The final design from each discipline must be consistent with z. For our study, we assume that the shared design variables are known beforehand for this type of problems.

The quasi-separable MDO problem specified by Eq. (10.2) is frequently encountered in MDO. It is part of the general multidisciplinary design optimization problem involving multiple disciplines that share some design variables and are linked by coupling variables. Figure 10.1 shows a two-disciplinary system, where x_1 and x_2 represent local variable vectors for disciplines 1 and 2, respectively. z represents the common variable vector for both disciplines. The corresponding objective functions for both disciplines are given by f_1 and f_2. For both disciplines, g_1 and g_2 represent the constraints. When the problem is considered as an entire system, f_{system} and g_{system} represent the system objective and constraint functions, respectively. Note that the f_{system} is a function of f_1 and f_2. The linkages or coupling of disciplines 1 and 2 are given by the coupling variables y_{12} and y_{21}. They are the result of a corresponding analysis of disciplines 1 and 2. They are also required for calculating the objective and constraint functions of disciplines 2 and 1, respectively.

For our study, we focus on MDO problems that are only coupled through shared design variables z, i.e. quasi-separable MDO problems. For a quasi-separable MDO problem, the original problem is decomposed along disciplines. In particular, duplicate variables are introduced to deal with the common design variables z. They are used in our proposed parallel cooperative CEAs for the optimization process. Our computational study will consider a particular design problem that assumes the

Fig. 10.1 A two-disciplinary system [32]

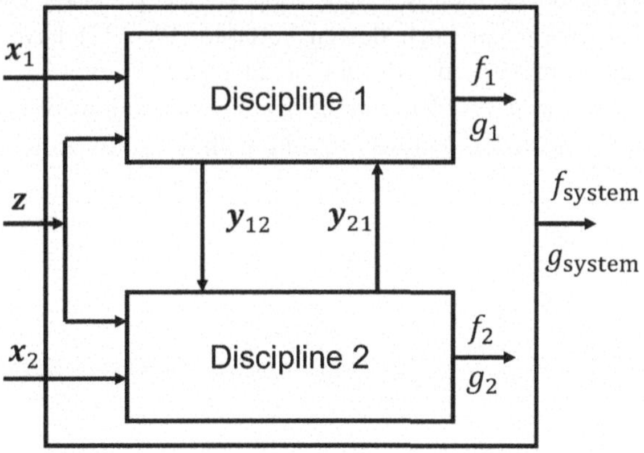

involvement of two disciplines, each having local design variables $\boldsymbol{x}_i,\ i = 1, 2$. The duplicate variables z^1 and z^2 are used to decompose the original problem in the form of Eq. (10.2) to one specified as follows [32]:

Subproblem 1

$$\begin{aligned} &\min_{\boldsymbol{x}_1, z^1} f_1(\boldsymbol{x}_1, z^1) \\ &\text{subject to} : z^1 = z^2. \end{aligned} \tag{10.3}$$

Subproblem 2

$$\begin{aligned} &\min_{\boldsymbol{x}_2, z^2} f_2(\boldsymbol{x}_2, z^2) \\ &\text{subject to} : z^2 = z^1. \end{aligned} \tag{10.4}$$

10.2.2 Design Problems for the Computational Studies

For our computational studies, we will consider the UEM design problem and a geometric programming problem [32]. We first present the general formulation of the UEM design problem that constitutes a widely used benchmark problem in the field of product family design [21, 46, 47]. In particular, two scenarios of the UEM design problems are considered. After that, we will present the specification of the geometric programming problem.

10.2.2.1 The Baseline UEM Design Problem

The UEM design problem involves 8 design variables represented by the design vector $\boldsymbol{x} = (N_c, N_s, A_{wf}, A_{wa}, I, r_o, t, L)$ [14]. The study [47] gives detailed descriptions of each design variable. We [32] have referenced the same study that formulates the design of an electric motor as a two-objective constrained optimization problem that have led to derivations of Eqs. (10.5)–(10.12) [47]. They are presented as follows, starting with the optimization problem formulation:

$$\min_{\boldsymbol{x}}\{\text{Mass}(\boldsymbol{x}), -\eta(\boldsymbol{x})\},$$

$$\boldsymbol{x} = (N_c, N_s, A_{wf}, A_{wa}, I, r_o, t, L),$$

$$\text{subject to}: \begin{cases} H(\boldsymbol{x}) & \leq 5000[\text{Amp} \cdot \text{turns/m}] \\ \frac{r_o}{t} & \geq 1 \\ \text{Mass}(\boldsymbol{x}) & \leq 2[\text{kg}] \\ \eta(\boldsymbol{x}) & \geq 15\% \\ P(\boldsymbol{x}) & = 300[\text{W}] \\ T(\boldsymbol{x}) & = a[\text{Nm}] \end{cases} \tag{10.5}$$

where Mass($\boldsymbol{x}$) and $\eta(\boldsymbol{x})$ denote the weight and efficiency calculation function, respectively. In Eq. (10.5), a denotes the user's requirement of the torque output $T(\boldsymbol{x})$, which is a real number and specified a priori. In the following, we describe in more details the formulations of $H(\boldsymbol{x})$, Mass($\boldsymbol{x}$), $\eta(\boldsymbol{x})$, $P(\boldsymbol{x})$, and $T(\boldsymbol{x})$.

First, the magnetizing intensity $H(\boldsymbol{x})$ is formulated as follows:

$$H(\boldsymbol{x}) = \frac{2 \cdot N_s \cdot I}{l_c + l_r + 2 \cdot l_{\text{gap}}}, \tag{10.6}$$

$$l_c = \pi \cdot \frac{(2 \cdot r_o + t)}{2},$$

$$l_r = 2 \cdot (r_o - t - l_{\text{gap}}).$$

Second, the Mass($\boldsymbol{x}$) is formulated as follows:

$$\text{Mass}(\boldsymbol{x}) = M_{\text{stator}} + M_{\text{armature}} + M_{\text{windings}}, \tag{10.7}$$

$$M_{\text{stator}} = \pi \cdot \left(r_o^2 - (r_o - t)^2\right) \cdot L \cdot \rho_{\text{steel}},$$

$$M_{\text{armature}} = \pi \cdot (r_o - t - l_{\text{gap}})^2 \cdot L \cdot \rho_{\text{steel}},$$

$$M_{\text{windings}} = \Big(\big(2 \cdot L + 4 \cdot (r_o - t - l_{\text{gap}})\big) \cdot N_c \cdot A_{wa} + \big(2 \cdot L + 4 \cdot (r_o - t)\big) \cdot 2 \cdot N_s \cdot A_{wf}\Big) \cdot \rho_{\text{copper}}.$$

Third, the power $P(\boldsymbol{x})$ is formulated as follows:

$$P(\boldsymbol{x}) = P_{\text{in}} - P_{\text{losses}}, \tag{10.8}$$

$$P_{\text{in}} = V \cdot I,$$

$$P_{\text{losses}} = P_{\text{brush}} + P_{\text{copper}},$$

$$P_{\text{brush}} = 2 \cdot I,$$

$$
\begin{aligned}
P_{\text{copper}} &= I^2 \cdot (R_a + R_s), \\
R_a &= \frac{\rho \cdot \left(2 \cdot L + 4 \cdot (r_o - t - l_{\text{gap}})\right) \cdot N_c}{A_{wa}}, \\
R_s &= \frac{\rho \cdot p_{\text{field}} \cdot \left(2 \cdot L + 4 \cdot (r_o - t)\right) \cdot N_s}{A_{wf}}.
\end{aligned}
$$

Fourth, the efficiency $\eta(\boldsymbol{x})$ is formulated as follows:

$$
\eta(\boldsymbol{x}) = \frac{P}{P_{in}} = \frac{P}{V \cdot I}. \tag{10.9}
$$

Lastly, the torque $T(\boldsymbol{x})$ is formulated as follows:

$$
T(\boldsymbol{x}) = K \cdot \phi \cdot I = \frac{N_c}{\pi} \cdot \phi \cdot I, \tag{10.10}
$$

$$
\begin{aligned}
K &= \frac{N_c}{\pi}, \\
\phi &= \frac{S}{R}, \\
S &= N_s \cdot I, \\
R &= R_s + R_r + 2 \cdot R_a, \\
R_s &= \frac{l_c}{2 \cdot \mu_{\text{steel}} \cdot \mu_0 \cdot t \cdot L}, \\
R_r &= \frac{l_r}{\mu_{\text{steel}} \cdot \mu_0 \cdot l_r \cdot L}, \\
R_a &= \frac{l_{\text{gap}}}{\mu_{\text{air}} \cdot \mu_0 \cdot l_r \cdot L},
\end{aligned}
$$

where

$$
\mu_{\text{steel}} = \begin{cases} -0.22791 \cdot H^2 + 52.411 \cdot H + 3115.8 & \text{if } H \leq 220, \\ 11633.5 - 1486.33 \cdot \ln(H) & \text{if } 220 < H \leq 1000, \\ 1000 & \text{if } H > 1000. \end{cases}
$$

The formulation of the UEM design problem in the form of a two-objective constrained optimization problem can be simplified by way of transformation to a single-objective optimization problem. This can be achieved if the preference between Mass($\boldsymbol{x}$) and $-\eta(\boldsymbol{x})$ is known and represented via weight coefficients associated with the two objectives. Then, one can take the weighted sum of these two objective functions that is expressed as follows:

$$f_{\text{weighted}}(\boldsymbol{x}) = w_1 \cdot (1 - \eta(\boldsymbol{x})) + w_2 \cdot \text{Mass}_{\text{normalized}}(\boldsymbol{x}), \tag{10.11}$$

where $\text{Mass}_{\text{normalized}}(\boldsymbol{x})$ represents the normalized weight that is derived by dividing the original weight with the maximum allowable weight $\text{Mass}_{\text{max}} = 2$ [kg]). The weight coefficients w_1 and w_2 are assumed to be equal, i.e. $w_1 = w_2 = 0.5$ [2].

In addition, the objective function can include a penalty term. The penalties with respect to the constraints in Eq. (10.5) that are associated with each design variable in $\boldsymbol{x}$ can be calculated. These penalties are summed and incorporated into the objective function. Their detailed calculations are given as follows:

$$\text{penalty} = p_H + p_{rt} + p_M + p_\eta + p_P + p_T, \tag{10.12}$$

where

$$p_H = \begin{cases} 0 & \text{if } H(\boldsymbol{x}) \leq 5000[\text{Amp} \cdot \text{turns/m}], \\ 1.0 + \left((H(\boldsymbol{x}) - 5000) \cdot 0.01\right)^2 & \text{otherwise,} \end{cases}$$

$$p_{rt} = \begin{cases} 0 & \text{if } r_o \geq t, \\ 1.0 + (t - r_o)^2 & \text{otherwise,} \end{cases}$$

$$p_M = \begin{cases} 0 & \text{if Mass}(\boldsymbol{x}) \leq 2[\text{kg}], \\ 1.0 + \left((\text{Mass}(\boldsymbol{x}) - 2) \cdot 100\right)^2 & \text{otherwise,} \end{cases}$$

$$p_\eta = \begin{cases} 0 & \text{if } \eta(\boldsymbol{x}) \geq 0.15, \\ 1.0 + \left((0.15 - \eta(\boldsymbol{x})) \cdot 100\right)^2 & \text{otherwise,} \end{cases}$$

$$p_P = \begin{cases} 0 & \text{if } |P(\boldsymbol{x}) - 300| \leq 5[W], \\ \left((|300 - P(\boldsymbol{x})| - 5) \cdot 0.1\right)^2 & \text{otherwise,} \end{cases}$$

$$p_T = \begin{cases} 0 & \text{if } |T(\boldsymbol{x}) - a| \leq 0.02 \cdot T(\boldsymbol{x}), \\ \left((|T(\boldsymbol{x}) - a| - 0.02 \cdot T(\boldsymbol{x})) \cdot 1000\right)^2 & \text{otherwise.} \end{cases}$$

In the calculations involved for the UEM design problem, there are various constants associated with materials among others. Table 10.1 gives the values for some of these constants that are used in the calculations.

Table 10.1 Values of constants for the UEM design problem [32]

Constant	Value	Constant	Value
V	115[V]	ρ	$1.69 \times 10^{-8}[\Omega \cdot \text{m}]$
l_{gap}	0.0007[m]	p_{field}	2
ρ_{steel}	$7850[\text{kg/m}^3]$	ρ_{copper}	$8960[\text{kg/m}^3]$
μ_{air}	1	π	3.14159

10.2.2.2 Scenario 1: Single Motor Design

In this ***single UEM design*** scenario, the aim is to design a motor with respect to one torque requirement a in Eq. (10.5). We will assume that the 8 design variables are decomposed into two groups $\{N_c, A_{wf}, I, t\}$ and $\{N_s, A_{wa}, r_o, L\}$. For each group, the respective design variables will be optimized simultaneously. This scenario is used as a case study for the monodisciplinary design problems. Furthermore, the single-objective optimization formulation is considered. In particular, we use the weighted sum function given by Eq. (10.11) as the objective function, and set both w_1 and w_2 to 0.5. The range for each design variable is set to be the same as used in [47]. They are specified in Table 10.2. Two test design problems are investigated by setting different a requirements, i.e. $a = 0.3$ for the first problem and $a = 0.5$ for the second problem [32].

10.2.2.3 Scenario 2: Overlapping Motors Design

In this scenario, two or more overlapping motors are needed to be designed, and that each motor has a different torque requirement a. Note that overlapping motors refer to different motors that share some design variables or parameters. The ***overlapping UEM design*** scenario is used as a case study for the quasi-separable MDO problems. This problem setting captures some of the important characteristics of designing a product family. As one example in the literature, the study [52] has considered designing multiple aircrafts, with each aircraft type having different mission but share a common wing design. In the context of the general UEM design problem, the overall objective for this scenario is specified as follows [32]:

$$
\begin{aligned}
&\min_{\boldsymbol{x},z} \sum_{i=1}^{m} f_{\text{weighted},i}(\boldsymbol{x}_i, z) \\
&= \min_{\boldsymbol{x},z} \sum_{i=1}^{m} \big(w_1 \cdot (1 - \eta(\boldsymbol{x}_i, z)) + w_2 \cdot \text{Mass}_{\text{normalized}}(\boldsymbol{x}_i, z)\big),
\end{aligned}
\tag{10.13}
$$

where m is the number of motors, $\boldsymbol{x}_i$ denotes the local design parameters of the ith motor, and z denotes the common design variables. We consider designing 2

Table 10.2 Range setting for motor parameters [32]

Parameter	Range
N_c	[100, 1500]
N_s	[1, 500]
A_{wf}	[0.01, 1.00] ([mm^2])
A_{wa}	[0.01, 1.00] ([mm^2])
I	[0.1, 6.0] (Amp)
r_o	[1, 10] (cm)
t	[0.5, 100] (mm)
L	[0.1, 20] (cm)

motors with different a values in Eq. (10.5) and assume they have 2 variables (t, L) in common. Three different test design problems are studied by specifying different sets of a values associated with the two different motors. They are $\{0.1, 0.125\}$, $\{0.1, 0.3\}$, and $\{0.05, 0.5\}$ for the first, second, and third problem, respectively [32].

10.2.2.4 The Geometric Programming Problem

The original ***geometric programming*** problem [3] is formulated as follows:

$$\min_{z\geq 0} z_1^2 + z_2^2$$

$$\text{subject to}: \begin{cases} z_3^{-2} + z_4^2 - z_5^2 & \leq 0, \\ z_5^2 + z_6^{-2} - z_7^2 & \leq 0, \\ z_8^2 + z_9^2 - z_{11}^2 & \leq 0, \\ z_8^{-2} + z_{10}^2 - z_{11}^2 & \leq 0, \\ z_{11}^2 + z_{12}^{-2} - z_{13}^2 & \leq 0, \\ z_{11}^2 + z_{12}^2 - z_{14}^2 & \leq 0, \\ z_1^2 - z_3^2 - z_4^{-2} - z_5^2 & = 0, \\ z_2^2 - z_5^2 - z_6^2 - z_7^2 & = 0, \\ z_3^2 - z_8^2 - z_9^{-2} - z_{10}^{-2} - z_{11}^2 & = 0, \\ z_6^2 - z_{11}^2 - z_{12}^2 - z_{13}^2 - z_{14}^2 & = 0. \end{cases} \tag{10.14}$$

However, the problem can be transformed using the equality constraints into the following formulation [32]:

$$\begin{aligned} &\min_{\boldsymbol{x}_i, z\geq 0} \sum_{i=1}^{2} f_i(\boldsymbol{x}_i, \boldsymbol{z}), \\ &\boldsymbol{x}_1 = \{z_4, z_8, z_9, z_{10}\},\ \boldsymbol{x}_2 = \{z_7, z_{12}, z_{13}, z_{14}\},\ \boldsymbol{z} = \{z_5, z_{11}\}, \\ &f_1(\boldsymbol{x}_1, \boldsymbol{z}) = z_4^{-2} + z_5^2 + z_8^2 + z_9^{-2} + z_{10}^{-2} + z_{11}^2, \\ &f_2(\boldsymbol{x}_2, \boldsymbol{z}) = z_5^2 + z_7^2 + z_{11}^2 + z_{12}^2 + z_{13}^2 + z_{14}^2 \\ &\text{subject to}: \begin{cases} (z_8^2 + z_9^{-2} + z_{10}^{-2} + z_{11}^2)^{-1} + z_4^2 - z_5^2 & \leq 0, \\ z_5^2 + (z_{11}^2 + z_{12}^2 + z_{13}^2 + z_{14}^2)^{-1} - z_7^2 & \leq 0, \\ z_8^2 + z_9^2 - z_{11}^2 & \leq 0, \\ z_8^{-2} + z_{10}^2 - z_{11}^2 & \leq 0, \\ z_{11}^2 + z_{12}^{-2} - z_{13}^2 & \leq 0, \\ z_{11}^2 + z_{12}^2 - z_{14}^2 & \leq 0. \end{cases} \end{aligned} \tag{10.15}$$

One can observe that the original problem in the form of Eq. (10.14) is transformed into a constrained quasi-separable MDO problem with common variables (z_5, z_{11}) as expressed in Eq. (10.15). This problem is used as another case study for the quasi-separable MDO problems. It is decomposed into two subproblems with duplicate variables $(z_{5,1}, z_{5,2})$ and $(z_{11,1}, z_{11,2})$ as follows [32]:

Subproblem 1

$$\min_{x_1, z \geq 0} z_4^{-2} + z_{5,1}^2 + z_8^2 + z_9^{-2} + z_{10}^{-2} + z_{11,1}^2$$

$$\text{subject to}: \begin{cases} (z_8^2 + z_9^{-2} + z_{10}^{-2} + z_{11,1}^2)^{-1} + z_4^2 - z_{5,1}^2 & \leq 0, \\ z_8^2 + z_9^2 - z_{11,1}^2 & \leq 0, \\ z_8^{-2} + z_{10}^2 - z_{11,1}^2 & \leq 0, \\ z_{5,1} - z_{5,2} & = 0, \\ z_{11,1} - z_{11,2} & = 0. \end{cases} \tag{10.16}$$

Subproblem 2

$$\min_{x_2, z \geq 0} z_{5,2}^2 + z_7^2 + z_{11,2}^2 + z_{12}^2 + z_{13}^2 + z_{14}^2$$

$$\text{subject to}: \begin{cases} z_{5,2}^2 + (z_{11,2}^2 + z_{12}^2 + z_{13}^2 + z_{14}^2)^{-1} - z_7^2 & \leq 0, \\ z_{11,2}^2 + z_{12}^{-2} - z_{13}^2 & \leq 0, \\ z_{11,2}^2 + z_{12}^2 - z_{14}^2 & \leq 0, \\ z_{5,1} - z_{5,2} & = 0, \\ z_{11,1} - z_{11,2} & = 0. \end{cases} \tag{10.17}$$

10.2.3 Use of Evolutionary Algorithms in MDO

Although an MDO problem can be solved as a single optimization problem, it would require a high integration of different disciplines that makes it not practical in some situations. One example is the situation whereby each disciplinary analysis is developed under a different specialized computer code and it becomes hard to combine these codes [51]. One general approach to solve MDO problems in such situations is distributed optimization methods based on decomposition that allow disciplinary autonomy. EAs are increasingly and widely employed as optimizers in MDO because they can be used to solve difficult optimization problems that are multimodal and nondifferentiable. In the following, we will briefly present three existing EA-based distributed optimization methods for MDO.

The first method is MORDACE that stands for Multidisciplinary Optimization and Robust Design Approaches Applied to Concurrent Engineering [15]. In MORDACE, each disciplinary optimization is performed independently. Upon the completion of a disciplinary optimization, MORDACE employs a method to deal with the common decision variables. This is because changes in common variable values due to compromise will make each disciplinary performance worse. As such, a robust design approach is used in each disciplinary optimization to alleviate the issue. A robust design approach would aim to optimize disciplinary objectives while at the same time minimize sensitivity of solution performance to variations in the common variables.

The second method is COSMOS that stands for Collaborative Optimization Strategy for Multi-Objective Systems [43]. COSMOS attempts to solve multi-objective problems in a multidisciplinary context. It performs a nested decomposition and two-level optimization, i.e. supervisor-level and disciplinary-level optimization. At the supervisor-level, the common variables are optimized to provide the values of the common variables to the disciplinary-level. Then, at the disciplinary-level, the disciplinary variables are optimized based on those provided values of common variables from the supervisor-level. The best function value is returned to the supervisor-level, which is then used as the performance value of the given common variables. One can observe that the optimization process at the disciplinary-level aims to compute the fitness of the given common variable values, and thus is a nested optimization. In COSMOS, the supervisor-level optimizer is based on the multi-objective Genetic Algorithm (MOGA) and is used to search the best common variable values based on the performance values returned from the disciplinary-level optimization.

The third method is ATC for Analytical Target Cascading [36], which constitutes a hierarchical multi-level methodology. It propagates system targets through a hierarchical structure and minimizes their unattainability [34]. When ATC is used to solve the MDO problem (see Fig. 10.1), copies of common design variables, coupling variables, and consistency constraints are created to ensure that common variables and coupling variables are consistent in the final design. In particular, ATC uses a penalty function to deal with the consistency constraints. Figure 10.2 illustrates the decomposition and coordination of ATC in solving the problem illustrated in Fig. 10.1.

In this problem setting, T represents the design targets. The duplicate common variables z^{u} are for system optimization while z^{11} and z^{12} are for disciplinary optimization. Note that at optimality, $z^{\mathrm{u}} = z^{11} = z^{12}$. $\mathrm{y}^{\mathrm{u}}_{12}$ and $\boldsymbol{y}^{\mathrm{u}}_{21}$, and $\boldsymbol{y}^{\mathrm{l}}_{12}$ and $\boldsymbol{y}^{\mathrm{l}}_{21}$ are duplicate coupling variables for system-level and disciplinary-level optimization, respectively. At optimality, one has $\boldsymbol{y}^{\mathrm{u}}_{12} = \boldsymbol{y}^{\mathrm{l}}_{12}$ and $\boldsymbol{y}^{u}_{21} = \boldsymbol{y}^{\mathrm{l}}_{21}$. System-level and discipline-level optimization procedures are implemented in turn in a cycle or loop. Each cycle is a sequence of system-level and discipline-level optimizations. Both optimization tasks communicate with each other upon completion by passing the required set of optimal values. In particular, the optimal f^{u}_1, f^{u}_2, z^{u}, $\boldsymbol{y}^{\mathrm{u}}_{12}$ and $\boldsymbol{y}^{\mathrm{u}}_{21}$ produced upon completion of the system-level optimization

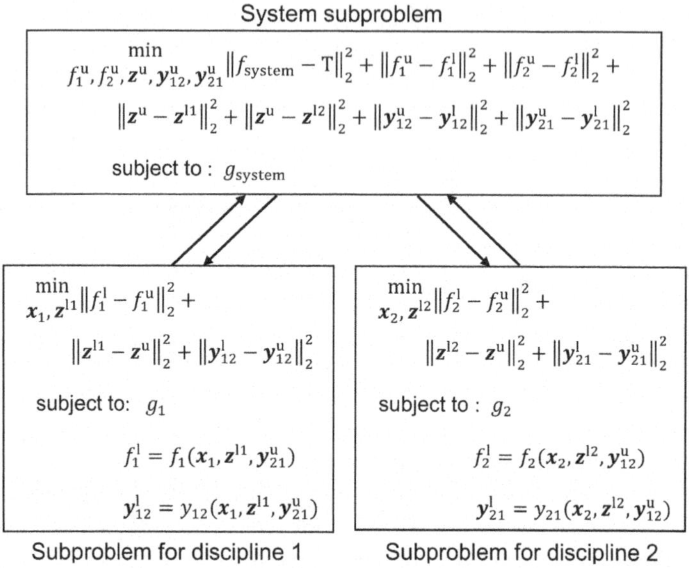

Fig. 10.2 ATC decomposition and coordination [32]

are passed to the discipline-level optimization, which in turn will pass the optimal f_1^l, f_2^l, z^{l1}, z^{l2}, y_{12}^l and y_{21}^l to the system-level optimization. The procedural optimizations that form this cycle can be iterated. In [36], the incorporation of penalty in the system optimization via ATC has a multi-objective formulation as follows [32]:

$$\begin{aligned}
&\min \| f_{system} - T\|_2^2, \\
&\min \| f_1^u - f_1^l\|_2^2 + \| f_2^u - f_2^l\|_2^2, \\
&\min \|z^u - z^{l1}\|_2^2 + \|z^u - z^{l2}\|_2^2 + \|y_{12}^u - y_{12}^l\|_2^2 + \|y_{21}^u - y_{21}^l\|_2^2, \\
&\text{subject to : } g_{system}.
\end{aligned} \tag{10.18}$$

NSGA-II [11] was applied to solve the multi-objective problem.

Cooperative CEAs have been employed to solve MDO problems [6, 7, 37, 53]. In general for this approach, the common variables of an MDO problem are decomposed into disjoint sets and assigned to different disciplines. The coupling variables are handled with duplicate variables or with the implicit iteration strategy [37]. The crucial aspect to note is that for cooperative CEA-based methods to work effectively, decisions have to be made on how the common variables should be decomposed. However, methods that could automatically perform such decomposition for general problems have not been fully developed. In [38], a decomposition approach for the common variables based on their main effects on the disciplinary objectives via orthogonal array-based experimental design is proposed and used before starting optimization process on MDO problems. In the next section, we will introduce our cooperative CEAs that are designed to solve two types of CE problems that include quasi-separable MDO problems.

10.3 Novelty-Driven Cooperative Coevolution

This section will introduce and describe in detail two cooperative CEAs that are developed to solve the two types of CE problems we have presented earlier in Sect. 10.2. We first motivate these new parallel cooperative CEAs in the next subsection. After that, we will present in turn the ***Novelty-Driven Cooperative Coevolution with Stochastic Selection*** (***NDCC-SS***) for the monodisciplinary design problems, and the ***Coevolutionary Concurrent Design Method*** (***CCDM***) for the quasi-separable MDO problems [32].

10.3.1 Novelty-Driven Cooperative CEA

In the cooperative CEA framework, the decision variables are first decomposed into several subproblems so that each subproblem is optimized separately in a genetically isolated population. Note that subproblems are also called subcomponents but in this chapter we use the term subproblem to better reflect decomposition along disciplines in CE. The optimization process of a cooperative coevolutionary Divide-and-Conquer process is procedurally implemented in the usual iterative manner with each iteration called a coevolutionary cycle (or simply cycle). In each cycle, the multiple populations with each representing candidate solutions for a specific subproblem undergo evolutionary optimization processes (e.g. with a specific EA) for a fixed number of iterations (or a fixed budget or number of function evaluations). Procedurally, these multiple populations can evolve sequentially or simultaneously, in the latter case the approach is called parallel cooperative CEA. Regardless of the two implementations, different populations communicate with each other through selected collaborative individuals from their respective current populations at the end of each cycle. These individuals are used to form the complete candidate solution of the original problem to allow evaluation of the individuals in the populations representing candidate solutions of subproblems in the next cycle. There are different methods to select such collaborative individuals [49] and the most commonly used method is to choose the best individual from each current population. Figure 10.3 illustrates a parallel cooperative CEA optimization process with some details. The decision variables are decomposed into two subcomponents. The objective function f requires as input the candidate solution for evaluation $\boldsymbol{x}_1$ ($\boldsymbol{x}_2$) and collaborative individual $\boldsymbol{x}_2^{\text{best},t_i}$ ($\boldsymbol{x}_1^{\text{best},t_i}$) selected from the respective population $\boldsymbol{P}_2^{t_i}$ ($\boldsymbol{P}_1^{t_i}$) over communication time points t_i, $i = 1, 2, 3, \ldots$. The

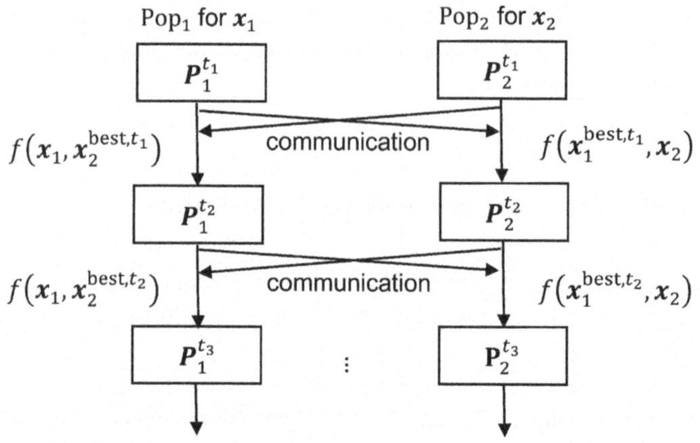

Fig. 10.3 Parallel cooperative coevolutionary optimization process [32]

communication interval denotes the difference between the generation numbers of two sequential communications in the parallel cooperative CEA as shown in Fig. 10.3. There will be more than two populations if there are more than two parts (disciplines) exist in the design, but the framework remains unchanged.

Despite the advantage offered through Divide-and-Conquer problem-solving capability, there are issues that affect cooperative CEAs from realizing their full potential. One issue involves the choice of collaboration selection strategy that is required as part of the fitness evaluation of candidate solutions in each population. In particular, the fitness evaluation depends on selected collaborative individuals obtained from other populations so as to form the complete solution. Such a fitness evaluation process can be misled by the selection of collaborators due to the presence of suboptimal equilibrium states (which is called mediocre stable states in coevolution literature) [18]. In an equilibrium state, any local change of a candidate solution with respect to a subproblem results in lower performance. There is a strong selection pressure to choose candidate solutions with respect to subproblems that form the complete solution representing that equilibrium state. Furthermore, cooperative CEAs could be affected by the issue of relative overgeneralization [40] as they identify local optima that have large basins of attraction [5] that do not correspond to the global optimum.

One approach to address the issue of cooperative CEA converging to such mediocre stable states is to incorporate ***novelty*** of individuals when scoring them as part of the fitness evaluation process [16, 18]. Such a ***novelty-based search*** in EA has been recently proposed to solve deceptive problems [28]. The original novelty search method scores individuals based on their behavioural novelty rather than fitness. This scheme encourages the evolutionary search process to continuously explore the regions of individuals with behavioural innovation rather than converging to one region. A novelty metric is proposed to measure how far an individual is from other individuals in the behaviour space [26]. This is used to calculate the novelty score of each individual as follows:

$$\mathrm{nov}(\boldsymbol{x}) = \frac{1}{k}\sum_{i=1}^{k} \mathrm{dist}(\boldsymbol{x}, \boldsymbol{\mu}_i) \tag{10.19}$$

where $\boldsymbol{\mu}_i$ is the ith nearest neighbour of the individual $\boldsymbol{x}$ according to the distance metric dist. The set of neighbours include the other individuals in the current population and optionally archived past individuals. Archived individuals could be composed of the most novel individuals [26, 29] or randomly selected individuals from every generation [27]. Novelty-based optimization has demonstrated better performance than some fitness-based optimization methods in many different applications.

As one might expect, selecting individuals based on novelty scores does not necessarily lead to high average fitness in the search outcomes [9]. This can be addressed by using both novelty and fitness objectives together to score individuals [18]. Different combination methods have been proposed in the literature [9, 17, 35]. In [9], the combination is done via a linear weighted sum of the normalized fitness $f_{\mathrm{norm}}(\boldsymbol{x})$ and novelty $\mathrm{nov}_{\mathrm{norm}}(\boldsymbol{x})$ as follows:

$$\begin{aligned} \mathrm{score}(\boldsymbol{x}) &= (1-\rho)\cdot f_{\mathrm{norm}}(\boldsymbol{x}) + \rho \cdot \mathrm{nov}_{\mathrm{norm}}(\boldsymbol{x}), \\ f_{\mathrm{norm}}(\boldsymbol{x}) &= \frac{f(\boldsymbol{x}) - f_{\min}}{f_{\max} - f_{\min}}, \\ \mathrm{nov}_{\mathrm{norm}}(\boldsymbol{x}) &= \frac{\mathrm{nov}(\boldsymbol{x}) - \mathrm{nov}_{\min}}{\mathrm{nov}_{\max} - \mathrm{nov}_{\min}}, \end{aligned} \tag{10.20}$$

where $f_{\min}$ and $f_{\max}$ are correspondingly the lowest and highest fitness values with respect to the current population, $\mathrm{nov}_{\min}$ and $\mathrm{nov}_{\max}$ are correspondingly the lowest and highest novelty scores, and ρ controls the importance of fitness and novelty and stays fixed during the optimization process. Another combination approach is to treat both novelty and fitness as two objectives to be optimized as in a multi-objective EA [35].

Both approaches have been incorporated into cooperative CEAs. The study [16] has employed the weighted sum approach and while the study [18] has used the multi-objective approach. However, novelty computations incur a high cost that makes the approach difficult for practical use in cooperative CEAs. In general, the time complexity of computing the novelty for a whole population of size P based on Eq. (10.19) without considering archived individuals is $O\big(\mathrm{P}^2 \cdot \mathrm{complexity(dist)} + \mathrm{P}^2 \log(\mathrm{P})\big)$ [32]. There is also the additional issue of employing a fixed weight ρ or multi-objective approach that attempts to balance both novelty and fitness objectives with respect to changing emphasis of the evolutionary optimization process between an exploration or exploitation phase.

10.3.2 *Novelty-Driven Cooperative Coevolution with Stochastic Selection*

NDCC-SS aims to prevent cooperative coevolutionary search to converge to suboptimal states by maintaining diversity in the selected individuals obtained from the selection process because the direction of cooperative coevolutionary search depends on these selected individuals. We design a new novelty metric in NDCC-SS that measures individuals' novelty scores based only on the individuals already selected for generating offsprings. In this manner, both novelty computation and individual selection are procedurally implemented together in NDCC-SS to improve computational efficiency. In particular, the novelty score of individual $\boldsymbol{x}$ is calculated as follows [32]:

$$\mathrm{nov}(\boldsymbol{x}) = \mathrm{dist}\left(\boldsymbol{x}, \frac{1}{k}\sum_{i=1}^{k}\boldsymbol{x}_{\mathrm{s}}^{i}\right) \tag{10.21}$$

with respect to k selected individuals $\boldsymbol{x}_{\mathrm{s}}^{i}$, $i = 1, \ldots, k$. Algorithm 10.1 procedurally describes the selection process **NOVSEL** that is based on the binary tournament selection. In particular, the cost of computing novelty has been reduced. In Algorithm 10.1, computational cost increases linearly with population size P with a time complexity of $O\big(\mathrm{P} \times \mathrm{complexity(dist)}\big)$ [32].

In addition, the procedure **NOVSEL** involves a stochastic selection process with an adaptive probability to make a trade-off between novelty and fitness objectives. This is achieved via a call to the procedure **COMPARISON**, which is detailed in Algorithm 10.2 [32]. The procedure **COMPARISON** compares two individuals $\boldsymbol{x}_1$ and $\boldsymbol{x}_2$ on the basis of their novelty $\mathrm{nov}(\boldsymbol{x}_1)$, $\mathrm{nov}(\boldsymbol{x}_1)$ and fitness $f(\boldsymbol{x}_1)$, $f(\boldsymbol{x}_2)$. The probability of selecting an individual based on novelty is p_n. This is implemented via a conditional that involves a call to the procedure **rand**(0, 1) that draws a random number in [0, 1] with uniform distribution. A schedule is developed to vary p_n with respect to the number of generations G. It is linearly decreased from p_0 to p_f within a specified number of generations and is kept constant as p_f thereafter. In particular, p_n^G is calculated as follows [32]:

$$p_n^G = \begin{cases} p_0 - (p_0 - p_f) \cdot \frac{G}{r \cdot \mathrm{MaxGen}}, & \text{if } G \leq r \cdot \mathrm{MaxGen}, \\ 0, & \text{otherwise}, \end{cases} \tag{10.22}$$

where the real parameter $r \in [0, 1]$ and MaxGen is the maximum number of generations. For monodisciplinary design problems that can be decomposed to two parts or subproblems that are known a priori, the parallel cooperative CEA framework as illustrated in Fig. 10.3 can be directly incorporated with our new novelty-driven method we have detailed here. In particular, the cooperative coevolutionary search process remains with the exception of the EA optimization

Algorithm 10.1 Novelty calculation and binary tournament selection [32]

Input:
X Current population of candidate solutions
P Population size
p_n Stochastic selection probability
Output:
Y Population of selected individuals for the next generation

procedure NOVSEL(X, P)
  $\boldsymbol{x}_s^1$:= **selBestInd**(X) ▷ $\boldsymbol{x}_s^1$ is best individual in X
  $Y[1].y := \boldsymbol{x}_s^1$
  $\boldsymbol{c} := \boldsymbol{x}_s^1$
  $k := 2$ ▷ Initialization
  while $k \leq$ P **do**
    $\boldsymbol{x}_1$:= **randSel**(X)
    $\boldsymbol{x}_2$:= **randSel**(X) ▷ Randomly select two individuals $\boldsymbol{x}_1, \boldsymbol{x}_2 \sim X$
    f_1 := **fitEval**($\boldsymbol{x}_1$)
    f_2 := **fitEval**($\boldsymbol{x}_2$) ▷ Evaluate fitness $f(\boldsymbol{x}_1), f(\boldsymbol{x}_2)$
    nov_1 := **novEval**($\boldsymbol{x}_1, \boldsymbol{c}$)
    nov_2 := **novEval**($\boldsymbol{x}_2, \boldsymbol{c}$) ▷ Evaluate novelty dist($\boldsymbol{x}_1, \boldsymbol{c}$), dist($\boldsymbol{x}_2, \boldsymbol{c}$)
    $\boldsymbol{x}_s^k$:= COMPARISON($\boldsymbol{x}_1, \boldsymbol{x}_2, f_1, \text{nov}_1, f_2, \text{nov}_2, p_n$) ▷ Select winner of $\boldsymbol{x}_1, \boldsymbol{x}_2$
    $Y[k].y := \boldsymbol{x}_s^k$
    $\boldsymbol{c} := \left((k-1)/k \star \boldsymbol{c}\right) + \left(1/k \star \boldsymbol{x}_s^k\right)$ ▷ + is for vector addition, ⋆ is scalar
    $k := k + 1$ ▷ multiplication of vector, / is scalar division
  end while
  return Y
end procedure

of the population representing the specific subproblem whereby individuals are selected based on either novelty or fitness scores according to Algorithm 10.2.

There are two advantages for employing the stochastic selection with an adaptive probability in NDCC-SS for balancing novelty and fitness objectives compared to the weighted sum and multi-objective approaches. First, stochastic selection provides a more direct and explicit scheme to determine how many comparisons are dominated by the novelty or fitness objective. In the weighted sum approach, the importance of novelty or fitness is unclear and that the parameter ρ might require tuning throughout the evolutionary process. Second, we employ an adaptive scheme that changes the emphasis of the objectives as evolutionary process proceeds. The linear schedule of decrease in p_n^G is designed to focus more on exploration in the early stage of evolutionary search and on exploitation in the later stage [32].

Algorithm 10.2 Comparison process between two individuals [32]

Input:

$\boldsymbol{x}_1$	First individual
$\boldsymbol{x}_1$	Second individual
f_1	$f(\boldsymbol{x}_1)$
nov_1	$\text{nov}(\boldsymbol{x}_1)$
f_2	$f(\boldsymbol{x}_2)$
nov_2	$\text{nov}(\boldsymbol{x}_2)$
p_n	Stochastic selection probability

Output:

$\boldsymbol{x}$	Winning individual

```
procedure COMPARISON(x1, x2, f1, nov1, f2, nov2, pn)
    if rand(0, 1) < pn then
        if nov1 > nov2 then                          ▷ nov(x1) > nov(x2)
            return x1                                ▷ nov(x1) wins
        else
            return x2                                ▷ nov(x2) wins
        end if
    else
        if f1 > f2 then                              ▷ f(x1) > f(x2)
            return x1                                ▷ nov(x1) wins
        else
            return x2                                ▷ nov(x2) wins
        end if
    end if
end procedure
```

10.3.3 *Coevolutionary Concurrent Design Method for Quasi-Separable MDO*

For our study, we consider quasi-separable MDO problems given by Eq. (10.2) that can be decomposed into two subproblems given by Eqs. (10.3) and (10.4) along two disciplines via duplicate variables z^1 and z^2 to deal with the common design variables z. After problem decomposition, CCDM optimizes each subproblem in a population with a specified EA simultaneously as illustrated in Fig. 10.3. At each communication step, the populations communicate with each other their best design information, i.e. z^i, $i = 1, 2$, in order to evaluate their separate designs.

CCDM deals with the equality constraints in Eqs. (10.3) and (10.4) by transforming them into an inequality constraint as follows [32]:

$$\frac{\sum_{k=1}^{n_c} |z_k^1 - z_k^2|}{n_c} \leq \delta, \tag{10.23}$$

where n_c is the number of common variables, z_k^1 and z_k^2 are the kth elements of z_1 and z_2, respectively, and δ is the tolerance parameter. An adaptive setting [33] is applied for δ as follows:

$$\delta(t+1) = \delta(t)/\hat{\delta}, \tag{10.24}$$

where $\delta(0)$ is set as the median of equality constraint violations, i.e. $\left(\sum_{k=1}^{n_c} |z_k^1 - z_k^2|\right)/n_c$ over all individuals in the initial population, and $\hat{\delta}$ is the rate of decay. In particular, $\hat{\delta}$ is set in a specific manner so that the δ value will decrease to a final value δ_f over a specified number of generations and then is kept fixed as δ_f thereafter [32].

A state-of-the-art constraint handling method based on stochastic ranking [45] is used to deal with the generated inequality constraints and other existing inequality constraints. In stochastic ranking, the individual with better objective value is selected when making comparisons among feasible individuals. Otherwise, the comparison is made based on the objective values of individuals with probability P_{stoch} and on the constraint violation values with probability $1 - P_{\text{stoch}}$. The constraint violation for the inequality constraint in Eq. (10.23) is defined as [32]

$$\phi(z) = \begin{cases} 0 & \text{if } \frac{\sum_{k=1}^{n_c} |z_k^1 - z_k^2|}{n_c} \leq \delta, \\ \delta - \frac{\sum_{k=1}^{n_c} |z_k^1 - z_k^2|}{n_c} & \text{otherwise.} \end{cases} \tag{10.25}$$

Note that a schedule is also developed to adaptively vary P_{stoch}, which is linearly decreased from $P_{\text{stoch}} = 0.475$ at the initial generation to $P_{\text{stoch}} = 0.25$ at the final generation. After optimization, the mean values of z^1 and z^2 in the best solutions are used as the final values of z in CCDM [32].

10.3.4 Communication Frequency Self-Adaptation

The best communication frequencies for the parallel cooperative CEA-based methods might vary with problem and communication cost. This will be studied in detail later in Sect. 10.4. Instead of performing parameter tuning on communication frequencies, we propose a ***Communication Frequency Self-Adaptation* (*CFS*)** method to automatically adjust communication frequency to a more optimal setting during the concurrent design process of NDCC-SS and CCDM [32]. The CFS method is based on the adaptation method [19] that is originally used to adapt the population size of an EA.

Algorithm 10.3 procedurally describes the CFS method. The main cooperative coevolutionary search procedure can make a call to **CFS** at the end of each cycle. Crucially, it can use the two main binary variables or parameters init and operate to control the internal implementation of **CFS**. If init $=$ true is passed to **CFS**, then the procedure will randomly initialize three communication intervals $\{p_0, p_1, p_2\}$ with positive integer values satisfying $p_1 < p_0 < p_2$. Note that in terms of design, p_0 is the main variable with $\{p_1, p_2\}$ representing its neighbours. During each operation cycle of the cooperative coevolutionary search process, in which

Algorithm 10.3 Communication frequency self-adaptation [32]

Input:

t	Current coevolutionary cycle (communication time point)
init	Initialization option flag
operate	Operation (true) or update (false) flag
sel	Communication interval selector
M	Communication intervals $\{p_0, p_1, p_2\}$ and performance improvements $\{I_0, I_1, I_2\}$
r	Setting for random integer generator

Output:

M	Updated $\{p_0, p_1, p_2\}$ and $\{I_0, I_1, I_2\}$

procedure CFS(t, init, operate, sel, **M**, **r**)
  if init = true **then**
    $\mathbf{M}[0].p := \mathbf{randInt}(\mathbf{r}[0], \mathbf{r}[2])$ ▷ Randomly generate communication
    $\mathbf{M}[1].p := \mathbf{M}[0] - \mathbf{randInt}(\mathbf{r}[0], \mathbf{r}[1])$ ▷ intervals p_0, p_1, p_2
    $\mathbf{M}[2].p := \mathbf{M}[0] + \mathbf{randInt}(\mathbf{r}[0], \mathbf{r}[1])$ ▷ satisfying $p_1 < p_0 < p_2$
    $\mathbf{M}[0].I := 0,\ \mathbf{M}[1].I := 0,\ \mathbf{M}[2].I := 0$
    return M
  else
    if operate = true **then** ▷ Use $\{p_0, p_1, p_2\}$ for a specified ▷ number of function evaluations
      $\mathbf{M}[\text{sel}].I :=$ PERIMPCAL(sel) ▷ Calculate performance improvement
      return M ▷ of selected $\{p_0, p_1, p_2\}$
    else
      case := PERIMPCMP(**M**, t) ▷ case $\in \{0, 1, 2, 3, 4\}$
      if case = 0 **then** ▷ p_0 results in largest improvement
        $q_0 := \mathbf{M}[0].p,\ q_1 := (\mathbf{M}[0].p + \mathbf{M}[1].p)/2,\ q_2 := (\mathbf{M}[0].p + \mathbf{M}[2].p)/2$
      else if case = 1 **then** ▷ p_1 results in largest improvement
        $q_0 := \mathbf{M}[1].p,\ q_1 := (\mathbf{M}[1].p)/2,\ q_2 := (\mathbf{M}[1].p + \mathbf{M}[0].p)/2$
      else if case = 2 **then** ▷ p_2 results in largest improvement
        $q_0 = \mathbf{M}[2].p,\ q_1 = (\mathbf{M}[0].p + \mathbf{M}[2].p)/2,\ q_2 := (\mathbf{M}[2].p) \times 2$
      else if case = 3 **then** ▷ All $p_i,\ i = 1, 2, 3$ have resulted in same improvement
        $q_0 := \mathbf{M}[0].p,\ q_1 := 2 \times \mathbf{M}[1].p - \mathbf{M}[0].p,\ q_2 := 2 \times \mathbf{M}[2].p - \mathbf{M}[0].p$
      else ▷ No improvement achieved from using any $p_i,\ i = 1, 2, 3$
        $q_0 := \mathbf{M}[1].p/2,\ q_1 := q_0/2,\ q_2 := (q_0 + \mathbf{M}[1].p)/2$
      end if
      $\mathbf{M}[0].p := q_0,\ \mathbf{M}[1].p = q_1,\ \mathbf{M}[2].p := q_2$
      return M
    end if
  end if
end procedure

case operate = true, each of $\{p_0, p_1, p_2\}$ is used as the communication interval for a specified number of function evaluations. Populations in NDCC-SS and CCDM communicate with each other every p_i generations, and then the obtained improvement is calculated and recorded by the appropriate setting of sel $\in \{0, 1, 2\}$ via a call to PERIMPCMP [32].

After the above process, the main procedure of the cooperative coevolutionary search can set init = false and operate = true so as to enable **CFS** to perform updates on $\{p_0, p_1, p_2\}$. These updates are based on greedy rules and are implemented in Algorithm 10.3 as five conditional cases $\{0, 1, 2, 3, 4\}$. The first three cases refer to situations where one of the communications intervals $\{p_0, p_1, p_2\}$ has resulted in the largest improvements. Procedurally, these three conditional cases correspond to the index used to label $\{p_0, p_1, p_2\}$. For example, case 0 is when p_0 has resulted in the largest improvement. For the updates to any of these three cases, p_0 is always updated with the value in $\{p_0, p_1, p_2\}$ that has led to the largest improvement. Then, two new neighbours $\{p_1, p_2\}$ in different directions are generated for p_0 and then the next cycle begins upon exiting **CFS**. Case 3 is for the situation where all the communication intervals have resulted in similar improvements, in which case the action of the greedy rule is to enlarge the neighbourhood of p_0. The last case, Case 4 is when there is no improvement achieved from using any of $\{p_0, p_1, p_2\}$. Here, the greedy rule is to enable more frequent communication between the populations [32].

Note that the procedure **PERIMPCMP** computes the improvement in performance as a result of using a particular communication frequency in the parallel cooperative coevolutionary search process. The degree of performance improvement is measured by the ratio of the improvement over either the best objective function value or the constraint violation value. This is normalized over the search effort (i.e. the number of function evaluations used). For NDCC-SS on the single motor design problem, the improvement measure is defined as follows [32]:

$$I = \frac{f^{t}_{\text{best}} - f^{t+1}_{\text{best}}}{f^{t}_{\text{best}} \cdot (\text{Evals}_{t+1} - \text{Evals}_{t})}, \tag{10.26}$$

where f^{τ}_{best} is the best objective function value at time τ, and Evals_{τ} is the number of function evaluations spent until time τ. Equation (10.26) also applies for CCDM on the overlapping motor design problems if the best solution satisfies the constraints. Otherwise, the ratio of improvement on the constraint violation value normalized by the used number of function evaluations is employed as the performance improvement measure.

Given the need to maintain a fixed budget or a number of function evaluations that can be consumed in total by any method to ensure fair comparison, we have to adjust the number of function evaluations that are consumed in one cycle of the parallel cooperative CEA for each p_i setting. This means that for each p_i, $i = 0, 1, 2$, the budget constitutes a multiple of $(p_i \times \text{N} \times \text{P} + \text{cost})$, where N is the number of populations, P is the population size, and cost is the communication cost [32]. For different p_i values, different multipliers have to be found and used to arrive at the same budget. This is a nontrivial task to be repeatedly solved for any combination of $\{p_0, p_1, p_2\}$ values.

The alternative to simply having a fixed number of functions evaluations for all p_i settings would introduce some biases. On the one hand, p_0 and p_1 settings would mean that the cooperative coevolutionary search process gets less function evaluations per cycle compared to when p_2 is used. The method might be compromised for problems associated with a preference towards small communication intervals.

On the other hand, setting too large a number of function evaluations per cycle would mean that the total budget is more quickly exhausted (particularly for the p_2 setting), in which case the number of cycles would be too small for effective self-adaptation. We employ the following mechanism to minimize the bias that consumes the following functions evaluations per cycle [32]:

1. p_2 setting: $(p_2 \times \mathrm{N} \times \mathrm{P} + \mathrm{cost})$.
2. $p_i,\ i = 0, 1$ setting: a random integer from $[a, b]$ with uniform distribution, where

$$a = \left\lfloor \frac{p_2 \times \mathrm{N} \times \mathrm{P} + \mathrm{cost}}{p_i \times \mathrm{N} \times \mathrm{P} + \mathrm{cost}} \right\rfloor \times (p_i \times \mathrm{N} \times \mathrm{P} + \mathrm{cost}),$$

$$b = \left\lceil \frac{p_2 \times \mathrm{N} \times \mathrm{P} + \mathrm{cost}}{p_i \times \mathrm{N} \times \mathrm{P} + \mathrm{cost}} \right\rceil \times (p_i \times \mathrm{N} \times \mathrm{P} + \mathrm{cost}).$$

10.4 Computational Studies

In this section, we present our computational studies [32] that are designed to systematically assess our proposed parallel cooperative CEA methods (i.e. NDCC-SS and CCDM) on two types of CE problems representing monodisciplinary design problems and quasi-separable MDO problems. In particular, we will compare our methods against some existing methods in terms of the quality of the best solution obtained with a fixed number of function evaluations. After that, we will conduct a systematic computational study to assess the impact of adapting communication frequency during the operations of NDCC-SS and CCDM via CFS.

For Scenario 1 involving the UEM design problem, the existing methods used for comparison with NDCC-SS include:

1. The original cooperative CEA without considering novelty (CC).
2. Novelty-driven CC with the weighted sum approach (NDCC-WS) [9].
3. Novelty-driven CC with the multi-objective approach (NDCC-MO) [18].

For NDCC-WS, three different weight settings $\rho \in \{0.25, 0.5, 0.75\}$ are considered. In addition, two different archive settings are applied in novelty computation: (i) Archive is not used, and (ii) archive is used with λ randomly selected individuals in the population added to the archive in every generation. Novelty computation is done with $k = 15$ and $\lambda = 6$ [9, 17]. This gives six different configurations of NDCC-WS: NDCC-WS(0.25), NDCC-WS(0.50), NDCC-WS(0.75), NDCC-WS-a(0.25), NDCC-WS-a(0.50), and NDCC-WS-a(0.75). Note that “a” means the archive is used for novelty computation. Similarly for NDCC-MO, two configurations are considered: (i) Archive is not used (NDCC-MO), (ii) archive is used (NDCC-MO-a). The binary tournament selection in both NDCC-MOs is based on the domination between two individuals, and that a random individual is selected between two nondominated individuals under comparison [32].

For Scenario 2, we compare against a baseline method that optimizes all decision variables as a whole without any decomposition. This lets us investigate the impact of problem decomposition and coordination of the common variables in CCDM for effective search. In addition, we compare CCDM with a particular COSMOS implementation that uses nested decomposition for single-objective system called COSSOS. For the geometric programming problem, the following existing methods are compared with CCDM [32]:

1. All-At-Once that optimizes all decision variables together (AAO) [22]
2. ATC method with the MO formulation (ATC-MO) [18]

We implement a real-coded GA with binary tournament selection and standard operators for simulated binary crossover and polynomial mutation [10] as the underlying EA for all the experiments to ensure fair comparisons. The crossover probability is set to 0.9, and the mutation probability is set to $\frac{1}{n}$, where n is the number of decision variables. The distribution indices [10] for the crossover and mutation operators are set to 15 and 20, respectively. For all parallel cooperative CEA-based methods, different populations communicate with each other at every generation. Note that we use the original results of AAO [22] and ATC-MO [18] to perform comparisons [32].

10.4.1 Results on Performance Comparisons

We start our presentation of results obtained from our computational studies for Scenario 1 (Single Motor Design). Given that all the methods are based on the parallel cooperative CEA framework, we set the size of each population to 50, and the maximum number of function evaluations is set to 80000. The additional parameters of NDCC-SS are set as follows: $p_0 = 0.45, p_f = 0, r = 0.4$. Each method is repeatedly and independently run 100 times. Then, we statistically pairwise compare results obtained from NDCC-SS with those from each of the other methods using Wilcoxon rank-sum test at a significance level of 0.05 [32].

Table 10.3 reports the means and standard deviations of the best function values over 100 repeated and independent runs obtained from each method, as well as the outcomes from statistical pairwise comparisons between NDCC-SS and each of the other methods on the two test problems $a \in \{0.3, 0.5\}$. The $-$ sign next to a particular entry in the table indicates that the associated method is statistically significantly worse compared to NDCC-SS. Results from Table 10.3 indicate that the proposed NDCC-SS performs the best among all methods under study on both test problems. This can be observed further in Fig. 10.4 that plots the evolutionary curves of CC, NDCC-WS-a(0.25), NDCC-MO, and NDCC-SS. In particular, NDCC-SS converges to lower average function values for the two minimization problems and requires less number of function evaluations [32].

Table 10.3 Comparisons of NDCC-SS against CC, NDCC-WS, and NDCC-MO for single motor design problems. The (−) sign denotes the method in that row is statistically significantly worse than NDCC-SS on the test problem in that column [32]

Method	$a = 0.3$	$a = 0.5$
CC	4.90e-01±1.41e-01 −	1.87e+01±6.00e+01 −
NDCC-WS(0.25)	8.73e-01±2.72e+00 −	4.56e+00±1.22e+01 −
NDCC-WS(0.50)	6.67e-01±9.93e-01 −	4.26e+00±1.37e+01 −
NDCC-WS(0.75)	2.93e+00±1.97e+01 −	6.35e+00±2.14e+01 −
NDCC-WS-a(0.25)	5.08e-01±5.45e-01 −	1.00e+00±1.86e+00 −
NDCC-WS-a(0.50)	1.14e+00±1.40e+00 −	1.56e+00±3.94e+01 −
NDCC-WS-a(0.75)	1.20e+01±2.36e+01 −	5.97e+01±9.01e+01 −
NDCC-MO	5.01e-01±4.96e-02 −	1.94e+00±3.55e+00 −
NDCC-MO-a	5.35e-01±6.93e-02 −	1.92e+00±4.38e+00 −
NDCC-SS	4.43e-01±3.59e-02	5.66e-01±3.72e-02

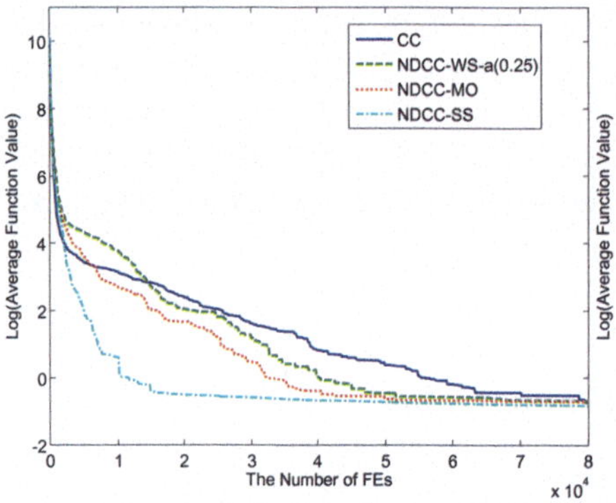

(a) $a = 0.3$

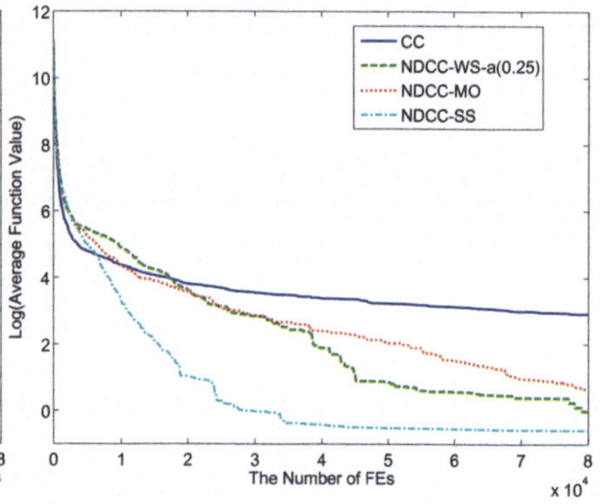

(b) $a = 0.5$

Fig. 10.4 Evolution curves of CC, NDCC-WS-a(0.25), NDCC-MO, and NDCC-SS on the single-objective case in Scenario 1 with specific a settings [32]

Next, we investigate whether the adaptive stochastic selection process and the new novelty calculation contribute to the good performance of NDCC-SS on the single motor design problems. In the first controlled experiment, we fixed the probability parameter p_n in the stochastic selection so that it is no longer adaptive in NDCC-SS. Three values for $p_n \in \{0.25, 0.50, 0.75\}$ are considered. In addition, we compare NDCC-SS with a variant that replaces the adaptive selection process with the multi-objective trade-off method that is labelled simply as "Multi-objective". Table 10.4 summarizes the results of this particular controlled experiment. In particular, NDCC-SS performs better than any other method listed in the table, thus demonstrating the advantage of the use of the adaptive stochastic selection process in NDCC-SS [32].

In the second controlled experiment, we replace the proposed novelty calculation in NDCC-SS with methods for the novelty calculation in [26] to create new variants

Table 10.4 Comparison between NDCC-SS and variants with fixed $p_n \in \{0.25, 0.50, 0.75\}$ and multi-objective trade-off method for single motor design problems. The (−) sign denotes the method in that row is statistically significantly worse than NDCC-SS on the test problem in that column [32]

Method	$a = 0.3$	$a = 0.5$
$p_n = 0.25$	5.13e-01±4.45e-02 −	6.14e-01±4.64e-02 −
$p_n = 0.50$	5.41e-01±4.68e-02 −	6.48e-01±5.29e-02 −
$p_n = 0.75$	1.54e+00±2.05e+00 −	7.73e+00±1.54e+01 −
Multi-objective	5.29e-01±4.14e-02 −	6.63e-01±1.41e-01 −
NDCC-SS	4.43e-01±3.59e-02	5.66e-01±3.72e-02

Table 10.5 Comparison between NDCC-SS and variants that employ methods specified in [26] for the novelty calculation. The (−) sign denotes the method in that row is statistically significantly worse than NDCC-SS on the test problem in that column [32]

Method Archive/No(p_0,p_f,r)	$a = 0.3$	$a = 0.5$
No(0.5,0.25,0.4)	1.32e+00±4.58e+00 −	3.96e+00±8.72e+00 −
No(0.5,0.25,0.8)	5.42e-01±3.41e-01 −	4.98e+00±1.16e+01 −
No(0.25,0,0.4)	5.45e-01±7.14e-01 −	1.57e+01±3.18e+01 −
No(0.25,0,0.8)	1.36e+00±7.09e+00 −	1.34e+01±3.53e+01 −
Archive(0.5,0.25,0.4)	4.98e-01±2.22e-01 −	1.15e+00±3.01e+00 −
Archive(0.5,0.25,0.8)	4.93e-01±1.07e-01 −	1.16e+00±2.33e+00 −
Archive(0.25,0,0.4)	5.77e-01±6.02e-01 −	1.66e+01±3.06e+01 −
Archive(0.25,0,0.8)	4.86e-01±1.06e-01 −	7.58e+00±1.95e+01 −
NDCC-SS	4.43e-01±3.59e-02	5.66e-01±3.72e-02

for comparison. These involve two archive implementation settings ("No/Archive" representing without/with the use of an archive) and used four parameter settings of p_0, p_f, r in the adaptive stochastic selection process. Table 10.5 indicates the superior search performance of NDCC-SS against these variants with lower computational cost due to the novel calculation as specified in Eq. (10.21) and implemented in Algorithm 10.1. The new novelty calculation method implemented in Algorithm 10.1 could maintain better diversity in the selected individuals than previous methods. In particular, the new method computes novelty based only on individuals in the selection process compared to previous methods that are based on the whole population before selection, i.e. as specified in Eq. (10.19). Given that the selected individuals decide the direction of the cooperative coevolutionary search, the new method might better preserve diversity in the population and prevent the coevolutionary process from converging to suboptimal states. In summary, our computational studies have demonstrated the efficacy of the new novelty calculation and the adaptive stochastic selection in NDCC-SS [32].

We now present our computational study for Scenario 2 (Overlapping Motors Design) whereby a systematic controlled experiment is done to compare CCDM with the baseline method that places all design variables in a single optimization

Table 10.6 Comparison between the baseline method, COSSOS and CCDM for overlapping motors design. The (−) sign and the (≈) sign denote the compared method is statistically significantly worse than and similar (no statistical significant difference) to CCDM, respectively [32]

a values	Baseline Method	COSSOS method	CCDM
$a_1 = 0.10, a_2 = 0.125$	3.69e+00±1.61e+01 −	2.13e+00±6.63e+00 −	6.87e-01±3.42e-01
$a_1 = 0.10, a_2 = 0.30$	4.30e+00±1.66e+01 −	3.68e+00±6.81e+00 −	2.47e+00±1.48e+01
$a_1 = 0.05, a_2 = 0.50$	1.57e+01±3.46e+01 ≈	1.51e+01±2.67e+01 −	1.51e+01±1.02e+02

process and COSSOS. For the baseline method, the population size is set to 100 and the maximum number of function evaluations is set to 32000. For CCDM, the settings involve a population size of 50 and a maximum number of function evaluations of 2 × 32000. This is because the baseline method has to evaluate both f_1 and f_2 in a function evaluation on the overlapping motor design problems with two motors during the optimization process, while in CCDM a single function evaluation is required due to the problem decomposition. Other parameter settings for CCDM are $\delta_f = 0.005, r = 0.8$. For COSSOS, the population size is 20 for the supervisor-level optimization and 50 for each disciplinary-level optimization. COSSOS evaluates only one f_i, $i = 1, 2$, in a function evaluation, in which case the maximum number of function evaluations is 2×32000 with 800 function evaluations restricted to each disciplinary-level optimization. Table 10.6 summarizes results from this controlled study and indicates that both decomposition-based methods (CCDM and COSSOS) outperform the baseline method, with CCDM obtaining better solutions on average on all test problems [32].

Last for this subsection, we present results from the computational study on the geometric programming problem. CCDM is implemented with the same setting of 0.055 for the maximum constraint violation tolerance compared to previous studies [22, 36]. The population size is 100, and the maximum number of function evaluations is 10000. Other parameter settings are $\delta_f = 0.01, r = 0.3$. The best solutions ($z_i : i = 1, \ldots, 14$) obtained from the three methods under study are (2.88, 3.00, 2.40, 0.73, 0.81, 2.77, 0.86, 0.99, 0.81, 0.76, 1.26, 0.97, 1.62, 1.59) for CCDM, (2.84, 3.09, 2.36, 0.76, 0.87, 2.81, 0.94, 0.97, 0.87, 0.8, 1.3, 0.84, 1.76, 1.55) for AAO [22] and (2.77, 3.14, 2.28, 0.76, 0.88, 2.86, 0.94, 0.96, 0.95, 0.85, 1.35, 0.84, 1.79, 1.58) for ATC-MO [36]. Table 10.7 shows the corresponding f values and the constraint violation values of g and h constraints for these three solutions. AAO is unable to satisfy all constraints (i.e. $h3$). CCDM is able to search for a solution with better f value compared to AAO and ATC-MO [32].

10.4.2 The Effect of Communication Frequency

Here, we present a systematic, controlled computational study [32] to assess the effect of communication frequency in parallel cooperative CEAs to solve CE

Table 10.7 Comparison between CCDM, AAO, and ATC-MO on the geometric programming problem [32]

	AAO [22]	ATC-MO [36]	CCDM
f	17.6137	17.5325	17.3364
g1	0.0002	0	0.0505
g2	0	0.0131	0.0473
g3	0.0078	0.0016	0.0393
g4	0.0128	0	0.0462
g5	0.0096	0.0356	0.0318
g6	0	0.0317	0.0193
h1	0.0078	0.0312	0
h2	0.0115	0.0220	0
h3	0.0550	0.0378	0
h4	0.0004	0.0490	0

problems. In particular, we investigate if there are best communication frequencies among populations, with and without communication costs, for the two scenarios of motor design problems and the geometric programming problem. We study the best communication frequencies for both NDCC-SS and CCDM that obtain solutions of the highest quality within a fixed number of function evaluations without considering communication costs. The methods are applied on the test problems with each communication interval independently for 100 times. The function values of the best solutions obtained using each communication interval over 100 runs from those methods are recorded. The parameter settings of NDCC-SS and CCDM are maintained except for the setting of communication frequencies. We consider these 12 different communication intervals $\{1, 2, 4, 5, 8, 10, 16, 20, 32, 40, 80, 160\}$ for the two scenarios of motor design problems and these 12 different communication intervals $\{1, 2, 4, 5, 8, 10, 20, 25, 40, 50, 80, 100\}$ for the geometric programming problem we used [32]. Note again that each number denotes the difference between the generation numbers of two sequential communication in the coevolutionary process.

We employ the Kruskal-Wallis test and a post hoc test of [8, 41] with significance level at 0.05 to determine which communication frequency (interval) leads to significantly good performance on each test problem. The Kruskal-Wallis test is a nonparametric multiple comparison test that compares more than two populations of random samples using rank sums. For our application, the function value of the best solution obtained in one run constitutes one sample. A population of 100 samples is obtained with respect to one communication frequency over 100 runs of the method for a test problem. The null hypothesis ($H0$) for the Kruskal-Wallis test states that all communication frequencies achieve equivalent performance. If the null hypothesis is rejected, the communication frequency with the minimum rank sum will be selected as the control frequency, and the post hoc test will be used to perform a comparison between the control frequency and every other communication frequency. We further apply Finner's adjustment of p-values [13] in this multiple hypothesis testing to control the family-wise error. Any communication

Table 10.8 Best communication intervals of CCDM on Scenario 2 and the geometric programming problem without communication cost [32]

Problem	Best communication intervals
$a_1 = 0.10, a_2 = 0.125$	1
$a_1 = 0.10, a_2 = 0.3$	1,2
$a_1 = 0.05, a_2 = 0.50$	1,2
Geometric programming	1,2,4,5,8,10,20,25,40,50,80,100

frequency whose associated performance is not significantly different from that of the control frequency will be recorded as the best communication frequency along with the control frequency [32].

For NDCC-SS on the single motor design problems ($a \in \{0.3, 0.5\}$), the best communication intervals to exchange information among populations are every one or two generations. For CCDM, the best communication intervals differ depending on the specific problems it is used to solve. Table 10.8 indicates communication in every generation or two for the overlapping motor design problems. However, CCDM performs similarly on every tested communication intervals for the geometric programming problem. In general, our computational study suggests that populations should communicate with each other as frequently as possible when applying the proposed parallel cooperative CEA methods to solve the test design problems when communication cost is not considered [32].

Next, we study the best communication frequency for the parallel cooperative CEA methods that now consider different communication costs. Here, the communication cost is considered as a fixed number of function evaluations. We vary the communication cost from $0 \times P$ to $40 \times P$ function evaluations for all test problems where P denotes the population size. These computational experiments are carried out using each communication interval from 1 to 20 generations for every communication cost considered. Tables 10.9 and 10.10 summarize the results of this computational study on the best communication intervals for NDCC-SS on the single motor design problems and CCDM on the overlapping motor design problems and the geometric programming problem, respectively. In general, there are sufficient experimental data to indicate that the best communication frequency is design problem and communication cost specific [32].

In comparing results for parallel cooperative CEAs in Tables 10.9 and 10.10 where communication costs are considered with those in Table 10.8 where there is no communication cost, we observe that the advantage of frequent communication to effective cooperative coevolutionary search is mitigated by the communication cost consideration. This is because more frequent communication will cause less time to be spent on the optimization process. However, the situation is method and problem specific. Table 10.9 indicates that one generation is always among the best communication intervals for NDCC-SS applied to Scenario 1 design problem with $a = 0.5$ even when the communication cost is increased to $40 \times P$ function evaluations. There might be a correlation between the decomposed subcomponents in which case it pays for the cooperative coevolution to communicate more frequently.

In addition, we observe that generally larger communication intervals are preferred when the communication cost becomes larger based on results in Tables 10.9 and 10.10. However, there is the trend that higher communication frequencies

Table 10.9 Best communication intervals of NDCC-SS on Scenario 1 with different communication costs [32]

Communication cost	$a = 0.3$	$a = 0.5$
2 × P	2–8	1–4
4 × P	1–20	1–8,10–11
6 × P	1–20	1–20
8 × P	1–20	1–20
10 × P	3–20	1–20
12 × P	1–20	1–20
14 × P	2,5–20	1–20
16 × P	4–20	1–20
18 × P	1–20	1–20
20 × P	3–20	1–20
40 × P	1–20	1–20

Table 10.10 Best communication intervals of CCDM on Scenario 2 and the geometric programming problem with different communication costs [32]

Cost	$a_1 = 0.10$ $a_2 = 0.125$	$a_1 = 0.10$ $a_2 = 0.30$	$a_1 = 0.05$ $a_2 = 0.50$	Geometric programming
2 × P	2–5	2–5	1–5	1–20
4 × P	2–7	2–7	4–5	1–20
6 × P	3–7	3–7	3–5	1–20
8 × P	4–8	2–8	2,4–9	1–20
10 × P	2–9,11	2–8	3–7,9	1–20
12 × P	3–11	3–6	4–8,10	1–20
14 × P	4–6, 9–10	2–7, 9–10	3–8, 12–13,20	1–20
16 × P	3–9,18,20	5,7,8,19,20	3–11,16,18–20	1–20
18 × P	8,17,20	3,5,20	3–5,7–10, 12,14–20	1–20
20 × P	5–9,12,14, 16–17,20	3–8,10,12, 13, 15–20	4–7,12, 14,16–20	1–20
40 × P	18–20	17–20	16–20	3–6, 8–20

become better again as communication cost becomes larger (e.g. one-generation communication is no longer the best at the cost of 16 × P but becomes best again at the cost of 18 × P as shown in Table 10.9 where $a = 0.3$). Since the maximum number of function evaluations is fixed for different communication cost settings, the actual available number of function evaluations for optimization reduces for increasing communication costs. This means that up to some point in communication interval settings one might observe the situation whereby frequent communication can lead to more effective search again as communication cost continues to increase. This is because frequent communication under high communication cost settings would compensate the reduced availability of function evaluation calls during the optimization phase [32].

More importantly, the results indicate that there might be opportunities to be exploited with regard to parameter tuning of communication intervals. As such, the next systematic and controlled computational study will be to assess the proposed CFS in both NDCC-SS and CCDM. A fixed communication cost of $N \times P$ is assumed. The CFS method is compared against two methods: random and fixed. In the random method, at each communication step, it randomly decides the next communication time based on one drawn with uniform distribution over a specified communication interval range $[1, 50]$. In the fixed communication method, a fixed communication interval of one generation is used. Similar settings for NDCC-SS and CCDM are used from their original performance studies. However, the maximum number of function evaluations MaxFEs is set to 2e5, with the number of generations of the evolutionary optimization process set to $\lfloor \text{MaxFEs}/(\text{N}\times\text{P}+\text{cost}) \rfloor$ [32].

Tables 10.11 and 10.12 summarize the results of the computational study where CFS is incorporated into NDCC-SS and CCDM, respectively. The CFS method performs better than both random and fixed methods in NDCC-SS on the single motor design problems. All pairwise comparisons indicate statistical significant results in favour of CFS except one involving CFS against fixed method on $a = 0.5$ problem. Similarly, the CFS method statistically significantly outperforms both random and fixed methods in CCDM on the overlapping motors design problems. However, no statistical significant results are observed comparing CFS with random and fixed methods in CCDM on the geometric programming problem. In general, the computational study suggests that the CFS method is effective in setting the appropriate communication frequency both NDCC-SS and CCDM when such information is unavailable in advance. The only exception is the case

Table 10.11 Comparison between the CFS method and the compared methods for NDCC-SS on Scenario 1. The (−) sign and the (≈) sign denote the compared method is statistically significantly worse than and similar (no statistical significant difference) to CFS, respectively [32]

Problem	Random method	Fixed method	CFS method
$a = 0.3$	1.34e+00±5.71e+00 −	4.87e-01±4.13e-02 −	4.54e-01±4.77e-02
$a = 0.5$	1.65e+01±3.62e+01 −	5.80e-01±3.94e-02 ≈	2.97e+00±1.33e+01

Table 10.12 Comparison between the CFS method and the compared methods for CCDM on Scenario 2 and the geometric programming problem. The (−) sign and the (≈) sign denote the compared method is statistically significantly worse than and similar (no statistical significant difference) to CFS, respectively [32]

Problem	Random method	Fixed method	CFS method
$a_1 = 0.10, a_2 = 0.125$	1.47e+00±3.83e+00 −	1.02e+00±2.04e+00 −	1.01e+00±1.89e+00
$a_1 = 0.10, a_2 = 0.3$	4.09e+00±1.42e+01 −	1.77e+01±1.17e+02 −	9.99e-01±2.38e+00
$a_1 = 0.05, a_2 = 0.5$	3.10e+01±2.59e+02 −	1.26e+01±6.42e+01 −	2.36e+00±1.33e+01
Geometric programming	4.20e+01±2.43e+01 ≈	4.97e+01±2.25e+01 ≈	4.78e+01±2.54e+01

of the geometric programming problem where the use of different communication intervals can lead to similar performance [32].

10.5 Discussion and Conclusion

This chapter has presented our study [32] that applies the principles of cooperative coevolution for CE design optimization. Our study has sought to leverage on the nature of CE practices whereby all design tasks are solved simultaneously and the existing methodologies for concurrent design that involves problem decomposition, optimization, and communication strategies among subproblems with that of the general framework of parallel cooperative CEA. In particular, two parallel cooperative CEA-based methods for concurrent design have been developed that are targeted on two different types of CE problems. First, the NDCC-SS method has been developed for monodisciplinary design problems with a complex structure. There are two main features in NDCC-SS: a computationally efficient novelty calculation approach and a stochastic selection process with an adaptive probability to adjust trade-off between novelty and fitness during cooperative coevolutionary search. Second, the CCDM method has been developed to address quasi-separable MDO problems. Here, duplicate variables and the stochastic ranking method are two main features in CCDM to deal with the common variables in such CE design optimization problems.

We have carried out systematic and controlled computational studies to assess the performance of these two parallel cooperative CEA-based concurrent design methods. For NDCC-SS, this has involved the UEM design problems. For CCDM, this has involved both the UEM design problems and the geometric programming problem. These design problems constitute standard benchmarks in CE studies. Our computational studies have demonstrated the effectiveness of these two new parallel cooperative CEA-based concurrent design methods when they are compared against existing methods. In the case of NDCC-SS, superior performance has been shown over existing novelty-driven cooperative CEA methods. Further computational studies have validated its two main features on novelty calculation and adaptive stochastic selection process. In the case of CCDM, superior performance is observed against baseline methods that optimize all design variables as whole and existing methods such as COSSOS in the UEM design problems involving overlapping motors and ATC-MO in the geometric programming problem.

In addition to performance evaluation and comparison with existing methods, we have carried out further investigations on one main issue in CE and cooperative coevolution that involves the setting of communication frequency between populations representing solutions for the subproblems or subcomponents. Although less frequent communication would increase parallelism and computational efficiency of the parallel cooperative CEA, it could also lead to an evolutionary optimization process of a particular population based on outdated information from other populations. The opposite might occur by having more frequent communication to ensure that the latest information from other populations is used for evolutionary optimization in a particular population, in which case this comes at a cost of computational efficiency of the parallel cooperative CEA. As such, there exists a trade-off being realized as the best communication frequency setting for NDCC-SS and CCDM. Our initial computational study has validated that the best frequency varies with the CE design problem and communication cost. Our response and motivation to address this issue further is to develop a self-adaptive method that can adjust the appropriate communication frequency automatically during the concurrent design process. This CFS method has been incorporated into both NDCC-SS and CCDM. Further computational study has verified the effectiveness of the CFS method whereby it performs better in general compared to one method that randomly sets a communication frequency over some range of values and another method that employs a fixed communication frequency of one generation.

Our study thus far has focused on those that can be formulated as single-objective design optimization problems on three standard benchmarks. In real-world settings, many CE problems could only be formulated as multi-objective design optimization problems. This provides opportunity for further research work to develop the appropriate parallel cooperative CEAs that can solve such multi-objective CE design problems. For example, more work is required to develop a different novelty calculation approach and common variable handling technique. The proposed novelty calculation in this study has prioritized computational efficiency rather than accuracy. As such, there is scope to improve on current novelty-driven cooperative CEAs.

References

1. Addo-Tenkorang, R.: Concurrent engineering: a review literature report. In: Proceedings of the World Congress on Engineering and Computer Science, pp. 1074–1080 (2011)
2. Akundi, S.V., Simpson, T.W., Reed, P.M.: Multi-objective design optimization for product platform and product family design using genetic algorithms. In: ASME 2005 International Design Engineering Technical Conferences and Computers and Information in Engineering Conference, pp. 999–1008 (2005)
3. Beightler, C.S., Phillips, D.T., Dembo, R.S., Reklaitis, G.V., Woolsey, R.E.D.: Applied Geometric Programming. Wiley, New York (1975)
4. Braun, R.D.: Collaborative optimization: an architecture for large-scale distributed design. Ph.D. thesis, Stanford University, Stanford (1996)
5. Bucci, A., Pollack, J.B.: On identifying global optima in cooperative coevolution. In: Proceedings of the 7th Annual Conference on Genetic and Evolutionary Computation (GECCO), pp. 539–544 (2005)
6. Chen, Q., Dai, J.: Distributed coevolutionary multidisciplinary design optimization: a flexible approach. In: The 9th AIAA/ISSMO Symposium on Multidisciplinary Analysis and Optimization, 4-6 September 2002, Altanta, Georgia (2002)
7. Chen, Q., Dai, J.: A distributed multi-objective evolutionary algorithm and application in missile design. In: The 2002 International Conference on Control and Automation (ICCA 2002), pp. 197–197 (2002)

8. Conover, W.J., Iman, R.L.: On multiple-comparisons procedures. Tech. rep., Los Alamos Scientific Laboratory, Los Alamos (1979)
9. Cuccu, G., Gomez, F.: When novelty is not enough. In: European Conference on the Applications of Evolutionary Computation, pp. 234–243. Springer, Berlin (2011)
10. Deb, K., Agrawal, R.B.: Simulated binary crossover for continuous search space. Complex Systems **9**(3), 1–15 (1994)
11. Deb, K., Pratap, A., Agarwal, S., Meyarivan, T.: A fast and elitist multiobjective genetic algorithm: NSGA-II. IEEE Trans. Evol. Comput. **6**(2), 182–197 (2002)
12. Dépincé, P., Guédas, B., Picard, J.: Multidisciplinary and multiobjective optimization: comparison of several methods. In: 7th World Congress on Structural and Multidisciplinary Optimization (2007)
13. Finner, H.: On a monotonicity problem in step-down multiple test procedures. J. Am. Stat. Assoc. **88**(423), 920–923 (1993)
14. Friedrich, T., Menzel, S.: A cascaded evolutionary multi-objective optimization for solving the unbiased universal electric motor family problem. In: Proceedings of the 2014 IEEE Congress on Evolutionary Computation (CEC'14), pp. 3184–3191 (2014)
15. Giassi, A., Bennis, F., Maisonneuve, J.J.: Multidisciplinary design optimisation and robust design approaches applied to concurrent design. Struct. Multidiscip. Optim. **28**(5), 356–371 (2004)
16. Gomes, J., Mariano, P., Christensen, A.L.: Avoiding convergence in cooperative coevolution with novelty search. In: Proceedings of the 2014 International Conference on Autonomous Agents and Multi-Agent Systems, pp. 1149–1156. International Foundation for Autonomous Agents and Multiagent Systems (2014)
17. Gomes, J., Mariano, P., Christensen, A.L.: Devising effective novelty search algorithms: a comprehensive empirical study. In: Proceedings of the 2015 Annual Conference on Genetic and Evolutionary Computation, pp. 943–950. ACM, New York (2015)
18. Gomes, J., Mariano, P., Christensen, A.L.: Novelty-driven cooperative coevolution. Evol. Comput. **25**(2), 275–307 (2017)
19. Hinterding, R., Michalewicz, Z., Peachey, T.C.: Self-adaptive genetic algorithm for numeric functions. In: Proceedings of the 4th Conference on Parallel Problem Solving from Nature–PPSN IV, pp. 420–429. Springer, Berlin (1996)
20. Kang, N., Kokkolaras, M., Papalambros, P.Y.: Solving multiobjective optimization problems using quasi-separable MDO formulations and analytical target cascading. Struct. Multidiscip. Optim. **50**(5), 849–859 (2014)
21. Khajavirad, A., Michalek, J.J., Simpson, T.W.: An efficient decomposed multiobjective genetic algorithm for solving the joint product platform selection and product family design problem with generalized commonality. Struct. Multidiscip. Optim. **39**(2), 187–201 (2009)
22. Kim, H.M., Michelena, N.F., Papalambros, P.Y., Jiang, T.: Target cascading in optimal system design. J. Mech. Des. **125**(3), 474–480 (2003)
23. Krishnan, V., Eppinger, S.D., Whitney, D.E.: A model-based framework to overlap product development activities. Manag. Sci. **43**(4), 437–451 (1997)
24. Kusiak, A.: Concurrent Engineering: Automation, Tools, and Techniques. Wiley, New York (1993)
25. Kusiak, A., Larson, N.: Decomposition and representation methods in mechanical design. J. Vib. Acoust. **117**(B), 17–24 (1995)
26. Lehman, J., Stanley, K.O.: Exploiting open-endedness to solve problems through the search for novelty. In: ALIFE, pp. 329–336 (2008)
27. Lehman, J., Stanley, K.O.: Efficiently evolving programs through the search for novelty. In: Proceedings of the 12th Annual Conference on Genetic and Evolutionary Computation Conference (GECCO), pp. 837–844 (2010)
28. Lehman, J., Stanley, K.O.: Abandoning objectives: evolution through the search for novelty alone. Evol. Comput. **19**(2), 189–223 (2011)
29. Liapis, A., Yannakakis, G.N., Togelius, J.: Constrained novelty search: a study on game content generation. Evol. Comput. **23**(1), 101–129 (2015)

30. Lin, J., Qian, Y., Cui, W., Miao, Z.: Overlapping and communication policies in product development. Eur. J. Oper. Res. **201**(3), 737–750 (2010)
31. Loch, C.H., Terwiesch, C.: Communication and uncertainty in concurrent engineering. Manag. Sci. **44**(8), 1032–1048 (1998)
32. Lu, X., Menzel, S., Tang, K., Yao, X.: Cooperative co-evolution-based design optimization: a concurrent engineering perspective. IEEE Trans. Evol. Comput. **22**(2), 173–188 (2018)
33. Mallipeddi, R., Suganthan, P.N.: Ensemble of constraint handling techniques. IEEE Trans. Evol. Comput. **14**(4), 561–579 (2010)
34. Martins, J.R., Lambe, A.B.: Multidisciplinary design optimization: a survey of architectures. Am. Inst. Aeronaut. Astronaut. J. **51**(9), 2049–2075 (2013)
35. Mouret, J.B.: Novelty-based multiobjectivization. In: New Horizons in Evolutionary Robotics, pp. 139–154. Springer, New York (2011)
36. Moussouni, F., Kreuawan, S., Brisset, S., Gillon, F., Brochet, P., Nicod, L.: Analytical target cascading for optimal design of railway traction system. In: International Conference on Engineering Optimization, pp. 1–10 (2008)
37. Nair, P.B., Keane, A.J.: Coevolutionary genetic adaptation–a new paradigm for distributed multidisciplinary design optimization. In: The 40th AIAA/ASME/ASCE/AHS/ASC Structures Structural Dynamics and Materials Conf., St. Louis, MO, AIAA Paper, 1999-1428 (1999)
38. Nair, P.B., Keane, A.J.: Coevolutionary architecture for distributed optimization of complex coupled systems. Am. Inst. Aeronaut. Astronaut. J. **40**(7), 1434–1443 (2002)
39. Panait, L.: Theoretical convergence guarantees for cooperative coevolutionary algorithms. Evol. Comput. **18**(4), 581–615 (2010)
40. Panait, L., Wiegand, R.P., Luke, S.: A visual demonstration of convergence properties of cooperative coevolution. In: International Conference on Parallel Problem Solving from Nature, pp. 892–901. Springer, Berlin (2004)
41. Pohlert, T.: The pairwise multiple comparison of mean ranks package (PMCMR). R package pp. 2004–2006 (2014)
42. Potter, M.A., De Jong, K.A.: Cooperative coevolution: an architecture for evolving coadapted subcomponents. Evol. Comput. **8**(1), 1–29 (2000)
43. Rabeau, S., Dépincé, P., Bennis, F.: Collaborative optimization of complex systems: a multidisciplinary approach. Int. J. Interact. Des. Manuf. **1**(4), 209–218 (2007)
44. Roemer, T.A., Ahmadi, R., Wang, R.H.: Time-cost trade-offs in overlapped product development. Oper. Res. **48**(6), 858–865 (2000)
45. Runarsson, T.P., Yao, X.: Stochastic ranking for constrained evolutionary optimization. IEEE Trans. Evol. Comput. **4**(3), 284–294 (2000)
46. Simpson, T.W.: Product platform design and customization: status and promise. AI EDAM: Artificial Intelligence for Engineering Design, Analysis and Manufacturing **18**(01), 3–20 (2004)
47. Simpson, T.W., Maier, J.R., Mistree, F.: Product platform design: Method and application. Res. Eng. Des. **13**(1), 2–22 (2001)
48. Sobieszczanski-Sobieski, J., Agte, J.S., Sandusky, R.R.: Bilevel integrated system synthesis. Am. Inst. Aeronaut. Astronaut. J. **38**(1), 164–172 (2000)
49. Stoen, C.: Various collaborator selection pressures for cooperative coevolution for classification. In: International Conference of Artificial Intelligence and Digital Communications (AIDC) (2006)
50. Tosserams, S., Etman, L., Rooda, J.: Augmented Lagrangian coordination for distributed optimal design in MDO. Int. J. Numer. Methods Eng. **73**(13), 1885–1910 (2008)
51. Tosserams, S., Etman, L.P., Rooda, J.: An augmented Lagrangian decomposition method for quasi-separable problems in MDO. Struct. Multidiscip. Optim. **34**(3), 211–227 (2007)
52. Willcox, K., Wakayama, S.: Simultaneous optimization of a multiple-aircraft family. J. Aircr. **40**(4), 616–622 (2003)
53. Xing, Y., Tang, S.: A distributed coevolutionary multidisciplinary design optimization algorithm. In: The 2010 International Joint Conference on Computational Science and Optimization (CSO), vol. 2, pp. 77–80 (2010)

54. Yang, Z., Tang, K., Yao, X.: Large scale evolutionary optimization using cooperative coevolution. Inf. Sci. **178**(15), 2985–2999 (2008)
55. Yassine, A., Braha, D.: Complex concurrent engineering and the design structure matrix method. Concurr. Eng. **11**(3), 165–176 (2003)
56. Yassine, A.A., Sreenivas, R.S., Zhu, J.: Managing the exchange of information in product development. Eur. J. Oper. Res. **184**(1), 311–326 (2008)
57. Zadeh, P.M., Toropov, V.V., Wood, A.S.: Metamodel-based collaborative optimization framework. Struct. Multidiscip. Optim. **38**(2), 103–115 (2009)
58. Zhang, X., Huang, H.Z., Wang, Z., Liu, Y., Li, Y.: A pareto set coordination method for analytical target cascading. Concurr. Eng. **21**(4), 286–295 (2013)

Part III
Coevolutionary Computation on Learning Problems

Chapter 11
Coevolutionary Learning in Noncooperative Games

11.1 Introduction to Coevolutionary Learning in Games

There are many problems in both natural and artificial settings that involve entities having to make decisions. These decision-making problems can be abstracted in the form of games. Perhaps not surprisingly, there is a close correspondence between the nature of what constitutes an iterative process to discover solutions to such decision-making problems and coevolution. Consider a ***multi-agent system*** that consists of a collection of agents that interact with each other in a pairwise manner. The setting of these interactions can be modelled by a specific game. In the case of a two-player game, rules constrain the possible behaviours the game-playing agent implements to ensure that only allowed (legal) moves are made in the course of the interaction. These decision-making behaviours (strategies) can be represented in a parametric form. Then, a specific choice of strategy representation and associated parameters prescribe the set of strategies for the game. Furthermore, the goal (objective) of the game is known and specified. Rewards in the form of payoffs are given to both players over their joint interaction (game-play). The payoffs accrued by both players determine the game outcome a player receives at the end of the game-play. As one example, a wide range of competitive games are zero-sum so that a positive outcome for a player means a negative outcome for the opposing player (e.g. the player that wins the game means the opposing player loses the game). Regardless, a detailed specification of a game means that the set of strategies is made available (e.g. to both players in the case of symmetric games) and that the outcomes for any pair of strategies adopted by the two players in the game-play are determined (i.e. they can be calculated and stored as in the case of normal form games or computed by actual and simulated game-play) [7].

In multi-agent coevolutionary systems, the collection of agents consists of an evolving population. In any one generation, agents in the population interact and ultimately compete for reproduction resources. In particular, the cumulative game outcomes an agent obtains determine in some manner the frequency it reproduces,

X. Yao, S. Y. Chong, *Coevolutionary Computation and Its Applications*,
Natural Computing Series, https://doi.org/10.1007/978-981-96-2841-4_11

i.e. how often it is selected to create offspring agents having similar or different but closely related strategies. The actual operation of replicating an offspring agent involves the application of variation whereby specific changes are made on the strategy representation parameters of the parent agent. Note that the specific parameter values not only describe the internal description of the strategy an agent implements but are also used with the strategy representation form to allow the strategy to express and mediate its specific decision-making behaviours as an input-output response function. This process is repeated for the new population in the next generation long enough that the subsequent population at the later generation would constitute agents implementing strategies that are superior compared to those in earlier generations with respect to some notions of goodness or optimality. There are two perspectives one can view such a coevolutionary process and the population of agents it produces. One could interpret such a simulated coevolution in the same way as the process of natural evolution whereby the focus is on the evolutionary success of certain strategies in the strategy set compared to others. One could also focus more on the agents and view coevolution as a learning process with which agents can adapt the strategies they choose to implement. It is with this view that ***coevolutionary learning*** is succinctly and typically described as consisting of a population of adaptive agents that learn behaviours through a reproduction process driven solely by their strategic interactions [10, 11].

A significant proportion of studies in coevolutionary learning are concerned with understanding the mechanisms of natural evolutionary processes and as a consequence develop learning procedures and systems that allow for the search of solutions for the problem at hand and in particular for decision-making problems abstracted as competitive games. For this problem-solving setting, coevolutionary learning is designed to encourage an iterative process that allocates reproduction resources towards parametric changes in the strategy representation of the agents that leads to a subset of high-performing game-playing strategies over time (generations). This is achieved through Coevolutionary Algorithms (CEAs) that search for these high-performing game-playing strategies via a generate-and-test process that optimizes the parameters associated with the strategy representation of the agents. This optimization process is guided by outcomes of competitive plays between agents in the evolving population. Successful coevolutionary learning comes about when the evolving population of competing agents enters into a sustained arms race dynamics, thereby generating agents implementing strategies with increasingly higher performance. Thus, it is crucial to note that coevolutionary learning differs in principle from classical settings in Machine Learning in a competitive game-play context. In comparison with supervised learning, there is no formal use of a training sample consisting of fixed agents implementing test strategies that is employed to evaluate performance of evolving agents in coevolutionary learning. In comparison with reinforcement learning, the adaptation process in coevolutionary learning is performed only at the end of game-plays whereby the final outcomes of games (e.g. wins) are used to guide the adaptation process of agents [7].

One example whereby coevolutionary learning has been applied to search for competitive game-playing strategies involves two-player board games of various difficulties from Checkers [7, 21], Othello [8, 9], Chess [22], and even Go [32]. Note that the approach that these studies have used is different from classical methods in Artificial Intelligence that solves the problem of two-player board games. The latter is concerned with developing effective and efficient procedures based on minimax principles to search in the game tree that generates possible alternating sequences (paths) of moves by the two players, the paths that lead to a win for the root player that initiates the search. The former still uses minimax search but only at a very shallow ply or depth (e.g. in single digit values) and typically for the purpose of ensuring that a legal move is chosen by the agent. Instead, the main motivation is to search in the very large set of strategies generated from the use of a complex, nonlinear function for board evaluations the subset of such game-playing strategies having superior performance. For example, the choice of representation for these board evaluation strategies typically involves the use of Artificial Neural Networks (ANNs) to capture complexities required in having to reason with possible patterns in the board representing current state of the game-play and then provide an appropriate response by selecting a move that favours the current player over its partner. Crucially, one of the arguments made in favour of the coevolutionary approach is that it offers the means with which adaptive agents are able to learn to play these board games to a high level with limited domain knowledge required to be incorporated into these agents. We [8, 9] have studied and observed that these coevolving ANN players exhibit strategic game-plays for Othello reminiscent of those used by high-level human players elsewhere for other board games [21, 22].

However, the coevolutionary approach is not just restricted to being employed for the purpose of searching for optimal game-playing strategies for competitive games that possess a subset of dominant strategies. In such a problem-solving setting, an agent that implements such a dominant strategy will always achieve the best outcome of the game independent to any alternate strategy implemented by the opposing agent. However, such a dominant strategy does not necessarily provide the optimal payoff. In other words, it is possible that payoffs can be simultaneously improved for all players in the game if other strategies are implemented [28]. This is not an issue for zero-sum games that are strictly competitive where the interest in these games coincides with the individual objective of a player that is about the strategy that maximizes the number of winning outcomes by ensuring better payoff compared to the partner in any encounter with an alternative strategy. There are other ***noncooperative games*** that are non-strictly competitive in the sense that there are both elements of competitiveness and cooperativeness in the interactions. In such ***nonzero-sum games***, one could interpret a competitive objective for the individual and another cooperative objective for the group of players. Games such as the ***Iterated Prisoner's Dilemma*** (***IPD***) [1–3] could model certain complex real-world interactions whereby the main interest of studies involve other notions of optimality or solution concepts such as the performance of the whole population of participating agents as a whole. For such settings, the multi-agent systems with learning approach to ***coevolutionary simulation*** enable one to conduct what-if analysis and answer questions such as establishing conditions in the interaction environments that promote adaptive agents to learn specific strategies [4, 14, 18–20, 25, 27].

This chapter is primarily concerned with this application of coevolutionary learning in noncooperative games as a simulation that we have carried out in [13]. In Sect. 11.2, we will first describe the IPD game in detail and present a brief literature review why the game has been widely used as the model of real-world interactions in various settings that range from biological, socio-economic, political, and to engineering settings. After that, we will present the more complex construction of the IPD game with more choices that better reflects real-world interactions. Section 11.3 presents the first part of our case study [13] in applying coevolutionary learning for IPD games to uncover one possible explanation why with the availability of more choices in IPD interactions, agents would learn to engage in interactions with lower cooperation levels that then lead to defection. It is not known why the mechanism of direct reciprocity (e.g. repeated encounters) that has been shown to promote the coevolutionary learning of cooperative strategies becomes less effective. Our simulated coevolution would show that agents are less able to resolve the intention of intermediate cooperative choices. They evolve to lower cooperation level plays that offer higher payoffs in the short-term view. Effectively, agents have more opportunities to exploit other agents with the availability of more choices to play. Despite this, it is known in the real-world human interactions that cooperation can be established through other mechanisms, in particular, those that involve indirect interactions (i.e. reputation). Section 11.4 presents the second part of our case study [13] whereby we incorporate reputation into IPD interactions for the coevolving agents. Results from our simulated coevolution let us uncover why and how reputation can help with agents learning to cooperate even for the IPD with more choices. In particular, agents learn to use reputation scores associated with other agents to quickly engage in mutual cooperative plays. Finally, Sect. 11.5 closes the chapter with a discussion of different implementations of reputation and their subsequent impact towards the evolution of cooperation of these adaptive agents.

11.2 Complex Iterated Prisoner's Dilemma for Real-World Interactions

In this section, we will provide an overview of the IPD games and a brief review of selected studies that have used this family of mathematical game models as abstractions of real-world interactions to develop more in-depth understanding of various phenomena that have been observed to arise from those interactions. We start with the classical IPD game involving two players who are repeatedly engaged, i.e. they have to decide which choice to play in every move over a finite number of times. Each player is given two choices to play for each move (cooperate and defect) in the classical IPD [1–3]. The payoff a player receives depends on the choices made in the joint-moves of both players. For the first player, all the payoffs are specified

$$\begin{array}{c} \\ \mathrm{C} \\ \mathrm{D} \end{array} \begin{array}{c} \mathrm{C}\,\mathrm{D} \\ \begin{pmatrix} R & S \\ T & P \end{pmatrix} \end{array}$$

Fig. 11.1 The payoff matrix for the two-player two-choice IPD game. The payoff listed is for the first or row player choosing the move. The payoff matrix for the second or the column player can be obtained by taking the transpose of the matrix illustrated in the figure

in the payoff matrix $\mathbf{G} = (g_{ij} : i, j = \mathrm{C}, \mathrm{D})$, where the index i represents the (row) first player's move and j represents the (column) second player's move.

For the classical two-player, two-choice IPD game, there are four payoff values that are associated with the four possible combinations of joint-moves. Here, $\mathbf{G} = (R, S, T, P) = (g_{\mathrm{CC}}, g_{\mathrm{CD}}, g_{\mathrm{DC}}, g_{\mathrm{DD}}),\ R, S, T, P \in \mathbb{R}_{\geq 0}$. The four payoffs R, S, T, P correspond to reward (R) for mutually cooperating, taking a sucker's (S) payoff when the other player defected, having the temptation (T) to defect, and punishment (P) for mutually defecting [3]. The payoffs in $\mathbf{G}$ must satisfy the following conditions: (1) $T > R$ and $P > S$ (defection pays more), (2) $R > P$ (mutual cooperation pays more than mutual defection), and (3) $R > (T + S)/2$ (alternating does not pay compared to mutual cooperation). The study in [4] uses $(R, S, T, P) = (3, 0, 5, 1)$, although other values can be used if they satisfy the IPD game conditions. In particular, we use $(R, S, T, P) = (4, 0, 5, 1)$ for our case studies. Note that the payoff matrix for the second player is simply $\mathbf{G}^{\mathsf{T}}$ as illustrated in Fig. 11.1.

The game is easily played. The game is usually simultaneously played whereby both players decide at the same time which choice to play in each move or round. For memory-based reactive strategies, and in particular the simple memory-one reactive strategies [12], we consider in the case study, each strategy includes a pregame archive. This archive is required to initiate the game, i.e. the play in the first round. One implementation is to store a joint-move and then feed this joint-move as input into the strategy's behavioural response function to obtain the outcome. The game is initiated when both players perform this initial move and thereafter use current joint-moves to obtain responses that serve as subsequent joint-moves. This gives a finite sequence of joint-moves that can be transformed to a finite sequence of payoffs for any player $(p_m : m = 1, \ldots, M),\ M \in \mathbb{N}$ and where each $p_m \in \{R, S, T, P\}$. The outcome for any player is obtained by taking the player's corresponding cumulative payoffs, e.g. taking the sum or average of the payoffs $(1/M)\sum_{m=1}^{M} p_m$. If average payoffs are used to decide a game outcome for a player, the outcome always takes a value that ranges over the interval $[S, T]$. Despite its simplicity in construction, this noncooperative game is able to capture the tension between rationality of individuals who are tempted to defect and rationality of the group where every individual is rewarded by mutual cooperation [2]. It is possible to simultaneously improve the outcomes for both players who mutually cooperate instead of mutually defect. However, such mutual cooperation means that cooperative players are vulnerable to exploitation by partners that implement the dominant strategy of defecting in order to gain the maximum individual payoff. The competitive-cooperative dilemma in this game seemingly extends to how one should go about reasoning the nature of its possible solutions.

The family of IPD games has become a popular model to study conditions with which the cooperative element in interactions between individuals is promoted and established against the backdrop of competitive element of the individual to self-maximize utility (payoff). These games have received much attention in many areas of study to model biological, social, economic, and political interactions in natural systems [3], as well as those involving artificial systems in engineering settings. The latter is prevalent in decentralized systems whereby entities have to decide among alternative choices, between those that minimize cost to themselves or to the group of interacting entities as a whole due to operational needs. Indeed, the ***Internet Service Provider (ISP) routing game*** provides one such compelling example [28]. Here, the interactions are directly interpretable in an engineering setting with the multi-agent system now representing a typical communication network system interconnecting more than one service provider. In the basic ISP routing game, the two players represent ISPs that need to interact with one another repeatedly, i.e. having to perform data communication across their networks due to service requests. In general, an ISP could decide to cooperate (which requires utilizing some of its own network resources) or defect (which compels nearly the full use of the other ISP's network resources). The payoff matrix of the ISP game is specified by the various costs (use of network resources) associated with routing choices. These values can be suitably transformed so that they satisfy the three IPD game conditions described earlier.

There have been many studies that have sought to identify conditions with which cooperative plays could form a viable strategy [1, 2]. These conditions could be a result of specifications imposed on the game interactions or the specific design choices implemented in the coevolutionary process. One example of conditions related to game-play interactions is the restriction of the set of strategies to those that are reactive with finite memory (i.e. the choice a strategy decides in the current move is based on previous historical moves). In another example on conditions related to the coevolutionary process, agents' interactions are assumed to be fully mixed whereby an agent interacts with all other agents in the population. It has been shown in general starting from a random population of IPD strategies that the population can be evolved towards mostly cooperative strategies. Furthermore, such an evolution of cooperation can persist for a long period of time [14] even though it is unstable [6]. For studies that have investigated why and how specific IPD strategies are successful in evolutionary settings via computer simulations, two broad approaches have been proposed. The first approach that employs ecological simulation emphasizing frequency dependent reproductions of predetermined strategies is more in line with classical evolutionary game studies [2, 5]. The second approach is based on the use of CEAs to develop multi-agent coevolutionary learning systems [4, 18–20, 25]. We use this approach as it provides a framework to study the learning effective strategies by adaptive agents through an adaptation process guided by their strategic interactions with respect to specific conditions in this multi-agent system.

The majority of early studies that have employed the coevolutionary learning approach for the IPD games [4, 14, 18–20, 25, 27] only consider those with two choices of cooperate and defect. There are several other studies that have extended the classical two-choice IPD game with more complex interactions that involve more choices. Two approaches for such ***IPD with more choices*** have been studied: (1) interactions involving a continuous level of cooperation [24] and (2) interactions involving multiple, discrete levels of cooperation [15–17]. In the second approach, the n-choice IPD is based on a simple linear interpolation of the classic two-choice IPD using the following equation [12]:

$$p_A = 2.5 - 0.5c_A + 2c_B, \quad -1 \leq c_A, c_B \leq 1, \tag{11.1}$$

where p_A is the payoff to player A, given that c_A and c_B are the cooperation levels of the choices that A (first player) and B (second player) make, respectively. The cooperation levels that c_A, c_B can take come from a set with n elements $\{\text{choice}_1, \ldots, \text{choice}_n\}$ corresponding to real values that are spaced out equally over the closed interval $[-1, 1]$, whereby $\text{choice}_1 = -1$ and $\text{choice}_n = 1$. For a four-choice IPD game, $c_A, c_B \in \{-1, -1/3, 1/3, 1\}$, where -1 represents full defection and 1 represents full cooperation.

To generate the payoff matrix for the n-choice IPD, the following conditions must be satisfied [12]:

1. For $c_A < c'_A$ and constant c_B: $p_A(c_A, c_B) > p_A(c'_A, c_B)$
2. For $c_A \leq c'_A$ and $c_B < c'_B$: $p_A(c_A, c_B) < p_A(c'_A, c'_B)$
3. For $c_A < c'_A$ and $c_B < c'_B$: $p_A(c'_A, c'_B) > \frac{1}{2}(p_A(c_A, c'_B) + p_A(c'_A, c_B))$

These conditions are analogous to those for the two-choice IPD: The first condition ensures that defection always pays more. The second condition ensures that mutual cooperation has a higher payoff than mutual defection. The third condition ensures that alternating between cooperation and defection does not pay in comparison to just playing cooperation. Figure 11.2 illustrates the payoff matrix for Player A for the four-choice IPD game. Note that any 2×2 submatrix in the generated matrix is a payoff matrix that satisfies the three IPD game conditions. Furthermore, if c_A, c_B take values with only two cooperation levels of $\{-1, 1\}$ that correspond to

$$\begin{array}{c} \\ +1 \\ +\frac{1}{3} \\ -\frac{1}{3} \\ -1 \end{array} \begin{array}{c} \begin{array}{cccc} +1 & +\frac{1}{3} & -\frac{1}{3} & -1 \end{array} \\ \begin{pmatrix} 4 & 2\frac{2}{3} & 1\frac{1}{3} & 0 \\ 4\frac{1}{3} & 3 & 1\frac{2}{3} & \frac{1}{3} \\ 4\frac{2}{3} & 3\frac{1}{3} & 2 & \frac{2}{3} \\ 5 & 3\frac{2}{3} & 2\frac{1}{3} & 1 \end{pmatrix} \end{array}$$

Fig. 11.2 The payoff matrix for the two-player four-choice IPD game. Each element of the matrix gives the payoff for the row player (Player A)

{defect, cooperate}, the generated payoff matrix $(R, S, T, P) = (4, 0, 5, 1)$ is the same as what is used for the classical IPD game.

All these studies that are based on the coevolution of agents implementing IPD strategies with more choices [15–17, 24] have employed ANNs as the main choice of strategy representation in the agents. In general, they have concluded that cooperation can still be evolved in the population when agents could implement IPD strategies with more choices. However, the evolution of cooperation is unstable [24] and more difficult to achieve compared to the game interaction setting of the classical two-choice IPD game [15–17]. This general observation is corroborated by other studies that considered different settings such as linear reactive strategies [33] as opposed to nonlinear reactive strategies that could be represented by ANNs, as well as more complex, localized agent interactions in a specific spatial configuration [23]. This raises the question as to why agents are less able to learn cooperative plays through coevolution when IPD interactions now involve more choices. One notes that any pair of agents still engages repeatedly in an IPD game-play or interaction despite the increased availability in choices to play in any one move or round. As in the case of the classical IPD with two choices, there are opportunities for cooperative plays in the population if agents reciprocate high levels of cooperation in their interactions.

As such, previous studies have sought to uncover specific conditions in the coevolutionary systems that could affect the outcome of their learning process. In particular, they have argued that certain conditions have to be met for the coevolutionary learning of cooperative agents. The coevolutionary simulations in [24] have shown that a minimum level of complexity in the strategy representation is required for the coevolution of cooperative plays. Studies that coevolve IPD strategies with multiple levels of cooperation [15–17] have shown that behavioural diversity of strategies in the population is the important factor for the coevolution of cooperative plays as opposed to genetic diversity. This is because population genetic diversity that reflects variability in parameter values of agents' strategy representations does not always correspond to population behavioural diversity that reflects variability in game-plays of agents. Agents that adopt seemingly different parameter values in their strategy representations could express similar strategic game-plays. These observations on conditions associated with the strategy representation for coevolutionary learning to realize certain behaviours (i.e. cooperative strategies) when interactions involve more choices are also made in other more complex interaction settings. We have shown that the choice of strategy representation is crucial because in some representations it is easier for the coevolutionary process to introduce and maintain behavioural diversity in the population of agents [12]. In particular, it is possible for coevolving agents to learn cooperative strategies even when their interactions are subjected to noise, in which case high levels of cooperative plays could be communicated with lower levels of cooperative plays.

However, it has been observed in complex human interactions that cooperation could be achieved and mediated by mechanisms other than direct interactions [29]. This might serve to explain the limitation and the challenge faced by the current coevolutionary learning framework for IPD games that have only involved direct interactions. The mechanism of indirect reciprocity is understood to mediate cooperation between current partners based on information how they behave in the past with others. Unlike the case of direct reciprocity where cooperation depends on repeated encounters between individuals (e.g. those modelled as IPD games), in the case of indirect reciprocity, an individual receives cooperation from third parties due to its cooperative behaviours to others. In [29], the two-player interaction is modelled as a single-round ***donation game***. Agents are randomly selected to be donors and recipients. Those who are selected to be donors are given two choices: cooperate (help) and defect (not help). The outcomes of any pair of donor-recipient interaction are (1) the donor pays a cost of c and the recipient receives the benefit b satisfying the condition $b > c$ if the donor cooperates, and (2) both donor and recipient receive zero payoff if the donor defects.

The study [29] has proposed the family of ***image scoring*** strategies and investigated how they mediate cooperative behaviours through indirect reciprocity. Here, each agent is given a value for reputation that is known to itself and its partners in the donation game interactions. There are two approaches for calculating reputation: ***image scores*** [30] and ***standing*** [26]. An agent's reputation score increases if they cooperate. The two approaches differ in terms of how reputation score is decreased when the agent defects. In image score, reputation decreases whenever the agent defects. In standing, reputation only decreases when the defection is unjustified (e.g. the recipient has a good reputation score). For image scoring strategies, their behavioural responses depend on reputation scores and some additional internal parameters used for decision-making via threshold h [26]. For example, one broad class of strategies considers only their own reputation scores for decision-making, i.e. cooperate if reputation score is at least h.

In this chapter, our case study [13] involving coevolutionary simulations of adaptive agents engaging in more complex interactions incorporating mechanisms that are believed to occur in real-world systems has very specific motivations. Section 11.3 presents the first part of our case study to better establish the manner with which the mechanism of direct reciprocity (e.g. repeated encounters) is less effective in promoting cooperation when the number of choices increases in the interactions of a coevolutionary learning system. This strongly motivates the second part of our case study that we present in Sect. 11.4, whereby we investigate the incorporation of the mechanism of indirect reciprocity (e.g. reputation) that is based on an agent's prior interactions with other agents [26, 30] as the means with which cooperation with the current and opposing agent can be established. This also extends the IPD games of almost all other studies that have used coevolutionary learning whereby the strategies only consider historical moves as the basis with which they decide on the choice to play for the current move. In particular for our case study, strategies now consider both previous moves and reputation as inputs for their decision-making. Indeed, both mechanisms of direct and indirect reciprocity exist together in complex interactions such as those found in human societies, with the mechanism of indirect reciprocity occurring as a result of direct reciprocity between other individuals [29]. As such, what our coevolutionary simulations could offer is to better reveal why and how such mechanisms could lead to agents' learning cooperative behaviours through their strategic interactions.

We have earlier investigated such ***IPD with reputation*** games in [35]. Our initial study has shown that cooperative strategies can be more easily learned through coevolution even in the case of IPD games with more choices and correspondingly shorter encounters or durations. For our implementation, each agent is tagged with a reputation or image score, which is calculated based on payoffs received from a small random sample of games that are played in the first phase. After that in the second phase, games between agents will consider both repeated encounters and reputation. Outcomes from these latter games contribute to reproduction resources of agents. In our study that we present later in Sect. 11.4, we will use actual choices that are played by the agent instead of the payoffs in order to calculate the reputation score. This new implementation better reflects the actual cooperativeness of the agent. We note that our studies are also different compared to other studies that have investigated the mechanism of indirect reciprocity in agent interactions. In particular, the game-playing strategies in [26, 30] encode reputation into their behavioural responses of strategies, i.e. only reputation is used for decision-making. Furthermore, they have employed an evolutionary game-theoretic or ecological approach whereby strategies are predetermined and their replacement depends on the proportions of existing strategies in the population.

Furthermore, our case study will address important issues with regard to the mechanism of indirect reciprocity that have been left open in our previous initial study in [35]. It is not known precisely why and how reputation helps with the learning of cooperative strategies through coevolution for the IPD games with more choices and also shorter game durations. Our study will be designed to better reveal how the mechanism of indirect reciprocity promotes cooperative plays in coevolving agents. We will also study the impact of different implementations of reputation in the coevolutionary process for the learning of cooperative strategies. Different methods for calculating reputation scores are proposed, which are based on different interpretations how an agent's reputation is estimated by other agents, thus affecting behaviours of future interactions. We focus on accurate reputation estimation in these methods that depends on two factors: (1) How memory of games from agents in previous generations can be incorporated in the calculation of reputation scores, and (2) how frequently reputation scores are updated. We will assess if increasing accuracy of reputation estimation has a significant and positive impact on the coevolutionary learning of cooperative strategies.

11.3 Challenges to Cooperation via Direct Reciprocity with More Choices

This section presents the first part of our case study [13] that employs a multi-agent coevolutionary learning system to more deeply investigate why with an increasing number of available choices for play it becomes more difficult for agents to learn highly cooperative strategies. This requires that agents implement strategies that are nice (i.e. fully cooperative) at the start of the interaction and continue to be mutually cooperative thereafter. In Sect. 11.3.1, we will present the specific design of our multi-agent coevolutionary learning system and the general details of the setup we use to run the coevolutionary simulations. Then, we will present results from coevolutionary simulations in 11.3.2 and our analysis in 11.3.3 why when there are more choices made available, strategies effectively have more opportunities to exploit others by playing lower cooperation levels. Unlike the case involving the classical IPD interactions where the mechanism of direct reciprocity allows agents to quickly reciprocate cooperation or punish defection. The addition of choices makes strategies less able to resolve the intention of an intermediate choice, e.g. whether it is a signal to engender further cooperation or a subtle exploitation. As a consequence, agents adapt to implement strategies that play lower cooperation levels to obtain higher payoffs in the short-term view when they are unable to resolve the intention of partners. In Sect. 11.3.4, we examine more closely the specific cooperative strategies that we observe in our coevolutionary simulation and analyze how they are learned by the population of coevolving agents.

11.3.1 Multi-Agent Coevolutionary Learning System for IPD

The multi-agent coevolutionary learning system we use to study the learning of game-playing strategies for the n-choice IPD game is based on the framework of coevolutionary learning of competitive games that has been employed in various studies such as [7–9, 21, 22, 32]. We start our presentation on the choice of strategy representation given that it would drive the design of other components in the coevolutionary learning system (e.g. variation operators). We use a fixed-architecture feedforward ANN based on our earlier study on coevolutionary learning of IPD games [12, 35]. We consider deterministic and reactive memory-one strategies (i.e. responds based on the moves made by the strategy and the partner in the previous round) to allow more direct comparisons with previous studies.

In particular, the ANN is a ***Multilayer Perceptron*** (***MLP***) with four input nodes, ten hidden nodes in the single hidden layer, and one output node. The four input nodes take in the following values [13]:

1. The agent's previous choice
2. The partner's previous choice
3. An input of $+1$ if the partner played a lower cooperation level compared to agent and 0 otherwise
4. An input of $+1$ if the agent played a lower cooperation level compared to the partner and 0 otherwise

The last two inputs are to enable the MLP to more easily compare cooperation levels of the agent against its partner. Each node (perceptron) is associated with input weights and a single bias. For studies on the n-choice IPD game, the MLP has a total of $N_w = 61$ connection weights. This comes about from $(4 + 1)10$ for the nodes in the single hidden layer and $(10 + 1)1$ for the single output node. The activation function is a tanh scaled to output values in $[-1, 1]$. After that, the MLP output is rounded to the nearest value defined by the choices of the n-choice IPD game.

Algorithm 11.1 General CEA for coevolutionary learning of IPD games [13]

Input:

Y	Population of candidate solutions
λ	Size of parent population
T	Number of generations
n	Number of choices in the IPD game

Output:

X	Population of evolved candidate solutions

procedure CEA(Y, λ, T)
  $t := 1$ ▷ Initialize time-stamp and loop
  $X :=$ **initialize**(Y) ▷ Initialize parent population in X
  while $t \leq T$ **do**
    $X :=$ **variation**(X) ▷ Apply variation operator on each $X[i],\ i = 1, \ldots, \lambda$ ▷ to generate child population $X[i],\ i = \lambda + 1, \ldots, 2\lambda$
    roundTour$(X, 2\lambda, n)$ ▷ Round robin tournament on population X to obtain ▷ fitness evaluation of agent $X[i].f,\ i = 1, \ldots, 2\lambda$
    $X :=$ **selection**(X) ▷ Apply selection operator on population X by sorting ▷ it according to $X[i].f$ in decreasing order
    $t := t + 1$
  end while
  return X
end procedure

After specifying the specific form of the strategy representation, we employ a CEA based on variations of real-valued weights of ANN [34] for the learning process of the adaptive agents. The CEA that we use here is based on the method we have studied earlier in [12]. In general, the CEA is a modification on self-adaptive $(\lambda + \lambda)$ Evolutionary Programming [36], with the main difference being that fitness evaluations of candidate solutions (agents) for the CEA are based on the cumulative game outcomes from interactions with all other agents in the population (e.g. round-robin tournament). A single-population competitive CEA is used given that the IPD games are symmetric, i.e. all agents can implement strategies from the same strategy set. Algorithm 11.1 specifies the procedural implementation of this CEA. The main procedure **CEA** takes in the population of agents $\boldsymbol{Y}$ that specifies for each agent $i = 1, \ldots, 2\lambda$ the data structure for its strategy representation (MLP) in the vector form $\boldsymbol{Y}[i].\boldsymbol{w}$ with associated methods for computing the input-output responses. Given the use of self-adaptive mutation, each agent is further equipped with the self-adaptive parameter vector $\boldsymbol{i Y.\sigma}$ The fitness of each agent is stored in $\boldsymbol{Y}[i].f$. Two other parameters that are passed to **CEA** include the size of the parent population $\lambda = 15$ and the number of generations $T = 600$. These parameters are chosen so that observations can be made on the coevolutionary simulations (e.g. persistent periods of cooperation or defection). Each experiment is repeated for 30 independent runs.

We provide more details of the other procedures in Algorithm 11.1. The initial parent population of agents is created via the procedure **initialize**. In particular, for each parent agent $\boldsymbol{X}[i]$, $i = 1, \ldots, \lambda$ [13]:

1. Vector of weights and biases $\boldsymbol{w} = (w_j : j = 1, \ldots, N_w)$: Each $w_j = u \sim U[-1, 1]$ is initialized with a random value drawn with uniform distribution over the interval of reals $[-1, 1]$.
2. Self-adaptive parameter vector $\boldsymbol{\sigma} = (\sigma_j : j = 1, \ldots, N_w)$: Each $\sigma_j = 0.05$ is initialized with the same fixed value of 0.05.
3. Pregame inputs (pregame_1, pregame_2): Each $\text{pregame}_k = \text{choice} \sim U\{\text{choice}_1, \ldots, \text{choice}_n\}$ is initialized with a random value drawn from a discrete uniform distribution over the set of n-choices $\{\text{choice}_1, \ldots, \text{choice}_n\}$.

The procedure **variation** is applied to the parent population $\boldsymbol{X}[i]$, $i = 1, \ldots, \lambda$, to produce the child population $\boldsymbol{X}[i]$, $i = \lambda + 1, \ldots, 2\lambda$. In particular, a ***self-adaptive Gaussian mutation*** operator is used to generate the additive perturbations on weights and biases of the MLP. The operator is applied to the associated parameters on each parent agent $\boldsymbol{X}[i]$ to produce the corresponding child agent $\boldsymbol{X}[\lambda + i]$. For each $i = 1, \ldots, \lambda$ [13],

$$\boldsymbol{X}[\lambda + i].\boldsymbol{\sigma}[j] = \boldsymbol{X}[i].\boldsymbol{\sigma}[j] \times \exp\big(\tau \times \textbf{randNorm}(0, 1)\big), \; j = 1, \ldots, N_w,$$
$$\boldsymbol{X}[\lambda + i].\boldsymbol{w}[j] = \boldsymbol{X}[i].\boldsymbol{w}[j] + \boldsymbol{X}[\lambda + i].\boldsymbol{\sigma}[j] \times \textbf{randNorm}(0, 1), \; j = 1, \ldots, N_w,$$

where $N_w = 61$, $\tau = (2(N_w)^{0.5})^{-0.5} \approx 0.2530$. The repeated calls to the procedure **randNorm**(0, 1) constitute sampling anew from the standard Gaussian distribution (zero mean and standard deviation of 1) $N(0, 1)$ for a random value in each call $j = 1, \ldots, N_w$. Each pregame input pregame_k is mutated separately. Mutation is of the form of adding the original value with a step value whose probability is described by a probability mass function that is approximately a standard Gaussian distribution. This is so that smaller step sizes in either increasing or decreasing direction are more likely to be chosen. The new value does not wrap around so that values greater than $+1$ are changed to $+1$, while values lower than -1 are changed to -1.

The procedure **roundTour** performs a round-robin tournament on the full population $\boldsymbol{X}$, and all agents engage in a pairwise manner game-play with all other agents (including itself). Every agent competes a total of 2λ games. The fitness value is simply the sum of the outcomes of all 2λ games and then is stored in $\boldsymbol{X}.f$. The procedure **selection** will take the best λ agents based on their fitness values. The implementation involves sorting $\boldsymbol{X}$ in a decreasing manner according to values of $\boldsymbol{X}[i].f$, $i = 1, \ldots, 2\lambda$, so that $\boldsymbol{X}[i]$, $i = 1, \ldots, \lambda$, could form the parent population for the next generation.

11.3.2 Revisiting the IPD with More Choices: Evolution of Defection

Here, we present our case study [13] that aims to reexamine the issue that agents are more likely to learn to defect through coevolution when there are more choices to play in the IPD game interactions. Most past studies that have employed coevolutionary simulations [15–17, 24] have suggested that the mechanism of direct reciprocity is less effective in promoting the coevolutionary learning of cooperative. We start with two sets of controlled experiments in our coevolutionary simulations. The first experiment set uses the 64-choice IPD, while the second experiment uses the 4-choice IPD. Both sets of experiments have the game duration of the IPD games being fixed to ten rounds. This means that the number of choices is larger than the number used for the game duration (e.g. 64-choices compared to 10 rounds) in the first experiment set. In the second experiment set, the opposite occurs (e.g. 4-choices compared to 10 rounds). These experiments allow us to study whether the mechanism of direct reciprocity in the coevolutionary learning of IPD games with significantly more choices is as effective as in the setting of IPD games with a small number of choices. Note that for analysis purposes we classify the run as one with (1) a mutual defection outcome if the final population average payoff is at most 1.5, (2) a mutual cooperation outcome if the final population average payoff is at least 3.5, and (3) an intermediate cooperation outcome if the population average payoff is between 1.5 to 3.5. In this way, we use the outcome labels in the legend of subsequent figures of "Defection", "Intermediate", and "Cooperation" to refer to the number of runs with mutual defection, intermediate, and mutual cooperation outcomes, respectively. Whenever we perform statistical testing for comparison, we employ a paired Wilcoxon Signed-Rank test at a 0.05 level of significance and then report the mean and standard deviation of the result for the respective experiment set.

Figure 11.3 summarizes the results of both sets of experiments. Our results have confirmed observations that have been made in past studies [15–17, 24] whereby populations are more likely to coevolve to mutual defection plays for IPD interactions with 64-choices where the number of choices is larger compared to the number used for the game duration. In particular, Fig. 11.3 shows that 25 out of 30 runs have resulted with mutual defection plays for 64-choice IPD game interaction setting. This stands in contrast to the setting with four choices whereby 9 out of 30 runs have resulted with such mutual defection plays in the population. The comparison of outcomes via the population average payoffs of individual runs from the two experiment sets indicates a statistical significant difference in the outcomes of the two experiments (2.27 ± 1.13 for 4-choices IPD against 1.37 ± 0.84 for 64-choices IPD) [13].

Next, we perform a deeper analysis on actual plays made by the coevolved agents by collecting statistics that describe behavioural responses collected from the best (i.e. having the highest fitness) agents in the final coevolved populations of all individual runs. Our results indicate more runs where coevolved agents implement

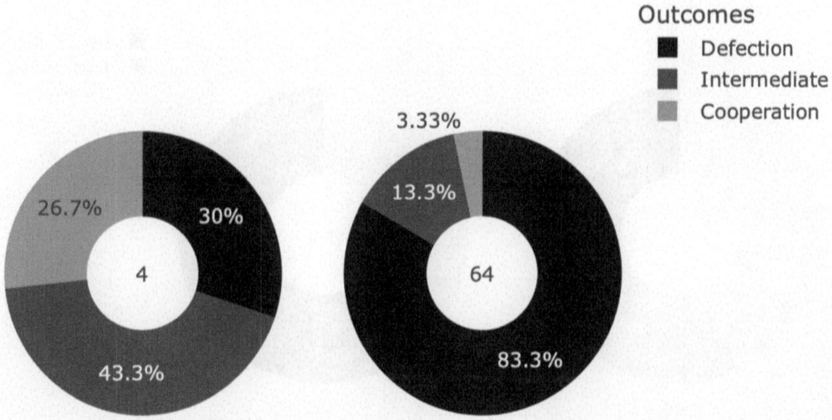

Fig. 11.3 Comparison of results from the experiments that used the IPD with 4-choices and 64-choices. Results taken at the end of the coevolutionary learning runs

strategies that are nasty (e.g. play lower cooperation levels at the start of the game) for the setting of the 64-choice IPD compared to the setting of the 4-choice IPD (i.e. 26 versus 19 runs). Furthermore, there are more runs where coevolved strategies respond with lower cooperation levels if previous moves indicated mutual cooperation play for the setting of the IPD game with more choices at 64 (e.g. 24 versus 13 runs). This indicates that highly cooperative play is more difficult to be achieved in the population because coevolved agents are not willing to continuously engage in mutual cooperation. However, the result also shows that these agents are not naive (e.g. respond with higher cooperation levels if previous moves indicated mutual defection play) in all the runs for both experiment sets. They implement strategies that respond with choices leaning towards defection if mutual defection occurred in the previous round [13].

We have performed close inspection of these agents in the final coevolved populations by individually examining the input-output responses of the strategies in the final coevolved population. In most runs where the population has coevolved towards defection for the 64-choice IPD experiment, we have observed that agents implement strategies that mostly start the game with low cooperation levels and then continue to respond with low cooperation levels. In addition, most of the agents in the population have also implemented similar strategies [13].

Our coevolutionary simulations have indicated that the mechanism of direct reciprocity is less effective in promoting the learning of cooperative behaviours when the number of choices increases. This is more apparent when the number of choices is more than the game duration. As such, we have conducted two more controlled experiment sets to clarify this point. We have modified Eq. (11.1) from $p_A = 2.5 - 0.5c_A + 2c_B$ to $p_A = 2.5 - 0.25c_A + 2.25c_B$. This is to reduce the penalty given for playing higher cooperation levels in the IPD payoffs and should encourage the coevolutionary learning of cooperative plays. Note that for the classical IPD,

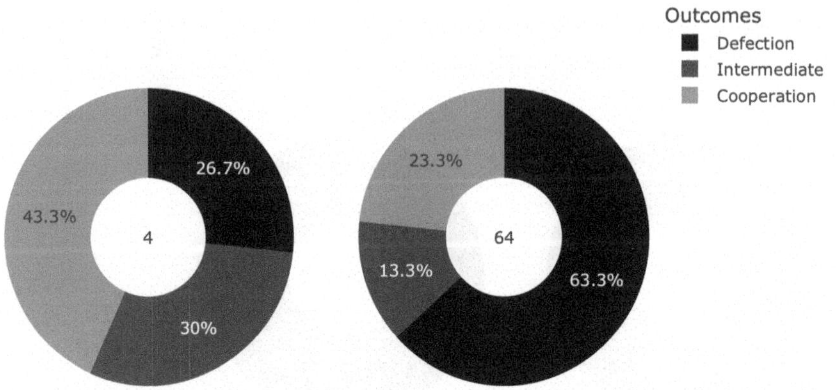

Fig. 11.4 Comparison of results from the experiments that used the IPD with 4-choices and 64-choices with payoffs that penalize higher cooperation levels less. Results taken at the end of the coevolutionary learning runs

the resulting payoffs are $(R, S, T, P) = (4.5, 0.5, 5, 1)$. Figure 11.4 summarizes results from these two experiment sets. Cooperation outcomes are more likely to be obtained for these new experiment sets, each corresponding to the IPD games with 64- and 4-choices. However, in general the majority of the runs in the coevolutionary simulations result in populations that coevolved towards low cooperation plays and leaning to defection [13].

11.3.3 More Choices Create More Opportunities to Exploit Partners

Our aim here is to investigate specifically why more choices lead to the coevolutionary learning of agents that implement strategies that play defection. We collect statistics how strategies that agents implement sample available choices of play during interactions in coevolutionary simulations involving 64-choice and 4-choice IPD games. First, we calculate the ***entropy*** on the distribution of sampled responses for play in an agent's implemented strategy behavioural response map. For an n-choice IPD game, there are $n_t = n^2$ elements in the behavioural response map of a memory-one reactive strategy that consider joint-move in the previous round of the partner and its own. The entropy is calculated as follows: [13]:

$$-\sum_{l}^{n_t}\left(\frac{m_l}{M}\cdot\mathbf{log}_2\frac{m_l}{M}\right), \tag{11.2}$$

where m_l/M is the frequency that the element l of the strategy map is sampled with respect to a total of $M = \sum_l^{n_t} m_l$ responses. Equation (11.2) is applied to the best agent in each generation so as to monitor how implemented strategies respond in game-plays during the coevolutionary learning process. We use unnormalized entropy values because our analysis is directed towards how much more responses are sampled, and hence more alternative choices are played. Normalized entropy would indicate the proportion of available choices that are played.

A plot on the average entropy values over 30 runs throughout 600 generations of the coevolutionary learning process of IPD games with 4- and 64-choices has been carried out [13]. Entropy as calculated using Eq. (11.2) provides a quantitative measure on how agents *invest* interaction opportunities so as to play alternative choices. Higher entropy values indicate more alternative responses to different combinations of previous moves and hence how different choices are sampled for play by a strategy that the agent implements. Our results show that the average entropy values over 30 runs taken at the final generation are 0.68 and 1.34 for the IPD with 4- and 64-choices, respectively. This suggests in general that more choices are played for the setting involving the 64-choice IPD game compared with the setting with 4-choice IPD game [13].

We note a very rapid decrease in average entropy value for the 64-choice IPD experiment set. Closer inspection of individual runs reveals that the populations coevolved to play lower cooperation levels more frequently after starting from initial random populations. This change from an initial, well-spread distribution of the frequency with which choices are sampled for play to a narrower distribution during the early period of the coevolutionary learning process can be observed in Fig. 11.5, which provide plots of these changes in distributions for four sample runs. Although more choices are sampled for play, they are all at low cooperation levels close to defection. This is in contrast with coevolutionary simulations involving the 4-choice IPD, whereby there are runs where populations have coevolved to play intermediate choices and mutual cooperation.

One would expect similar opportunities for encouraging cooperation and exploiting partners when more choices are made available in the IPD interactions. Although payoffs are higher for exploiting a cooperative partner than mutual cooperation in the short term (e.g. for that particular round itself), there is the possibility agents can learn to implement strategies that leverage repeated interactions and use the mechanism of direct reciprocity to reciprocate cooperation and punish defection as the means to establish mutual cooperative plays. However, deeper analysis of our coevolutionary simulations has shown that this is not true. Mutual defection outcomes are always more likely through coevolutionary learning process compared to mutual cooperation outcomes when there are more choices to play. Note that the coevolutionary learning process is driven by strategic interactions of agents based on the cumulative IPD payoffs they receive from of all games they have played. As such, results from the coevolutionary simulations would suggest that there are more incentives for agents to adapt their strategic responses to exploit their partners rather than to engender cooperation. In other words, there are effectively more opportunities to exploit partners for interactions with more choices available for play. This makes the mechanism of direct reciprocity less effective in promoting the coevolutionary learning of cooperative behaviours [13].

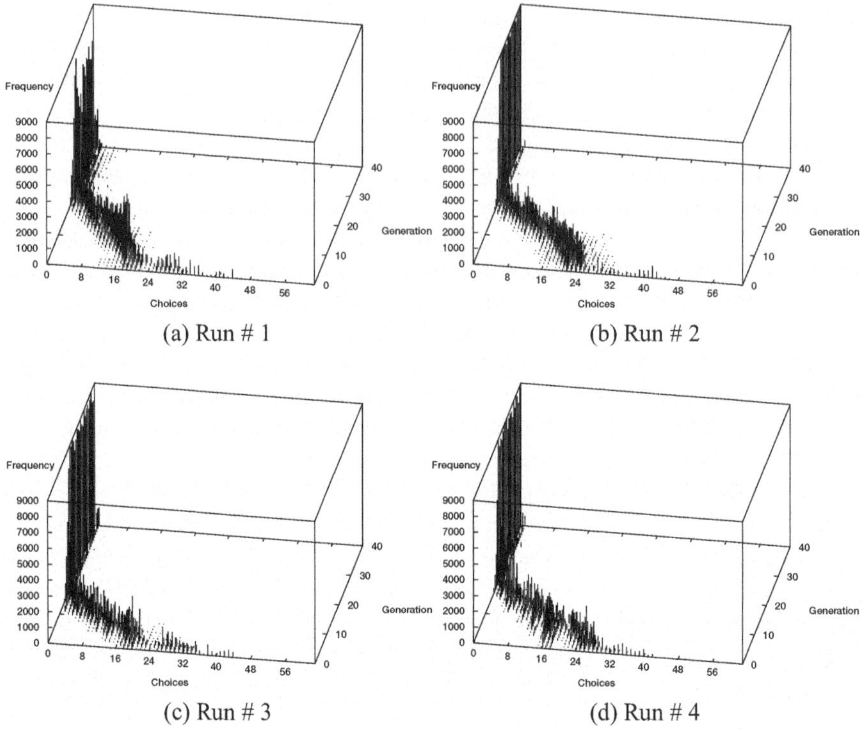

Fig. 11.5 Sampling distribution of choices of the populations of the 64-choice IPD experiment. Choices are represented from 0 (full defection) to 63 (full cooperation). A higher vertical line for a particular choice and a particular generation indicates a higher sampling frequency for the choice [13]

One explanation for such a decrease in the effectiveness of the mechanism of direct reciprocity in coevolutionary learning is that strategies are less able to resolve an intermediate choice whether as a signal to engender further cooperation or a subtle exploitation in an interaction setting with more choices available for play. This is despite having repeated encounters because a moderate increase in the number of choices leads to a substantially larger increase in the possible sequences of choices agent plays, which makes resolving the intention of intermediate choices via the game outcome (cumulative payoff) more difficult. As such, we have conducted additional experiments that could verify this explanation by increasing game durations for the 64-choice IPD. Results from our coevolutionary simulations in Fig. 11.6 indicate that with longer game durations, the tendency for the learning of strategies with low cooperation levels is reduced. The number of runs where populations have coevolved to defection has reduced from 25 to 15 after game duration is increased to 30 (note: no further decrease is observed for game durations lasting 150 rounds). In particular, the average payoffs and standard deviations of the experiments involving 30 and 150 rounds are 1.89 ± 0.94 and 1.80 ± 0.91,

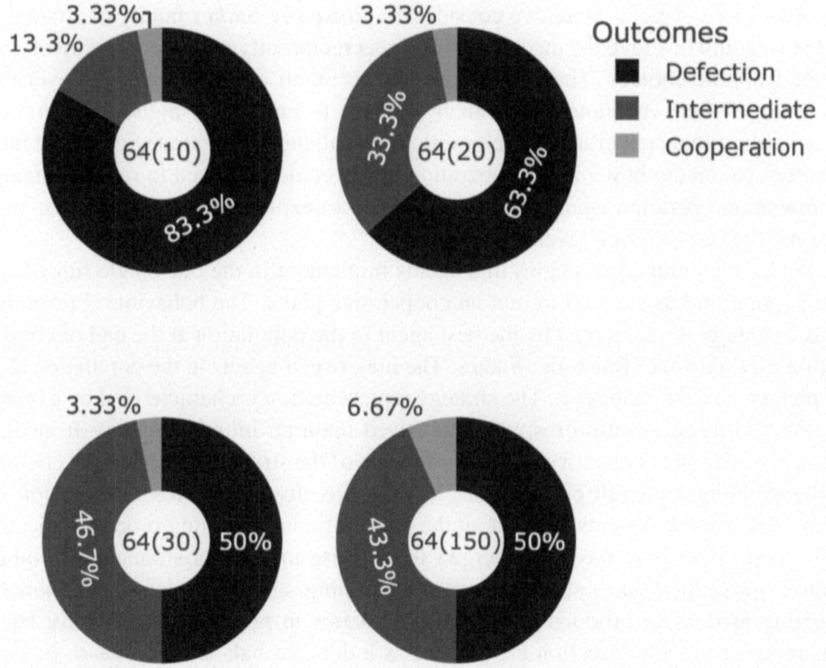

Fig. 11.6 Comparison of results from the experiments that used the 64-choice IPD with longer game durations. Results taken at the end of the coevolutionary learning runs

respectively. These are compared against those obtained when ten rounds are used in the games where the result is 1.37 ± 0.84 [13]. However, the reduction in mutual defection outcomes in the populations is replaced by those that engage in intermediate cooperation levels. This is further corroborated by analysis on how agents have sampled alternative choices for play, with higher average entropy measurements on the coevolving populations with longer game duration settings. In particular, the average entropy values over 30 runs taken at the final generation are 1.34, 1.96, 2.16, and 2.02 for the 64-choice IPD with game duration of 10, 20, 30, and 150, respectively [13].

11.3.4 Is Cooperative Play Still Possible?

One might ask at this point whether it is still possible for agents to learn to engage in highly cooperative plays through the mechanism of direct reciprocity. Unlike ***Tit-for-Tat*** family of strategies in the classical IPD game setting that has properties to enable populations of agents to learn to engage in mutual cooperation from initial defection plays, some other strategies are required when the interaction setting involves more choices. Here, we consider the ***Raise-the-Stakes*** family of strategies [31] that could leverage the mechanism of direct reciprocity in interactions involving more available choices. These strategies increase their cooperation levels over the duration of their encounters with their partners to encourage higher cooperative play until mutual cooperation is achieved. The challenge these strategies would face involves balancing how much cooperation levels could be raised to rapidly engage in mutual cooperation against being vulnerable to exploitation by others that stop raising their cooperation levels midway [31].

We have examined strategies that agents implement in the one single run where the population has evolved to mutual cooperative plays. The behavioural response of the strategy implemented by the best agent in the population at the end of coevolution uses a form of Raise-the-Stakes. The majority of agents in the population also implement similar strategies. The strategy representation is characterized by a broad region of full cooperation responses achieved upon an initial steep rise from full defection, which is indicative of some tolerance of the strategy to alternative choices by responding with full cooperation. Despite this discovery, closer inspection of responses made by agents throughout the coevolutionary learning process indicates that cooperative play has not emerged from Raise-the-Stakes strategies. In other words, Raise-the-Stakes strategies are not stepping stones for the coevolutionary learning process to produce cooperative strategies in populations that have been engaging in mutual defection [13]. There is a delicate balance that has to be met in how fast a strategy can raise the level of cooperation without making itself vulnerable to exploitation [31].

Figure 11.7 plots the three most indicative properties (e.g. the first move, the response to mutual cooperation, and the response to mutual defection) of strategies implemented by the coevolving agents. The transition from low to higher cooperation level plays occurs around generation 256 whereby strategies start their interactions being nice (i.e. play with very high cooperation levels). Full cooperation response as a result to previous mutual cooperation play is learned in a very short time nine generations later (middle graph of Fig. 11.7). These strategies are more similar to Tit-for-Tat. Around 60 generations later, we observe Raise-the-Stakes strategies similar to that we have discussed above appearing in the population. Such a rapid transition from lower to higher cooperative plays is also observed in plots of the sampling distribution of the frequency of choices played by the coevolving population of agents [13]. Our coevolutionary simulations have indicated that cooperative behaviours have emerged from the population of agents that have been engaging in mutual defection with agents adopting strategies that mutually engage in highly cooperative plays right at the start of their interaction. Other studies such as [33] have reported similar observations, with the suggestion that offering high cooperation levels at the start of the interaction is important for the evolution of cooperative plays because it avoids the need having to build up trust by raising the stakes.

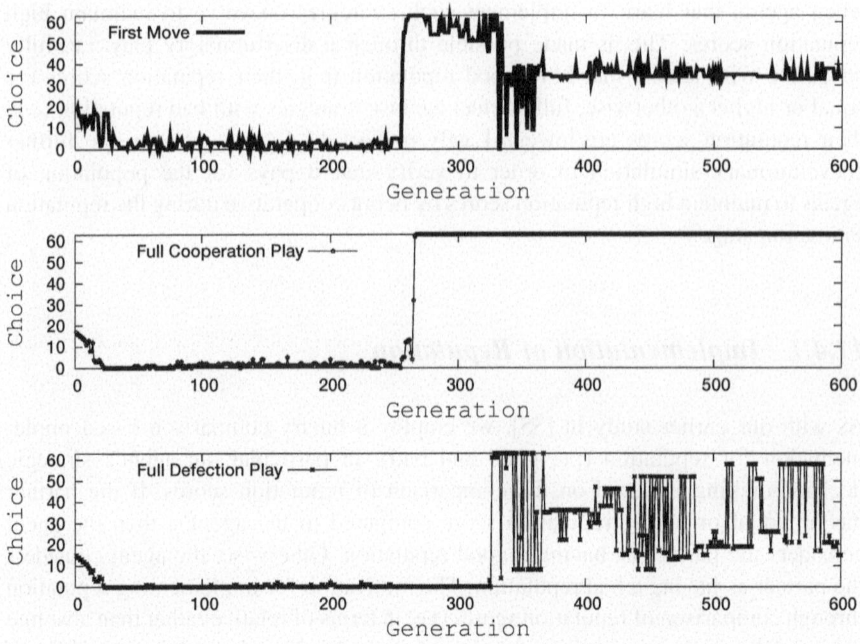

Fig. 11.7 Properties of strategies implemented by the best agents in the run that has coevolved to cooperation. The top graph plots the choices played for the first move. The middle graph plots the choices played if previous moves are mutual cooperation (full cooperation play of both the strategy and the partner). The lower graph plots the choices played if previous moves are mutual defection (full defection play of both the strategy and the partner) [13]

11.4 Incorporating the Mechanism of Indirect Reciprocity in Coevolution

In the second part of our case study [13], we aim to demonstrate why and how the addition of reputation in the IPD game promotes the coevolutionary learning of cooperative behaviours. Our coevolutionary simulations would reveal that agents learn strategies that use reputation as a mechanism to estimate behaviours of future partners and elicit cooperation right from the start of their interactions. First in Sect. 11.4.1 we will present the specific implementation of reputation for strategic decision-making in the coevolving agents that also engage in repeated interactions via IPD game-plays. In Sect. 11.4.2, we present results of our coevolutionary simulations indicating mutual cooperation outcomes are possible in the populations through the IPD game with reputation. Next in Sect. 11.4.3, we present our analysis to determine why and how mutual cooperative behaviours in the population of agents are possible. In particular, we observe that agents implement strategies that effectively engage in binary choices (e.g. full cooperation and full defection) despite the option to play other intermediate choices. Cooperative outcomes are obtained when agents that learn to implement such strategies coevolve to maintain high reputation scores. This is made possible through a discriminatory play, i.e. fully cooperate with agents that have good reputation (e.g. their reputation scores are equal or higher), otherwise, fully defect against strategies with bad reputation (e.g. their reputation scores are lower). Lastly in Sect. 11.4.4, we present one further coevolutionary simulation in order to verify that it pays for the population of agents to maintain high reputation scores by being cooperative during the reputation estimation stage.

11.4.1 Implementation of Reputation

As with our earlier study in [35], we employ a binary comparison-based implementation for reputation (i.e. good and bad). In particular, an agent's strategic decision-making is based on the comparison of reputation scores. If the partner has an equal or higher reputation score compared to the agent's, then the agent considers the partner as having a good reputation. Otherwise, the agent considers the partner as having a bad reputation. The motivation for implementing reputation through comparison of reputation scores (i.e. in terms of relative rather than absolute values) follows the motivation of earlier studies [26, 30]. Here, the comparison of reputation scores provides a well-defined and efficient means for agents to identify partners who have earlier paid the cost (e.g. leaving themselves open to exploitation) through highly cooperative plays to obtain a large reputation score, which otherwise would be more difficult if the implementation is based on the use of absolute values of reputation scores. Crucially, our implementation links both the mechanisms of direct and indirect reciprocity.

Given that an agent takes as inputs both previous moves and reputation scores for decision-making, then reputation scores of agents that could be involved in any interaction have to be determined a priori. We consider a ***two-stage procedure*** we have first proposed in [35] for the ***IPD with more choices and reputation***. In the first (reputation estimation) stage, an agent's reputation score is calculated based on the choices the agent has played in some random n-choice IPD games with other agents of the population in the current generation. Then in the second stage, agents interact through the n-choice IPD with reputation. Each agent will receive the n-choice IPD game payoffs from interactions in both stages. As such, the random selection in the first stage is done without replacement. We note here that such a two-stage procedure for estimating reputation of agents is simplistic. Later, we will examine more realistic procedures and determine their impact on the coevolutionary learning of strategic behaviours by the agents.

The procedure for estimating the reputation score of an agent based on IPD games with other agents randomly sampled from the population is described next. Although some agents might play more games than others in the first stage, our procedure performs normalization by using the proportions of choices played over all involved games in the calculation. Let $\boldsymbol{p} = (p_1, \ldots, p_n)$ be the empirical discrete distribution of an agent's responses in all games involved for reputation estimation, where p_i is the proportion with which the agent samples or plays the ith choice of the n-choice IPD game. Let $\boldsymbol{x} = (x_1, \ldots, x_n)$ specify the weights associated with n choices. In particular, we use $(x_1, \ldots, x_n) = (\text{choice}_1, \ldots, \text{choice}_n)$. This means the weight for full defection is $x_1 = \text{choice}_1 = -1$, while the weight for full cooperation is $x_n = \text{choice}_n = 1$. Then, reputation score is calculated as follows [13]:

$$R = \sum_{i=1}^{n} x_i \cdot p_i \tag{11.3}$$
$$= \frac{\text{Sum of all choices made}}{\text{Total number of moves made}}.$$

Note that an agent's reputation score is calculated based on the choices it plays in this first stage, instead of the payoffs it receives that we have used earlier in [35]. In this approach, the reputation score would more accurately reflect the actual cooperativeness of an agent. The use of payoffs as an indication of cooperativeness can be misleading, e.g. higher average payoffs could be due to mutual cooperation or from the agent exploiting its partners.

Algorithm 11.1 that procedurally implements the coevolutionary learning process for the IPD game with more choices can be easily adapted for the interaction setting that involves the addition of reputation. First, the agent's data structure now is $\boldsymbol{X}[i] = \left(\boldsymbol{w}, \boldsymbol{\sigma}, (\text{pregame}_1, \text{pregame}_1), \boldsymbol{p}, \boldsymbol{x}, R, f\right)_i$. Note that the MLP for the strategy representation of the n-choice IPD game with reputation is updated with one additional input node that takes in the following value: an input of $+1$ if the partner has an equal or higher reputation score compared to the agent's and -1 otherwise. During the reputation estimation stage, the fifth input node takes in a 0 value. Altogether, there are $N_w = (5+1)10+(10+1)1 = 71$ weights and biases. As such, we have $\boldsymbol{w} = (w_j : j = 1, \ldots, N_w)$, $\boldsymbol{\sigma} = (\sigma_j : j = 1, \ldots, N_w)$, and the self-adaptation parameter $\tau = (2(N_w)^{0.5})^{-0.5} \approx 0.2436$. All the various and relevant procedures are updated to ensure that the data with regard to reputation would be suitably processed, e.g. initializations. Another major update would involve the original procedure **roundTour** that now has to incorporate two-stage game-play procedure, which is now implemented as **ROUNDTOURREP** is procedurally described in Algorithm 11.2. The procedural implementation described here takes advantage that the games are symmetric, and as such there are $(\lambda^2 + \lambda)/2$ unique paired interactions in total. The probability that a game involving a pair of agents being selected for reputation estimation is set by the parameter r_p with a value at most 0.5. It should be noted that selection of strategies is based on combined payoffs of games from both stages (reputation estimation and n-choice IPD with reputation). This procedural implementation explicitly models the mechanism of indirect reciprocity that acts on the current pair of agents as a result of direct reciprocal interactions of these agents with others in the past [29].

Algorithm 11.2 Round-robin tournament with two-stage game [13]

Input:

X	Full population of candidate solutions
2λ	Size of full population
n	Number of choices in the IPD game
r_p	Probability a paired interaction selected for reputation estimation

Output:

X	Full population with updated fitness values

```
procedure ROUNDTOURREP(X, 2λ, n, r_p)
    X := initializeRep(X)          ▷ Initialize reputation score X[i].R = 0, i = 1, ..., 2λ
    for i = 1 : 2λ do                      ▷ Stage 1: Games for Reputation Estimation
        for k = i : 2λ do
            if rand(0, 1) < r_p then   ▷ Paired interaction selected for reputation estimation
                selStage[i][k] := true
                IPDREPEST(X[i], X[k], n)   ▷ Update pair of agents' reputation scores and
                                           ▷ fitness values X[i].R, X[i].f, X[k].R, X[k].f
            else
                selStage[i][k] := false
            end if
        end for
    end for
    for i = 1 : 2λ do                      ▷ Stage 2: IPD with Reputation Games
        for k = i : 2λ do
            if ¬(selStage[i][k]) then              ▷ Paired interaction not selected
                                                   ▷ for reputation estimation
                IPDREP(X[i], X[k], n)              ▷ Update pair of agents' fitness
            end if                                 ▷ values X[i].f, X[k].f
        end for
    end for
    return X
end procedure
```

11.4.2 Reputation Promotes Cooperative Behaviours

First, we confirm that reputation promotes coevolutionary learning of cooperative behaviours for the IPD games with more choices and even those with short durations. As such, we consider the interaction setting involving the 64-choice IPD with a game duration of ten rounds. For the interaction setting that further involves reputation, we investigate the outcomes for the following set of values $r_p \in \{0.05, 0.15, 0.25, 0.50\}$. Figure 11.8 summarizes the results of the coevolutionary simulations for the experiments with and without reputation for comparison. In general, our results show incorporating reputation into the IPD interactions promotes the coevolutionary learning of cooperative behaviours in the agents. In particular, the populations are able to learn to be mutually cooperative even for the setting involving IPD games with 64-choices and very short duration of ten rounds, which is challenging for the coevolving population to establish mutual cooperation (see Sect. 11.3.2). Our results indicate a tradeoff in setting r_p on the coevolutionary

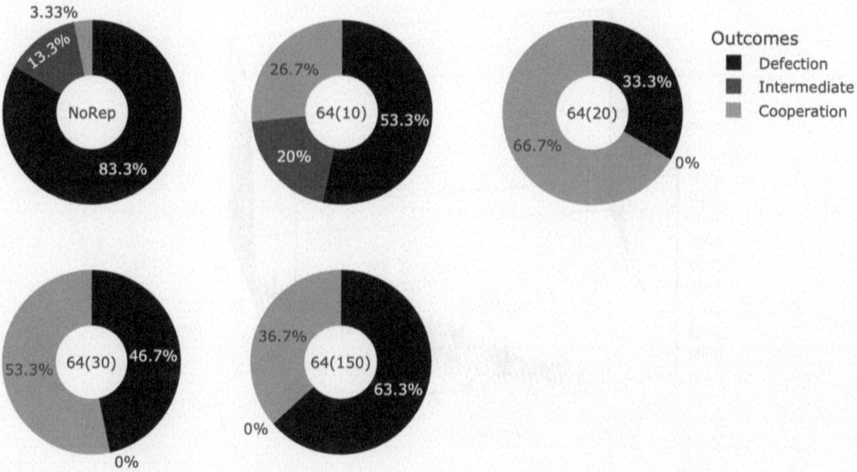

Fig. 11.8 Comparison of results of the experiments that consider the 64-choice IPD game with and without reputation. Results taken at the end of the coevolutionary learning runs

outcomes. An initial increase of r_p from 0.05 to 0.15 leads to more cooperative outcomes, which then decreases as r_p is increased to 0.50. Statistical analysis indicates that both the experiment sets involving r_p settings of 0.15 and 0.25 give significantly different results when compared against the baseline experiment set without reputation. In particular, the results involve 2.93 ± 1.40 ($r_p = 0.15$) and 2.60 ± 1.51 ($r_p = 0.25$) against 1.37 ± 0.84 (baseline with no reputation). Given that payoffs of games in both stages are used for the fitness evaluations of agents, the tradeoff might reflect the situation of having to balance between having enough number of games to reliably estimate reputation scores of agents and then having a large enough number of games for which reputation can be leveraged by agents to engage in mutual cooperative plays for which they can receive favourable outcomes in terms of fitness values for reproduction [13].

11.4.3 Why and How Reputation Helps with Cooperation

Here, we delve deeper into the coevolutionary learning process to better understand why and how reputation promotes the population of agents to engage in mutual cooperative plays. We start with the analysis on the sampling distribution of plays on agents' behavioural maps and compare results obtained for the interaction settings of IPD games with and without reputation. The average entropy values over the population of agents as they coevolve for both experiment sets with reputation ($r_p = 0.05$) and without reputation decrease to 1.34 and 0.40, respectively. Similar trends are observed when other r_p settings are used. Our coevolutionary simulation results indicate that agents play less alternative choices by sampling

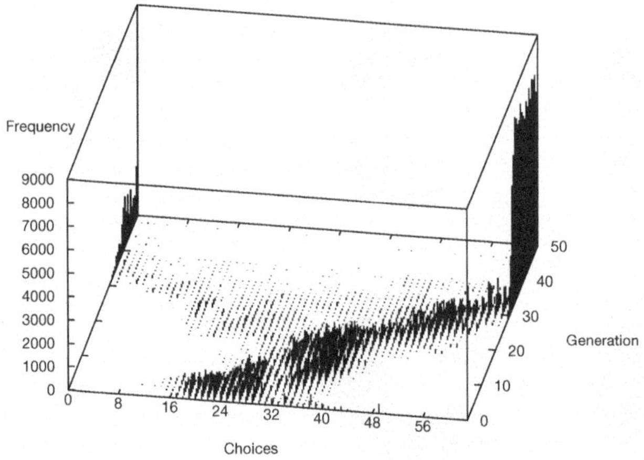

Fig. 11.9 Sampling distribution of choices of the population in the case of the IPD with 64-choices and reputation throughout the coevolutionary process of a typical run that converged to cooperation. Choices are represented from 0 (full defection) to 63 (full cooperation). A higher vertical line for a particular choice and a particular generation indicate a higher sampling frequency for the choice. The graph plots the sampling distributions from generations 1 to 50 to show the transition from mostly alternative choices play to that of binary choices play [13]

smaller portions of their behavioural maps for plays when the interaction setting further includes reputation. Closer inspection on individual runs reveals agents learning to play binary choices, i.e. full cooperation and full defection. Figure 11.9 plots the sampling distribution of choices played by the population of agents in a typical run. One can observe the population coevolving to play mostly binary choices in less than 100 generations. The higher frequency of full cooperation play compared to that of full defection play indicates a cooperative outcome for the population [13].

At this point, we are able to determine why reputation promotes cooperative plays in the coevolving population, that is, agents learn to play binary choices of full cooperation and full defection only. Next, we perform an analysis to facilitate our understanding of how agent behaviours change during the coevolutionary learning process that lead to binary choices of play. We perform observations on the learning process of three properties involving the best agent in the coevolving population: (1) the first move, (2) the response to mutual cooperation in the previous move, and (3) the response to mutual defection in the previous move. These observations are made for these agents as they engage in game-playing interactions for reputation estimation stage and then later in the second stage for partners with good reputation and bad reputation.

Our observation on runs with mutual cooperation outcomes reveals that agents coevolve to be naive and highly cooperative for the reputation estimation stage. One such typical run involves agents coevolving to be nice and respond with full cooperation regardless of what have been played in the previous joint-move, i.e.

whether it is mutual cooperation or mutual defection. For the second stage that considers reputation, these agents have coevolved to engage in highly cooperative play with partners having good reputation. However, the same agents engage in the opposite manner for partners with bad reputation. Close inspection indicates that these agents have coevolved to be unforgiving and engage with defection, i.e. start with full defection, and then continue to respond with full defection regardless of the case of mutual cooperation or mutual defection in the previous joint-move [13].

Lastly, we closely inspect strategies that these agents have learned in order to understand better how reputation helps with mutual cooperative plays. We note that the coevolved agents implement a relatively simple strategy: (Stage 1) It is nice (starts with high cooperation levels) and continues to be highly cooperative. (Stage 2) It employs ***discriminatory*** play mostly with respect to the partner's reputation, i.e. fully cooperate if the partner has good reputation, else fully defect against the partner with bad reputation. In other words, these strategies only consider reputation for behavioural responses, and that previous joint-moves are effectively superfluous. In addition, we note that such strategies are implemented by most of the population. This strategy is successful in proliferating throughout the population because it is able to engage in full mutual cooperation play throughout the entire game with similar strategies, thus enabling agents to obtain the highest joint payoffs. The strategy also quickly punishes those that do not engage in similar highly cooperative play during the interaction setting that involves more choices and reputation.

For these interactions, reputation acts as a sort of signalling mechanism for agents to engage in mutual cooperation. High reputation scores reflect an agent's willingness to engage in highly cooperative play in future interactions, while the opposite is the case for agents with lower reputation scores. A comparison of reputation scores allows agents implementing discriminatory strategies to know in advance which future partners to trust to engage in mutual cooperation and others that are punished with uncompromising defection play. A deep inspection on the MLPs' input-output responses reveals significantly higher connection weight associated with the fifth input for reputation than all the other four inputs for previous moves. Such structures of MLP weights enable discriminatory play based only on comparison of reputation scores. Next, we examine these generated MLPs from different experiment sets. In particular, MLPs with such weight structures are observed in coevolutionary learning processes whereby $r_p < 0.5$ is set. One could argue for the case of higher selection pressure in coevolution on the trait that is connected to discriminatory play, e.g. substantially higher connection weight for the fifth input (reputation). This high selection pressure on reputation input can be explained by considering that there are more games where an agent uses reputation (second stage) that contributes to its fitness compared to games without reputation (first stage) when $r_p < 0.5$. We have not observed such discriminatory play in the coevolutionary learning process with $r_p = 0.5$ set. Given that the number of games for both stages are similar on average, there would not be significant selection pressure on the fifth input of MLPs for which discriminatory play can be learned through coevolution [13].

11.4.4 Does It Pay to Maintain Good Reputation?

At this point, one immediate question one would ask is whether it pays for coevolving agents to maintain good reputation. The answer to this question is important because the additional input for reputation introduces a subtle and complex relationship between responses in current interaction to gain immediate favourable payoffs and those for future interactions that are about gaining probable favourable payoffs from other agents in the population. These two sets of responses are linked by the mechanism whereby the present choices inform how reputation scores are calculated, which in turn through their comparison form one crucial consideration for future strategic decision-making. Recall that an agent will process its partner as having a good reputation if the partner's reputation score is equal to or higher than its reputation score. Our coevolutionary simulations reveal that in all our runs whereby the populations have learned cooperative plays, the best agent has high reputation scores. Observation made on a typical run with cooperative outcome reveals that the reputation scores of successful agents coincide with the state of cooperation (using average population payoff) in the population. For runs whereby populations engage in defection play, the reputation score of the best agents in the population fluctuates between -1 (for playing full defection) and 0 (for not being sampled for the reputation estimation stage) [13].

Close inspection reveals that the population consists of agents with high reputation scores of 1, which is the same as the reputation score of the best agent. Such a high reputation score of 1 can only be achieved when agents play full cooperation levels during the reputation estimation stage, e.g. refer to Eq. (11.3). We note that these agents implement strategies that have broad response surfaces corresponding to full cooperation play for games in the reputation estimation stage. Later in interactions in the second stage that include reputation, these agents are able to elicit mutual cooperative play from other agents having comparable, high reputation scores, and implementing similar strategies. They engage in discriminatory play and, as a consequence, defect against agents with lower reputation scores. In this manner, reputation helps to promote cooperation because it discourages agents from playing other alternative choices of lower cooperation levels [13]. This is especially important in interaction settings with more choices and shorter durations that have been earlier established in Sect. 11.3.3 to be challenging for populations of adaptive agents to learn cooperative strategies through coevolution.

Unlike previous studies [26, 30] that have focused only on the mechanism of indirect reciprocity, one concern with the approach we have employed here is the possibility of having cheaters, i.e. agents that obtain high reputation scores so as to be perceived as having good reputation only for them to exploit those that cooperate with high reputation agents. However, our implementation discourages agents to learn such cheating behaviours. There are two reasons for this with the first related to the variation operator and the second related to the selection operator. First, we note that it is functionally very difficult for MLP-based agents to be rewired to implement such a cheating behavioural response after having inherited those that express cooperative plays for the reputation estimation stage and discriminatory play for the second stage. In particular, the use of self-adaptive mutation means that the population of agents at this state of learning in the coevolutionary process would have had small σ parameter values. Second, we observe that such a cheater does not do well against other cheaters (including self-play). Furthermore, there might be scope in the coevolutionary learning process to generate more discriminatory play from the population, i.e. those that are not simply naive cooperators in the second stage but instead employ Tit-for-Tat that is known to be more robust against strategies that play defection.

11.5 Discussion and Conclusion

In this chapter, we have presented our case study [13] that applies the principles of competitive coevolution to develop a multi-agent coevolutionary system for which detailed simulations can be conducted to establish why and how some conditions promote the learning of specific behaviours when agents interact in complex settings. The first part of our case study has specifically aimed to uncover why the coevolving population of agents is less likely to learn cooperative behaviours when more choices are available for play in their IPD interactions. Our coevolutionary simulations have revealed that agents effectively have more opportunities to exploit others by playing lower cooperation levels because they are less able to resolve the intention of an intermediate choice, which could signal a call to engender further cooperation or a subtle exploitation. As a consequence, agents adapt to play defection to obtain higher payoffs in the short term. In summary, the mechanism of direct reciprocity that has been proposed as the means with which agents engage in mutual cooperative plays when their interactions consist of repeated encounters has become less effective. In two-player IPD games, a moderate increase in the number of choices leads to a substantially larger increase in the possible sequences of choices that agent plays. Agents are less able to resolve the intention of intermediate choices based on game outcomes in the form of cumulative IPD payoffs.

However, there are other mechanisms of which the one based on indirect reciprocity has been shown to be effective in promoting cooperative behaviours in complex human interactions. The mechanism of indirect reciprocity can be explained in terms of reputation, whereby cooperative behaviours between two individuals depend on their prior interactions with other individuals. This motivates the second part of our case study where we have incorporated reputation into the IPD interactions in our multi-agent coevolutionary learning system. This is also in line with arguments made in previous studies that indirect reciprocity is a result of direct reciprocity that occurs between other individuals in complex, real-world interactions. As such, the coevolutionary learning of agents implementing strategies for IPD games with multiple choices and reputation explicitly connects these two mechanisms. In particular, reputation scores of agents are estimated from choices that they have made in a small number of interactions that do not involve reputation in the population, which are then used in latter interactions that involve reputation with the other remaining agents. Our coevolutionary simulations reveal that incorporating reputation into the IPD interactions promotes the learning of cooperative behaviours in the coevolving population. The agents implement strategies that are cooperative for the reputation estimation stage to obtain high reputation scores and then adopt discriminatory play for the second stage whereby they cooperate with those that have good reputation but defect against those that have bad reputation.

It should be noted that our simple two-stage approach for which reputation can be incorporated into the IPD interactions between agents might not be realistic because of the two main assumptions we have made about how reputation is estimated. First, an agent's reputation is static for the period with which the agent interacts with others in the population until the next update, which in the current implementation setting only occurs in the next generation. Second, an agent's reputation score is always calculated anew in every generation because there is no memory of those games. Before closing this chapter, we study two alternative implementations of reputation that aim to address these two issues we raise here. In general, we view these alternative implementations from the perspective of the accuracy of reputation estimation of agents based on their prior interactions with other agents. In particular, these two alternative implementations replace the two-stage procedure with a one-stage procedure so that any interaction between two agents involves IPD games and reputation simultaneously. Note that the fifth input of the MLP now only takes $+1$ or -1. Next, we enable agents to store choices they have played in previous generations that are used to calculate reputation scores. This implementation effectively incorporates memory of games played in previous interactions including those in previous generations of the coevolutionary learning process. As such, we can investigate how accuracy of reputation reflects an agent's behaviour to its future partners is related to how much memory of previous games (e.g. those in past generations) is incorporated and how frequent reputation score is updated [13].

We first discuss results of our coevolutionary simulations with the first alternative implementation of reputation. Note that an agent's empirical discrete distribution of choices played $\boldsymbol{p} = (p_1, \ldots, p_n)$ is initialized to $(0, \ldots, 0)$ and inheritable by the child agent. Reputation score for each agent is calculated using Eq. (11.3) at the start of every generation, which remains unchanged for interactions within the same generation. At the start of coevolution, agents are arbitrarily assigned reputation scores from a uniform distribution $U[-1, 1]$. This ensures on average that the behavioural responses of implemented strategies for partner's having good and bad reputation will be sampled for play. Results of our coevolutionary simulations indicate that cooperative outcomes are more likely to occur when the populations of coevolving agents engage in such IPD interactions with more choices and reputation. In addition, we have conducted additional experiments whereby a parameter p_u is used to specify the probability that a particular game is not used to update an agent's $\boldsymbol{p}$. Results indicate a small reduction in the number of outcomes for which populations coevolved towards cooperation [13].

Closer inspection of strategies implemented by coevolved agents reveals that they predominantly respond with binary choices of either full cooperation or full defection. In particular, these agents are highly cooperative for partners with good reputation. They are also cooperative with partners having bad reputation but will respond to punish them with defection plays if the partners have played close to defection choices. In this regard, the strategies that agents implement for coevolutionary simulations with a one-stage implementation differ with that observed in those that have employed a two-stage implementation. We have earlier studied and established that the reputation implementation based on two-stage games promotes discriminatory strategies due to selection pressure on the input for reputation and especially for those with low r_p settings where there are more games played for the second stage that involve IPD interactions with reputation. This is further confirmed with close inspections on the MLPs whereby connection weights associated with the input for reputation are not substantially higher [13].

Next, we discuss results of our coevolutionary simulations for which the second alternative implementation of reputation has been employed. The main difference compared to the first alternative implementation is that reputation scores are updated after every game played. Results indicate a positive impact on cooperative outcomes. In general, these cooperative outcomes produced from the coevolutionary learning process are stable or robust to the various p_u settings we have investigated. As for the coevolution of agents' behavioural responses, we have observed similar strategies that are learned with the first alternative implementation. That is, agents have typically learned to implement strategies that play mostly binary choices of full cooperation or full defection. They are highly cooperative with both good and bad reputation partners that are cooperative in previous moves. Although these agents are not discriminatory on the basis of reputation consideration, they are less tolerant (e.g. response with full defection for partners that have played lower cooperation levels) to bad reputation partners as compared to good reputation partners. In other words, these strategies do not discriminate based entirely on reputation alone. Close inspections on the weights of MLPs associated with these strategies have not revealed significant differences between input connection weights for previous moves and reputation. These strategies employ more subtle discrimination that is based on how interactions unfold in which previous moves and reputation are considered for decision-making [13].

In summary, our coevolutionary simulations have indicated the importance of accurate reputation estimation for the further learning of cooperative strategies by the population of coevolving agents. Factors that have an impact to such coevolutionary learning outcomes are memory of games of prior interactions (including those from previous generations) for reputation estimation and the frequency with which reputation scores are updated. Before we close the chapter, we remark that in complex real-world interactions the mechanism of indirect reciprocity might be subjected to other effects. We have investigated one effect whereby not all prior games are attributed to an agent's reputation score. However, there are other effects namely that of noise in the interactions that could affect the coevolutionary learning process of adaptive agents. There is a need to determine the specific types of noise in the interactions (e.g. whether one that involves a change in perception or an actual change to what was played) and their levels that could present more significant challenges to the coevolutionary learning of cooperative behaviours.

References

1. Axelrod, R.: Effective choice in the prisoner's dilemma. J. Confl. Resolut. **24**(1), 3–25 (1980)
2. Axelrod, R.: More effective choice in the prisoner's dilemma. J. Confl. Resolut. **24**(3), 379–403 (1980)
3. Axelrod, R.: The Evolution of Cooperation. Basic Books, New York (1984)
4. Axelrod, R.: The evolution of strategies in the iterated prisoner's dilemma. In: L.D. Davis (ed.) Genetic Algorithms and Simulated Annealing, chap. 3, pp. 32–41. Morgan Kaufmann, New York (1987)
5. Axelrod, R., Hamilton, W.D.: The evolution of cooperation. Science **211**, 1390–1396 (1981)
6. Batali, J., Kitcher, P.: Evolution of altruism in optional and compulsory games. J. Theor. Biol. **175**, 161–171 (1995)
7. Chellapilla, K., Fogel, D.B.: Evolution, neural networks, games, and intelligence. Proc. IEEE **87**(9), 1471–1496 (1999)
8. Chong, S.Y., Ku, D.C., Lim, H.S., Tan, M.K., White, J.D.: Evolved neural networks learning Othello strategies. In: Proceedings of the 2003 Congress on Evolutionary Computation (CEC'03), pp. 2222–2229. IEEE Press, Piscataway, NJ (2003)
9. Chong, S.Y., Tan, M.K., White, J.D.: Observing the evolution of neural networks learning to play the game of Othello. IEEE Trans. Evol. Comput. **9**(3), 240–251 (2005)
10. Chong, S.Y., Tiňo, P., Ku, D.C., Yao, X.: Improving generalization performance in co-evolutionary learning. IEEE Trans. Evol. Comput. **16**(1), 70–85 (2012)
11. Chong, S.Y., Tiňo, P., Yao, X.: Measuring generalization performance in co-evolutionary learning. IEEE Trans. Evol. Comput. **12**(4), 479–505 (2008)
12. Chong, S.Y., Yao, X.: Behavioral diversity, choices, and noise in the iterated prisoner's dilemma. IEEE Trans. Evol. Comput. **9**(6), 540–551 (2005)
13. Chong, S.Y., Yao, X.: Multiple choices and reputation in multiagent interactions. IEEE Trans. Evol. Comput. **11**(6), 689–711 (2007)
14. Darwen, P., Yao, X.: On evolving robust strategies for iterated prisoner's dilemma. In: Progress in Evolutionary Computation. Lecture Notes in Artificial Intelligence, vol. 956, pp. 276–292 (1995)
15. Darwen, P., Yao, X.: Does extra genetic diversity maintain escalation in a co-evolutionary arms race. Int. J. Knowledge-Based Intell. Eng. Syst. **4**(3), 191–200 (2000)
16. Darwen, P., Yao, X.: Why more choices cause less cooperation in iterated prisoner's dilemma. In: Proceedings of the 2001 Congress on Evolutionary Computation (CEC'01), pp. 987–994. IEEE Press, Piscataway, NJ (2001)
17. Darwen, P., Yao, X.: Co-evolution in iterated prisoner's dilemma with intermediate levels of cooperation: Application to missile defense. Int. J. Comput. Intell. Appl. **2**(1), 83–107 (2002)
18. Fogel, D.B.: The evolution of intelligent decision making in gaming. Cybern. Syst. An Int. J. **22**, 223–236 (1991)
19. Fogel, D.B.: Evolving behaviors in the iterated prisoner's dilemma. Evol. Comput. **1**(1), 77–97 (1993)
20. Fogel, D.B.: On the relationship between the duration of an encounter and the evolution of cooperation in the iterated prisoner's dilemma. Evol. Comput. **3**(3), 349–363 (1996)
21. Fogel, D.B.: Blondie24: Playing at the Edge of AI. Morgan Kaufmann, San Francisco, CA (2002)

22. Fogel, D.B., Hays, T.J., Hahn, S.L., Quon, J.: A self-learning evolutionary chess program. Proc. IEEE **92**(12), 1947–1954 (2004)
23. Frean, M.R., Abraham, E.R.: A voter model of the spatial prisoner's dilemma. IEEE Trans. Evol. Comput. **5**(2), 117–121 (2001)
24. Harrald, P.G., Fogel, D.B.: Evolving continuous behaviors in the iterated prisoner's dilemma. BioSystems: Special Issue on the Prisoner's Dilemma **37**, 135–145 (1996)
25. Julstrom, B.A.: Effects of contest length and noise on reciprocal altruism, cooperation, and payoffs in the iterated prisoner's dilemma. In: Proceedings of the 7th International Conference on Genetic Algorithms (ICGA'97), pp. 386–392. Morgan Kauffman, San Francisco, CA (1997)
26. Leimar, O., Hammerstein, P.: Evolution of cooperation through indirect reciprocity. Proc. R. Soc. Lond. Ser. B **268**, 745–753 (2001)
27. Lindgren, K.: Evolutionary phenomena in simple dynamics. In: C.G. Langton, C. Taylor, J.D. Farmer, S. Rasmussen (eds.) Artificial Life II, pp. 295–312. Addison-Wesley, New York (1991)
28. Nisan, N., Roughgarden, T., Tardos, E., Vazirani, V.V. (eds.): Algorithmic Game Theory. Cambridge University, New York (2007)
29. Nowak, M.A., Sigmund, K.: The dynamics of indirect reciprocity. J. Theor. Biol. **194**, 561–574 (1998)
30. Nowak, M.A., Sigmund, K.: Evolution of indirect reciprocity by image scoring. Nature **393**, 573–577 (1998)
31. Roberts, G., Sherratt, T.N.: Development of cooperative relationships through increasing investment. Nature **394**, 175–179 (1998)
32. Runarsson, T.P., Lucas, S.M.: Coevolution versus self-play temporal difference learning for acquiring position evaluation in small-board go. IEEE Trans. Evol. Comput. **9**(6), 540–551 (2005)
33. Wahl, L.M., Nowak, M.A.: The continuous prisoner's dilemma: I. linear reactive strategies. J. Theor. Biol. **200**(3), 307–321 (1999)
34. Yao, X.: Evolving artificial neural networks. Proc. IEEE **87**(9), 1423–1447 (1999)
35. Yao, X., Darwen, P.: How important is your reputation in a multi-agent environment. In: Proceedings of the 1999 Conference on Systems, Man, and Cybernetics (SMC'99), pp. 575–580. IEEE Press, Piscataway, NJ (1999)
36. Yao, X., Liu, Y., Lin, G.: Evolutionary programming made faster. IEEE Trans. Evol. Comput. **3**(2), 82–102 (1999)

Chapter 12
Generalization in Coevolutionary Learning

12.1 Introduction to Performance Analysis in Competitive Coevolution

Competitive coevolution has been applied successfully to solve the problem of games [2, 10, 11, 46, 55, 57] and other related problem-solving formulations such as those in the context of cooptimization [39] and classification [41, 60] through a learning process. However, ***competitive coevolutionary learning*** is known to suffer from pathological issues that can negatively impact the search performance of coevolutionary learning systems [9, 24, 26, 28, 30, 49, 54, 59]. Earlier in Chap. 5, we have introduced and discussed how specific cycle structures in the pairwise relations of solutions that come about from intransitivity give rise to these pathological issues. We now have some deeper understanding about these coevolutionary pathologies through formal studies about coevolutionary problem structures [6, 7, 12]. They are fundamental towards how competitive coevolution operates since pairwise comparisons of solutions constitute the basic primitives with which the relative fitness values are defined, calculated, and then used in the selection process.

There are deeper connections between such cycle structures in coevolutionary problems with cycle structures in the dynamics of coevolutionary processes. Current theoretical studies [29, 58] have only begun to uncover these complex dynamics describing the rich trajectories of population configurations or states as coevolution operates that arise even in simple interactions in the form of two-player games of two pure strategies. Much of what we understand and in settings where we could offer some explanations about coevolutionary search come about from quantitative analysis with respect to measurements made on coevolutionary systems. In the literature of competitive coevolution, cyclic dynamics might occur as a result of the population cycling between states of overspecialized population members corresponding to a particular solution to other overspecialized population states each corresponding to other specific solution, and that underlying these individual solu-

X. Yao, S. Y. Chong, *Coevolutionary Computation and Its Applications*,
Natural Computing Series, https://doi.org/10.1007/978-981-96-2841-4_12

tions are intransitive relations between them (e.g. Rock-Paper-Scissors dynamics) [9, 24, 26, 28, 59].

There are broader implications in the design of such competitive coevolutionary learning system to ensure effective search performance that might be achieved other than solely about mitigating pathological issues. In particular, coevolutionary search performance is dependent on other components in the general framework of Coevolutionary Algorithms (CEAs). For example, the choice of solution representation [16] and variation operators [16, 24, 25] can affect effectiveness of coevolutionary search because they change the nature of the search space. Indeed, systematic and controlled computational studies are still being employed in studies of competitive coevolution even when equipped with theoretical insights [4, 12]. This necessitates other studies specifically towards the methodology that measures and is able to monitor the progress of coevolutionary search of more innovative and sophisticated solutions (e.g. arms race dynamics) when not much are known about the structures of the problem and search processes [3, 18, 31, 56]. Another approach is to consider a global view of increasing performance in coevolutionary learning. This is because tools that are designed to monitor progress of arms race dynamics are inappropriate for such a global view analysis because they only provide relative performance information of solutions between different generations.

In this chapter, we present a series of studies [13–15] where such a global view analysis has been employed in competitive coevolution setting. Our approach is based on the notion of ***generalization*** from the field of Machine Learning that is concerned with a global view of performance in learning systems. Generalization refers to the ability of the learning system to produce solutions in the form of input-output mappings that best predict the required output for any new input that has not been seen during the training process. As such, a learning system that produces solutions with good generalization performance is those that realize the underlying properties of the problem from a small sample of training data [5]. Like other learning systems, competitive coevolution uses a small sample of test cases during the evolutionary (training) process to guide search for solutions to the problem at hand. However, coevolutionary learning differs from learning systems in general and those constructed under a supervised setting in particular in one main aspect. In coevolution, the training samples are not fixed but could change in response to the evolving solutions. Indeed, these specific interactions between solutions and test cases leading to their respective coadaptations constitute the hallmark of competitive learning in coevolution [53, 54].

Despite this, the generalization framework can be applied to competitive coevolutionary learning systems to provide a global view in performance analysis. Although the notion of generalization is well understood in Machine Learning, there is a lack of theoretical framework that specifies what generalization performance of a coevolutionary learning system is with respect to the problem it is used to solve. Past studies [20, 64] have used a large sample of randomly obtained test cases to estimate the generalization performance of competitive coevolution. However, there is no notion what accuracy of an estimated generalization performance entails when using random test samples given that the precise definition of the true generalization performance has not been formulated. To address this issue, we first develop a theoretical framework of generalization performance for competitive coevolution setting whereby interactions between solutions and test cases are modelled as competitive game-plays [14]. We focus on symmetric two-player games between solutions and test cases for simplicity in presentation. Note that for the rest of this chapter we will use a simplified terminology whereby we refer to agents implementing pure game-playing strategies directly as (test) strategies. We define generalization performance of a strategy produced by competitive coevolution as its expected or average performance against test strategies drawn from the strategy space. It follows that a game strategy that generalizes well to the game means that it defeats a large number of opponent test strategies that exist. However, note that the theoretical framework is general given that there are many problems that involved cooptimization settings whereby performance comparisons of solutions require test-based evaluations instead of formulated objective functions [48, 50].

The remainder of this chapter is organized into two main technical sections. Section 12.2 formally introduces the generalization performance framework for competitive coevolution. We propose an estimation procedure that computes the average performance against a small sample of random test strategies and derive confidence bounds with respect to some precision levels that are distribution-free. We demonstrate how various generalization performance measurements can be constructed relevant to analysis of coevolutionary systems through controlled computational studies. In Sect. 12.3, we will demonstrate how the original distribution-free formulation of generalization performance can be improved by taking advantage of the near-Gaussian nature of average game outcomes and provide tighter bounds based on parametric testing. The formal analysis requires use of higher (second and third) order moments of the game outcomes with respect to the underlying distribution with which random test strategies are drawn. They are used to estimate the speed (in terms of sample sizes) with which the distribution of average game outcomes converges to the standard Gaussian. The chapter will close with brief remarks how these quantitative analysis tools have been used to aid with the study to determine the relationship between generalization and diversity in competitive coevolutionary learning.

12.2 Generalization Performance in Coevolutionary Learning

In this section, we present our study [14] where we have developed the theoretical framework of generalization performance in competitive coevolution, demonstrated how statistical estimators for generalization performance measurements can be constructed, and subsequently applied those estimators to provide rigorous quantitative analysis of coevolutionary learning. Section 12.2.1 will formally define generalization in competitive coevolution, its statistical estimation using a sample of random test strategies, and then how a distribution-free confidence bound can be constructed that lets one make statistical claims of the accuracy of generalization performance estimates. This is then followed by constructions of various useful performance measurements such as generalization with respect to a biased sample of high-performing test strategies and generalization performance measure of an ensemble of strategies, Sect. 12.2.2. Then Sect. 12.2.3 will present various computational studies to verify the framework and also to illustrate how these estimators can be used to provide rigorous quantitative analysis of coevolutionary learning for competitive games.

12.2.1 Theoretical Framework of Generalization Performance

In competitive coevolutionary learning, the quality of a solution is determined with respect to its performance against test cases. The interaction between a solution and a test case is formulated as a two-player game. Assume that the interaction (game-play) is symmetric so that both strategy i and test (opponent) strategy j are taken from the finite set of distinct strategies $\mathcal{S}$. Let the game outcome of i against j be $G_i(j)$ (conversely, one can define the game outcome $G_j(i)$ of j against i). The absolute quality of a strategy i is measured with respect to its expected performance in solving tests provided by opponent strategies j.

Let the selection of individual test strategies from $\mathcal{S}$ be represented by a random variable J taking on values $j \in \mathcal{S}$ with probability $\mathbb{P}_{\mathcal{S}}(j)$. Some strategies are favoured over others, or all strategies are considered with equal probability a priori. We define the ***true generalization performance*** G_i of a strategy i as the mean of the random variable $G_i(J)$, $G_i = \mathbb{E}_{\mathbb{P}_{\mathcal{S}}}[G_i(J)]$, which is calculated as [14]

$$G_i = \sum_{j=1}^{M} \mathbb{P}_{\mathcal{S}}(j)\, G_i(j). \tag{12.1}$$

For example, when all strategies are equally likely to be selected, we have

$$G_i = \frac{1}{M} \sum_{j=1}^{M} G_i(j).$$

There are three related difficulties applying Eq. (12.1) directly as a generalization performance measure in practice. First, a closed-form formula for G_i is very difficult to obtain even for simple games since the game outcome $G_i(j)$ is unknown in advance. Second, the probability distribution $\mathbb{P}_{\mathcal{S}}$ on $\mathcal{S}$ is usually unknown and that in practice one only can sample from $\mathcal{S}$. The sampling is dependent on the specific strategy representation that one uses. As such, some strategies j exist but are not included in the computation of G_i because $\mathbb{P}_{\mathcal{S}}(j) = 0$. Third, it is not practical to directly compute G_i via brute-force even for simple games. For example, consider the n-choice Iterated Prisoner's Dilemma (IPD) game with deterministic and reactive memory-one strategies. The total number of unique strategies is given by n^{n^2+1} [16], which gives a faster than exponential growth. One has $2^{2^2+1} = 2^5 = 0.32 \times 10^2$, $3^{3^2+1} = 3^{10} \approx 0.59 \times 10^4$, $4^{4^2+1} = 4^{17} \approx 0.17 \times 10^{10}$, and so forth. Computing G_i directly quickly becomes a computationally infeasible task.

Instead, we exploit the formulation of $G_i = \mathbb{E}_{\mathbb{P}_S}[G_i(J)]$ and employ a procedure to estimate its true value using a random sample S_N of N test strategies drawn iid from $\mathcal{S}$. The ***estimated generalization performance*** $\hat{G}_i(S_N)$ of strategy i with respect to the sample S_N is calculated as [14]

$$\hat{G}_i(S_N) = \frac{1}{N} \sum_{j \in S_N} G_i(j). \tag{12.2}$$

For the rest of this chapter, we use the simplified notation $\hat{G}_i$ to represent $\hat{G}_i(S_N)$. We also refer $\hat{G}_i(S_N)$ as the ***statistical estimator*** of the generalization performance of strategy i.

The issue of accuracy in the ***estimation*** of generalization performance is to determine how close $\hat{G}_i$ is in comparison to G_i. For a particular ***level of precision*** given by a small positive number $\epsilon > 0$, this means determining if the absolute difference between the estimated and true values $|\hat{G}_i - G_i|$ is less than ϵ. If $|\hat{G}_i - G_i| < \epsilon$ is true, then one can claim that $\hat{G}_i \approx G_i$ within the specific level of precision. However, $|\hat{G}_i - G_i|$ is unknown because G_i itself is unknown. The alternative then is to find out how likely $\hat{G}_i$ is close enough to G_i. To make such a statistical claim, one needs to determine what is the probability $|\hat{G}_i - G_i| \geq \epsilon$ for which a random sample S_N is used to obtain $\hat{G}_i$. This question can be answered with ***Chebyshev's Theorem*** [36].

Note that the difference $D_N = \hat{G}_i - G_i$ is a random variable since $\hat{G}_i$ is a random variable. From the definition of D_N, we obtain the following [14]:

$$\mathbb{P}_N(|\hat{G}_i - G_i| \geq \epsilon) \leq \frac{\sigma_i^2}{N \cdot \epsilon^2} \tag{12.3}$$

for a strategy i. We use the simplified notation σ^2 in place of σ_i^2 for the rest of this chapter. Furthermore,

$$\begin{aligned}
\sigma^2 &= \mathbb{E}_{P_S}\Big[\big(G_i(J) - G_i\big)^2\Big] \\
&= \mathbb{E}_{\mathbb{P}_S}\Big[\big(G_i(J) - \mathbb{E}_{\mathbb{P}_S}[G_i(J)]\big)^2\Big] \\
&= \mathrm{var}_{\mathbb{P}_S}[G_i(J)].
\end{aligned}$$

Although the second moment σ^2 is also unknown, we exploit the fact that its value can be bounded from above. In general, for the random variable $G_i(J)$

distributed over the interval $[G_{\min}, G_{\max}]$, where $G_{\max}$ and $G_{\min}$ are the maximum and minimum of $G_i(J)$, respectively, and the maximum variance $\sigma^2_{\max}$ is when half of the mass is at $G_{\min}$ and the other half is at $G_{\max}$. One obtains the mean $(G_{\min} + G_{\max})/2$ and standard deviation $\sigma_{\max} = (G_{\max} + G_{\min})/2$ [40]. Let $R = G_{\max} - G_{\min}$ be the range of the random variable $G_i(J)$. Then, $\text{var}_{\mathbb{P}_S}[G_i(J)]$ being upper bounded by $\sigma^2_{\max} = R^2/4$ is used to prove the following theoretical result (Lemma 12.1).

Lemma 12.1 ([14]) *For a strategy i, let $\hat{G}_i$ be the estimated generalization performance with respect to N random test strategies and G_i be the true generalization performance. Consider the absolute difference $|\hat{G}_i - G_i|$, which is a random variable with distribution $\mathbb{P}_N$ taken on a compact interval $[G_{\min}, G_{\max}]$ of length $R = G_{\max} - G_{\min}$. Then, for any positive number $\epsilon > 0$,*

$$\mathbb{P}_N(|\hat{G}_i - G_i| \geq \epsilon) \leq \frac{R^2}{4N \cdot \epsilon^2}. \tag{12.4}$$

Note that ***Chebyshev's bounds*** given by Eq. (12.4) are ***distribution-free***. In particular, no specific distribution of $G_i(J)$ is assumed to derive Chebyshev's bound. Statistical claims can be made on the accuracy of an estimate $\hat{G}_i(S_N)$ computed using a random sample of N test strategies. Taking the complement of Eq. (12.4), one then claims with probability of at least $p = 1 - R^2/4N\epsilon^2$ or confidence level at $100p\%$ that $|\hat{G}_i - G_i| < \epsilon$ for a precision level $\epsilon > 0$ [14].

We can illustrate Chebyshev's bound as follows. Let $|D_N|' = |D_N|/R = |\hat{G}_i - G_i|/R$ and $\epsilon' = \epsilon/R$. This normalization allows us to express Eq. (12.4) in the simpler form

$$\mathbb{P}_N(|D_N|' \geq \epsilon') \leq \frac{1}{4N \cdot \epsilon'^2}, \tag{12.5}$$

and then illustrate Chebyshev's bound as a plot of $\mathbb{P}_N(\epsilon') = 1/(4N\epsilon'^2)$ as points in the $\mathbb{P}$-ϵ' space. Figure 12.1 shows these plots $\mathbb{P}_N$ for Chebyshev's bounds for different Ns over a range of ϵ's in $[0.01, 0.1]$. Each plot $\mathbb{P}_N$ gives the upper bound for $\mathbb{P}_N(|D_N|' \geq \epsilon')$. Note that $\mathbb{P}_N(\epsilon')$ is inversely proportional to N and ϵ'. For any ϵ', an increasingly larger number for N are required to obtain lower Chebyshev's bounds (Fig. 12.1). In practice, one specifies N to be as large as computationally possible to obtain $\hat{G}_i$. Then, one makes a trade-off between the confidence and precision levels.

More importantly, we note the generality of this framework by observing the following three points [14]:

1. The framework is **independent of the complexity of the game**, since it is independent of the size of the strategy space and independent of the strategy distribution in the strategy space.
2. Both $G_{\max}$ and $G_{\min}$ are often known a priori as they are specified by the game, which means that we can always obtain an upper bound on (12.4).

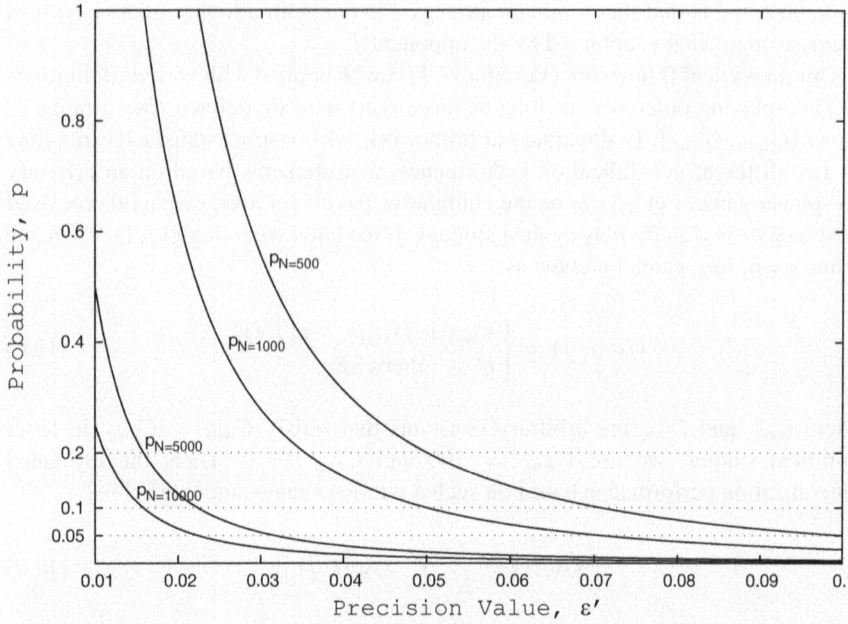

Fig. 12.1 Figure showing various plots of Chebyshev's bounds $\mathbb{P}_N$ for different sample sizes N in the range of precision value ϵ' in $[0.01, 0.1]$ [14]

3. The framework is **independent of learning algorithms** since the bound holds for any strategy in the strategy space.

12.2.2 Constructions of Generalization Performance Measurements

Here, we demonstrate the construction of different generalization performance measurements that can be applied to provide different aspects of quantitative analysis in competitive coevolutionary learning systems. We consider three different aspects of these performance measurements. First, the notion of performance in competitive coevolutionary setting depends on the specific definition of the interaction outcome $G_i(j)$ between the solution and test case. In two-player competitive games, one can define performance in the context of a win-lose game outcome. Note that classic game theory [45] defines zero-sum games to have the sum of outcomes between any pair of strategies implemented by the two competing players to be zero. In such a setting, a win of $G_i(j)$ results in a lose of $G_j(i)$ and draws automatically result with zero for both players. For nonzero-sum games, one specific notion of a winning game outcome is that the dominant strategy i is one with a higher game payoff in comparison to what is obtained by the opponent j.

Our theoretical framework (Lemma 12.1) can be applied with various definitions of game-playing outcomes, as long as those outcomes are defined over a range of values $[G_{\min}, G_{\max}]$. To illustrate our framework, we construct statistical estimators for two different generalization performance measurements based on an arbitrary two-player game. Let $g(i, j)$ be the cumulative payoff (or average payoff) received by strategy i in a game-play against strategy j, the latter receiving $g(j, i)$. Then, we define a win-lose game outcome as

$$G_{\mathrm{W}}(i, j) = \begin{cases} C_{\mathrm{win}} \text{ if } g(i, j) > g(j, i), \\ C_{\mathrm{lose}} \text{ otherwise}, \end{cases} \tag{12.6}$$

where C_{win} and C_{lose} are arbitrary constants that satisfy $C_{\mathrm{win}} > C_{\mathrm{lose}}$. In latter empirical studies, we use $C_{\mathrm{win}} = 100$ and $C_{\mathrm{lose}} = 0$. Then, the estimated generalization performance based on such a win-lose game outcome is [14]

$$\hat{G}_{\mathrm{W}}(i) = \frac{1}{N} \sum_{j \in S_N} G_{\mathrm{W}}(i, j). \tag{12.7}$$

Note that $G_{\mathrm{W}}(i)$ denotes the true generalization performance.

However, one can define generalization performance in terms of game payoff. Assume nonnegative game payoff $g(i, j) \geq 0$ for any pair of strategies $i, j \in \mathcal{S}$. Then, the estimated generalization performance based on aggregation of game payoffs is [14]

$$\hat{G}_{\mathrm{A}}(i) = \frac{1}{N} \sum_{j \in S_N} G_{\mathrm{A}}(i, j). \tag{12.8}$$

As before, $G_{\mathrm{A}}(i)$ correspondingly denotes the true generalization performance. One example for illustration is to consider the IPD games. Here, the game outcome is the average payoff per move where $G_{\mathrm{A}}(i, j) = g(i, j) \in [G_{\min}, G_{\max}]$, where $G_{\max} = T$ and $G_{\min} = S$ represent the payoffs for joint-moves associated with temptation to defect and payoff for a sucker's play, respectively.

The second aspect of generalization performance measurements has to do with the issue of computing performance with respect to a ***biased sample*** of test strategies. This is related to the probability distribution $\mathbb{P}_{\mathcal{S}}$ on the strategy space $\mathcal{S}$, which is typically unknown, and might change depending on the competition settings that one considers. Past studies have employed *ad hoc* methods to obtain such a biased sample. This includes the use of expert human knowledge [2, 35] that is typically unavailable. Furthermore, these human expert biased samples are typically small-sized and as such when used to estimate generalization performance they could introduce a scalability issue and make those measurements coarse because of large differences in competence levels between the test strategies and those produced via learning systems because of their typical small sizes.

The alternative is to sample from the strategy space to obtain these biased test samples. However, this is very difficult in practice because there is no notion of goodness metric in the strategy space where one can locate and then sample high-performing test strategies with higher probabilities for most games other than abstract ones that are specifically constructed with such a metric space. We propose a procedure based on a simple heuristic to obtain a biased test sample. The ***multiple partial enumerative search*** extends our earlier approach in [19]. The procedure is described as in Algorithm 12.1.

Algorithm 12.1 Multiple partial enumerative search [14]

Input:

Y	Population of candidate test strategies
C	Sample size of biased test strategies
N	Population size of single partial enumerative search

Output:

S	Sample of biased test strategies

```
procedure MULPARSEARCH(Y, C, N)
    r := 1                                   ▷ Initialize run count
    S := initialize(Y)                       ▷ Initialize S
    while r ≤ C do
        S_r := PARSEARCH(N)                  ▷ Single partial enumerative search
        r := r + 1
    end while
    return S
end procedure

procedure PARSEARCH(N)
    Q := ((q, f)_i, i = 1, ..., N)    ▷ Sample q ∈ S randomly, initialize scores fs to zeroes
    roundrobinTour(Q)     ▷ Round-robin tournament on Q to obtain cumulative scores fs
    ascendingSort((Q, f))                    ▷ Sort (q, f)_i in order of increasing scores f
    return Q[N].q
end procedure
```

We require the population size N in each run of the partial enumerative search to be larger than the maximum number of unique strategies that can be obtained from a coevolutionary run, i.e. $N >$ (generation $\times$ population size of coevolution). The heuristic argument is that a learning system is likely more effective in searching the strategy space for a higher performing strategy than one performed by direct sampling. The procedure of repeating the partial enumerative search C independent times is aimed at obtaining a more diverse sample of high-performance test strategies compared to the earlier approach in [19] that obtains a sample of C test strategies from a single run. Repeating the partial enumerative search might address the issue of obtaining lower performing test strategies due to a particular poor sample of the population of a partial enumerative search.

The third and last aspect of generalization performance measurements that we introduce and discuss has to do with individual and population level measurements. In most cases, analysis is focused on the best-performing strategy in the coevolving population. However, there is no prerequisite to choose the best coevolved strategy of the population to be the sole representative, especially in game-playing settings where one might suspect intransitivity to occur even between top performing strategies (i.e. they form a strong component). One can choose the best U coevolved strategies from the population, $\mathrm{SPOP}_U = \{\mathrm{spop}_1, \ldots, \mathrm{spop}_U\}$ (e.g. ordered according to their relative fitness values). Here, we introduce three different measurements of generalization performance for coevolutionary learning systems [14]:

(i) ***Best***

$$\mathrm{Best}(G_{\mathrm{SPOP}_U}) = \hat{G}_{\mathrm{spop}_1}. \tag{12.9}$$

(ii) ***Average***

$$\mathrm{Avg}(G_{\mathrm{SPOP}_U}) = \frac{1}{U}\left(\sum_{l}^{U} \hat{G}_{\mathrm{spop}_l}\right). \tag{12.10}$$

(iii) ***Ensemble***

$$\mathrm{Ens}(G_{\mathrm{SPOP}_U}) = \frac{1}{N}\left(\sum_{j \in S_N} \min\left(\sum_{l}^{U} G_{\mathrm{spop}_l}(j), G_{\max}\right)\right). \tag{12.11}$$

The first two measurements are straightforward. $\mathrm{Best}(G_{\mathrm{SPOP}_U})$ refers to the generalization performance of the best strategy relative to the subset of the population SPOP_U. $\mathrm{Avg}(G_{\mathrm{SPOP}_U})$ refers to the average of generalization performances of $\{\mathrm{spop}_1, \ldots, \mathrm{spop}_U\}$. However, the ***ensemble*** measurement given by $\mathrm{Ens}(G_{\mathrm{SPOP}_U})$ assumes that the ensemble of SPOP_U coevolved strategies is equipped with a perfect gating mechanism [22]. The ensemble always chooses the best coevolved strategy in SPOP_U that best performs against each of the test strategies. For example, consider the case involving five test strategies. If spop_1 outperforms the first two test strategies and spop_2 outperforms the last three test strategies, then the ensemble of these two evolved strategies outperforms all test strategies, i.e. it will choose spop_1 for the first two test strategies and spop_2 for the others. Note that this measurement is only relevant for $G_{\mathrm{W}}(i)$. This measurement is used to determine whether the population as a whole generalizes better compared to an individual strategy.

12.2.3 *Computational Studies of Chebyshev's Bound for Competitive Coevolution*

In this subsection, we present various computational studies we have conducted in [14] to demonstrate the applications of Chebyshev's bound for competitive coevolutionary learning systems. First, we present a computational study that is aimed at verifying Chebyshev's bound and to illustrate the mechanics of the proposed statistical estimator for generalization performance based on the use of a random test sample S_N. We consider the IPD game, specifically with three-choice IPD with deterministic and reactive memory-one strategies. There are various strategy representations that have been employed, e.g. a look-up table with bit-string encoding [2], Finite State Machines [32–34], and Artificial Neural Networks (ANNs) [16, 23, 35, 38]. Here, we use the ***direct look-up table*** strategy representation we have introduced in [16]. Figure 12.2 illustrates the direct look-up table representation for the agent implementing the three-choice IPD strategy that considers one past joint-moves of its own and the opponent's, where m_{ij}, $i, j = 1, 2, 3$, specifies the choice to be made with the indices i and j representing the agent's or player's previous choice and the opponent's previous choice, respectively. The first move m_{fm} play is specified independently. For three choices, the table elements each take a value from $\{-1, 0, +1\}$.

The main advantage of using the direct look-up table representation is that it induces a finite, discrete strategy space $\mathcal{S}$, which simplifies and allows direct investigation on the generalization performance of coevolutionary learning. We assume a uniform probability distribution $\mathbb{P}_{\mathcal{S}}$ on the strategy space $\mathcal{S}$. It is possible to compute the true generalization performance G_i since the strategy space has $M = 59049$ unique strategies. It is not possible for the four-choice IPD game, while it is trivial for the two-choice IPD game. We use this set of values $T = 5$, $R = 4$, $P = 1$, and $S = 0$ and employ the following linear interpolation to generate the n-choice IPD game payoff matrix:

$$p_{\text{A}} = 2.5 - 0.5c_{\text{A}} + 2c_{\text{B}}, \quad -1 \leq c_{\text{A}}, c_{\text{B}} \leq 1, \tag{12.12}$$

where p_{A} is the payoff to player A, given that c_{A} and c_{B} are the cooperation levels of the choices that players A and B make, respectively. The discrete cooperation

$$\begin{array}{c} \\ +1 \\ 0 \\ -1 \end{array} \begin{array}{c} \begin{array}{ccc} +1 & 0 & -1 \end{array} \\ \begin{pmatrix} m_{11} & m_{12} & m_{13} \\ m_{21} & m_{22} & m_{23} \\ m_{31} & m_{32} & m_{33} \end{pmatrix} \end{array}$$

Fig. 12.2 The direct look-up table representation for the deterministic and reactive memory-one IPD strategies that consider three choices. The reactive part is represented by the illustrated 3×3 matrix. This first move m_{fm} is not shown in the figure

levels correspond to the number of choices n, with full cooperation being $+1$ and full defection being -1 [16, 23–25, 63]. For example, the payoff matrices for the three-choice and four-choice IPD games are given by Figs. 12.3 and 12.4 [16], respectively. All IPD games involve a fixed game length of 150 iterations.

In the following experiments, the estimated $\hat{G}_i(S_N)$ and true G_i values of generalization performances of 4000 random strategies drawn from S have been computed. The proportion of these 4000 strategies where $|D_N|' > \epsilon'$ is determined. This proportion corresponds with the actual probability $\mathbb{P}_N P(|D_N|' \geq \epsilon')$. Experiments have been repeated for 50 samples of random test strategies S_N for different Ns (rough % of M) at 10 (0.02%), 50 (0.1%), 100 (0.2%), 500 (1%), 1000 (2%), 5000 (10%), and 10000 (20%) that are drawn identically and independently. Note that for each sample S_N, no strategy is obtained more than once.

Figure 12.5 shows the empirical results for the generalization performance defined by $G_W(i)$ and estimated with $\hat{G}_W(i)$. One can observe that the proportion

Fig. 12.3 The payoff matrix for the two-player three-choice IPD game. Each element of the matrix gives the payoff for the row player (Player A)

$$\begin{array}{c} \\ +1 \\ 0 \\ -1 \end{array}\begin{array}{c} \begin{array}{ccc} +1 & 0 & -1 \end{array} \\ \left(\begin{array}{ccc} 4 & 2 & 0 \\ 4\frac{1}{2} & 2\frac{1}{2} & \frac{1}{2} \\ 5 & 3 & 1 \end{array}\right) \end{array}$$

Fig. 12.4 The payoff matrix for the two-player four-choice IPD game. Each element of the matrix gives the payoff for the row player (Player A)

$$\begin{array}{c} \\ +1 \\ +\frac{1}{3} \\ -\frac{1}{3} \\ -1 \end{array}\begin{array}{c} \begin{array}{cccc} +1 & +\frac{1}{3} & -\frac{1}{3} & -1 \end{array} \\ \left(\begin{array}{cccc} 4 & 2\frac{2}{3} & 1\frac{1}{3} & 0 \\ 4\frac{1}{3} & 3 & 1\frac{2}{3} & \frac{1}{3} \\ 4\frac{2}{3} & 3\frac{1}{3} & 2 & \frac{2}{3} \\ 5 & 3\frac{2}{3} & 2\frac{1}{3} & 1 \end{array}\right) \end{array}$$

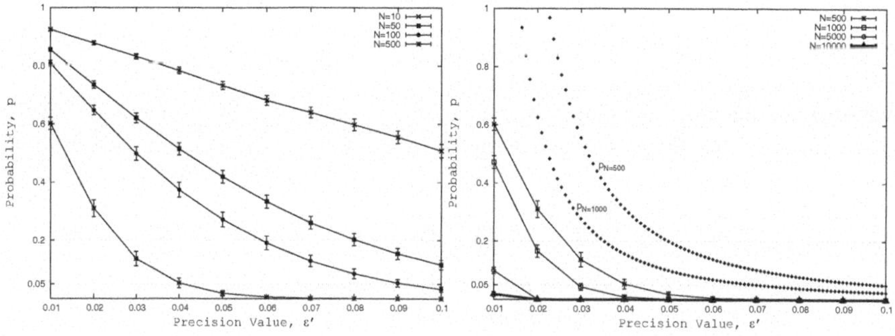

(a) $N = \{10, 50, 100, 500\}$ (b) $N = \{500, 1000, 5000, 10000\}$

Fig. 12.5 Figures showing the probability that $|D_N|' > \epsilon'$ for ϵ' in [0.01, 0.1] where the sample of random test strategies S_N of size N is used to compute $\hat{G}_W(i)$. Each graph is obtained by averaging from results of using 50 independent samples S_N with error bars representing 95% confidence interval. P_N gives the curves for the Chebyshev's bounds with different Ns [14]

of the 4000 values of $\{|D_N|' > \epsilon'\}_i$, $i = 1, \ldots, 4000$, reduces as N increases from 10 to 10000 for ϵ' in $[0.01, 0.1]$. The empirical probability curves are observed to be lower than those given by Chebyshev's bounds $\mathbb{P}_N(|D_N|' \geq \epsilon')$. Then, we analyze the empirical distribution (histogram) of $G_i(j) - G(i)$ for individual strategies (which we randomly picked) against a sample of test strategies (which we also randomly picked from many samples S_Ns). A closer inspection reveals that for four strategies, the empirical distributions are always bimodal (i.e. having two separate peaks) since a strategy i either wins or loses to a strategy j. Crucially, one can observe that the empirical distributions are similar as sample sizes N increase. There are two implications for these results. First, estimates based on using a smaller S_N are as accurate as those based on using larger S_Ns. Second, the estimations are stable in terms of varying sample sizes starting from a small S_N. This is not surprising given that Chebyshev's bounds $\mathbb{P}_N(|D_N|' \geq \epsilon')$ assume the worst case that $\sigma_i^2 = R^2/4$ [14].

Next, we illustrate how the generalization performance framework can be applied for rigorous quantitative analysis of competitive coevolutionary learning systems. In particular, we focus on simple IPD games, e.g. those with *two* and *three* choices where the true generalization performance of a strategy can be directly computed. This computational study lets us demonstrate that a small sample S_N can provide good estimates of generalization performances of strategies produced by a coevolutionary learning system. Crucially, various generalization performance measurements can be made on the coevolving population that aid with analysis of the learning process. For the following experiments, we consider a single-population competitive coevolution that is based on Evolutionary Programming with tournament selection with a full population of 2λ (see Sect. 1.2.4). Algorithm 12.2 describes this ***Classical Coevolutionary Learning*** (**CCL**) [16]. A simple mutation operator is used to generate offsprings directly from parents where **mutation** replaces the original choice of a particular element in the direct look-up table (m_{fm} and m_{ij}) with one of the other remaining $n - 1$ possible choices with an equal probability of $1/(n-1)$. A fixed mutation rate p_{m} is used. A strategy's fitness $X[i].f$ is the average game payoff from all games played in round-robin tournaments. A fixed and sufficiently long duration for the evolutionary process (e.g. $\text{b}_{\text{gen}} = 300$) is used to observe changes in the generalization performance of CCL. All experiments are repeated for 30 independent runs.

Some observations can be made on the results in [14] for the generalization performance of CCL with respect to $G_{\text{W}}(i, j)$ using all three measurements (best, average, and ensemble). Note that ensemble measurements are always higher compared to best and average measurements for CCL of both IPD games. However, the difference is not substantial indicating that the coevolving population consists of mostly similar strategies. The generalization performance of CCL differs for the two-choice and three-choice IPD games. For CCL of the two-choice IPD, $G_{\text{W}}(i)$ measurements decrease as evolutionary process continues. For CCL of the three-choice IPD, $G_{\text{W}}(i)$ measurements remain around similar levels throughout the evolutionary process. For the simpler two-choice IPD game, the population of CCL is known to overspecialize to more cooperative strategies (e.g. all cooperate

Algorithm 12.2 Classical coevolutionary learning (CCL) [16]

Input:
Y Population of λ candidate strategies
b_{gen} Termination criterion fixed number of generation
Output:
X Population of λ coevolved strategies

```
procedure CCL(Y, b_gen)
    t := 0                                        ▷ Initialize time-stamp
    X_p := initialize(Y)                          ▷ Initialize parent population X_p
    while t < b_gen do
        X_o := variation(X_p)                     ▷ Generate offspring population
                                                  ▷ for i = 1, ..., λ, X_o[i].x := mutation(X_p[i].x)
        X := combine(X_p, X_o)                    ▷ X = X_p ∪ X_o
        roundrobinTour(X)                         ▷ Round-robin tournament on X to obtain
                                                  ▷ cumulative game payoffs X[i].fs
        X_p := selectionTour(X)                   ▷ Tournament selection: choose best half of X
        t := t + 1
    end while
    return X_p
end procedure
```

strategies) that do not generalize well when $G_W(i)$ measurements are used. In particular, the game outcome only registers a win for strategy i if $g(i, j) > g(j, i)$. Cooperative strategies can be exploited by those that defect and thus will obtain lower IPD game payoffs compared to their opponents $g(i, j) < g(j, i)$ leading to poor generalization performance $G_W(i)$ [14].

The higher tendency of CCL to overspecialize to naive cooperators for IPD games with fewer choices can be explained by noting that the strategy space $\mathcal{S}$ consists of a higher proportion of naive cooperators when there are less choices to play. For all cooperate in the two-choice IPD, it plays full cooperation for the first move $m_{fm} = +1$ and plays full cooperation regardless of the previous joint-move itself and the opponent made. This is reflected by the following direct look-up table configuration:

$$\begin{pmatrix} +1 & +1 \\ * & * \end{pmatrix},$$

where $*$ can be $+1$ or -1 in the two-choice IPD. If $m_{fm} = +1$ and that the choices in the first row are all $+1$, then such a strategy will always play full cooperation since it will only access elements in the first row. There are 2^2 combinations of such tables, which gives one the proportion of effectively all cooperate strategies in $\mathcal{S}$ as $2^2/2^5$.

The same analysis can be applied for the three-choice IPD case, where the all cooperate strategy has $m_{fm} = +1$ and a direct look-up table configuration:

$$\begin{pmatrix} +1 & +1 & +1 \\ * & * & * \\ * & * & * \end{pmatrix},$$

where $*$ can be $+1$, 0, or -1 in the three-choice IPD. The proportion of all cooperate in $\mathcal{S}$ is $3^6/3^{10}$. For the n-choice IPD, the proportion of all cooperate is given by $n^{(n^2-n)}/n^{(n^2+1)} = 1/n^{(n+1)}$. The proportion of all cooperate strategies decreases exponentially with n. For the two-choice IPD, the proportion is 12.5%. For the three-choice IPD, the proportion is approximately 1%.

Various previous studies have found that coevolution that employs cumulative or average game payoffs calculated from round-robin tournaments as fitness values are more likely to produce populations of mutually cooperative strategies. This is also the case for CCL [16]. In a population of mutual cooperators, CCL cannot distinguish between different strategies that are mutually cooperating because they have similar fitness values. The population is more likely to drift to naive cooperators when CCL fails to distinguish between mutually cooperating strategies. This is reflected in our results that show on average, around 12% of the population is all cooperate for the two-choice IPD case compared to around 0.2% for the three-choice IPD case. Regardless, this computational study has demonstrated that the generalization performance measures could provide quantitative analysis that corroborates with those from deep structural analysis of the strategy space $\mathcal{S}$, the combinatorics of cooperative plays, and previous understanding that has been developed for the coevolution towards cooperative plays.

Lastly for this subsection, we compare the generalization performance of biased and unbiased samples of test strategies in the case of CCL of the three-choice IPD. We have implemented Algorithm 12.1 and made calls to the procedure **MULPARSEARCH**(Y, C, N) with $C = 20$ to produce a sample of 20 test strategies. We have investigated different population sizes $N \in \{1, 4, 10, 50, 100, 10000\}$. Only $N = 10000$ satisfies the requirement of having searched more strategies compared to what CCL can be best achieved (i.e. $10000 > \mathrm{b_{gen}} \times 2\lambda = 300 \times 20 = 6000$). This choice of N is to investigate its impact on multiple partial enumerative search procedure in producing a biased and diverse sample of high-performing test strategies that is challenging for coevolved strategies. We refer $\hat{G}_i^{\mathrm{B}}$ to denote the estimated generalization performance using a biased sample of test strategies. Results from experiments are summarized in Fig. 12.6. In general, increasing the population size N in calls for the procedure **MULPARSEARCH**(Y, C, N) produces a sample of high-performing test strategies that are harder to defeat. For example, the same coevolved strategies have lower $\hat{G}_i^{\mathrm{B}}$ measurements when compared to G_i measurements. Differences between the two measurements are more pronounced especially for cases where $\hat{G}_i^{\mathrm{B}}$s are obtained with respect to the biased sample produced using **MULPARSEARCH**(Y, C, N) with larger population sizes N as shown in Fig. 12.6b [14].

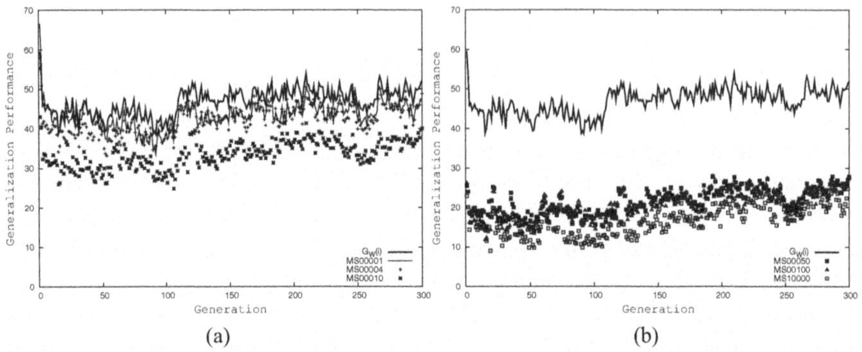

Fig. 12.6 Comparison between $G_W(i)$ and $\hat{G}^B_W(i)$s for CCL across the evolutionary process, averaged over 30 independent runs ("Best" measurements only). (**a**) Multiple partial enumerative search with population sizes of one, four, and ten. (**b**) Multiple partial enumerative search with population sizes of 50, 100, and 10000 [14]

12.3 Improving Generalization Performance Estimates in Coevolutionary Learning

This section introduces theoretical studies we have made in [13, 14] to improve statistical estimators for generalization performance in competitive coevolution, especially with regard to their computational costs that are directly proportional to the number of test strategies used to compute those estimates. First in Sect. 12.3.1 we show how the statistical estimator can be improved by performing a ***second estimation of the variance*** σ_i^2 that could well be lower than the worst case of $R^2/4$ used in Chebyshev's bound. Crucially in Sect. 12.3.2, we will directly address this issue in Chebyshev's bound and demonstrate how an underlying property of these statistical estimators (i.e. the near-Gaussian nature of average game outcomes) can be exploited to provide ***tighter bounds based on parametric testing***. We can exploit this formalization to perform statistical testing for comparison of strategies, which we present in Sect. 12.3.3. Then Sect. 12.3.4 will present various computational studies to verify these theoretical studies and demonstrate how these improved estimators can be used for analysis of coevolutionary learning of more complex competitive games.

12.3.1 *Obtaining Tighter Chebyshev's Bound by Estimating the Variance* σ^2

Recall that the generalization performance G_i of a strategy $i \in \mathcal{S}$ is defined as the expected mean performance (game outcome) $\mathbb{E}_{\mathbb{P}_S}[G_i(J)]$ against the random test strategy J taking on value $j \in \mathcal{S}$. We have that $G_i = \mathbb{E}_{\mathbb{P}_S}[G_i(J)]$ with $G_i(J)$

taking on value from a finite interval. In the following, we will demonstrate how Chebyshev's bound can be improved by a second estimation on the variance of $G_i(J)$ that we have formulated in [14]. The argument can be technically involved, so to simplify our presentation we now consider $G_i(J)$ as the random variable X with the underlying distribution $\mathbb{P}_X$ taken from a compact interval of real numbers $X \in [a, b]$. For N realizations $x_1, x_2, \ldots, x_N$, the empirical mean $\hat{\mu}_N$ of X is given as

$$\hat{\mu}_N = \frac{1}{N} \sum_{j=1}^{N} x_j,$$

whereas the true mean $\mathbb{E}_{\mathbb{P}_X}[X]$ and true variance σ^2 are given as

$$\mathbb{E}_{\mathbb{P}_X}[X] = \mu,$$
$$\sigma^2 = \mathbb{E}_{\mathbb{P}_X}\left[\left(X - \mathbb{E}_{\mathbb{P}_X}[X]\right)^2\right].$$

The maximum variance of a random variable X over $[a, b]$ is $\sigma^2_{\max} = R_X^2/4$ where $R_X = b - a$. For a positive ϵ, this lets us obtain Chebyshev's bound:

$$\mathbb{P}_N(|\hat{\mu}_N - \mu| \geq \epsilon) \leq \frac{R_X^2}{4N \cdot \epsilon^2}.$$

We want to obtain a tighter upper bound compared to Chebyshev's bound. If $\mu = \mathbb{E}_{\mathbb{P}_X}[X]$ is known, we could obtain an unbiased estimate of σ^2 that is given as

$$\hat{\sigma}_N^2 = \frac{1}{N} \sum_{j=1}^{N} (x_j - \mu)^2 = \frac{1}{N} \sum_{j=1}^{N} y_j,$$

where $y_j = (x_j - \mu)^2$, $j = 1, \ldots, N$, are realizations of a new random variable $Y = (X - \mathbb{E}_{\mathbb{P}_X}[X])^2$. Indeed, for N realizations $y_1, y_2, \ldots, y_N$, the empirical mean is $\hat{\sigma}_N^2$, whereas the true mean $\mathbb{E}_{\mathbb{P}_Y}[Y]$ is

$$\mathbb{E}_{\mathbb{P}_Y}[Y] = \mathbb{E}_{\mathbb{P}_X}\left[(X - \mathbb{E}_{\mathbb{P}_X}[X])^2\right] = \sigma^2.$$

We apply Chebyshev's Theorem a second time and specifically for Y. Then, for a positive δ, we obtain [14]

$$\mathbb{P}_N(|\hat{\sigma}_N^2 - \sigma^2| \geq \delta) \leq \frac{\mathrm{var}_{\mathbb{P}_Y}[Y]}{N \cdot \delta^2},$$

where $\mathrm{var}_{\mathbb{P}_Y}[Y]$ is the variance of Y. For a finite $y_j = (x_j - \mu)^2$, its minimal possible value is $y_{\min} = 0$ if $x_j = \mu$, and its maximal possible value is $y_{\max} = (b - a)^2$ if

$x_j = a \wedge \mu = b$. Then, the range for Y is $R_Y = y_{\max} - y_{\min} = (b-a)^2$. This gives an upper bound of $\mathrm{var}_{\mathbb{P}_Y}[Y]$ via $\mathrm{var}_{\max}[Y] \leq R_Y^2/4 = (b-a)^4/4 = R_X^4/4$. As such, we can obtain the following [14]:

$$\mathbb{P}_N(|\hat{\sigma}_N^2 - \sigma^2| \geq \delta) \leq \frac{R_X^4}{4N \cdot \delta^2}.$$

We summarize the two points above as follows. First, we have $|\hat{\mu} - \mu| \leq \epsilon$ with probability of at least $c_1 = 1 - R_X^2/(4N\epsilon^2)$ with respect to generation of many N-tuples $(x_1, \ldots, x_N)$ from $\mathbb{P}_X^N$. Second, we have $|\hat{\sigma}_N^2 - \sigma^2| \leq \delta$ with probability of at least $c_2 = 1 - R_X^4/(4N\delta^2)$. We would like to use $\hat{\sigma}_N^2 + \delta$ as an upper bound for σ^2 when the latter is lower than $R_X^4/4$. However, we have assumed that we know μ. In practice, μ is unknown.

We examine the correction required that $\hat{\mu}$ instead of μ is used to calculate the variance $\hat{\sigma}_N^2$ [14]. We have that

$$\hat{\hat{\sigma}}_N^2 = \frac{1}{N}\sum_{j=1}^{N}(x_j - \hat{\mu}_N)^2.$$

We know $|\hat{\mu} - \mu| \leq \epsilon$ with probability at least $c_1 = 1 - R_X^2/(4N\epsilon^2)$. The worst case in upper bounding σ^2 is when $\hat{\hat{\sigma}}_N^2$ underestimates $\hat{\sigma}_N^2$ because $\hat{\mu}_N$ is closer to x_j than μ. There are two such cases, and we provide the corresponding correction: (a) For $x_j < \hat{\mu}_N < \mu$, we calculate $(\hat{\mu}_N + \epsilon - x_j)^2$. (b) For $\mu < \hat{\mu}_N < x_j$, we calculate $(x_j - \hat{\mu}_N - \epsilon)^2$.

Note that in both cases since the terms are squared, we can use $(x_j - \hat{\mu}_N - \epsilon)^2 = \left(x_j - (\hat{\mu}_N + \epsilon)\right)^2$. Then, we have $(x_j - \hat{\mu}_N - \epsilon)^2 \geq (x_j - \mu)^2$ with probability $c_1 = 1 - R_X^2/(4N\epsilon^2)$ with respect to generation of many N-tuples $(x_1, \ldots, x_N)$ from $\mathbb{P}_X^N$. That is, with probability c_1, we can be sure that

$$\hat{\sigma}_{N,U}^2 = \frac{1}{N}\sum_{j=1}^{N}(x_j - \hat{\mu}_N - \epsilon)^2 \geq \hat{\sigma}_N^2.$$

Then, with probability $c_2 = 1 - R_X^4/(4N\delta^2)$, we have the upper bound $\sigma^2 \leq \hat{\sigma}_N^2 + \delta$.

What we need to do now is to combine the two probabilities c_1, c_2 for the two underlying events in a simple way, i.e. they are independent [14]. This requires two sets of N-tuples of observations. The first set of $x_{1,1}, x_{1,2}, \ldots, x_{1,N}$ is used to estimate $\hat{\mu}_{1,N} = \frac{1}{N}\sum_{j=1}^{N} x_{1,j}$. The second set $x_{2,1}, x_{2,2}, \ldots, x_{2,N}$ is used to estimate $\hat{\mu}_{2,N} = \frac{1}{N}\sum_{j=1}^{N} x_{2,j}$ and $\hat{\sigma}_{N,U}^2 = \frac{1}{N}\sum_{j=1}^{N}(x_{2,j} - \hat{\mu}_{2,N} - \epsilon)^2$. Then, with probability

$$c_1 \cdot c_2 = \left(1 - \frac{R_X^2}{4N \cdot \epsilon^2}\right) \cdot \left(1 - \frac{R_X^4}{4N \cdot \delta^2}\right),$$

we have $\sigma^2 \leq \hat{\sigma}_{N,U}^2 + \delta$ given that $\sigma^2 \leq \hat{\sigma}_N^2 + \delta$ and $\hat{\sigma}_N^2 \leq \hat{\sigma}_{N,U}^2$.

As such, one can claim with probability at least $c_1 c_2$ that the following inequality holds:

$$\mathbb{P}_N(|\hat{\mu}_{1,N} - \mu| \geq \epsilon) \leq \frac{\hat{\sigma}_{N,U}^2 + \delta}{N \cdot \epsilon^2}.$$

For this inequality to be of use, we require that $\sigma^2 \leq \hat{\sigma}_{N,U}^2 + \delta < \sigma_{\max}^2 = R_X^2/4$. Our theoretical formulation above is summarized as the proof argument for the following Lemma 12.2 for an improved statistical estimator for generalization performance of competitive coevolution based on the second estimation of the variance [14].

Lemma 12.2 ([14]) *For a strategy i, consider two independent non-overlapping sets of N test strategies: T_1 and T_2, where $T_1 \cap T_2 = \emptyset$ and $|T_1| = |T_2| = N$. The first set is used to estimate the generalization performance $\hat{G}_i(T_1) = \frac{1}{N}\sum_{j \in T_1} G_i(j)$. The second set is used to estimate the variance $\hat{\sigma}_{N,U}^2 = \frac{1}{N}\sum_{j \in T_2} \left(G_i(j) - \hat{G}_i(T_2) - \epsilon\right)^2$, for some positive number $\epsilon > 0$, where $\hat{G}_i(T_2) = \frac{1}{N}\sum_{j \in T_2} G_i(j)$. Then, for $\delta > 0$ with probability at least $c_1 c_2 = (1 - R^2/(4N\epsilon^2))(1 - R^4/(4N\delta^2))$, the following inequality holds:*

$$\mathbb{P}_N(|\hat{G}_i(T_1) - G_i| \geq \epsilon) \leq \frac{\hat{\sigma}_{N,U}^2 + \delta}{N \cdot \epsilon^2}. \tag{12.13}$$

12.3.2 Error Estimations for Gaussian-Distributed Generalization Estimates

Although we are now able to obtain a tighter bound compared to Chebyshev's that assumes the worst case, this comes at twice the computational cost since a second non-overlapping set of random test strategies is required. Here, we will address this issue of improving generalization performance estimates directly and exploit an underlying property of generalization estimates that being averages of game outcomes are near-Gaussian for large enough realizations of those outcomes. Our subsequent formal analysis we have made in [13] employs higher (second and third) order moments of the game outcomes against the random test strategies to estimate the speed (in terms of sample sizes) with which the distribution of average game outcomes (i.e. generalization performance) converges to the standard Gaussian.

In particular, selection of the sample S_N of test strategies can be formalized through a random variable $\mathcal{S}_N$ on $\mathcal{S}^N$ endowed with the product measure induced by $\mathbb{P}_{\mathcal{S}}$. Estimates of the generalization performance of strategy i can be viewed as realizations of the random variable $\hat{G}_i(\mathcal{S}_N)$. Since game outcomes $G_i(J)$ have finite mean and variance, by the ***Central Limit Theorem***, for large enough N, $\hat{G}_i(\mathcal{S}_N)$ is Gaussian-distributed. Claims regarding the speed of convergence of the (cumulative) distribution of $\hat{G}_i(\mathcal{S}_N)$ to the (cumulative) distribution of a Gaussian can be made quantitative using the ***Berry-Esseen Theorem*** [44].

To demonstrate this formally, we need to normalize $\hat{G}_i(\mathcal{S}_N)$ to zero mean and unit standard deviation [13]. First, we normalize $G_i(J)$ to zero mean:

$$X_i(J) = G_i(J) - G_i.$$

Denote the variance of $G_i(J)$ (and hence the variance of $X_i(J)$) by σ_i^2. Since $G_i(J)$ takes on values in a finite domain, the third absolute moment

$$\rho_i = \mathbb{E}_{\mathbb{P}_{\mathcal{S}}}[|X_i(J)|^3]$$

of $X_i(J)$ is finite. Second, we normalize $\hat{G}_i(\mathcal{S}_N)$ to zero mean:

$$Y_i(\mathcal{S}_N) = \hat{G}_i(\mathcal{S}_N) - G_i = \frac{1}{N} \sum_{j \in \mathcal{S}_N} X_i(j).$$

Third, we normalize $\hat{G}_i(\mathcal{S}_N)$ to unit standard deviation:

$$Z_i(\mathcal{S}_N) = \frac{Y_i(\mathcal{S}_N)}{\frac{\sigma_i}{\sqrt{N}}}.$$

The Berry-Esseen Theorem states that the cumulative distribution function (cdf) F_i of $Z_i(\mathcal{S}_N)$ converges (pointwise) to the cdf Φ of the standard normal distribution $N(0, 1)$: For any $x \in \mathbb{R}$,

$$|F_i(x) - \Phi(x)| \leq \frac{0.7975}{\sqrt{N}} \frac{\rho_i}{\sigma_i^3}. \tag{12.14}$$

Only σ_i and ρ_i are required to make an estimate on the pointwise difference between cdf's of $Z_i(\mathcal{S}_N)$ and $N(0, 1)$. Since the theoretical moments σ_i and ρ_i are unknown in practice, we use their empirical estimates instead. To ensure that the cdf's of $Z_i(\mathcal{S}_N)$ and $N(0, 1)$ do not differ pointwise by more than $\epsilon_{\text{cdf}} > 0$, we need at least $N_{\text{cdf}}(\epsilon)$ test strategies given as follows [13]:

$$N_{\text{cdf}}(\epsilon) = \frac{0.7975^2}{\epsilon_{\text{cdf}}^2} \frac{\rho_i^2}{(\sigma_i^2)^3}. \tag{12.15}$$

We now show how by employing parametric testing, the new statistical estimator could potentially improve on the previous estimator via a reduction in computational cost through reduced sample size N requirements [13]. Assume that the generalization estimates are Gaussian-distributed (e.g. using the analysis above, we gather enough test points to make the means *almost* Gaussian distributed). Denote by $z_{\alpha/2}$ the upper $\alpha/2$ point of $N(0, 1)$, i.e. the area under the standard normal density for $(z_{\alpha/2}, \infty)$ is $\alpha/2$, and for $[-z_{\alpha/2}, z_{\alpha/2}]$ it is $(1 - \alpha)$. For large strategy samples S_N, the estimated generalization performance $\hat{G}_i(S_N)$ has standard error of $\sigma_i/\sqrt{N}$, and the $100(1 - \alpha)\%$ error margin of $\hat{G}_i(S_N)$ is $z_{\alpha/2}\sigma_i/\sqrt{N}$. Since σ_i is generally unknown, the standard error can be estimated as [13]

$$\frac{\hat{\sigma}_i(S_N)}{\sqrt{N}} = \sqrt{\frac{\sum_{j \in S_N} \left(G_i(j) - \hat{G}_i(S_N)\right)^2}{N(N-1)}} \tag{12.16}$$

to give the $100(1 - \alpha)\%$ error margin $\Upsilon_i(\alpha, N)$ of $\hat{G}_i(S_N)$ that we obtain as

$$\Upsilon_i(\alpha, N) = z_{\alpha/2} \cdot \sqrt{\frac{\sum_{j \in S_N} \left(G_i(j) - \hat{G}_i(S_N)\right)^2}{N(N-1)}}.$$

This allows us to formulate the following expression whereby to require that the error margin be at most $\delta_{\mathrm{em}} > 0$ leads to samples of at least [13]:

$$N_{\mathrm{em}}(\delta) = \frac{z_{\alpha/2}^2 \, \sigma_i^2}{\delta_{\mathrm{em}}^2}. \tag{12.17}$$

In other words, we need $N_{\mathrm{em}}(\delta)$ test strategies to be $100(1-\alpha)\%$ sure that $|\hat{G}_i(S_N) - G_i| < \delta_{\mathrm{em}}$. Then, the $100(1 - \alpha)\%$ confidence interval for the true generalization performance of strategy i is

$$(\hat{G}_i(S_N) - \Upsilon_i(\alpha, N), \hat{G}_i(S_N) + \Upsilon_i(\alpha, N)). \tag{12.18}$$

There is one more property of generalization estimates being Gaussian-distributed that we can exploit. In particular, the same reasoning we employ making use of the Central Limit Theorem can be applied to sample estimates of σ_i whose true value is unknown in practice [13]. For m iid test strategy samples S_N^r, $r = 1, 2, \ldots, m$, each of size N, the generalization performance estimates

$$\hat{G}_i(S_N^r) = \frac{1}{N} \sum_{j \in S_N^r} G_i(j)$$

are close to being Gaussian-distributed with mean G_i and standard deviation $\sigma_i/\sqrt{N}$ (for large enough N). These generalization estimates can be used to estimate the confidence interval for σ_i.

The sample variance V_m^2 of those estimates $\hat{G}_i(S_N^r)$, $r = 1, 2, \ldots, m$, is [13]

$$V_m^2 = \frac{\sum_{r=1}^{m}(\hat{G}_i(S_N^r) - \Gamma_i)^2}{m-1},$$

$$\Gamma_i = \frac{1}{m}\sum_{r=1}^{m}\hat{G}_i(S_N^r).$$

Crucially, the normalized sample variance U_m^2 of a Gaussian-distributed $\hat{G}_i(S_N^r)$

$$U_m^2 = \frac{(m-1)\,V_m^2}{\frac{\sigma_i^2}{N}}$$

is χ^2-distributed with $m-1$ degrees of freedom. Note that there is a trivial inequality for the third moment (page 210 of [17]) that however leads to rather broad bounds.

As such, the $100(1-\alpha)\%$ confidence interval for $\sigma_i/\sqrt{N}$ is

$$\left(V_m \cdot \sqrt{\frac{m-1}{\chi^2_{\alpha/2}}},\; V_m \cdot \sqrt{\frac{m-1}{\chi^2_{1-\alpha/2}}}\right),$$

where χ^2_β is the value such that the area to the right of χ^2_β under the χ^2 distribution with $N-1$ degrees of freedom is β. This then finally gives us the $100(1-\alpha)\%$ confidence interval for σ_i

$$\left(V_m \cdot \sqrt{\frac{N(m-1)}{\chi^2_{\alpha/2}}},\; V_m \cdot \sqrt{\frac{N(m-1)}{\chi^2_{1-\alpha/2}}}\right),$$

which we rewrite in the following form for direct computation [13]:

$$\left(\sqrt{\frac{N \cdot \sum_{r=1}^{m}(\hat{G}_i(S_N^r) - \Gamma_i)^2}{\chi^2_{\alpha/2}}},\; \sqrt{\frac{N \cdot \sum_{r=1}^{m}(\hat{G}_i(S_N^r) - \Gamma_i)^2}{\chi^2_{1-\alpha/2}}}\right). \qquad (12.19)$$

12.3.3 Statistical Testing for Comparison of Strategies

The knowledge that $\hat{G}_i(S_N)$ is Gaussian distributed allows us to conduct ***statistical testing for comparison of strategies*** [13]. First, one might be interested in strategies with true generalization performance greater than some threshold $\tilde{G}$. We can test whether i is a low-performing strategy by testing for $G_i < \tilde{G}$. The hypothesis H_1 that $G_i < \tilde{G}$ is substantiated at significance level of $\alpha\%$ (against the null hypothesis

H_0 that $G_i = \tilde{G}$) if the test statistic

$$Z'_i(S_N, \tilde{G}) = \frac{\hat{G}_i(S_N) - \tilde{G}}{\sqrt{\frac{\sum_{j \in S_N} \left(G_i(j) - \hat{G}_i(S_N)\right)^2}{N(N-1)}}} \tag{12.20}$$

falls below $-z_\alpha$, i.e. if $Z'_i(S_N, \tilde{G}) \leq -z_\alpha$. Alternatively, the hypothesis that strategy i is an acceptable strategy, i.e. $G_i > \tilde{G}$, is accepted (against H_0) at significance level of $\alpha\%$, if $Z'_i(S_N, \tilde{G}) > z_\alpha$. We can also perform a nondirectional test for $G_i \neq \tilde{G}$ that requires $|Z'_i(S_N, \tilde{G})| > z_{\alpha/2}$.

Crucially, we can compare two strategies $i, j \in S$ for their relative performance [13]. This can be important in (co)evolutionary learning setting when constructing a new generation of strategies. Let both strategies i and j play against the same set of N test strategies $S_N = \{t_1, t_2, \ldots, t_N\}$. Statistical tests regarding the relation between the true generalization performances of i and j can be made using paired tests. One computes a series of performance differences on S_N

$$D(n) = G_i(t_n) - G_j(t_n), \ n = 1, 2, \ldots, N.$$

The performance differences are analyzed as a single sample. At significance level of $\alpha\%$, strategy i appears to be better than strategy j by more than a margin $\tilde{D}$ (against the null hypothesis that i has better performance than j exactly by the margin $\tilde{D}$), if $Z''_i(S_N, \tilde{D}) \geq z_\alpha$, where

$$Z''_i(S_N, \tilde{D}) = \frac{\hat{D}(S_N) - \tilde{D}}{\sqrt{\frac{\sum_{n=1}^{N}(D(n) - \hat{D}(S_N))^2}{N(N-1)}}}, \tag{12.21}$$

$$\hat{D}(S_N) = \frac{1}{N}\sum_{n=1}^{N} D(n). \tag{12.22}$$

For simply testing whether strategy i outperforms strategy j, we set the margin $\tilde{D} = 0$. Analogously, strategy i is lower performing than strategy j at significance level of $\alpha\%$ provided that $Z''_i(S_N, 0) \leq -z_\alpha$. Finally, strategies i and j appear to be different at significance level of $\alpha\%$, if $|Z''_i(S_N, 0)| \geq z_{\alpha/2}$ [13].

It is very crucial to take note that statistical comparison of strategies $i, j \in S$ is done through a set of test strategies S_N. This is different compared to pairwise comparison of strategies i, j via game-play between the two. Intransitivity in the game can offer misleading results if one is not careful which comparison one refers to. For example, one could claim strategies are equivalent in the sense they belong to the same strong component in their tournament graph (refer to Sect. 5.3.1). Here, strategy j has a higher score that implies a higher generalization performance compared to i. However, it is still possible that i defeats j in a game-play directly.

12.3.4 Computational Studies of Improved Generalization Estimates in Coevolutionary Learning

Here, we present several computational studies in [13, 14] to demonstrate how generalization performance estimates can be improved and then applied. First, we show that in practice Chebyshev's bound that assumes the worst-case $\sigma^2_{\max} = R^2/4$ can be improved with a second estimation of σ^2 with the assumption that $\hat{\sigma}^2_{N,U} + \delta < \sigma^2_{\max}$. Note that to apply the tighter Chebyshev's bound given by Eq. (12.13) with high probability c_1c_2, $c_1 = 1 - R^2/(4N\epsilon^2)$, $c_2 = 1 - R^4/(4N\delta^2)$ claims, we want c_2 to be as high as possible since this is where the tighter bound is useful because it is the case for strategy i that σ^2 is small and that $\hat{\sigma}^2_{N,U} + \delta < \sigma^2_{\max}$ even for larger δ.

Our computational study has applied Chebyshev's Theorem again and this time for the random variable $|\hat{\sigma}^2_N - \sigma^2|$ to investigate empirically how good is the estimate of σ^2 using a sample of test strategies for the three-choice IPD game since we can actually compute the true value of σ^2:

$$\mathbb{P}_N(|\hat{\sigma}^2_N - \sigma^2| \geq \delta) \leq \frac{R^4}{4N \cdot \delta^2}.$$

We can simplify the above by letting $|D(\hat{\sigma}^2_N)|' = |\hat{\sigma}^2_N - \sigma^2|/R^2$ and $\delta' = \delta/R^2$ so that we now obtain

$$\mathbb{P}_N(|D(\hat{\sigma}^2_N)|' \geq \delta') \leq \frac{1}{4N \cdot \delta'^2}.$$

Experiments have been conducted for $G_W(i)$ to determine the empirical probability (proportion) of 4000 random strategies where $|D(\hat{\sigma}^2_N)|' > \delta'$ for 50 samples of S_N of sizes $N \in \{10, 50, 100, 500, 1000, 5000, 10000\}$ that are drawn uniformly and independently. Note that we actually compute $\hat{\hat{\sigma}}^2_N$ rather than $\hat{\sigma}^2_N$. However, our results will show that in most cases $\hat{\hat{\sigma}}^2_N$ overestimates σ^2. More importantly, corrected estimates $\hat{\sigma}^2_{N,U}$ can be made and then used to obtain a tighter Chebyshev's bound.

Table 12.1 presents the $\hat{\sigma}^2_N$ for various N and also the true value σ^2 for some random strategies for $G_W(i)$. Results of $|\hat{\sigma}^2_N - \sigma^2|$ for $G_W(i)$ can be derived from the table. If we consider a large precision value of $\delta' = 0.04$ ($\delta = 400$), we know from Chebyshev's bound for example that for $N = 10000$, the probability that $|\hat{\sigma}^2_N - \sigma^2| \leq \delta$ is $1 - 1/(4 \times 10000 \times 0.04^2) > 0.98$. However, the actual estimation of σ^2 is better than the theoretical value given by Chebyshev's bound. For example, $|\hat{\sigma}^2_N - \sigma^2|$ for $N = 10000$ is much smaller than 400 for a probability greater than 0.98. Note that $\hat{\sigma}^2_N$ in Table 12.1 for a strategy can fluctuate (higher or lower) around the true value σ^2. Erratic results can be obtained for very small sample size (e.g. $N = 10$, $\hat{\sigma}^2_N = 0$). Our results do not indicate a monotonically decreasing $|D_N|$ as

Table 12.1 Comparison of $\hat{\sigma}_N^2$ for various N with the true value σ^2 for ten strategies for $G_W(i)$ where $\sigma^2 \in [0, 2500]$ [14]

N	# 1	# 2	# 3	# 4	# 5	# 6	# 7	# 8	# 9	# 10
10	2500	2400	2500	2400	2400	2100	0	2400	2500	1600
50	2464	2496	2244	1476	2496	1924	1204	2356	2464	2484
100	2304	2484	2059	1344	2464	1924	1275	2436	2496	2379
500	2436	2500	2224	1539	2491	1895	1317	2477	2497	2375
1000	2419	2498	2131	1470	2484	1875	1218	2482	2500	2278
5000	2432	2492	2108	1481	2490	1849	1341	2476	2499	2220
10000	2434	2497	2118	1515	2493	1816	1321	2473	2500	2231
TRUE	**2436**	**2496**	**2126**	**1475**	**2491**	**1816**	**1293**	**2477**	**2499**	**2228**

N is increased because the computed σ^2 are better than what Chebyshev's bound gives [14].

A tighter Chebyshev's bound can be found if $\hat{\sigma}_N^2 + \delta < \sigma_{\max}^2$. For a particular strategy (# 7 in Table 12.1), $\sigma^2 = 1293$, while $\hat{\sigma}_N^2 = 1321$. The corrected estimate is found to be $\hat{\sigma}_{N,U}^2 = 1337$ (where the probability that $\hat{\sigma}_{N,U}^2 \geq \hat{\sigma}_N^2$ is $c_1 = 1 - 1/(4 \times 10000 \times 0.04^2) > 0.98$). In this case, we know that $\hat{\sigma}_N^2$ overestimates σ^2. Regardless, with probability $c_1 c_2 = (1 - 1/(4 \times 10000 \times 0.04^2)) \times (1 - 1/(4 \times 10000 \times 0.04^2)) > (0.98) \times (0.98) > 0.95$), we know that $\sigma^2 \leq \hat{\sigma}_{N,U}^2 + \delta$ with $\delta = 400$. Here, we obtain $\hat{\sigma}_{N,U}^2 + \delta = 1337 + 400 < \sigma_{\max}^2 = 2500$ [14].

The observation that estimates of σ^2 are better than what Chebyshev's bound predict naturally motivates the question whether small sample sizes can be used reliably to estimate generalization performance. Indeed, we have already answered this theoretically in Sect. 12.3.4. We do not have to perform detailed studies how the game governs the spread of game outcomes of a strategy $i \in \mathcal{S}$ against a random sample S_N. Nevertheless, our principled approach exploits the near-Gaussian nature of generalization estimates and finds out the rate at which the distribution of generalization estimates converges to a Gaussian as the random sample of test strategies S_N used to compute generalization estimates grows in size N.

In the following computational study [13], we use the three-choice IPD game to illustrate our approach. We first collect a random sample of 50 base strategies $i \in \mathcal{S}$. We also collect 1000 independent samples S_N to compute 1000 estimates $\hat{G}_i(S_N)$. Each random sample S_N consists of N test strategies drawn iid from $\mathcal{S}$. In all sampling cases we use a uniform distribution $\mathbb{P}_{\mathcal{S}}$. For this game, we directly compute G_i so as to be able to compute $X_i(J) = G_i(J) - G_i$. Then, we compute estimates of the variance and the third absolute moment of $X_i(J)$ with respect to S_N, i.e. we have a 1000-sample estimate of $\hat{\sigma}_i^2$ and another 1000-sample estimate of $\hat{\rho}_i$.

From the Berry-Esseen Theorem (Eq. (12.15)) [44], we can compute for each base strategy i the deviation ϵ_{cdf} from the Gaussian:

$$\epsilon_{\text{cdf}} = \frac{0.7975 \cdot \hat{\rho}_i}{\sqrt{N} \cdot \hat{\sigma}_i^3} \tag{12.23}$$

given different sample sizes of S_N. We can observe how fast the distribution of generalization estimates converges to a Gaussian by systematically computing the error ϵ_{cdf} for $N = \{50, 100, 200, 300, 400, 500, 1000, 2000, 3000, 4000, 5000, 10000, 50000\}$. We take a pessimistic estimate of ϵ_{cdf} since its true value is unknown. Both 1000-sample estimates of $\hat{\sigma}_i^2$ and $\hat{\rho}_i$ are first rank-ordered in an ascending order. A pessimistic estimate of ϵ_{cdf} would be to take a smaller value (2.5%-tile) of $\hat{\sigma}_i^2$ and a larger value (97.5%-tile) of $\hat{\rho}_i$.

Although we can directly compute the quantile intervals for $\hat{\sigma}_i^2$ from the χ^2-distribution, we have loose bounds for $\hat{\rho}_i$ (based on the inequality from page 210 of [17]), which would result in unnecessarily larger values in our pessimistic estimate of ϵ_{cdf}. Our comparison of quantiles from a 1000-sample estimate of $\hat{\sigma}_i^2$ between empirical estimates and estimates obtained directly from the χ^2-distribution (Eq. (12.19)) indicates an absolute difference around 0.03 when $N = 50$ (which is the smallest sample size we consider) and is smaller for larger values of N on average. As such, we could use quantiles for $\hat{\sigma}_i^2$ and $\hat{\rho}_i$ obtained empirically for subsequent experiments. Note that uniform approximation of the true distribution function by an empirical distribution function based on sample values is guaranteed, e.g. by the Glivenko-Cantelli Theorem [17, 51].

Table 12.2 lists out ϵ_{cdf} for different Ns for ten strategies i. Although results indicate that ϵ_{cdf} decreases when N increases, there is a trade-off between more robust estimation of generalization performance and increasing computational cost. In particular, results indicate that ϵ_{cdf} decreases rapidly when N increases from 50 to 1000 but starts to level off from around $N = 1000$ onwards [13]. Table 12.2 suggests that S_N at $N = 2000$ would provide a sufficiently robust estimate of generalization performance for a reasonable computational cost since one would need a fivefold increase of N to 10000 to reduce ϵ_{cdf} by half. Note that a pessimistic estimate of ϵ_{cdf} could be greater than the real one for small sample sizes (e.g. strategy # 7 at $N = 50$)—this could be corrected by simply taking the maximum value of 1. Furthermore, $\hat{\sigma}_i^2$ and $\hat{\rho}_i$ in Eq. (12.23) are finite moments bounded away from 0. Even for non-pathological strategies (i.e. those demonstrating large variation of game outcomes when playing against a wide variety of opponent test strategies), ϵ_{cdf} is dominated by the term $N^{-1/2}$ for larger N. This implies that the trade-off between more robust estimations of generalization and computational cost is roughly the same for most of the strategies [13].

Assume for the moment that estimates $\hat{G}_i(S_N)$ for a base strategy i are Gaussian-distributed. We use Eq. (12.17) to obtain the following error:

$$\delta_{\text{Gauss}} = \frac{z_{\alpha/2}\,\hat{\sigma}_i}{\sqrt{N}}.$$

Table 12.2 Pessimistic estimates of ϵ_{cdf} from Eq. (12.23) for ten strategies i. The results show a trade-off between more robust estimation of generalization performance and increasing computational cost. Increasing the sample size N leads to decreasing error ϵ_{cdf} but soon levels off around $N = 1000$ onwards [13]

N	# 1	# 2	# 3	# 4	# 5	# 6	# 7	# 8	# 9	# 10
50	0.153908	0.12059	0.277445	0.791636	0.125334	0.40297	1.095542	0.132867	0.11612	0.233136
100	0.09876	0.083803	0.155298	0.357576	0.08611	0.227821	0.452723	0.090216	0.081356	0.138341
200	0.066866	0.058456	0.097296	0.200709	0.059497	0.136423	0.240982	0.06164	0.057204	0.086697
300	0.053277	0.047474	0.073737	0.141732	0.048203	0.099911	0.170278	0.049558	0.046587	0.067476
400	0.045463	0.041004	0.063622	0.114299	0.041509	0.082325	0.135654	0.042775	0.040304	0.057132
500	0.040469	0.036562	0.054709	0.100387	0.036949	0.071225	0.11657	0.037963	0.036012	0.049744
1000	0.027898	0.025662	0.036199	0.06219	0.025904	0.045869	0.072467	0.026473	0.025397	0.033765
2000	0.019337	0.018072	0.024639	0.039612	0.018206	0.031049	0.046544	0.018536	0.017922	0.023025
3000	0.015663	0.014728	0.019851	0.031357	0.014826	0.024533	0.036414	0.015082	0.014621	0.018436
4000	0.013501	0.012742	0.016939	0.02645	0.012814	0.020883	0.030649	0.013013	0.012655	0.015841
5000	0.012039	0.011384	0.015053	0.023489	0.011444	0.018533	0.026889	0.011622	0.011315	0.014013
10000	0.008441	0.008033	0.010422	0.015988	0.008074	0.012733	0.018249	0.008178	0.007994	0.009747
50000	0.003740	0.003583	0.004536	0.006792	0.003597	0.005481	0.007667	0.003634	0.003571	0.004266

Table 12.3 Statistics of $\{|\epsilon_{\text{cdf}} - \delta_{\text{Gauss}}|\}_i$ for 50 strategies i [13]

N	Mean	Std. Dev.
50	0.199785	0.359618
100	0.093783	0.168228
200	0.043777	0.068606
300	0.030180	0.046173
400	0.024052	0.036315
500	0.020268	0.030288
1000	0.011935	0.016976
2000	0.007458	0.010326
3000	0.005749	0.007831
4000	0.004822	0.006521
5000	0.004219	0.005665
10000	0.002825	0.003728
50000	0.001173	0.001510

We compute the pessimistic estimate of δ_{Gauss}, taking 97.5%-tile of $\hat{\sigma}_i^2$ from the rank-ordered 1000-sample estimates. Our results for δ_{Gauss} also indicate a trade-off between more robust estimation and increasing computational cost and suggest that S_N at $N = 2000$ would provide a sufficiently robust estimate of generalization performance for a reasonable computational expense. Table 12.3 shows results indicating that $\{|\epsilon_{\text{cdf}} - \delta_{\text{Gauss}}|\}_i$ becomes smaller for larger sample sizes, e.g. at $N > 1000$ the absolute difference is less than 0.01, with similar trade-off suggestion [13].

We have verified how statistical estimators for generalization performance in coevolutionary learning can be improved. Now, we demonstrate how such an improved statistical estimator can be applied in the coevolutionary learning of the more complex Othello game [13]. ***Othello*** is a deterministic, perfect information, zero-sum, board game played by two players (black and white) that alternatively place their (the same coloured) pieces on an eight-by-eight board. The game starts with each player having two pieces already on the board as shown by Fig. 12.7. The black player starts the game by making the first move. A legal move is one where the new piece is placed adjacent horizontally, vertically, or diagonally to an opponent's existing piece so that at least one of the opponent's pieces between the player's new and existing pieces can be flipped over to become the player's pieces to complete the move (e.g. Fig. 12.7 (right)). A player that could not make a legal move forfeits and passes the move to the opponent. The game ends when all the squares of the board are filled with pieces or when neither player is able to make a legal move successively [11].

Strategy representations for the coevolutionary learning of Othello strategies (in the form of a board evaluation function that is then fed to a minimax search of the game tree) are ***Weighted Piece Counters*** (***WPCs***) [43] and ANNs [11, 42]. We use WPCs primarily for simplicity in representation, computational cost consideration, and that one could make the argument the finite-precision representation of each

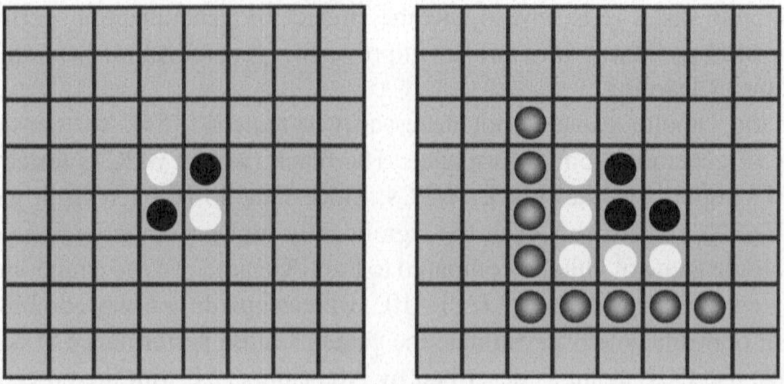

Fig. 12.7 Figure illustrates basic Othello moves. (Left) The positions of respective players' pieces at the start of the game. (Right) The possible legal moves for the black player (which are indicated by gray circles) at a later point of the game. The black move is initiated with a selected a legal move and completed where surrounded white pieces are flipped over to become black pieces

element in WPC means that WPC-representation induces a very large but finite strategy space S, the latter requiring the use of statistical estimators for generalization performance measurements. The WPC representing the board evaluation function of an Othello game strategy can take the form of a vector of 64 weights, indexed as w_{rc}, $r = 1, \ldots, 8$, $c = 1, \ldots, 8$, $w_{rc} \in [-10, 10]$, where r and c represent the position indexes for rows and columns of an eight-by-eight Othello board, respectively. Let the Othello board state be correspondingly represented by x_{rc}, $r = 1, \ldots, 8$, $c = 1, \ldots, 8$, $x_{rc} \in \{-1, 0, +1\}$, where the values of $+1$, -1, or 0 represent black, white, and empty piece, respectively [43]. Then, the output of WPC for a board state $\boldsymbol{x}$ input is calculated as

$$\text{WPC}(\boldsymbol{x}) = \sum_{r=1}^{8} \sum_{c=1}^{8} w_{rc} \cdot x_{rc}, \tag{12.24}$$

where the more positive (negative) value of WPC($\boldsymbol{x}$) indicates WPC's interpretation that the board state $\boldsymbol{x}$ is more favorable if WPC is a black (white) player [43].

We use a simple additive mutation operator to generate WPC of the offspring's w'_{rc} from the parent's w_{rc}:

$$w'_{rc} = w_{rc} + k \cdot U_{rc}(-1, 1), \quad r = 1, \ldots, 8, c = 1, \ldots, 8, \tag{12.25}$$

where k is a scaling constant ($k = 0.1$) and $U_{rc}(-1, 1)$ is a real number randomly drawn from $[-1, 1]$ with a uniform distribution and resampled for every combination of r and c (total of 64 weights). The simple mutation operator can provide sufficient variation to the Othello game strategy represented in the form of a WPC evaluation function. These choices for various parameters in strategy representation and variation operator are not optimized. The main emphasis of our subsequent computational study is to investigate the impact of generalization performance estimates used as fitness measure to improve the generalization performance of coevolutionary learning.

Since the Othello game is not necessarily symmetric [54], we consider two estimates of generalization performance: The black (white) WPC is tested against a random sample of white (black) WPCs. Since Othello is a zero-sum game, we employ $\hat{G}_W(S_N)$ measure. Given the significantly higher computational cost executing a single game of Othello compared to the IPD game, we use a random sample of 50000 test WPCs, with $w_{rc} \sim U_{rc}[-10, 10]$ resampled for every combination of r and c, of opposite colour to estimate the generalization performance of coevolved WPCs. We use CCL again as described by Algorithm 12.2 with the procedure call **CCL**($\boldsymbol{Y}$, b_{gen}) made with the full population $2\boldsymbol{Y} = 2\lambda$ and $b_{gen} = 200$ generations as the baseline. We have compared the generalization performances of CCL with a coevolutionary learning that directly uses $\hat{G}_i(S_N)$ as the fitness measure that we call ***Improved Coevolutionary Learning*** (***ICL***) and CCL-ICL hybrids that employ a mixture of $\hat{G}_i(S_N)$ and round-robin tournament scores as fitness. In particular, results for coevolution of IPD games indicate that there is no significant difference in using such hybrids against ICL, but that both are superior in the context of improved generalization performances against CCL [13]. Thus in the following, we compare CCL and ICL. We repeat all experiment settings twice: one for black WPC and one for white WPC. All experiments are repeated in 30 independent runs to allow for statistical analysis.

CCL uses $2\lambda = 50$, while ICLs use $2\lambda = 20$ and S_Ns of different sample sizes $N = \{50, 500, 1000, 5000, 10000, 50000\}$ to estimate $\hat{G}_i(S_N)$. The sample S_N, $N = 50000$ is to provide a ground truth estimate close to the generalization performance estimates with which to compare results for cases where much smaller samples are used. Note that different samples S_N are used to estimate $\hat{G}_i(S_N)$ as the fitness measure in ICL and to estimate the generalization performance of ICL for analysis. Figure 12.8 summarizes our results with each graph plotting the $\hat{G}_i(S_N)$ ($N = 50000$) of the top performing strategy of the population throughout the evolutionary process for all 30 independent runs. Results for the CCL of black and white WPCs indicate a search process with large fluctuations in the generalization performance of strategies throughout coevolution (Fig. 12.8a and b) [13].

However, we observe that ICL can search WPCs with higher generalization performance although small fluctuations can be seen during coevolution when estimates $\hat{G}_i(S_N)$ are computed using a small sample of 50 test strategies. Further increase in sample size leads to further improvements in generalization performance of ICL, e.g. when $N = 500$ (Fig. 12.8c and d). Despite this, there is a trade-off whereby a significant increase in the sample size does not bring about a significant increase in generalization performance of ICL. For example, results for ICL-N5000 are similar to that of ICL-N50000 (Fig. 12.8e and f). This is consistent with results from experiments to find the required number of test strategies for robust estimation of generalization performance with the trade-off occurring around $N = 5000$ for S_N [13].

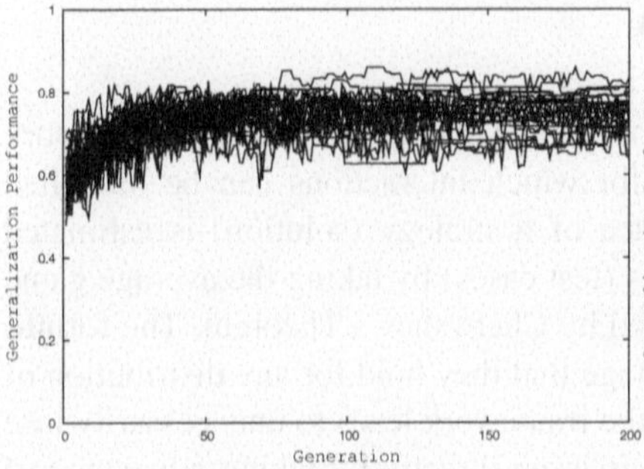

(a) CCL black WPC

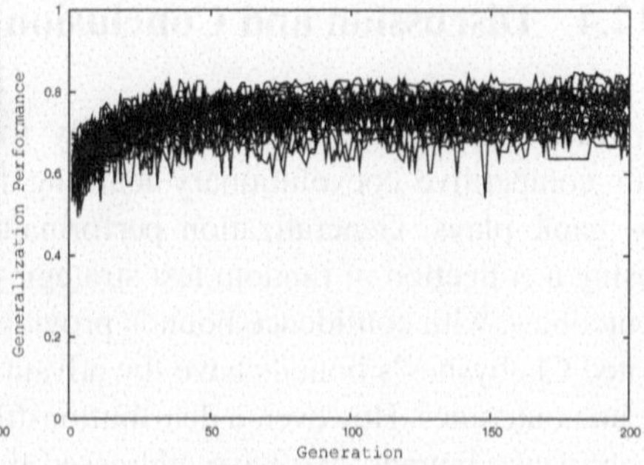

(b) CCL white WPC

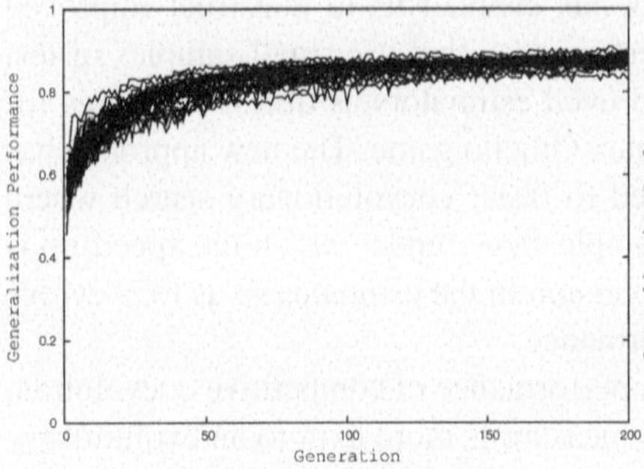

(c) ICL-N500 black WPC

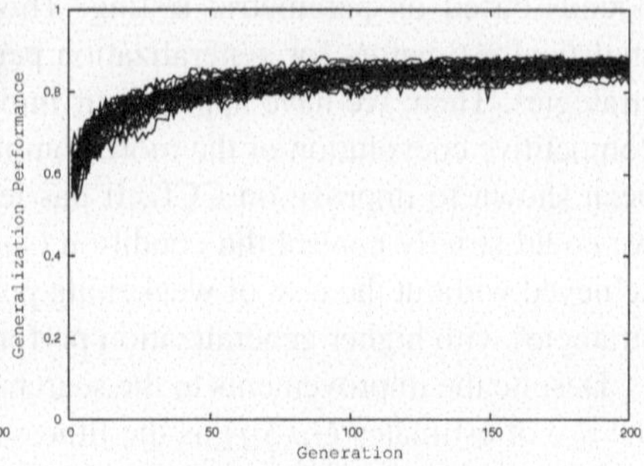

(d) ICL-N500 white WPC

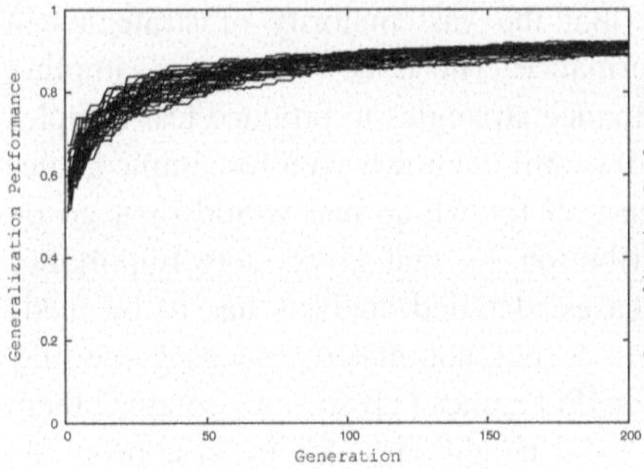

(e) ICL-N50000 black WPC

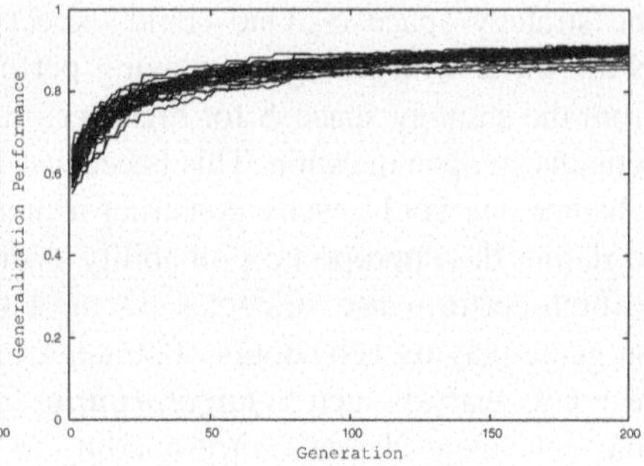

(f) ICL-N50000 white WPC

Fig. 12.8 Comparison of CCL and different ICLs for the Othello game: (**a**) CCL black WPC, (**b**) CCL white WPC, (**c**) ICL-N500 black WPC, (**d**) ICL-N500 white WPC, (**e**) ICL-N50000 black WPC, and (**f**) ICL-N50000 white WPC. Shown are plots of the estimated generalization performance $\hat{G}_i(S_N)$ (with $N = 50000$) of the top performing strategy of the population throughout the evolutionary process for all 30 independent runs [13]

12.4 Discussion and Conclusion

In this chapter, we have theoretically formulated the generalization framework for competitive coevolutionary learning for which interactions can be modelled as game-plays. Generalization performance of a strategy (solution) is estimated using a collection of random test strategies (test cases) by taking the average game outcomes, with confidence bounds provided by Chebyshev's Theorem. The formulated Chebyshev's bounds have the advantage that they hold for any distribution of game outcomes. However, a distribution-free framework leads to unnecessarily lose confidence bounds. We have addressed this issue directly by taking advantage of the near-Gaussian nature of average game outcomes for which we provide tighter bounds based on parametric testing. This has enabled us to construct improved statistical estimators for generalization performance that use small samples of test strategies. Then, we have applied our improved estimators as fitness measures for competitive coevolution of the more complex Othello game. The new approach has been shown to improve on CCL. It has led to faster coevolutionary search where we could strictly control the condition (sample sizes) under which the speed-up is achieved without the cost of weakening precision in the estimates so as to coevolve strategies with higher generalization performance.

Despite the improvements to the search performance of competitive coevolution, the use of estimates $\hat{G}_i(S_N)$ as the fitness measure is more akin to an evolutionary learning approach. More importantly, this is in the context of a generalization performance measure with respect to a uniform probability distribution $\mathbb{P}_{\mathcal{S}}$ on the strategy space $\mathcal{S}$. One could expect that the vast majority of strategies in $\mathcal{S}$ are those with low game-playing performance. The issue of directly sampling from the strategy space $\mathcal{S}$ for high-performance strategies to produce test samples remains an open question. This is because it is still unknown even for simple games whether one could easily construct a metric $\mathcal{S}$ for which one would then go on to define the appropriate probability distribution $\mathbb{P}_{\mathcal{S}}$ that gives more importance to high-performance strategies. In most cases, detailed analysis has to be made on game-playing behaviours of strategies in $\mathcal{S}$ (e.g. automated strategy game-play response analysis such as ***fingerprinting*** for IPD games [1]) so as to organize them into equivalent classes (corresponding to local neighbourhoods in an appropriate metric $\mathcal{S}$) that one then hopes can be organized in an ascending order of performance because the game is solvable (i.e. there is a dominant subset of strategies superior to the rest of the strategies in $\mathcal{S}$). This is related to the issue of representation in that one could envision a powerful representation that in effect gives a compact $\mathcal{S}$ whereby all equivalent strategies are reduced to their single representative strategies as single points of this compact $\mathcal{S}$ whose pairwise relations are essentially transitively chained.

As such, there remains significant scope and motivation to design and work on heuristics in the hope of improving search performance of competitive coevolutionary systems. We close this chapter with one such important area of research in competitive coevolutionary systems. In particular, we have earlier conducted a detailed computational study to investigate the issue of whether there is a relationship between generalization performance and diversity in coevolutionary learning [15]. This is a long-standing issue in the literature that requires having a better grasp and understanding since the lack of diversity in the population is commonly cited as a cause of various coevolutionary pathologies than then have negative impacts to coevolutionary search performance. Our computational study has sought to systematically examine various diversity levels and generalization performances of competitive coevolutionary systems augmented with various heuristic-based diversity maintenance mechanisms. We have employed 4-choice IPD game for our case studies as we are able to take quantitative measurements of both generalization performances and diversity levels of the coevolving population.

For ***diversity measurements***, we have adapted those studied earlier in [8, 37] that have conducted systematic analysis of various diversity measurements for GP. We will briefly describe next those we have used. Let an IPD game-play strategy represented in the form of a direct look-up table be encoded as the string $\mathbf{s}_i = (i_k : k = 1, \ldots, n_e)$, $n_e = n^2 + 1$, where n_e is the total number of elements of the direct look-up table and n is the number of choices of the IPD game with discrete levels of cooperation. Then, the ***edit distance*** between strategies i and j is associated with the minimum number of transformations required to change a strategy's direct look-up table $\mathbf{s}_i$ to $\mathbf{s}_j$:

$$\text{dist}_{\text{ed}}(\mathbf{s}_i, \mathbf{s}_j) = \frac{1}{n_e} \sum_{k=1}^{n_e} d(i_k, j_k), \tag{12.26}$$

where $d(i_k, j_k)$ is a distance metric on i_k and j_k. For the distance metric $d(i_k, j_k)$, we use the Levenshtein distance [47]:

$$d(i_k, j_k) = \begin{cases} 0, \ i_k = j_k \\ 1, \ i_k \neq j_k. \end{cases}$$

A transformation for a particular element is made only as a substitution from one choice to another. For example, if two strategies i and j differ only in the choice they make for the first move, e.g. $+1$ for strategy i and -1 for strategy j, only a single transformation is required. The edit distance between these two strategies is $1/n_e$.

As a consequence, one can then define a ***genotypic diversity measurement*** based on edit distance on a coevolving population which can be obtained by taking the average value of pairwise distances of strategies in the population [27]. However, a ***phenotypic diversity measurement*** can be defined as well to quantify the level of variety in the phenotypic space. Such a diversity measure is useful because for games one might expect a many-to-one relation between genotypes and phenotypes in which case many different strategies genotypic-wise could actually behave similarly against various opponents. The methodologies for such measurements range from sophisticated automated analysis based on fingerprints [1] to simpler phenotypic diversity measurements. We use the latter, specifically one based on the ***entropy*** diversity measurement [8, 52]

$$-\sum_k p_k \cdot \log p_k, \tag{12.27}$$

where p_k refers to the proportion of a particular fitness value in the population. In our case studies, the fitness value is the average IPD game payoff obtained from a round-robin tournament with the entire population of coevolving strategies.

We have been motivated to answer two related questions in our studies: (1) What is the right form of diversity that would be beneficial for generalization, and (2) what amount of the right form of diversity is needed for good generalization. This is so that specific conditions relating to diversity might be identified with potential implications for design choices in coevolutionary search algorithms to improve generalization performance. Our case studies employ CCL with a fixed mutation rate of $p_m = 0.05$ as the baseline—we refer to the baseline as CCL-PM05. Then, we consider both explicit (via mutation [23]) and implicit (via selection) diversity maintenance mechanisms. For explicit mechanisms, we increase mutation rates in steps of 0.05 until $p_m = 0.25$ (CCL-PM25). For implicit mechanisms, we consider both competitive (CFSD and CFSR) and implicit fitness sharing mechanisms [21]. The sampling procedure for the opponents to obtain fitness of coevolving strategies can be made with a simple random approach (CFSR and IFSR) or a ***disassortative sampling*** approach that successively chooses an opponent that is more different than the first one in the context of edit distance (CFSD and IFSD). Pareto coevolution (PAR) and reduced virulence (RV) are also investigated.

Detailed analysis we have conducted [15] of these results that include correlation analysis has indicated that the introduction and maintenance of diversity in the population of coevolutionary learning does not necessarily lead to a significant increase in the generalization performance. There are cases such as coevolution with reduced virulence where a positive relationship between generalization performance and diversity could not be positively established. However, we have observed that diversity maintenance that led to a speciated population (specifically implicit fitness sharing) did lead the coevolved population to better cover the strategy space. This can be observed in the results of [15] from the nonsignificant difference against the baseline for the Best(G_{SPOP_U}) and Avg(G_{SPOP_U}) measurements, but significantly better Ens(G_{SPOP_U}) measurements. This is because for the later an ideal gating mechanism is assumed that always chooses among the population one that best performs against the test strategy. Similar observations can be made for pareto coevolution. We have also obtained similar results when a biased test strategy is employed to compute generalization performances [15].

Our theoretical and computational studies on generalization performance have shown that rigorous quantitative analysis can be made on any competitive coevolutionary system with respect to a large class of problems that can be formulated as two-player strategic games. When these generalization performance estimates are used with care, particularly with the understanding that performance here refers to the average outcomes against test cases, then one could obtain some meaningful qualitative results as well. This is important for analysis of competitive coevolutionary systems that are applied to solve very complex real-world problems. In such cases, one might have at the start limited understanding about the problem structures. At this point having quantitative tools that provide rigorous analysis and yet can be easily setup is crucial to practitioners and designers of competitive coevolution so that the right choices can be made and discovered to implement the specific learning systems targeted towards the problem at hand.

Before closing, we note that despite limited theoretical studies of competitive CEAs compared to EAs, there has been significant progress over the years. For example, there have been studies on cooptimization as a further generalization over classical black box optimization. One would note that the earlier theoretical study [61] has established the ***No Free Lunch*** result in some settings within traditional optimization whereby informally all algorithms are similar in terms of their average performances across all problems. This was a follow-up in a competitive coevolution setting in [62] for which ***Free Lunch*** can be established in the more restricted self-play setting. More recently in [50], a theoretical framework for cooptimization algorithm aggregate performance analysis over all functions in the problem domain is developed and then refined to include crucial aspects such as budget considerations and search histories that are typically encountered in real-world practice. Then, the study [48] has established close links between competitive CEAs, cooptimization, and supervised learning.

References

1. Ashlock, D., Kim, E.Y.: Fingerprinting: Visualization and automatic analysis of prisoner's dilemma strategies. IEEE Trans. Evol. Comput. **12**(5), 647–659 (2008)
2. Axelrod, R.: The evolution of strategies in the iterated prisoner's dilemma. In: L.D. Davis (ed.) Genetic Algorithms and Simulated Annealing, chap. 3, pp. 32–41. Morgan Kaufmann, New York (1987)
3. Bader-Natal, A., Pollack, J.B.: Towards metrics and visualizations sensitive to coevolutionary failures. In: Coevolutionary and Coadaptive Systems Workshop: 2005 AAAI Symposium. Washington DC (2005)
4. Bari, A.G.: Interactive fitness domains in competitive coevolutionary algorithm. Ph.D. thesis, University of South Florida, Florida (2019)
5. Bishop, C.M.: Neural Networks for Pattern Recognition. Oxford University, Oxford (1995)
6. Bucci, A.: Emergent geometric organization and informative dimensions in coevolutionary algorithms. Ph.D. thesis, Brandeis University, Massachusetts (2007)
7. Bucci, A., Pollack, J.B.: Focusing versus intransitivity geometrical aspects of co-evolution. In: Proceedings of the 2003 Genetic and Evolutionary Computation Conference (GECCO'03), pp. 250–261 (2003)
8. Burke, E.K., Gustafson, S., Kendall, G.: Diversity in genetic programming: an analysis of measures and correlation with fitness. IEEE Trans. Evol. Comput. **8**(1), 47–62 (2004)
9. Cartlidge, J., Bullock, S.: Combating coevolutionary disengagement by reducing parasite virulence. Evol. Comput. **12**(2), 193–222 (2004)

10. Chellapilla, K., Fogel, D.B.: Evolution, neural networks, games, and intelligence. Proc. IEEE **87**(9), 1471–1496 (1999)
11. Chong, S.Y., Tan, M.K., White, J.D.: Observing the evolution of neural networks learning to play the game of Othello. IEEE Trans. Evol. Comput. **9**(3), 240–251 (2005)
12. Chong, S.Y., Tiňo, P., He, J., Yao, X.: A new framework for analysis of coevolutionary systems—directed graph representation and random walks. Evol. Comput. **27**(2), 195–228 (2019)
13. Chong, S.Y., Tiňo, P., Ku, D.C., Yao, X.: Improving generalization performance in co-evolutionary learning. IEEE Trans. Evol. Comput. **16**(1), 70–85 (2012)
14. Chong, S.Y., Tiňo, P., Yao, X.: Measuring generalization performance in co-evolutionary learning. IEEE Trans. Evol. Comput. **12**(4), 479–505 (2008)
15. Chong, S.Y., Tiňo, P., Yao, X.: Relationship between generalization and diversity in coevolutionary learning. IEEE Trans. Comput. Intell. AI Games **1**(3), 214–232 (2009)
16. Chong, S.Y., Yao, X.: Behavioral diversity, choices, and noise in the iterated prisoner's dilemma. IEEE Trans. Evol. Comput. **9**(6), 540–551 (2005)
17. Chung, K.L.: A Course in Probability Theory, 3rd edn. Academic Press, San Diego (2001)
18. Cliff, D., Miller, G.F.: Tracking the red queen: measurements of adaptive progress in co-evolutionary simulations. In: Advances in Artificial Life: Proceeding of the Third European Conference on Artificial Life. Lecture Notes in Computer Science, vol. 929, pp. 200–218. Springer, Berlin (1995)
19. Darwen, P.J.: Co-evolutionary learning by automatic modularization with speciation. Ph.D. thesis, University of New South Wales, Sydney (1996)
20. Darwen, P.J., Yao, X.: On evolving robust strategies for iterated prisoner's dilemma. In: Progress in Evolutionary Computation. Lecture Notes in Artificial Intelligence, vol. 956, pp. 276–292. Springer, Berlin (1995)
21. Darwen, P.J., Yao, X.: Every niching method has its niches: fitness sharing and implicit fitness sharing compared. In: The 4th International Conference on Parallel Problem Solving from Nature PPSN IV. Lecture Notes in Computer Science, vol. 1141, pp. 398–407. Springer, Berlin (1996)
22. Darwen, P.J., Yao, X.: Speciation as automatic categorical modularization. IEEE Trans. Evol. Comput. **1**(2), 101–108 (1997)
23. Darwen, P.J., Yao, X.: Does extra genetic diversity maintain escalation in a co-evolutionary arms race. Int. J. Knowledge-Based Intell. Eng. Syst. **4**(3), 191–200 (2000)
24. Darwen, P.J., Yao, X.: Co-evolution in iterated prisoner's dilemma with intermediate levels of cooperation: Application to missile defense. Int. J. Comput. Intell. Appl. **2**(1), 83–107 (2002)
25. Darwen, P.J., Yao, X.: Why more choices cause less cooperation in iterated prisoner's dilemma. In: Proceedings of the IEEE 2001 Congress on Evolutionary Computation (CEC'01), pp. 987–994. Seoul, Korea (2002)
26. de Jong, E.D., Pollack, J.B.: Ideal evaluation from coevolution. Evol. Comput. **12**(2), 159–192 (2004)
27. Ekárt, A., Németh, S.: Maintaining the diversity of genetic programs. In: Proceedings of the 5th European Conf. on Genetic Programming EuroGP 02. Lecture Notes in Computer Science, vol. 2278, pp. 162–171. Springer, Berlin (2002)
28. Ficici, S.G.: Solution concepts in coevolutionary algorithms. Ph.D. thesis, Brandeis University, Massachusetts (2004)
29. Ficici, S.G., Melnik, O., Pollack, J.B.: A game-theoretic and dynamical-systems analysis of selection methods in coevolution. IEEE Trans. Evol. Comput. **9**(6), 580–602 (2005)
30. Ficici, S.G., Pollack, J.B.: Pareto optimality in coevolutionary learning. In: Advances in Artificial Life: 6th European Conference (ECAL'01). Lecture Notes in Computer Science, vol. 2159, pp. 316–325. Springer, Prague (2001)
31. Floreano, D., Nolfi, S.: God save the red queen! competition in co-evolutionary robotics. In: J.R. Koza, K. Deb, M. Dorigo, D.B. Fogel, M. Garzon, H. Iba, R.L. Riolo (eds.) Genetic Programming 1997: Proceedings of the Second Annual Conference, pp. 398–406. Morgan Kaufmann, Stanford University, CA, USA (1997)

32. Fogel, D.B.: The evolution of intelligent decision making in gaming. Cybern. Syst. An Int. J. **22**, 223–236 (1991)
33. Fogel, D.B.: Evolving behaviors in the iterated prisoner's dilemma. Evol. Comput. **1**(1), 77–97 (1993)
34. Fogel, D.B.: On the relationship between the duration of an encounter and the evolution of cooperation in the iterated prisoner's dilemma. Evol. Comput. **3**(3), 349–363 (1996)
35. Franken, N., Engelbrecht, A.P.: Particle swarm optimization approaches to coevolve strategies for the iterated prisoner's dilemma. IEEE Trans. Evol. Comput. **9**(6), 562–579 (2005)
36. Gnedenko, B.V., Gnedenko, G.V.: Theory of Probability. Taylor & Francis, Routledge (1998)
37. Gustafson, S.: An analysis of diversity in genetic programming. Ph.D. thesis, University of Nottingham, Nottingham (2004)
38. Harrald, P.G., Fogel, D.B.: Evolving continuous behaviors in the iterated prisoner's dilemma. BioSystems: Special Issue on the Prisoner's Dilemma **37**, 135–145 (1996)
39. Hillis, W.D.: Co-evolving parasites improve simulated evolution as an optimization procedure. Phys. D **42**, 228–234 (1990)
40. Jacobson, H.I.: The maximum variance of restricted unimodal distributions. Ann. Math. Stat. **40**(5), 1746–1752 (1969)
41. Juille, H., Pollack, J.B.: Co-evolving the "ideal" trainer: application to the discovery of cellular automata rules. In: J.R. Koza, W. Banzhaf, K. Chellapilla, K. Deb, M. Dorigo, D.B. Fogel, M.H. Garzon, D.E. Goldberg, H. Iba, R. Riolo (eds.) Proceedings of the Third Annual Conference of Genetic Programming 1998, pp. 519–527. Morgan Kaufmann, Wisconsin (1998)
42. Kim, K.J., Choi, H., Cho, S.B.: Hybrid of evolution and reinforcement learning for Othello players. In: Proceedings of the 2007 IEEE Symposium on Computational Intelligence and Games (CIG 2007), pp. 203–209. IEEE Press, New York (2007)
43. Lucas, S., Runarsson, T.P.: Temporal difference learning versus co-evolution for acquiring Othello position evaluation. In: Proceedings of the 2006 IEEE Symposium on Computational Intelligence and Games (CIG 2006), pp. 52–59. IEEE Press, New York (2006)
44. Manoukian, E.B.: Modern Concepts and Theorems of Mathematical Statistics. Springer, New York (1986)
45. Nisan, N., Roughgarden, T., Tardos, E., Vazirani, V.V. (eds.): Algorithmic Game Theory. Cambridge University, New York (2007)
46. Noble, J.: Finding robust Texas Hold'em poker strategies using pareto coevolution and deterministic crowding. In: Proceedings of the 2002 International Conference on Machine Learning and Applications (ICMLA'02), pp. 233–239. Nevada, USA (2002) The International Journal of Intelligent Control and Systems
47. O'Reilly, U.M.: Using a distance metric on genetic programs to understand genetic operators. In: Proceedings of the IEEE 1997 International Conference on Systems, Man, and Cybernetics, pp. 4092–4097. IEEE Press, Piscataway, NJ (1997)
48. Popovici, E.: Bridging supervised learning and test-based co-optimization. J. Mach. Learn. Res. **18**(38), 1–39 (2017)
49. Popovici, E., Jong, K.D.: Relationships between internal and external metrics in co-evolution. In: Proceedings of the IEEE 2005 Congress on Evolutionary Computation (CEC'05), pp. 2800–2807. Edinburgh, UK (2005)
50. Popovici, E., Winston, E.: A framework for co-optimization algorithm performance and its application to worst-case optimization. Theor. Comput. Sci. **567**, 46–73 (2015)
51. Resnick, S.I.: A Probability Path. Birkhäuser, Boston (1998)
52. Rosca, J.P.: Entropy-driven adaptive representation. In: Proceedings of the Workshop Genetic Programming: From Theory to Real-world Applications, pp. 23–32. Tahoe City, CA (1995)
53. Rosin, C.D., Belew, R.K.: A competitive approach to game learning. In: COLT'96: Proceedings of the Ninth Annual Conference on Computational Learning Theory, pp. 292–302 (1996)
54. Rosin, C.D., Belew, R.K.: New methods for competitive coevolution. Evol. Comput. **5**(1), 1–29 (1997)
55. Runarsson, T.P., Lucas, S.M.: Coevolution versus self-play temporal difference learning for acquiring position evaluation in small-board go. IEEE Trans. Evol. Comput. **9**(6), 540–551 (2005)

56. Stanley, K.O., Miikkulainen, R.: The dominance tournament method of monitoring progress in coevolution. In: Bird of a Feather Workshop: Proceedings of the 2002 Genetic and Evolutionary Computation Conference (GECCO'02), pp. 242–248 (2002)
57. Stanley, K.O., Miikkulainen, R.: Competitive coevolution through evolutionary complexification. J. Artif. Intell. Res. **21**, 63–100 (2004)
58. Tiňo, P., Chong, S.Y., Yao, X.: Complex co-evolutionary dynamics—structural stability and finite population effects. IEEE Trans. Evol. Comput. **17**(2), 155–164 (2013)
59. Watson, R.A., Pollack, J.B.: Coevolutionary dynamics in a minimal substrate. In: Proceedings of the 2001 Genetic and Evolutionary Computation Conference (GECCO'01), pp. 702–709 (2001)
60. Werfel, J., Mitchell, M., Crutchfield, J.P.: Resource sharing and coevolution in evolving cellular automata. IEEE Trans. Evol. Comput. **4**(4), 388–393 (2000)
61. Wolpert, D.H., Macready, W.G.: No free lunch theorems for optimization. IEEE Trans. Evol. Comput. **1**(1), 67–82 (1997)
62. Wolpert, D.H., Macready, W.G.: Coevolutionary free lunches. IEEE Trans. Evol. Comput. **9**(6), 721–735 (2005)
63. Yao, X., Darwen, P.J.: How important is your reputation in a multi-agent environment. In: Proceedings of the IEEE 1999 Conference on Systems, Man, and Cybernetics (SMC'99), pp. 575–580. Tokyo, Japan (1999)
64. Yao, X., Liu, Y., Darwen, P.J.: How to make best use of evolutionary learning. In: R. Stocker, H. Jelinck, B. Burnota, T. Bossomaier (eds.) Complex Systems - From Local Interactions to Global Phenomena, pp. 229–242. IOS Press, Amsterdam (1996)

Chapter 13
Impact of Problem Structures in Coevolution

13.1 Introduction of Coevolutionary Problem Structures and Their Effects on Search

Coevolutionary Computation has been applied to a wide variety of problems across different settings such as strategic decision-making [1, 11, 12, 25], Machine Learning problems (e.g. classification) [26, 65], optimization [26, 34, 44, 56, 66], and Artificial Life simulations [23, 60]. They have been employed in different problem domains (see Chap. 4). Given the breadth of problem setting scope with which Coevolutionary Algorithms (CEAs) have been applied so that it becomes almost necessary to have a close literature review of CEAs employed on similar problem settings, nevertheless, practitioners and algorithm designers could reasonably ask whether there are some general principles that guide the specification and customization of CEAs to solve the problem at hand. Or, if one turns the question around, what are the relevant and crucial problem structures that affect the performance of coevolutionary search. This chapter aims to introduce and demonstrate formal studies we have made of relevant but general problem structures that could arise and how they are specifically linked to coevolutionary search processes so that some general guiding principles can be extracted to inform about performance of CEAs.

This requires that one first sets the scope of the coevolutionary problem at hand so as to better interpret these results we demonstrate later. As we have introduced earlier in Sect. 2.3, there are two general but distinct settings CEAs are used: competitive and cooperative. Briefly, competitive CEAs are applied to test-based problems whereby test cases are required to assess the quality of generated candidate solutions, while cooperative CEAs are applied to compositional problems whereby components are assembled to construct candidate solutions [53]. Note that in compositional problems, the evaluation of candidate solutions could involve the use of test cases for the more challenging and complex problems, or the objective function is available. It is crucial that one first distinguishes between problems for

X. Yao, S. Y. Chong, *Coevolutionary Computation and Its Applications*,
Natural Computing Series, https://doi.org/10.1007/978-981-96-2841-4_13

which test-case evaluations must be used and those where one could query from a function to provide specific objective quality measurements on candidate solutions.

In particular, the main principle underlying cooperative CEAs is a Divide-and-Conquer problem solving with a parallel search approach. As we have discussed earlier in Sect. 3.3.2, this requires optimization problems to admit problem decompositions that result in simpler problem subcomponents individually solved by genetically isolated populations. The advantage or strength with which cooperative search approach could be fully demonstrated is that these populations interact to enable a search that improves the subcomponents in tandem. In the case of continuous optimization problems [45], one of the crucial problem structures is that of variable interactions. For example, nonseparable problems exhibit various degrees of variable interactions. The study in[54] has demonstrated how such variable interactions affect cooperative CEA performances.

In particular, a detailed case study is employed to illustrate how nonseparability in two-dimensional continuous optimization problems can have effects on design choices in cooperative CEAs such as the use of elitism and setting of population sizes. The problem setting here automatically forces a cooperative CEA to employ a two-population search (each handling a specific variable) for which there exist interactions. These interactions are captured as best response curves (i.e. the partial derivative with respect to the variable of the optimization function). Assuming minimization, the best response curve with respect to x_2 is the function of type $\mathbb{R}_{x_2} \rightarrow \mathbb{R}_{x_1}$ defined by $g_{\text{best}}(x_2) = \text{argmin}_{x_1 \in \mathbb{R}_{x_1}} f(x_1, x_2)$. The other best response curve with respect to x_1 can be defined in the same manner. Note that the study has chosen optimization functions such that the best response curves are always one to one.

More importantly, the geometric structures of the two best response curves and their relation have useful interpretations. Together, an idealized coevolutionary search process can be constructed. This can be visualized by first having these two best response curves of type $\mathbb{R}_{x_2} \rightarrow \mathbb{R}_{x_1}$ and $\mathbb{R}_{x_1} \rightarrow \mathbb{R}_{x_2}$ embedded in the $\mathbb{R}^2$ plane. In the idealized case, the end result of an evolutionary search cycle would mean that only points from the corresponding best response curve are sampled. Over a single coevolutionary cycle, the cooperative CEA search that is represented by best-of-generation (i.e. employing best collaborator strategy) trajectory is visualized as first taking a sample point from a particular best response curve and translating it over to the other best response curves (horizontally or vertically depending on the two curves' geometric relation) and then repeating the process one more time so as to map back to a point in the first curve. The full trajectory is simply connected sequences of such jumps in $\mathbb{R}^2$. For the simpler ***Off Axis Quadratic*** function, trajectories appear similar to one where the global optimum is an attractor with a negative feedback [24]. However, variable interactions and nonlinearities in a complex nonseparable problem such as the ***Rosenbrock*** could have negative impact on such a cooperative coevolutionary search performance. The best response curves are non-linear, and computer-generated trajectories could in effect lead to a cooperative search where the full candidate solutions $\boldsymbol{x} = (x_1, x_2)$ could actually stay away from the global optimum.

In this chapter, our main focus is on the analysis of problem structures in competitive coevolution and their effects on coevolutionary search based on our studies in [13, 14]. The mathematical tools that we require are thus specific for a large number of competitive coevolutionary problems with solution-test-case interactions modelled as two-player strategic games. We have already introduced ***coevolutionary digraph*** in Sect. 5.3.1 that captures the complete, relevant structures of such problems. The remainder of this chapter is structured as follows. Section 13.2 will first formulate ***abstract coevolution*** as ***random walks*** (***finite state Markov chains***) on coevolutionary digraphs that are then applied to develop key theoretical results providing crucial qualitative insights on what makes coevolutionary problems challenging for search. Our theory of Markov chains on coevolution is also used to develop quantitative analysis tools that can be used to characterize the speed of coevolutionary search for a given problem (cycle) complexity. Section 13.3 presents a further theoretical study that establishes a deep connection between ***PageRank*** and abstract coevolution. PageRank has been used to characterize the importance of a network of webpages, and we establish that the notion PageRank authorities can be applied to indicate the importance (performance) of vertices in coevolutionary digraphs. Crucially, PageRank authorities have natural and second interpretation as visitation probabilities of coevolutionary search on digraphs with restart. Section 13.4 closes the chapter with a brief remark about the diverse nature of theoretical tools that have been used to better understand coevolutionary systems and the links between them.

13.2 Random Walks on Digraphs as Models of Coevolutionary Search Processes

In this section, we [14] present a theoretical formulation of population-one competitive coevolutionary systems that will allow us to study in depth how structures in coevolutionary problems completely captured as the pairwise relations of vertices in associated digraphs affect the search process. Section 13.2.1 will introduce the abstract coevolution as random walks on coevolutionary digraphs that model coevolutionary search processes naturally as sampling processes on digraphs. These Markov chains allow us to study how specific structures in coevolutionary digraphs can affect the dynamics of coevolutionary search processes in Sect. 13.2.2. They provide crucial interpretable insights as to what specific structures in these competitive problems that make them challenging for coevolutionary search. Then in Sect. 13.2.3, we will demonstrate how our theory of Markov chains on coevolution can be applied to develop quantitative analysis tools that can be used to determine the speed with which coevolution can find the dominant solution of solvable problems with respect to specific problem (cycle) structures.

13.2.1 Coevolutionary Search Processes as Random Walks on Digraphs

We have defined a walk on the digraph $D = (V, A)$ as the alternating sequence of vertices and arcs $v_1 a_1 v_2 a_2 v_3 \ldots a_{k-1} v_k$, where $v_1, \ldots, v_k \in V$ and $a_1, \ldots, a_{k-1} \in A$ earlier in Sect. 5.3.1. It is crucial to take note that a walk must satisfy $v_i a_i v_{i+1}$ for all i. For at least one v_i, there might be several ***out-neighbours*** us given by the set of ***out-neighbourhood*** $N_D^+(v_i) = \{u \in V \backslash \{v_i\} : v_i \to u \in A\}$. This means that there might be more than one strategy u defeating v_k. As such, there could be more than one distinct walk between two vertices v_1, v_k in the digraph. One could extend having these collections of walks between any pair of vertices in the digraph. The notion of a ***random walk*** can thus be informally viewed as a sampling process on the digraph. More specifically, there is a process that samples from collections of walks defined between two vertices in the digraph with some underlying distribution.

Random walks can be procedurally implemented. Algorithm 13.1 describes a ***standard random walk*** on ***labelled digraphs*** D (labelled here refers to the fact that the prior information of out-neighbourhood is known for every vertex in D). It starts at a random vertex and in each subsequent step randomly moves to an out-neighbour of the current vertex with a uniform probability. This is procedurally described in Algorithm 13.1. In practice, one would usually implement a random walk on ***unlabelled digraphs*** such as the ***(1+1)CEA*** (Algorithm 13.2). In the (1+1)CEA, the random move in each step is made through a two-stages process of variation (generates another random offspring from the vertex set V) and selection (moves to the offspring vertex if it dominates the parent vertex). As such, there is a chance the (1+1)CEA stays at the current vertex in a step. This only occurs in the standard random walk when it reaches the dominant vertex of D. Note as well that in practice a (1+1)CEA can apply an arbitrary stopping criterion (a finite number of steps as in Algorithm 13.2).

Algorithm 13.1 Random walk on Digraph D [14]

Input:
D Coevolutionary Digraph $D = (V, A)$
u Current vertex representing a game-playing strategy
procedure RWALKD(D, u)
  $v :=$ **uniSamp**($D.V$) ▷ $v \in V$ randomly chosen with uniform $\mathbb{P}_V$
  repeat
    $N_D^+(v) :=$ **outNeigh**(v) ▷ $N_D^+(v) = \{u \in V \backslash \{v\} : v_i \to u \in A\}$
    if $N_D^+(v) \neq \emptyset$ **then** ▷ $N_D^+(v) \subset V$ consists of strategies defeating v
      $u :=$ **uniSamp**($N_D^+(v)$) ▷ $u \in N_D^+(v)$ chosen with probability $1/|N_D^+(v)|$
    end if ▷ Else u remains as it is
    $v := u$
  until Forever
end procedure

Algorithm 13.2 (1+1)CEA on Digraph D [14]

```
D       Coevolutionary Digraph D = (V, A)
u       Current vertex representing a game-playing strategy
b_gen   Termination criterion fixed number of generation
procedure CEA(D, u, b_gen)
    t := 0                          ▷ Initialize time-stamp
    v_t := uniSamp(D.V)             ▷ v ∈ V randomly chosen with uniform P_V
    u := v_t                        ▷ Initialize parent as a single vertex u
    while t ≤ b_gen do
        v_t := mutate(u)            ▷ Generate offspring via uniSamp(V\{u})
        if v_t ← u then
            u := v_t
        end if                      ▷ Selection chooses winning strategy in {u, v_t}
        t := t + 1
    end while
end procedure
```

Regardless of how random walks on digraphs are procedurally implemented, our focus in this section is on the underlying ***stochastic processes*** that generate those walks. More specifically, we would like to model these population-one coevolutionary search processes on digraphs as a ***discrete-time, finite state Markov chains*** [47, 50]. Our ***abstract coevolution*** [14] as random walks on coevolutionary digraphs $\boldsymbol{\Phi}$ is a collection of $\{\Phi_t : t \in \mathbb{N}_{\geq 0}\}$ of random variables with each Φ_t taking values in a countable set X at time t, with the ***state space*** X corresponding to the finite set of vertices $V(D)$ of the coevolutionary digraph $D = (V, A)$. However, constructing a general abstract coevolution as a discrete-time stochastic process is very technically involved as one needs to work on probability spaces. In general, one first prescribes the structural definition for the sample path behaviours and the dynamics of steps on the digraph through time to define the probability laws that govern the evolution of the chain $\boldsymbol{\Phi}$. Even in the usual case where sequences $\{\Phi_0, \Phi_1, \Phi_2, \ldots\}$ take *discrete* values in X, developing a complete, functional description of the distribution of the random walk on most digraphs other than those with structures that strictly constrain its sample path behaviours (e.g. acyclic and regular tournaments) is very difficult.

Instead, we use the observation that a large family of restricted coevolutionary search processes satisfy the ***Markov property***, i.e. the step to a future state only depends on the current state. For the example of (1+1)CEA, the probability that it steps from a particular vertex $x_t = u \in V(D)$ to an out-neighbour $x_{t+1} = v \in N_D^+(u)$ is independent of the time (generational) step t, i.e. the probability is the same whether it is $x_0 = u$ at the start or $x_t = u$, which is t generations later. We exploit this Markov property to formulate and define the Markov chain for a coevolutionary process on digraphs as follows. Let $D(V_{S(n)}, A) \in \mathcal{D}(V_{S(n)})$ be a coevolutionary digraph having a finite vertex set $V_{S(n)}$ of n vertices. A ***Coevolutionary Markov Chain*** (***CMC***) [14] is a random walk on $D(V_{S(n)}, A)$ with initial distribution $\boldsymbol{\mu}$ and Markov transition matrix $\boldsymbol{P}$ on the state space $X = V_{S(n)}$ satisfying:

1. $\boldsymbol{\mu} = (\mu_x : x \in X)$, where $0 \leq \mu_x \leq 1$ and $\sum_{x \in X} \mu_x = 1$.
2. $\boldsymbol{P} = (\mathbb{P}(x, z) : x, z \in X)$, where every row is a distribution whereby $0 \leq \mathbb{P}(x, z) \leq 1$ and $\sum_{y \in X} \mathbb{P}(x, y) = 1$.

Since $V_{S(n)}$ has a finite number of elements and by definition is a countable set, guarantees are provided for the construction of such a CMC $\boldsymbol{\Phi} = \{\Phi_t : t \in \mathbb{N}_0\}$ defined on the state space $X = V_{S(n)}$, initial distribution $\boldsymbol{\mu}$, and transition matrix $\boldsymbol{P}$ by the Existence Theorem of Markov chains on countable state space [18].

The distribution describing a CMC is calculated from $\boldsymbol{\mu}$ and $\boldsymbol{P}$. We start the Markov chain with the initial distribution being fixed (point mass) at $x_0 \in X$. Subsequently, for each $x_0 \in X$, we define $\mathbb{P}_{x_0}$ of the Markov chain inductively:

$$\mathbb{P}_{x_0}(\Phi_0 = x_0) = 1,$$

$$\mathbb{P}_{x_0}(\Phi_0 = x_0, \Phi_1 = x_1) = \mathbb{P}(x_0, x_1),$$

$$\mathbb{P}_{x_0}(\Phi_0 = x_0, \Phi_1 = x_1, \Phi_2 = x_2) = \mathbb{P}(x_0, x_1)\mathbb{P}(x_1, x_2),$$

and so forth, using $\boldsymbol{P}$ that gives the chain's one-step transition probabilities $\mathbb{P}(x_t, x_{t+1})$ [47]. Other useful information can be made, for example, the probabilities $\mathbb{P}_\mu(\Phi_n = y \mid \Phi_0 = x) = \mathbb{P}^n(x, y)$ starting from x after n steps that the chain is at y. First, set $\boldsymbol{P}^0 = \mathbf{I}$ (the identity matrix), which corresponds to the initial conditions $\mathbb{P}^0(x, x) = 1$ for all $x \in X$. Then, the *n-step transition matrix* given by the iterates $\boldsymbol{P}^n = (\mathbb{P}^n(x, y) : x, y \in X)$ for $n \geq 0$ can be defined inductively through $\boldsymbol{P}^{n+1} = \boldsymbol{P}^n \boldsymbol{P}$ using the following expression [14]:

$$\mathbb{P}^{n+1}(x, z) = \sum_{y \in X} \mathbb{P}^n(x, y)\mathbb{P}(y, z). \tag{13.1}$$

Next, we provide some concrete examples of CMC constructions. We start with two examples of CMCs on transitive tournaments. The initial distribution can be arbitrary so we just need to construct the appropriate Markov transition matrix. The first example is the random walk on labelled transitive coevolutionary tournaments. Let $T(V_{S(m)}, A)$ be a transitive coevolutionary tournament with vertices $(v_1, v_2, \ldots, v_m)$ indexed according to its score sequence $(s_1, s_2, \ldots, s_m)$. (Note: The score of a vertex v of a coevolutionary digraph is the number of incoming arcs, which corresponds to the number of vertices dominated by v.) We simplify our notation, and let $\boldsymbol{P}_{m \times m} = (p_{ij} : i, j \in \{1, 2, \ldots, m\})$ be the ***Markov transition matrix*** for this CMC on $T(V_{S(m)}, A)$.

$\boldsymbol{P}_{m \times m}$ is defined inductively as follows. The basis $(n = 0)$ is $T(V_{S(m)}, A)$ of order $m = n + 1 = 1$. Since $T(V_{S(m)}, A)$ has one vertex that is also its winning vertex, $v_m = v_1$, it has a corresponding transition probability corresponding to the self-loop $p_{mm} = 1$. As such, $\boldsymbol{P}_{1 \times 1} = [p_{11}] = [1]$. The induction step $(n + 1)$ adds an additional vertex so that the tournament is of order $m = n + 1$. This additional vertex is indexed v_1. There are $n = m - 1$ vertices indexed from v_2 to v_m that dominates v_1 (see Fig. 13.1a). This gives equal probability of transitions from v_1 to

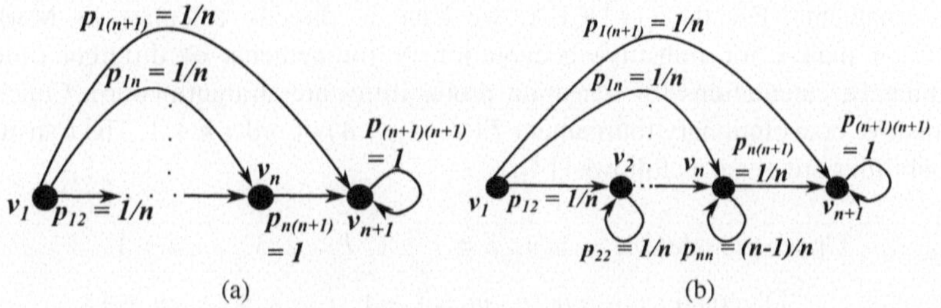

Fig. 13.1 The CMC for a (**a**) random walk on labelled transitive tournaments $T(V_{S(n+1)}, A)$ and (**b**) (1+1)CEA on transitive tournaments $T(V_{S(n+1)}, A)$ [14]

each of $\{v_2, v_3, \ldots, v_m\}$ so that

$$\begin{aligned} p_{11} &= 0, \\ p_{1j} &= 1/(m-1),\ j = 2, 3, \ldots, m, \\ p_{i1} &= 0,\ i = 2, 3, \ldots, m. \end{aligned}$$

We can then construct the Markov transition matrix for the random walk on labelled transitive tournaments $T(V_{S(n+1)}, A)$ of order $n+1$ inductively on $m = 1, 2, \ldots, n+1$

$$\begin{aligned} p_{ii} &= 0,\ i = m \\ p_{ij} &= 1/(n+1-m),\ i = m,\ j = i+1, i+2, \ldots, n+1, \\ p_{ij} &= 0,\ i = j+1, j+2, \ldots, n+1,\ j = m, \end{aligned}$$

and $p_{(n+1)(n+1)} = 1$. This will result in

$$\boldsymbol{P}_{(n+1)\times(n+1)} = \begin{pmatrix} 0 & \frac{1}{n} & \frac{1}{n} & \ldots & \frac{1}{n} & \frac{1}{n} & \frac{1}{n} \\ 0 & 0 & \frac{1}{n-1} & \ldots & \frac{1}{n-1} & \frac{1}{n-1} & \frac{1}{n-1} \\ 0 & 0 & \ddots & \ddots & \vdots & \vdots & \vdots \\ \vdots & \vdots & \vdots & \ddots & \ddots & \frac{1}{3} & \frac{1}{3} \\ 0 & 0 & 0 & \ldots & 0 & \frac{1}{2} & \frac{1}{2} \\ 0 & 0 & 0 & \ldots & 0 & 0 & 1 \\ \hline 0 & 0 & 0 & \ldots & 0 & 0 & 1 \end{pmatrix}, \tag{13.2}$$

where the last row indicates transition probabilities for the tournament's single dominant vertex, with a drawn horizontal line for emphasis [14].

The next example is the (1+1)CEA on unlabelled transitive coevolutionary tournament. Previously, the Markov transition matrix has a nice inductive structure (the lower right submatrix induces a random walk on the corresponding labelled subtournament). For the (1+1)CEA, we have to directly calculate its Markov transition matrix for transitive coevolutionary tournaments of different orders. Fortunately, calculations for transition probabilities are straightforward. Consider a transitive coevolutionary tournament $T(V_{S(n+1)}, A)$ of order $n+1$. The transition probabilities are given as follows [14]:

$$\text{Upper triangle: } p_{ij} = 1/n,\ i < j+1,\ i = 2, 3, \ldots, n+1,$$

$$\text{Diagonal : } p_{ii} = m/n,\ m = i-1,\ i = 2, 3, \ldots, n+1,$$

$$\text{Lower Triangle: } p_{ij} = 0,\ i > j-1,\ i = 2, 3, \ldots, n+1.$$

$$\boldsymbol{P}_{(n+1)\times(n+1)} = \left(\begin{array}{ccccccc} 0 & \frac{1}{n} & \frac{1}{n} & \cdots & \frac{1}{n} & \frac{1}{n} & \frac{1}{n} \\ 0 & \frac{1}{n} & \frac{1}{n} & \cdots & \frac{1}{n} & \frac{1}{n} & \frac{1}{n} \\ 0 & 0 & \frac{2}{n} & \ddots & \vdots & \vdots & \vdots \\ \vdots & \vdots & \vdots & \ddots & \frac{1}{n} & \frac{1}{n} & \frac{1}{n} \\ 0 & 0 & 0 & \cdots & \frac{n-2}{n} & \frac{1}{n} & \frac{1}{n} \\ 0 & 0 & 0 & \cdots & 0 & \frac{n-1}{n} & \frac{1}{n} \\ \hline 0 & 0 & 0 & \cdots & 0 & 0 & 1 \end{array}\right), \tag{13.3}$$

with the corresponding Markov chain illustrated in Fig. 13.1b.

We now consider two examples of CMCs on regular tournaments (labelled and unlabelled). Let $T(V_{S(n)}, A)$ be a regular coevolutionary tournament of odd order $n \geq (3+2m)$, $m = 0, 1, 2, \ldots$. Its ***adjacency matrix*** $\mathbf{M}_{\text{adj}}$ can be constructed as follows [7]:

$$\mathbf{M}_{\text{adj}} = \mathbf{L}_n + \mathbf{L}_n^2 + \mathbf{L}_n^3 + \ldots + \mathbf{L}_n^{(n-1)/2},$$

where $\mathbf{L}_n$ is the permutation matrix that is obtained from the permutation vector $(2, 3, 4, \ldots, n, 1)$. The permutation vector $(r_1, r_2, r_3, \ldots, r_m)$ specifies the row permutation operations on a $m \times m$ identity matrix. For the first row, 1 in the first column is exchanged with 0 in the r_1th column, and so forth, ending with the mth row where 1 in the mth column is exchanged with 0 in the r_mth column. For example, the permutation matrix for a permutation vector (2, 3, 1) is

$$\begin{bmatrix} 0 & 1 & 0 \\ 0 & 0 & 1 \\ 1 & 0 & 0 \end{bmatrix}.$$

The value $\text{adj}(i, j) = 1$ in the adjacency matrix specifies the arc $i \rightarrow j$ of the corresponding digraph. The Markov transition matrix requires that we calculate from the digraph's ***domination matrix*** $\mathbf{M}_{\text{dom}}$ where $\text{dom}(i, j) = 1$ means $i \leftarrow j$. Note that $\mathbf{M}_{\text{dom}}$ is the transpose of $\mathbf{M}_{\text{adj}}$, $\mathbf{M}_{\text{dom}} = \mathbf{M}_{\text{adj}}^T$. We can then compute the Markov transition matrix for a random walk on labelled regular $T(V_{S(n)}, A)$ as [14]

$$P = \frac{2}{(n-1)} \mathbf{M}_{\text{dom}}, \tag{13.4}$$

while for $(1+1)$CEA on unlabelled regular $T(V_{S(n)}, A)$, it is [14]

$$P = \frac{1}{2}\mathbf{I} + \frac{1}{(n-1)} \mathbf{M}_{\text{dom}}. \tag{13.5}$$

These examples highlight cases where we can do the constructions formally and then to perform analysis in the next subsection without resorting to sampling and large-scale computations as is often the case in real-world problems.

13.2.2 Characterization of Coevolutionary Markov Chains

The notion of ***communicating classes*** in a Markov chain is important towards its characterization. Let x, y be two distinct states $x, y \in X = V_{S(n)}$ of the Markov chain. We have $x \rightsquigarrow y$ (x ***leads to*** y) if and only if there is a positive probability $p_{xy}^{(n)} > 0$ for some $n \geq 0$ that starting at x the Markov chain reaches y after n-steps. The relation $\rightsquigarrow$ is reflexive ($x \rightsquigarrow x$) by definition. It is transitive from the property of the n-step transition probability (as specified by Eq. (13.1)) since $p_{xz}^{(m+n)} \geq p_{xy}^{(m)} p_{yz}^{(n)} > 0$ for some $m \geq 0$ and $n \geq 0$. Then, $x \leftrightsquigarrow y$ (x ***communicates*** with y) if and only if $x \rightsquigarrow y$ and $y \rightsquigarrow x$. The relation $\leftrightsquigarrow$ being symmetric and transitive forms an equivalence relation on X and allows one to partition X into disjoint subsets that are called communicating (equivalent) classes [18]. Let $C(x)$ be the class containing $x \in X$. A state space X consisting of a single class $C(x) = X$ (and the Markov chain operating on X) is said to be irreducible.

In [14], we formally demonstrate the links between all three digraph theoretic notions of strong connectivity, game theoretic notion of solvability, and now the Markov chain notion of communication. This is summarized in the following Lemmas 13.1 and 13.2. They form the basis with which all the three notions are unified, allowing us to make qualitative statements about competitive coevolution setting that cover all aspects starting from problem structures (and hence the structures of solutions via their pairwise relations) and then structures in coevolutionary search processes. In qualitatively characterizing competitive coevolution, the dichotomy is transferred from problem structures to search processes. In particular, an ***irreducible CMC*** operates on a ***strongly connected*** (irreducible) coevolutionary digraph that represents a nonsolvable coevolutionary problem. Otherwise, the ***absorbing CMC*** has a single absorbing class corresponding to the ***dominant subset*** of the coevolutionary digraph where ***solutions*** of the ***solvable*** coevolutionary problem reside.

Lemma 13.1 ([14]) *A CMC is irreducible if it operates on an irreducible coevolutionary digraph.*

Lemma 13.2 ([14]) *For a reducible CMC, there is only one absorbing class. The absorbing class consists of states corresponding to dominant solutions in the coevolutionary digraph.*

In addition to the general qualitative characterization of abstract coevolution we have introduced above, one might require more detailed quantitative characterization of abstract coevolution when applied to specific problems. One relevant analysis on the behaviour of such a CMC is concerned with its distributions at certain ***random times*** as it evolves on X. In the following, we will precisely introduce and describe such quantities. In particular, one such quantity refers to the random times with which a CMC $\mathbf{\Phi}$ visits certain states in X. Let $\mathcal{A} \subset \mathcal{B}(X)$, where $\mathcal{B}(X)$ is the set of all subsets of X. The ***first hitting time*** of $\mathcal{A}$ is the random variable $\tau_{\mathcal{A}} : X \to \mathbb{N}_0 \cup \{\infty\}$ defined as

$$\tau_{\mathcal{A}} = \inf\{t \geq 0 : \Phi_t \in \mathcal{A}\}. \tag{13.6}$$

The first hitting time of a state $x \in X$ is simply $\tau_x = \inf\{t \geq 0 : \Phi_t = x\}$. The first hitting time of an empty set $\emptyset$ is defined as ∞ by convention [50]. One can focus on starting states outside of $\mathcal{A}$ so as to define the first visit to $\mathcal{A}$ as $\tau_{\mathcal{A}}^{+} = \inf\{t \geq 1 : \Phi_t \in \mathcal{A}\}$ [47].

As in the context of analysis of EAs [31], studying the mean hitting time of an absorbing CMC $\mathbf{\Phi}$ is crucial. Absorbing CMCs have two properties that are not always found in EA settings. First, its state space is always partitioned into two classes—a closed absorbing class of recurrent states and another consisting of transient states only. Second, complete connectivities in the coevolutionary digraph mean $\tau_{\mathcal{A}}^{+} = 1$ although it is a rare event for $\mathbf{\Phi}$ to make a one-step walk to $\mathcal{A}$. Nevertheless, we can formally define the ***expected hitting time*** $\mathbb{E}_x(\tau_{\mathcal{A}})$ of non-empty, reachable $\mathcal{A}$ starting from $x \in X$ as

$$\mathbb{E}_x(\tau_{\mathcal{A}}) = \sum_{t<\infty} t\mathbb{P}_x(\tau_{\mathcal{A}} = t), \tag{13.7}$$

where $\mathbb{P}_x(\tau_{\mathcal{A}} = t)$ is the probability that starting for state $x \in X$ that $\mathbf{\Phi}$ hits $\mathcal{A}$ at time t [50]. This formulation takes $\mathbb{P}_x(\tau_{\mathcal{A}} = \infty) = 0$ by definition in assuming non-empty, reachable $\mathcal{A}$. We use expected hitting times $\mathbb{E}_x(\tau_{\mathcal{A}})$ as a quantitative measure for the speed with which coevolution searches for the dominant solution in $\mathcal{A}$ [14].

Before calculations of $\mathbb{E}_x(\tau_{\mathcal{A}})$ can be made, one first reindexes the states in the reducible CMC and makes the appropriate simplification to produce the ***canonical form*** of its Markov transition matrix $\boldsymbol{P}^*$ [46]

$$\boldsymbol{P}^* = \begin{bmatrix} \mathbf{I} & \mathbf{0} \\ \mathbf{R} & \mathbf{Q} \end{bmatrix}.$$

Here, **R** represents transition probabilities from transient states to the absorbing class. **Q** represents transition probabilities among transient states. **0** is a row vector of zeroes. Then, we [14] obtain the following theoretical result that allows for the direct calculation of $\mathbb{E}_i(\tau_{\mathcal{A}})$ starting from any transient state $i \in Q$.

Lemma 13.3 ([14]) *Consider a reducible CMC with Markov transition matrix in its canonical form $\boldsymbol{P}^*$. Let $i \in Q$ be the transient states. Let the column vector $\mathbf{h} = [h_i]_{i \in Q}$ where $h_i = \mathbb{E}_i(\tau_{\mathcal{A}})$ is the expected hitting time starting from a transient state i to the absorbing class $\mathcal{A}$. Then,*

$$\mathbf{h} = (\mathbf{I} - \mathbf{Q})^{-1}\mathbf{1}, \tag{13.8}$$

where **1** *is a column vector of ones.*

The technical proof makes use of the fact that every reducible CMC is always an absorbing Markov chain with one absorbing class (for which then Theorem 3.2 in [36] applies). For an absorbing Markov chain, the matrix $(\mathbf{I} - \mathbf{Q})^{-1}$ exists and is called the ***fundamental matrix*** of the Markov transition matrix $\boldsymbol{P}$. One can perform direct computation of the expected hitting time of any reducible CMC by first computing its fundamental matrix (Theorem 11.4 in [29]):

$$(\mathbf{I} - \mathbf{Q})^{-1} = \mathbf{I} + \mathbf{Q} + \mathbf{Q}^2 + \mathbf{Q}^3 + \cdots .$$

This series converges given that $\mathbf{Q}^n \to \mathbf{0}$ as $n \to \infty$ (Theorem 11.3 in [29]).

Now, we are able to provide analytic expression for the expected hitting time calculations for specific CMCs on transitive coevolutionary tournaments. Note that the Markov transition matrices given by Eqs. (13.2) and (13.3) are in the form of an upper triangle matrix. We relabel the indices in the respective $\boldsymbol{P}$ so that it becomes a lower triangle matrix. For both of these specific CMCs, **Q**s are also in the form of a lower triangle matrix. Crucially, each of the $n \times n$ matrices $(\mathbf{I} - \mathbf{Q})$ has a $rank(\mathbf{I} - \mathbf{Q}) = n$ so its inverse exists [46].

It is straightforward to obtain the following for the random walk on labelled transitive coevolutionary tournament of order $n + 1$ [14]

$$\mathbf{I} - \mathbf{Q} = \begin{pmatrix} 1 & 0 & 0 & \dots & 0 & 0 \\ -\frac{1}{2} & 1 & 0 & \dots & 0 & 0 \\ -\frac{1}{3} & -\frac{1}{3} & \ddots & \ddots & \vdots & \vdots \\ \vdots & \vdots & \ddots & \ddots & \ddots & \vdots \\ -\frac{1}{n-1} & -\frac{1}{n-1} & \ddots & -\frac{1}{n-1} & 1 & 0 \\ -\frac{1}{n} & -\frac{1}{n} & \dots & -\frac{1}{n} & -\frac{1}{n} & 1 \end{pmatrix}. \tag{13.9}$$

The fundamental matrix $\boldsymbol{P}$ has an inductive structure that we exploit. In particular, we apply the ***Gauss-Jordan*** method to obtain the $m \times m$ fundamental matrix, and then use this result to partially fill the $m \times m$ submatrix of the $(m + 1) \times (m + 1)$

fundamental matrix. Starting from $m = 2$ and applying this iteratively, we obtain the following form [14]:

$$(\mathbf{I}-\mathbf{Q})^{-1} = \begin{pmatrix} 1 & 0 & 0 & \dots & 0 & 0 \\ \frac{1}{2} & 1 & 0 & \dots & 0 & 0 \\ \frac{1}{2} & \frac{1}{3} & \ddots & \ddots & \vdots & \vdots \\ \vdots & \vdots & \ddots & \ddots & \ddots & \vdots \\ \frac{1}{2} & \frac{1}{3} & \ddots & \frac{1}{n-1} & 1 & 0 \\ \frac{1}{2} & \frac{1}{3} & \dots & \frac{1}{n-1} & \frac{1}{n} & 1 \end{pmatrix}. \tag{13.10}$$

The inductive structure can be demonstrated as follows. Note that as the fundamental matrix increases in size, a new row is added to the bottom and a new column is added to the left. Observe that the element in the first column and last row of $(\mathbf{I}-\mathbf{Q})^{-1}$ always add to $1/2$. For 2×2, it is obviously $1/2$ due to Gauss-Jordan operations. Increasing the matrix in size,

$$\begin{aligned}
3 \times 3: \quad & \tfrac{1}{3} \times \tfrac{1}{2} + \tfrac{1}{3} \times 1 = \tfrac{1}{3} \times (\tfrac{3}{2}) \\
&= \tfrac{1}{2} \\
4 \times 4: \quad & \tfrac{1}{4} \times \tfrac{1}{2} + \tfrac{1}{4} \times \tfrac{1}{2} + \tfrac{1}{4} \times 1 = \tfrac{1}{4} \times (\tfrac{4}{2}) \\
&= \tfrac{1}{2} \\
5 \times 5: \quad & \tfrac{1}{5} \times \tfrac{1}{2} + \tfrac{1}{5} \times \tfrac{1}{2} + \tfrac{1}{5} \times \tfrac{1}{2} + \tfrac{1}{5} \times 1 = \tfrac{1}{5} \times (\tfrac{5}{2}) \\
&= \tfrac{1}{2}.
\end{aligned}$$

For $(n+1) \times (n+1)$, we note that

$$\begin{aligned}
& \overbrace{\frac{1}{n+1} \times \frac{1}{2} + \frac{1}{n+1} \times \frac{1}{2} + \dots + \frac{1}{n+1} \times \frac{1}{2}}^{n-1} + \tfrac{1}{n+1} \times 1 \\
&= \tfrac{1}{n+1} \times (\tfrac{n+1}{2}) \\
&= \tfrac{1}{2}.
\end{aligned}$$

A simple but laborious calculation lets us verify that $(\mathbf{I}-\mathbf{Q})(\mathbf{I}-\mathbf{Q})^{-1} = \mathbf{I}$. We apply Lemma 13.3 to calculate the expected hitting time and then obtain [14]

$$\mathbf{h} = (\mathbf{I}-\mathbf{Q})^{-1}\mathbf{1} = \left(1, \frac{1}{2}+1, \frac{1}{2}+\frac{1}{3}+1, \dots, \sum_{i=1}^{n} \frac{1}{n}+1\right)^T. \tag{13.11}$$

For the transitive tournament, the reindexing in $\boldsymbol{P}^*$ means that $\mathbf{h} = [h_i]_{i \in Q} = [h_i]_{i=1}^n$ where the index $i = 1, \dots, n$ refers to vertices in the reverse order of the score sequence $(s_j : j = 1, \dots, n)$, which is written in an ascending manner. As such, $[h_1] = [1]$ is the expected hitting time $\mathbb{E}_{s_n}(\tau_{s_{n+1}})$ of the vertex s_n (expressed in the original index following the score sequence) with score $n-1$ to the dominant vertex s_{n+1} in a tournament of order $n+1$ having a score of n. The expected hitting time starting from the weakest vertex to the dominant vertex is then $\mathbb{E}_{s_1}(\tau_{s_{n+1}}) = [h_n] = [\sum_{i=1}^{n} 1/n + 1]$.

Although we are able to derive analytic expressions for the case of the (1+1)CEA on transitive coevolutionary tournament of order $n+1$ and then by direct calculation obtain $\mathbf{h} = [h_i]_{i \in Q} = [n, n, n, \ldots, n]^T$, further theoretical work on absorbing (1+1)CEAs has revealed a surprising general result. This is stated in the following Theorem 13.1. The technical proof exploits a unique invariant property that all absorbing (1+1)CEAs possess in their associated probability transition matrices, i.e. all the diagonal elements of $\mathbf{Q}$ have the same value of $\frac{1}{2}\frac{n-1}{n}$.

Theorem 13.1 ([14]) *Consider a (1+1)CEA operating on a reducible coevolutionary tournament* $T(V_{S(n+n_d)}, A) \in \mathcal{T}(V_{S(n+n_d)})$. *Let* $V_{S(n+n_d)}$ *consist of two disjoint subsets* $V^1_{S(n)}$ *and* $V^2_{S(n_d)}$, *whereby* $V^1_{S(n)} \cap V^2_{S(n_d)} = \emptyset$. *Let* $V^2_{S(n_d)} \Leftarrow V^1_{S(n)}$. *Let the CMC be defined with Markov transition matrix in its canonical form* $\boldsymbol{P}^*$. *Let* $i \in Q$ *be the transient states and that* $|Q| = n$. *Let the column vector* $\mathbf{h} = [h_i]_{i \in Q}$ *where* $h_i = \mathbb{E}_i(\tau_{\mathcal{A}})$ *is the expected hitting time starting from a transient state i to the absorbing class* $\mathcal{A} = V^2_{S(n_d)}$. *Then,* $\mathbf{h} = [h_i]_{i \in Q} = [n, n, n, \ldots, n]^T$.

In the following, we illustrate by example the simplified proof argument we have made in [14] for Theorem 13.1 that showcases the technical arguments. Consider the (1+1)CEA on an unlabelled coevolutionary tournament $T(V_{S(n+1)}, A)$ with a single dominant vertex and a single dominated strong component that consists of a regular tournament of odd order $n \geq (3+2k)$, $k = 0, 1, 2, \ldots$. We then have the following:

$$\boldsymbol{P}^* = \begin{pmatrix} 1 & 0 & 0 & 0 & 0 & \cdots & 0 & 0 \\ \frac{1}{n} & \frac{1}{2}\frac{n-1}{n} & 0 & \frac{1}{n} & 0 & \cdots & 0 & \frac{1}{n} \\ \frac{1}{n} & \frac{1}{n} & \frac{1}{2}\frac{n-1}{n} & 0 & \frac{1}{n} & \cdots & \frac{1}{n} & 0 \\ \frac{1}{n} & 0 & \frac{1}{n} & \frac{1}{2}\frac{n-1}{n} & 0 & \cdots & 0 & \frac{1}{n} \\ \vdots & \vdots & \vdots & \vdots & \ddots & \ddots & \vdots & \vdots \\ \vdots & \vdots & \vdots & \vdots & \vdots & \ddots & \ddots & \vdots \\ \frac{1}{n} & \frac{1}{n} & 0 & \frac{1}{n} & 0 & \cdots & \frac{1}{2}\frac{n-1}{n} & 0 \\ \frac{1}{n} & 0 & \frac{1}{n} & 0 & \frac{1}{n} & \cdots & \frac{1}{n} & \frac{1}{2}\frac{n-1}{n} \end{pmatrix}. \tag{13.12}$$

We can write $\mathbf{Q}$ with elements substituted with these two equalities $a = \frac{1}{2}\frac{n-1}{n}$ and $b = \frac{1}{n}$:

$$\mathbf{Q} = \begin{pmatrix} a & 0 & b & 0 & \cdots & 0 & b \\ b & a & 0 & b & \cdots & b & 0 \\ 0 & b & a & 0 & \cdots & 0 & b \\ \vdots & \vdots & \vdots & \ddots & \ddots & \vdots & \vdots \\ \vdots & \vdots & \vdots & \vdots & \ddots & \ddots & \vdots \\ b & 0 & b & 0 & \cdots & a & 0 \\ 0 & b & 0 & b & \cdots & b & a \end{pmatrix}.$$

We observe that the $(i+1)$th row is a right shift of the ith row of $\mathbf{Q}$. Furthermore, each of the ith row sum Q_i has the same value and is given by $a+(\frac{n-1}{2})b = \frac{1}{2}\frac{n-1}{n} + (\frac{n-1}{2})\frac{1}{n} = \frac{n-1}{n}$. This follows from the definition of a regular tournament whereby each vertex i has the same number of incoming and outgoing arcs $d^+_D(i) = d^-_D(i)$. $\mathbf{Q}^1_{n\times n}\mathbf{1} = [\frac{n-1}{n}, \ldots, \frac{n-1}{n}]^T$. Obviously, one can obtain the same result by taking $\mathbf{1} - \mathbf{R1}$ since $\mathbf{R} = [\frac{1}{n}, \ldots, \frac{1}{n}]^T$ and that $(\mathbf{R1} + \mathbf{Q1}) = \mathbf{1}$.

We need to show that the row sums of $(\mathbf{I}-\mathbf{Q})^{-1} = \mathbf{I}+\mathbf{Q}+\mathbf{Q}^2+\mathbf{Q}^3+\cdots$ follow a geometric series. We have that $\mathbf{Q}^m_{n\times n}\mathbf{1} = [Q^m_i : i = 1, \ldots, n]^T$, $Q^m_i = (\frac{n-1}{n})^m$ inductively on m. Then, each element of $\mathbf{h} = (\mathbf{I} - \mathbf{Q})^{-1}\mathbf{1}$ follows the geometric series $1 + \frac{n-1}{n} + (\frac{n-1}{n})^2 + (\frac{n-1}{n})^3 + \cdots = 1/(1 - \frac{n-1}{n}) = n$, in which case $\mathbf{h} = [h_i]_{i\in Q} = [n, n, n, \ldots, n]^T$. One can verify the calculations, noting that

$$\mathbf{I} - \mathbf{Q} = \begin{pmatrix} 1-a & 0 & -b & 0 & \cdots & 0 & -b \\ -b & 1-a & 0 & -b & \cdots & -b & 0 \\ 0 & -b & 1-a & 0 & \cdots & 0 & -b \\ \vdots & \vdots & \vdots & \ddots & \ddots & \vdots & \vdots \\ \vdots & \vdots & \vdots & \vdots & \ddots & \ddots & \vdots \\ -b & 0 & -b & 0 & \cdots & 1-a & 0 \\ 0 & -b & 0 & -b & \cdots & -b & 1-a \end{pmatrix}.$$

Each of the ith element of $(\mathbf{I} - \mathbf{Q})\mathbf{h}$ is exactly one, i.e. $(1-a)n + (\frac{n-1}{2})(-b)n = (1 - \frac{1}{2}\frac{n-1}{n})n + (\frac{n-1}{2})(-\frac{1}{n})n = \frac{2n-n+1}{2} - \frac{n-1}{2} = \frac{2}{2} = 1$.

More importantly, both theoretical results as stated in Theorem 13.1 and Eq. (13.11) give us a crucial insight how structures in a coevolutionary problem structure represented by its associated digraph can affect the search process modelled as random walks operating on the digraph. In particular, the information represented by labelling of digraphs specified as the set of out-neighbourhoods of vertices $N^+_D(v_i)$, $i = 1, \ldots, n+1$ can be interpreted as a form of ***domain knowledge***. Incorporating such information into coevolutionary search results in the construction of a specific CMC (i.e. standard random walk) that has a fast search process. For example, a random walk on transitive tournaments has $O(\ln n)$ time complexity to obtain the dominant solution v_{n+1} [20]. (Note: The expected hitting time starting from v_1 to v_{n+1} takes the form of a ***Harmonic Series*** H_n as specified in Eq. (13.11).) Theorem 13.1 will give a $O(n)$ time complexity for a (1+1)CEA on transitive tournaments. However, analysis of CMCs in the context of expected hitting times starting from v_1 to v_{n+1} only gives a partial understanding of the disparity between standard random walk and (1+1)CEA. If one considers expected hitting times from each vertex $v_1, \ldots, v_n$, then the actual difference in the speed of which the respective CMCs find v_{n+1} becomes really apparent. Specifically, there is no change whether one starts at v_1 or even v_n for (1+1)CEA. The lack of domain knowledge in the form of $N_D^+(v_i)$, $i = 1, \ldots, n+1$ means that (1+1)CEA behaves as though it is just uniformly sampling the vertex set $V_{S(n+1)}$. If we make the full use interpreting Theorem 13.1, then one can have different structures in the dominated set $V_{S(n+1)} \setminus \{v_{n+1}\}$ and speed with which (1+1)CEA obtains the dominant solution stays the same. In other words, (1+1)CEA is performing a blind search on any reducible coevolutionary tournament.

13.2.3 Computational Studies on Quantitative Measures of Coevolutionary Markov Chains

We conclude this section with a computational study on quantitative measures based on expected hitting times for the analysis of CMCs. Previously, we are able to derive analytical expressions that then let us calculate these hitting times for specific CMCs of reducible coevolutionary tournaments with simple structures, e.g. transitive tournaments. In practice, real-world reducible coevolutionary tournaments would be more complex than transitive tournaments. As such, another aspect of this computational study is to investigate how random coevolutionary tournaments can be generated in a controlled manner. The potential application here is to use these random tournaments as benchmarks to study other methodologies estimating quantities of interest in more complex and realistic CMCs. Note that the simple approach of randomly orienting edges of the underlying graphs would usually generate irreducible random tournaments [48]. There is more than 50% chance to generate an irreducible random tournament of order five, which increases to more than 99.9% for random tournaments of order 16, beyond which such generated random tournaments would almost certainly be irreducible. Another approach is to first generate all valid score sequences [32] that are then used to reconstruct the corresponding tournaments [48].

Here, we introduce a method to generate a hierarchy of random coevolutionary tournaments between two complexity extremes—irreducible tournaments and tournaments with prominent transitive structures—with only one control parameter. The method is motivated by network growth methodologies via preferential attachment in studies of ***complex networks*** [58]. Algorithm 13.3 describes a general procedural implementation. The tournament can be seeded with an initial, small-sized tournament in a random or arbitrary manner (e.g. acyclic ordering) via the procedure

Algorithm 13.3 Growth-based algorithm to generate random tournaments $T(V_{S(n)}, A)$ [14]

```
Input:
  T        Coevolutionary tournament T(V, A)
  n_init   Size of initial tournament
  n_final  Size of final tournament
Output:
  T(V, A)  Coevolutionary tournament T(V_S(n_final), A)
procedure GROW(T, n_init, n_final)
    n := 0
    T := initTournament(n_init)      ▷ Initialize coevolutionary tournament T(V_S(n_init), A)
    n := n_init
    while n ≤ n_final do
        T_t := reindex(T)            ▷ For T_t(V_t, A_t), V_t := V, A_t := A, then reindexed
                                     ▷ according to its score sequence (s_1, s_2, s_3, ..., s_n)
        p := attachPr(T_t)           ▷ Compute attachment probabilities for V_S(n), p = {P_i^-}_1^n
        T.V := addVertex(T.V_t, v_{n+1})                    ▷ V := V_t ∪ v_{n+1}
        i := 0
        for all i ≤ n do
            T.A := addArc(T.A_t, v_{n+1}, v_i, p[i])   ▷ A := A_t ∪ v_{n+1} → v_i with P_i^-,
                                                      ▷ otherwise A := A_t ∪ v_i → v_{n+1}
            i := i + 1
        end for
        n := n + 1
    end while
    return T
end procedure
```

initTournament(n_{init}). For each iterate in $n_{\text{init}} \leq n \leq n_{\text{final}}$, a new node (vertex) is added to the existing tournament of order n. Then, the $n+1$ edges from the new node to those in the existing tournament are oriented. This orientation is done randomly and independently but can be biased in a controlled manner towards the existing vertices with higher scores. This requires computing the ***attachment probabilities***

$$\mathbb{P}_i^- = \frac{1}{1 + e^{-\beta(x_i - \overline{x})}}, \tag{13.13}$$

where $\{x_i\}_1^n$ are the relevant statistics of the tournament. Parameter $\beta > 0$ controls the steepness of the sigmoid link function of the Bernoulli distribution [14]. A lower value of β reduces the steepness of the function with $\mathbb{P}_i^- \to 1/2$ for $\beta \to 0$ (which is the same as the case of random orientation discussed above).

The method is motivated by phenomena described in ***statistical physics***. Each node of the existing tournament has an ***energy*** related to its score (in degree). The new incoming node $n+1$ will be more likely to orient its edges towards high-energy node in the existing tournament. Parameter β plays the role of inverse temperature. For high-temperature regimes, the energy of nodes begins to equalize, and the probability of edge orientation is the same for any node in the existing tournament. In contrast, cooling the system (increasing β) leads to orientations of edges from the new node $n + 1$ towards high-energy nodes of the existing tournament to become more persistent.

The following computational study requires measurements that can characterize complexity of cycle structures in coevolutionary tournaments. We formally construct such an index as follows. Let a coevolutionary tournament $T(V_{S(n)}, A)$ of order n with vertices indexed as $v_1, v_2, v_3, \ldots, v_n$ according to the score sequence $(s_1, s_2, s_3, \ldots, s_n)$. Let the two integers $i, j = 1, 2, 3, \ldots, n$. $T(V_{S(n)}, A)$ is associated with two sequences $S = (s_1, \sum_{i=1}^{2} s_i, \sum_{i=1}^{3} s_i, \ldots, \sum_{i=1}^{n} s_i)$ and $L = (0, \binom{2}{2}, \binom{3}{2}, \ldots, \binom{n}{2})$. We write S_i to denote the ith element of S (similarly, for L we write L_i). Then, we define the Landau's index ν as follows [14]:

$$\nu = \sum_{i=1}^{n} \mathbb{I}\{S_i = L_i\}, \tag{13.14}$$

$$\mathbb{I}\{S_i = L_i\} = \begin{cases} 1 \text{ if } \sum_{j=1}^{i} s_j = \binom{i}{2} \\ 0 \text{ if } \sum_{j=1}^{i} s_j \neq \binom{i}{2}. \end{cases}$$

The index ν ranges from 1 to n. Essentially, it is a count of the number of strong components in the coevolutionary tournament and provides a useful macro-scale indication of cycle structure complexity in coevolutionary tournaments. $\nu = 1$ indicates an irreducible tournament, while $\nu = n$ indicates a transitive tournament. A value of ν closer to n indicates the presence of many shorter length cycles that can be grouped separately into different subtournaments. When ν is close to one, the tournament has longer length cycles between $n/2$ to $n - 1$. The index has a linear computation cost with the size of the tournament with precomputed Ls.

We [14] have performed initial experiments to determine appropriate choices for the parameters involved. We initialize the graph growth with transitive tournaments. For the statistics $\{x_i\}_1^n$, we use the node ranks $(i)_1^n$ with $\overline{x} = n/2$. We have $\beta \in [0, 4]$. Tournaments of order $n_{\text{final}} = 1000$ are generated from initial ones of order $n_{\text{init}} \in \{10, 100\}$. Each experiment set is repeated 100 times, thus generating 100 tournaments randomly and independently. The results are summarized in Fig. 13.2, with results indicating generated tournaments with a larger number of strong components when the procedure **Grow** is initialized with larger tournament in Fig. 13.2b.

We next introduce into our 1000-node tournaments an additional single dominant (1001th) node that has all the edges from the original 1000 vertices oriented towards it. This construction makes the tournaments reducible. Then, we compute the expected hitting time from the weakest node x_1 having the lowest score to the dominant (absorbing) node x_{1001}. This lets us study the impact of increasing transitive structures, which is measured by Landau's index ν, on the speed with which coevolution searches for the dominant node via its expected hitting time. For coevolution, we employ the standard random walk. We observe signatures

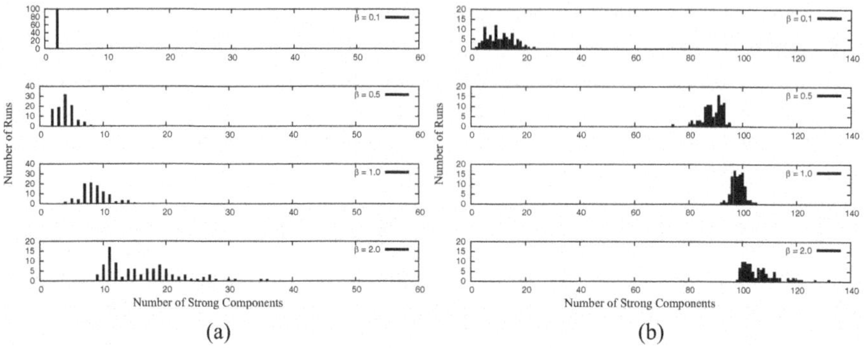

Fig. 13.2 Histograms of a number of runs generating random tournaments having a specific number of strong components (Landau's Index). Corresponding experiment sets where initial transitive tournaments are of order (**a**) 10 and (**b**) 100 [14]

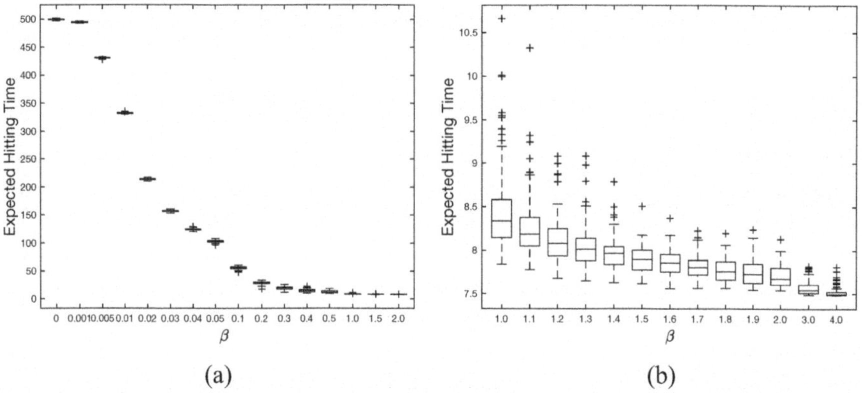

Fig. 13.3 Boxplots with centerline representing the median, the two edges corresponding to the 25th and 75th percentiles, and outliers are plotted as "+" using β set to a range in (**a**) $[0, 2]$ and (**b**) $[1, 4]$ [14]

of higher and lower temperature regimes in Fig. 13.3a. In the high-temperature regime ($\beta \approx [0, 0.001]$), all the base 1000-node tournaments are irreducible; hence, there is only one strong component. Together with the dominant 1001th node, there are two strong components. There is little variability of hitting times (around 492–497), which reflects limited variability of the internal structure of the irreducible, dominated 1000-node base tournaments. For comparison, the expected hitting time is exactly 500 for a regular 999-node base tournament. In the low-temperature regime ($\beta \approx [1, 4]$), prominent transitive structure dominates the base tournaments and the hitting times range in $[7.486, 10.659]$. For comparison, the expected hitting times would be slightly less than 10 for a vertex pancyclic 1000-node base tournament (which has the longest transitive structure) and around 7.4855 for a transitive 1000-node base tournament.

As in most complex systems, in between these two temperature regimes sits an intermediate temperature regime that exhibits the most interesting behaviour. Full spectrum of possible values of Landau's index and expected hitting times can be observed. Indeed, this is the case where transitive structures coexist with a wide range of hitting times, reflecting different degrees of cycle structures in the strong components. Figure 13.3 illustrates the relationship between the (inverse) temperature β and the hitting times of the dominant 1001th node. As expected, extremely high temperatures lead to irreducible base tournaments with intricate cycle structures and long hitting times, while very low-temperature regimes lead to base tournaments with prominent trivial transitive structures that yield minimal hitting times. Furthermore, one can expect on average (with respect to generation of many random tournaments) that the expected hitting times will decrease with decreasing temperature that induces more transitive structures in the base tournaments. The most complex and interesting regimes lie in between these two extremes at $\beta \approx (0.005, 0.5)$.

We have repeated the experiment with (1+1)CEA and obtain results that verify what Theorem 13.1 will indicate. Computations for the expected hitting times are dependent on the probability transition matrix associated with the CMC under study. Different constructions of coevolutionary search processes could lead to different outcomes in the expected hitting time analysis. The expected hitting time has a useful interpretation of being a measure of speed with which coevolution can search for the dominant solution. However, one needs to obtain prior, qualitative knowledge of the underlying structures of these digraphs to be able to use this quantitative measure appropriately. Although hitting times can be formally defined with respect to a starting vertex and then ending vertex within an irreducible digraph, here we have proposed hitting times as a measure of how fast coevolution searches for the dominant solution. It is only the case of CMC operating on reducible tournaments that the measure can be meaningfully interpreted.

13.3 Duality of Solution Performance and Visitation Probability in Competitive Coevolution

In this section, we present a study [13] on the deep connection between CMC and PageRank to address the issue where quantitative analysis of abstract coevolution requires prior qualitative knowledge about the underlying structure of digraphs. We develop a principled approach to measure and rank the performance (importance) of solutions (vertices) in a given coevolutionary digraph. In PageRank formalism, v transfers part of its authority to u if u dominates v, i.e. $v \rightarrow u \in A$ in the digraph $D = (V, A)$. Thus, PageRank authority indicates the importance of a vertex. First in Sect. 13.3.1, we will formulate PageRank authorities in the context of competitive coevolution settings. Then in Sect. 13.3.2, we will demonstrate formally that PageRank authorities with suitable normalization have a dual and natural interpretation of long-term visitation probabilities over the digraph by the coevolutionary random walk with restart. We can precisely quantify changes to these authorities as restart probability is varied. Lastly in Sect. 13.3.3, we conduct computational studies to demonstrate how PageRank authorities are used to characterize coevolutionary digraphs with different underlying structures.

13.3.1 Invariant Measures and PageRank Authority for Abstract Coevolution

Earlier in Sect. 13.2, we introduced and studied the expected hitting times $\mathbb{E}_x(\tau_{\mathcal{A}})$ starting from any transient state $x \in Q$ to the absorbing class $\tau_{\mathcal{A}}$ as a quantitative measure for analysis of reducible CMCs. Transient states are associated with dominated vertices, while the absorbing class is associated with the dominant subset of the coevolutionary digraph. Are there relevant quantitative characterizations of CMCs operating on both reducible and irreducible digraphs? We first introduce the notion of an ***invariant measure*** on Markov chains [50]. Let the row vector $\boldsymbol{\pi} = (\pi_x : x \in X)$ be a probability measure on X and $\boldsymbol{P}$ be the Markov transition matrix. Then, $\boldsymbol{\pi}$ is said to be ***invariant*** if $\boldsymbol{\pi P} = \boldsymbol{\pi}$. A $\boldsymbol{\pi}$ defined on X that satisfies this invariant property would represent the long-term limiting distribution of the associated Markov chain.

In the context of abstract coevolution [14], such a ***distribution*** $\boldsymbol{\pi}$ can be described in equivalent terms as follows:

(i) ***Stationary*** - If $(\Phi_n)_{n\geq 0}$ is a CMC with $\boldsymbol{\mu} = \boldsymbol{\pi}$ and $\boldsymbol{P}$, then $(\Phi_{m+n})_{n\geq 0}$ having $\mathbb{P}(\Phi_m = x) = (\boldsymbol{\pi P}^m)_x = \pi_x$ for all $x \in X$ is also a CMC with $\boldsymbol{\mu} = \boldsymbol{\pi}$ and $\boldsymbol{P}$ (Theorem 1.7.1 in [50]).
(ii) ***Equilibrium*** - For all $x \in X$, assuming complete connectivities of coevolutionary digraphs, $\mathbb{P}^n(x, y) \to \pi_y$ as $n \to \infty$ for all $y \in X$ (Theorem 1.7.2 in [50]).

It is known that for an irreducible Markov chain $\boldsymbol{\Phi}$ on countable state space X, its stationary distribution can be calculated from the ***expected return times*** $\mathbb{E}_x(\varrho_x)$ [42]. The general intuition is that if one breaks up the trajectory of $\boldsymbol{\Phi}$ into identically distributed segments through visits to and from some arbitrary states, the average fraction of time that is spent on a state $x \in X$ by each segment would be similar to the long-term fraction of time spent on x by $\boldsymbol{\Phi}$. Furthermore, starting from any arbitrary distribution $\boldsymbol{\mu}$, the chain would converge to its stationary distribution in the long run [42, 50]. There is an intimate connection between qualitative characterizations of global connectivity structures in D_C and quantitative characterizations of $\boldsymbol{\Phi}$ [14]. A random walk on a reducible D_C leads to an absorbing $\boldsymbol{\Phi}$. In the same manner, an irreducible $\boldsymbol{\Phi}$ operates on a strongly connected D_C. One can obtain the long-term limiting distribution of $\boldsymbol{\Phi}$ that satisfies the invariant property $\boldsymbol{\pi P} = \boldsymbol{\pi}$. For absorbing $\boldsymbol{\Phi}$, the probability mass is concentrated on the absorbing class with all other states having zero probabilities [47]. We have the following results for CMCs.

Lemma 13.4 ([14]) *Consider an irreducible CMC with Markov transition matrix* $\boldsymbol{P}$. *Then, there is a unique invariant probability distribution* $\boldsymbol{\pi} = (\pi_x : x \in X)$ *on* X *where* $0 \leq \pi_x \leq 1$ *and* $\sum_{x \in X} \pi_x = 1$ *such that* $\boldsymbol{\pi P} = \boldsymbol{\pi}$. *Furthermore,* $\pi_x = 1/\mathbb{E}_x(\varrho_x)$.

Lemma 13.5 ([14]) *Consider a reducible CMC with Markov transition matrix* $\boldsymbol{P}$. *For all inessential states* $y \in X$, $\pi_y = 0$. *Then, there is a unique invariant probability distribution* $\boldsymbol{\pi}$ *concentrated on the absorbing class* C *of essential states in* X.

The stationary distribution of a CMC can be formulated, but the associated quantitative analysis and interpretation of subsequent results require prior knowledge of the digraph's underlying structure. For example, the entire probability mass will be concentrated on the absorbing class for a CMC operating on a reducible digraph, but the immediate consequence is that one would have no knowledge about the structures underlying the dominated subset of vertices. Here, we have refined the invariant measure $\boldsymbol{\pi}$ to incorporate some knowledge of coevolutionary digraph structures. This involves establishing a direct connection between abstract coevolution and PageRank (used originally to measure webpage importance [6, 40]), which constitutes one of the large network analysis methodologies [27]. In the PageRank formalism, strategies transfer part of their authorities to those that dominate them at pairwise interactions level, and these authorities give indication of strategies' performances. These authorities are linked to the underlying structures of the digraph.

We reformulate the PageRank authority [5] using digraph-theoretic notations to make explicit its link to digraph structures. Let $u, v \in V_S$, the in-neighbourhood of u is $N_D^-(u) = \{v \in V_S \backslash \{u\} : vu \in A_R\}$, and the out-degree (the number of outgoing arcs) of v is $d_D^+(v) = |\{(u, v) \in A_R : v \in V_S \backslash \{u\}\}|$. The ***PageRank authority*** of u is

$$\varphi_u = \alpha + (1 - \alpha) \sum_{v \in N_D^-(u)} \frac{\varphi_v}{d_D^+(v)},$$

where $\alpha \in (0, 1)$. The parameter α has several interpretations depending on the context in which PageRank is formulated. Here, it is used for the convex combination of the free authority that each vertex (node) has and those as a result of connectivity structures to calculate PageRank authorities [5].

Let vertices in a coevolutionary digraph $D_C(V_{S(n)})$ be labelled as $v_1, v_2, v_3, \ldots, v_n$. Then, the PageRank authorities of $D_C(V_{S(n)})$ are given by

$$\boldsymbol{\varphi} = \alpha \mathbf{e} + (1 - \alpha) \boldsymbol{\varphi} \mathbf{M}, \tag{13.15}$$

where $\mathbf{e} = [1, 1, 1, \ldots, 1]$ and the matrix $\mathbf{M} = (m_{ij} : i, j \in \{1, 2, 3, \ldots, n\})$ with $m_{ij} = \frac{1}{d_D^+(v_i)}$ if $i \to j$, $m_{ij} = 0$ if $i \nrightarrow j$. $\boldsymbol{\varphi} = [\varphi_1, \ldots, \varphi_n]$ (where we use a simplified notation in place of φ_{v_i} as the ith element φ_i of $\boldsymbol{\varphi}$) [13]. The influence of the digraph structure can be seen from non-zero entries of $\mathbf{M}$ that corresponds to the adjacency matrix associated with $D_C(V_{S(n)})$.

13.3.2 Connection Between PageRank and Coevolutionary Markov Chains

Actual PageRank computation requires reformulation as an eigensystem or an equivalent linear system if $\boldsymbol{\psi}$ is suitably normalized ($\boldsymbol{\psi}\mathbf{e}^{\mathsf{T}} = 1$) [27]. ***PageRank*** is a Markov chain with a primitive probability transition matrix $\boldsymbol{P}_{\mathrm{PR}}$ (nonnegative, irreducible matrix having one eigenvalue $r = \rho(\boldsymbol{P}_{\mathrm{PR}})$ on its spectral circle [46]). Setting the usual adjustment $\boldsymbol{P}_{\mathrm{PR}} = \alpha\mathbf{e}^{\mathsf{T}}\mathbf{s} + (1-\alpha)\boldsymbol{P}$ where $\mathbf{s}$ is a general probability vector (e.g. uniform $\frac{1}{n}\mathbf{e}$) [41] makes any CMC irreducible (e.g. an original absorbing CMC will then become irreducible) and as a Markov chain similar to PageRank on $D_C(V_{S(n)})$. Crucially, this setting has the interpretation of introducing restart into CMC with probability α. It is this simple refinement that then makes PageRank authorities applicable to all coevolutionary digraphs (i.e. reducible and irreducible).

Next, the setting of $\boldsymbol{P}$ indicates which coevolutionary search process is used, e.g. $\boldsymbol{P} = \mathbf{M}$ means that the CMC is a standard random walk. Note that PageRank authorities of the coevolutionary digraph $\boldsymbol{\psi}$ are the solution to $\boldsymbol{\psi}\boldsymbol{P}_{\mathrm{PR}} = \boldsymbol{\psi}$. This gives two natural interpretations to $\boldsymbol{\psi}$: (1) It indicates how coevolution measures the importance (authority) of individual $v \in V_{S(n)}$ and provides the means to rank them. (2) It gives long-term visitation probabilities over the digraph by coevolutionary search modelled by the specific CMC.

In the following, we will present several theoretical results connecting CMC and PageRank we have established in [13], which are reflected by the relationship between the stationary distributions $\boldsymbol{\pi}$ and $\boldsymbol{\psi}_\alpha$ associated with CMC and PageRank (CMC with restart probability α), respectively [37]. We start with an existence guarantee of a PageRank vector associated with a CMC, which is summarized as Lemma 13.6. The technical proof involves direct calculations on $\left(\beta\mathbf{I} + (\mathbf{I} - \boldsymbol{P})\right)^{-1}$, which is made possible by employing a lazy random walk. The latter modifies the probability transition matrix $\mathbf{Z} = (\mathbf{I} + \boldsymbol{P})/2$ but maintains the original stationary distribution $\boldsymbol{\pi}$ since $\boldsymbol{\pi} = \boldsymbol{\pi}\mathbf{Z}$. More importantly, the modification introduced in the lazy random walker ensures that $\mathbf{Z}$ is aperiodic [16]. This lets us claim that $\boldsymbol{\pi}$ is the limiting invariant distribution of a CMC operating on a strongly connected digraph $D_C(V_{S(n)})$ that is aperiodic, which follows from the Perron-Frobenius Theorem. Lastly, $\mathbf{W} = \beta\mathbf{I} + (\mathbf{I} - \boldsymbol{P})$ is a strictly dominant diagonal matrix and is invertible [35].

Lemma 13.6 ([13]) *Let $D_C(V_{S(n)}) \in \mathcal{D}_C(V_{S(n)})$ be a coevolutionary digraph. Let $\boldsymbol{P}$ be the probability transition matrix associated with a CMC operating on $D_C(V_{S(n)})$ and $\mathbf{Z} = (\mathbf{I} + \boldsymbol{P})/2$ its lazy version. The row vector $\mathbf{s}$ is the probability distribution over the set vertices $V_{S(n)}$. Given $\alpha \in (0, 1)$ and $\beta = \frac{2\alpha}{1-\alpha}$, the personalized PageRank vector is the unique solution to the linear system defined as*

$$\boldsymbol{\psi}_\alpha(\mathbf{s}) = \alpha\mathbf{s} + (1 - \alpha)\boldsymbol{\psi}_\alpha(\mathbf{s})\mathbf{Z} \tag{13.16}$$

and can be computed as

$$\boldsymbol{\psi}_\alpha(\mathbf{s}) = \beta\mathbf{s}\big(\beta\mathbf{I} + (\mathbf{I} - \boldsymbol{P})\big)^{-1}. \tag{13.17}$$

Next, we employ a linear systems formulation to uncover several properties related to the ***coefficient matrix*** $\mathbf{W}$, which we summarize in Lemma 13.7. Let $\mathbf{R}^{n\times n}$ be the set of real, square $n \times n$ matrices. $\mathbf{A} \in \mathbf{R}^{n\times n}$ is an M-matrix with the form $\mathbf{A} = c\mathbf{I} - \mathbf{B}$, where $\mathbf{B} \geq \mathbf{0} = (b_{ij} \geq 0 : 1 \leq i, j \leq n)$ and $c \geq \rho(\mathbf{B})$ with $\rho(\cdot)$ denoting the spectral radius [4]. $||\mathbf{A}||_\infty$ is the ∞-norm on $\mathbf{A}$ and $\kappa_\infty(\mathbf{A}) = ||\mathbf{W}||_\infty||\mathbf{W}^{-1}||_\infty$ is the condition number with respect to the ∞-norm for $\mathbf{A}$ [46]. The main idea of the technical proof makes use of $\mathbf{W} = c\mathbf{I} - \mathbf{B}$ being an M-matrix with $c = 1 + \beta$ and $\mathbf{B} = \boldsymbol{P}$ a stochastic matrix. As such, $\mathbf{W}$ is ***inverse positive***, i.e. $\mathbf{W}^{-1}$ exists and $\mathbf{W}^{-1} \geq \mathbf{0}$ (via Theorem 2.3, [4]).

Lemma 13.7 ([13]) *Let $\boldsymbol{\psi}_\alpha(\mathbf{s})\big(\beta\mathbf{I} + (\mathbf{I} - \boldsymbol{P})\big) = \beta\mathbf{s}$ be the formulation of the PageRank problem as a linear system. The coefficient matrix $\mathbf{W} = \beta\mathbf{I} + (\mathbf{I} - \boldsymbol{P})$ has the following properties:*

1. *$\mathbf{W}$ is an M-matrix.*
2. *$\mathbf{W}$ is nonsingular.*
3. *The row sums of $\mathbf{W}$ are β.*
4. *$||\mathbf{W}||_\infty = 2 + \beta$.*
5. *$\mathbf{W}^{-1} \geq \mathbf{0}$.*
6. *The row sums of $\mathbf{W}^{-1}$ are $\frac{1}{\beta}$.*
7. *$||\mathbf{W}^{-1}||_\infty = \frac{1}{\beta}$.*
8. *$\kappa_\infty(\mathbf{W}) = \frac{2+\beta}{\beta} = \frac{1}{\alpha}$.*

$\kappa_\infty(\mathbf{W})$ in Lemma 13.7 indicates the sensitivity of the solution to PageRank to perturbations in $\mathbf{W}$. It quantifies the extent $\mathbf{W}$ is ill-conditioned with respect to machine precision when direct computation based on Gaussian elimination is used. For a machine precision m_p, this quantity is taken to be $m_p\kappa_\infty(\mathbf{W})$ (Chapter 3, [28]). For example, $m_p\kappa_\infty(\mathbf{W}) \leq 1$ would indicate a loss of a single decimal digit of precision for $\alpha = 0.1$. Establishing the quantity $\kappa_\infty(\mathbf{W})$ is important in practice because iterative methods are typically used for PageRank computation (see Theorem 2.2 [27] for error characterizations of these methods).

Now, we can proceed to investigate precisely how introducing restart to a CMC and then varying α will impact the long-term limiting distribution of CMC. We start with a qualitative result that is stated in Theorem 13.2. The technical proof is by direct calculation to which we refer the interested reader to the Appendix section of [13]. The proof employs results involving ***digraph Laplacian*** of random walks generalized to digraphs in [43], which we formally define next. Let $\boldsymbol{P}$ be associated with a CMC operating on irreducible $D_C(V_{S(n)})$ with stationary vector $\boldsymbol{\pi} = (\pi_1, \pi_2, \pi_3, \ldots, \pi_n)$ where $\boldsymbol{\pi P} = \boldsymbol{\pi}$. Let the diagonal matrix be $\boldsymbol{\Pi} = \text{diag}(\pi_i)$. The normalized digraph Laplacian is $\tilde{\mathcal{L}} = \boldsymbol{\Pi}^{\frac{1}{2}}(\mathbf{I} - \boldsymbol{P})\boldsymbol{\Pi}^{-\frac{1}{2}}$, with its individual element given by

$$\tilde{\mathcal{L}}_{ij} = \begin{cases} 1 - p_{ii} & \text{if } i = j \\ -\pi_i^{1/2} p_{ij} \pi_j^{1/2} & \text{if } (i, j) \in A \\ 0 & \text{otherwise.} \end{cases}$$

The Green's function for digraphs $\tilde{\mathcal{Z}} = \tilde{\mathcal{L}}^+$ is the Moore-Penrose pseudoinverse of $\tilde{\mathcal{L}}$ with $\tilde{\mathcal{Z}}\tilde{\mathcal{L}} = \tilde{\mathcal{L}}\tilde{\mathcal{Z}} = \mathbf{I} - \tilde{\mathcal{J}}$, where $\tilde{\mathcal{J}} = (\boldsymbol{\pi}^{1/2})^T \boldsymbol{\pi}^{1/2}$.

Theorem 13.2 ([13]) *Let $\boldsymbol{\pi}$ be the stationary vector associated with a CMC operating on an irreducible $D_C(V_{S(n)}) \in \mathcal{D}_C(V_{S(n)})$. The personalized PageRank vector is given by*

$$\boldsymbol{\psi}_\alpha(\mathbf{s}) = \mathbf{s}\big(\mathbf{I} - \boldsymbol{\Pi}^{-\frac{1}{2}} \tilde{\mathcal{L}}(\beta\mathbf{I} + \tilde{\mathcal{L}})^{-1} \boldsymbol{\Pi}^{\frac{1}{2}}\big). \tag{13.18}$$

Furthermore,

$$\lim_{\beta \to 0} \tilde{\mathcal{L}}(\beta\mathbf{I} + \tilde{\mathcal{L}})^{-1} = \tilde{\mathcal{L}}(\tilde{\mathcal{L}})^+ = \tilde{\mathcal{L}}\tilde{\mathcal{Z}},$$

where $\beta = \frac{2\alpha}{1-\alpha}$ and $\tilde{\mathcal{Z}} = \tilde{\mathcal{L}}^+$ is the Moore-Penrose pseudoinverse of $\tilde{\mathcal{L}}$. For small values of α (subsequently, small values of β), the PageRank vector is approximated by the stationary vector, i.e. $\boldsymbol{\psi}_\alpha(\mathbf{s}) \approx \boldsymbol{\pi}$ with equality ($\boldsymbol{\psi}_0(\mathbf{s}) = \boldsymbol{\pi}$) for any $\mathbf{s}$.

The digraph Laplacian explicitly makes the connection between structures in D_C and quantities arising from random walks on D_C. Theorem 13.2 shows the direct relationship between $\boldsymbol{\pi}$ and $\boldsymbol{\psi}_\alpha(\mathbf{s})$ as a result of introducing restart in CMC, expressed in terms of both the digraph Laplacian $\tilde{\mathcal{L}}$ and its pseudoinverse in the form of the Green's function $\tilde{\mathcal{Z}}$. The latter incorporates information on the restart probability α. Informally, Theorem 13.2 indicates that for small $\alpha > 0$, the long-term fraction of time the random walker spends on each vertices is redistributed according to the perturbed normalized digraph Laplacian $(\beta\mathbf{I} + \tilde{\mathcal{L}})^{-1}$.

The following Lemma 13.8 quantitatively establishes the difference between $\boldsymbol{\pi}$ and $\boldsymbol{\psi}_\alpha(\mathbf{s})$ due to restart with probability α by providing a closed-form expression on the difference. The proof by direct calculation makes use of results from perturbation theory on finite Markov chains (specifically, Theorems 1 and 2 [59]). However, the following Corollaries 13.1 and 13.2 are of more practical importance. The first result states that the difference which is expressed in Eq. (13.19) is monotonic with respect to α, i.e. $||\boldsymbol{\pi} - \boldsymbol{\psi}_\alpha(\mathbf{s})||$ increases as α increases. The second result gives an upper bound on the difference that is intuitively appealing.

The largest difference comes about when restart redistributes $\boldsymbol{\pi}$ so that it becomes uniform $\mathbf{s} = \frac{1}{n}\mathbf{e}$ (note that we adopt a centrality measure approach [27] and take s to be uniform for a global digraph analysis).

Lemma 13.8 ([13]) *Let a CMC operating on irreducible $D_C(V_{S(n)}) \in \mathcal{D}_C(V_{S(n)})$ be associated with a probability transition matrix $\boldsymbol{P}$, stationary distribution $\boldsymbol{\pi}$, and the fundamental matrix Z. Correspondingly, the personalized PageRank on $D_C(V_{S(n)})$ is a CMC with probability transition matrix $\boldsymbol{P}_{\mathrm{PR}}$ and stationary distribution $\boldsymbol{\psi}_\alpha(\mathbf{s})$. Then*

$$\begin{aligned}\boldsymbol{\pi} - \boldsymbol{\psi}_\alpha(\mathbf{s}) &= -\boldsymbol{\psi}_\alpha(\mathbf{s})\big(\mathbf{I} - \boldsymbol{P}\big)Z \\ &= \beta(\boldsymbol{\pi} - \mathbf{s})\big(\beta\mathbf{I} + (\mathbf{I} - \boldsymbol{P})\big)^{-1}.\end{aligned}$$

Corollary 13.1 ([13]) *For $\boldsymbol{\pi}$ and $\boldsymbol{\psi}_\alpha(\mathbf{s})$ associated with CMC on an irreducible $D_C(V_{S(n)}) \in \mathcal{D}_C(V_{S(n)})$, the following inequality is given for restart probabilities $\alpha_1 \le \alpha_2$*

$$||\boldsymbol{\pi} - \boldsymbol{\psi}_{\alpha_1}(\mathbf{s})|| \le ||\boldsymbol{\pi} - \boldsymbol{\psi}_{\alpha_2}(\mathbf{s})|| \tag{13.19}$$

where $\alpha_1, \alpha_2 \in (0, 1)$.

Corollary 13.2 ([13]) *Associated to each irreducible $D_C(V_{S(n)}) \in \mathcal{D}_C(V_{S(n)})$ is CMC with stationary distribution $\boldsymbol{\psi}_\alpha(\mathbf{s})$ for $\alpha \in (0, 1)$ and $\boldsymbol{\pi}$ with the following perturbation bound*

$$||\boldsymbol{\pi} - \boldsymbol{\psi}_\alpha(\mathbf{s})||_\infty \le ||\boldsymbol{\pi} - \mathbf{s}||_\infty \tag{13.20}$$

with equality when $\alpha = 1$.

Although characterizing $||\boldsymbol{\pi} - \boldsymbol{\psi}_\alpha(\mathbf{s})||_1$ against changes in α is more useful, tight bounds are difficult to obtain (e.g. $\frac{2}{\alpha}$) [5, 41, 49]. We use the coupling approach [49] with digraph-theoretic arguments to improve the bound to

$$||\boldsymbol{\pi} - \boldsymbol{\psi}_\alpha(\mathbf{s})||_1 \le \frac{2}{1-\alpha}\left(1 - \left(\pi_{v_n} - \frac{1}{n}\right)\right), \tag{13.21}$$

for irreducible D_C of order n having specific cycle structures (this is described in detail in the Appendix of [13]) with v_n having the highest score. Let D_C be vertex pancyclic with the least number of 3-cycles [10]. It has a specific vertex partition of three disjoint subsets $\{v_1\}$, $V_{S(n-2)}$, and $\{v_n\}$. Note that the subtournament $T(V_{S(n-2)})$ is transitive. Then, the long-term limiting visitation probabilities of v_1 and v_n are the same and given by [13]

$$\pi_{v_n} = \pi_{v_1} = \frac{1}{2}\left(1 - \frac{\mathbb{E}(\eta_{V_{S(n-2)}})}{2 + \mathbb{E}(\eta_{V_{S(n-2)}})}\right), \tag{13.22}$$

$$\mathbb{E}(\eta_{V_{S(n-2)}}) = \frac{1}{n-2}\sum_{m=1}^{n-2}\mathbb{E}_{x_m}(\tau_{v_n}), \tag{13.23}$$

where $\mathbb{E}(\eta_{V_{S(n-2)}})$ is the fraction of time on average the random walker spends in $V_{S(n-2)}$ as it repeatedly cycles through $v_1 \rightsquigarrow V_{S(n-2)} \rightsquigarrow v_n \rightsquigarrow v_1$. As such, $\mathbb{E}(\eta_{V_{S(n-2)}})$ is just the average of expected hitting times starting from each vertex in $V_{S(n-2)}$ to v_n

$$\mathbb{E}_{v_{n-1}}(\tau_{v_n}) = \mathbb{E}_{x_1}(\tau_{v_n}) = H_1 = 1$$

$$\mathbb{E}_{v_{n-2}}(\tau_{v_n}) = \mathbb{E}_{x_2}(\tau_{v_n}) = H_2 = 1 + \frac{1}{2}$$

$$\vdots$$

$$\mathbb{E}_{v_2}(\tau_{v_n}) = \mathbb{E}_{x_{n-2}}(\tau_{v_n}) = H_{n-2} = 1 + \frac{1}{2} + \frac{1}{3} + \cdots + \frac{1}{n-2},$$

where $H_n = \sum_{k=1}^{n}\frac{1}{k}$ (nth Harmonic number). Other general upper bound of $||\boldsymbol{\pi} - \boldsymbol{\psi}_\alpha(\mathbf{s})||_1$ can be obtained for probability vectors associated with random walks on labelled (isomorphic) tournaments [48, 61], taking into account their unilateral and directional duals [3, 9], which we have formulated as follows [13]:

$$||\boldsymbol{\pi}_1 - \boldsymbol{\pi}_2||_1 \leq 2\Big(1 - \frac{1}{n}\Big).$$

13.3.3 Computational Studies in PageRanking Competitive Coevolution

Here, we present a systematic computational study we have performed in [13] on CMCs operating on coevolutionary tournaments $T_C(V_{S(n)})$ having specific structures [3, 14, 48]. They are selected based on known internal cycle structures that are informative, and let us better evaluate their impact on the visitation probabilities of vertices by CMCs using quantitative measures we have developed earlier. We also compute differences on PageRank orderings [30] that we describe in detail later. We employ the standard ***Power Method*** approach [41] to compute the PageRank vectors $\boldsymbol{\psi}_\alpha(\mathbf{s})$, using the lazy version of the CMCs with uniform teleportation vector $\mathbf{s}$.

For this study, we use the following set of coevolutionary tournaments $T_C(V_{S(n)})$:

(a) **Irreducible**: We will use the ***pancyclic (maximal transitive subtournament)*** $T_C(V_{S(n)})$ with the least number of 3-cycles [10]. They are obtained from transitive tournaments of order n indexed by its score sequences and then just reverse the arc $v_1 \rightarrow v_n$ to $v_n \rightarrow v_1$. Another vertex pancyclic (or simply pancyclic) $T_C(V_{S(n)})$ that we use are those where we further reverse the arc $v_2 \to v_{n-1}$ to $v_{n-1} \to v_2$.

(b) **Reducible**: Two approaches are used to generate reducible $T_C(V_{S(n)})$ with various degrees of cycle structures in a controlled manner. The first approach exploits known digraph-theoretic structures, i.e. reducible $T_C(V_{S(n)})$, $n \geq 2$ has a strong decomposition (refer to Sect. 5.3.1 for details). In addition to transitive tournaments, we generate reducible $T_C(V_{S(n)})$ with an odd number of components, where each odd-numbered-ith component is a single vertex and even-numbered-ith component is a strong component of known structures, e.g. vertex pancyclic (maximal transitive subtournament) and regular [3]. The second approach is the same implementation of Algorithm 13.3 for a network-growth-based methodology via preferential attachment we have employed earlier in Sect. 13.2.3 to generate reducible tournaments.

(c) **Random**: Random $T_C(V_{S(n)})$ is generated by orienting each edge $\{v_i, v_j\} \in E$ of its underlying, complete graph $G_C(V_{S(n)}, E)$ randomly with equal probability for each direction. Although a random $T_C(V_{S(n)})$ is likely to be irreducible for the sizes we consider for this study [48], nevertheless we employ an irreducibility test on random $T_C(V_{S(n)})$s using their score sequences and select those that are irreducible [3].

First, we study the impact of the restart probability α on max norm $||\boldsymbol{\pi} - \boldsymbol{\psi}_\alpha(\mathbf{s})||_\infty$ and one norm $||\boldsymbol{\pi} - \boldsymbol{\psi}_\alpha(\mathbf{s})||_1$ as the structure and size of $T_C(V_{S(n)})$ are varied in a controlled manner. Figure 13.4 shows the results for $||\boldsymbol{\pi} - \boldsymbol{\psi}_\alpha(\mathbf{s})||_\infty$. For reducible $T_C(V_{S(n)})$, they all have one absorbing state v_n so that $\boldsymbol{\pi} = (0, 0, 0, \ldots, 1)$. As such, Theorem 13.2 implies that $\boldsymbol{\psi}_\alpha(\mathbf{s})$ is a redistribution of $\boldsymbol{\pi}$ and amounts to leakages of the probability mass concentrated on v_n, i.e. $||\boldsymbol{\pi} - \boldsymbol{\psi}_\alpha(\mathbf{s})||_\infty = |\pi_{v_n} - \psi_{v_n}|$. Since there are leakages to a maximal subset $V_{S(n)} \backslash \{v_n\}$ of $n-1$ vertices, $||\boldsymbol{\pi} - \boldsymbol{\psi}_\alpha(\mathbf{s})||_\infty$ for reducible cases would be larger in comparison with the irreducible cases where there are no leakages. $||\boldsymbol{\pi} - \boldsymbol{\psi}_\alpha(\mathbf{s})||_\infty$ becomes larger and gets closer to the upper bound $(1 - \frac{1}{n})$ (bold line in Fig. 13.4) as a result of an increasing number of 3-cycles that leads to fewer transitive components in the reducible $T_C(V_{S(n)})$. This is achieved by setting the dominated strong components from pancyclic with the least number of 3-cycles in B to regular with the most number of 3-cycles [48] in C [13].

For irreducible cases, both stationary and PageRank vectors will be nonnegative (all entries > 0) so that $||\boldsymbol{\pi} - \boldsymbol{\psi}_\alpha(\mathbf{s})||_\infty$ are substantially lower than $(1 - \frac{1}{n})$. As the number of 3-cycles increases, i.e. towards tournaments with regular cycle structures that have a maximum number of 3-cycles and uniform $\boldsymbol{\pi}$ [14], $||\boldsymbol{\pi} - \boldsymbol{\psi}_\alpha(\mathbf{s})||_\infty$ will decrease. This can be observed by comparing results in D for pancyclic $T_C(V_{S(n)})$ with the least number of 3-cycles and results in E and particularly in F where random $T_C(V_{S(n)})$ becomes increasingly more regular-like as the number of vertices increases (inset of F) [13].

Computational results indicate that $||\boldsymbol{\pi} - \boldsymbol{\psi}_\alpha(\mathbf{s})||_\infty$ is monotonic with α (see the higher-valued dotted lines in Fig. 13.4 representing results where larger α values are used for reducible and irreducible cases). $||\boldsymbol{\pi} - \boldsymbol{\psi}_\alpha(\mathbf{s})||_\infty$ gets closer to the upper

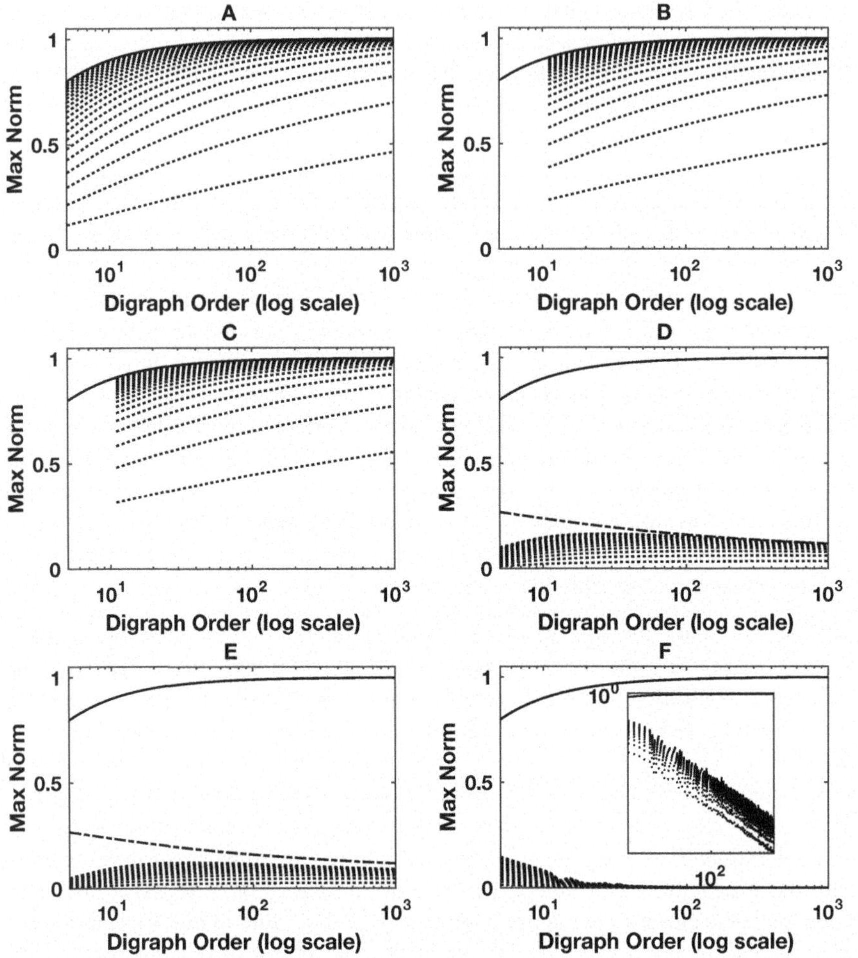

Fig. 13.4 $||\pi - \psi_\alpha(\mathbf{s})||_\infty$ of CMCs operating on tournaments that are (**a**) transitive, (**b**) with pancyclic (maximal transitive subtournament) components of order 9, (**c**) with regular components of order 9, (**d**) pancyclic (maximal transitive subtournament), (**e**) pancyclic, and (**f**) random tournaments of odd order $n \in [5, 999]$. All plots are semilog except the inset of F, which is log-log. Dotted lines indicate $||\pi - \psi_\alpha(\mathbf{s})||_\infty$ for $\alpha \in [0.05, 0.95]$ in steps of 0.05. Upper bounds: $(1 - \frac{1}{n})$ (bold lines) and $(\frac{a_1}{a_2+H_{m+1}} - \frac{1}{m+2})$ (dash-dot lines in D-E) [13]

bound $||\boldsymbol{\pi} - \mathbf{s}||_\infty \leq ||\boldsymbol{\pi}_{\text{tran}} - \mathbf{s}||_\infty = (1 - \frac{1}{n})$ for the reducible A-C cases and $||\boldsymbol{\pi} - \mathbf{s}||_\infty$ for the irreducible D-E cases. Although this property was proved for irreducible CMCs in Corollary 13.1, our computational study indicates that $||\boldsymbol{\pi} - \boldsymbol{\psi}_\alpha(\mathbf{s})||_\infty$ is monotonic for reducible CMCs from the point of leakages and setting a uniform $\mathbf{s}$ [13].

Sharp bounds can be obtained for CMCs operating on $T_C(V_{S(n)})$ with specific structures in D that consist of pancyclic (maximal transitive subtournament). We note that $||\boldsymbol{\pi} - \boldsymbol{\psi}_\alpha(\mathbf{s})||_\infty = |\pi_{v_1} - \psi_{v_1}| \leq ||\boldsymbol{\pi} - \mathbf{s}||_\infty = |\pi_{v_1} - \frac{1}{n}|$. This coincides with our understanding how this structure affects PageRank calculation. Redistribution affects vertex v_1, which shares most of the probability mass with v_n. As in [5], the PageRank authority of v_1 is obtained from a single link $v_n \to v_1$ even though v_n has the highest authority. We can calculate this bound directly using digraph-theoretic arguments. We have these identities: $H_n = \sum_{k=1}^{n} \frac{1}{k}$ (nth Harmonic number), $\sum_{i=1}^{n} H_i = (n+1)(H_{n+1} - 1)$ [19], and $\mathbb{E}(\eta_{V_{S(n-2)}}) = \frac{1}{n-2} \sum_{i=1}^{n-2} H_i$. We apply these identities with $m = n - 2, n \geq 3$ on Eq. (13.22) noting that $\pi_{v_1} = \pi_{v_n}$ to obtain [13]

$$
\begin{aligned}
\pi_{v_1} &= \frac{1}{2}\left(1 - \frac{\frac{1}{m}\sum_{i=1}^{m} H_i}{2 + \frac{1}{m}\sum_{i=1}^{m} H_i}\right) \\
&= \frac{1}{2}\left(1 - \frac{\frac{1}{m}(m+1)(H_{m+1} - 1)}{2 + \frac{1}{m}(m+1)(H_{m+1} - 1)}\right) \\
&= \frac{a_1}{a_2 + H_{m+1}},
\end{aligned}
$$

where $a_1 = \frac{m}{m+1}$ and $a_2 = \frac{m-1}{m+1}$. Then,

$$
||\boldsymbol{\pi} - \boldsymbol{\psi}_\alpha(\mathbf{s})||_\infty \leq \left(\frac{a_1}{a_2 + H_{m+1}} - \frac{1}{m+2}\right).
$$

For these tournaments with large but a finite number of vertices $n = m+2$, the right-hand side of the inequality is dominated by the reciprocal of H_{m+1} (since H_{m+1} grows logarithmically) and is bounded away from zero [19]. Equality is obtained only for $m = 1$ where the right-hand side of the inequality is zero. This corresponds to the isomorphic pancyclic tournament of $n = 3$ vertices, which is regular and has $||\boldsymbol{\pi} - \boldsymbol{\psi}_\alpha(\mathbf{s})|| = 0$.

Figure 13.5 shows the results for $||\boldsymbol{\pi} - \boldsymbol{\psi}_\alpha(\mathbf{s})||_1$ for the same set of experiments. The upper bound $||\boldsymbol{\pi} - \boldsymbol{\psi}_\alpha(\mathbf{s})||_1 \leq 2(1 - \frac{1}{n})$ is plotted as a bold line, while $||\boldsymbol{\pi} - \boldsymbol{\psi}_\alpha(\mathbf{s})||_1 \leq \frac{2}{1-\alpha}\left(1 - \left(\pi_{v_n} - \frac{1}{n}\right)\right)$ is plotted as a dash-dot line in Fig. 13.5. For reducible A-C cases with one absorbing state v_n, we can verify numerically that $||\boldsymbol{\pi} - \boldsymbol{\psi}_\alpha(\mathbf{s})||_1 = 2||\boldsymbol{\pi} - \boldsymbol{\psi}_\alpha(\mathbf{s})||_\infty$. Any leakage that is measured by the max norm would be redistributed over all other states. The one norm cumulatively adds these leakages in addition to that given by the max norm for v_n. Crucially, this

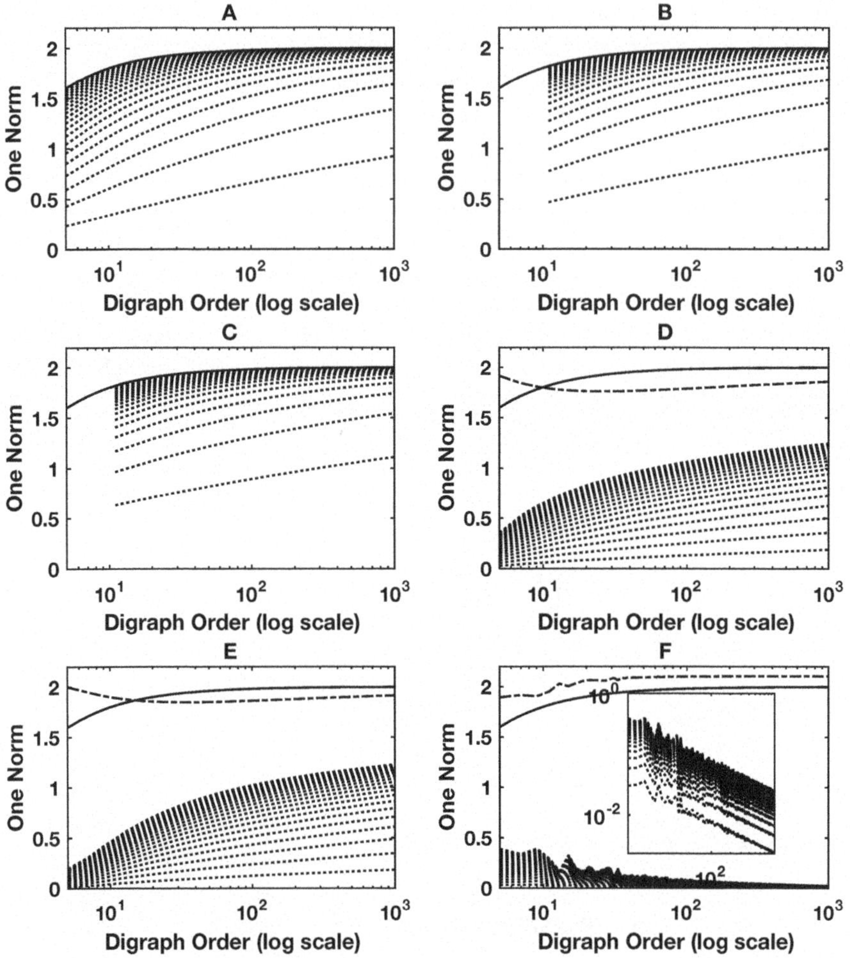

Fig. 13.5 $||\pi - \psi_\alpha(\mathbf{s})||_1$ of CMCs operating on tournaments that are (**a**) transitive, (**b**) with pancyclic (maximal transitive subtournament) components of order 9, (**c**) with regular components of order 9, (**d**) pancyclic (maximal transitive subtournament), (**e**) pancyclic, and (**f**) random tournaments of odd order $n \in [5, 999]$. All plots are semilog except the inset of (**f**), which is log-log. Dotted lines indicate $||\pi - \psi_\alpha(\mathbf{s})||_1$ for $\alpha \in [0.05, 0.95]$ in steps of 0.05. Upper bounds: $2(1 - \frac{1}{n})$ (bold lines) and $\frac{2}{1-\alpha}\left(1 - \left(\pi_{v_n} - \frac{1}{n}\right)\right)$ at $\alpha = 0.05$ setting (dash-dot lines in (**d**)–(**f**)) [13]

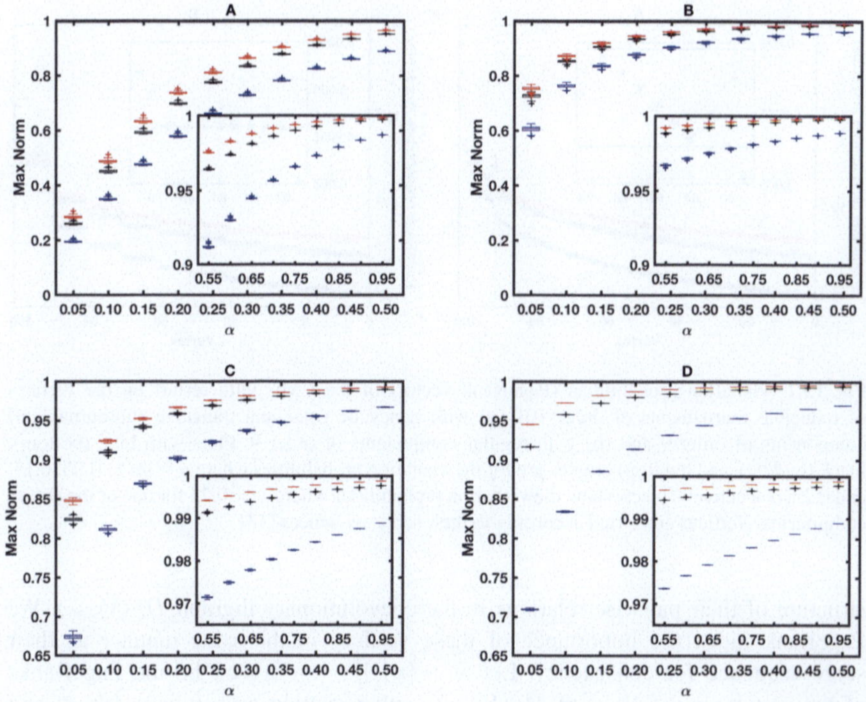

Fig. 13.6 $||\boldsymbol{\pi} - \boldsymbol{\psi}_{\alpha}(\mathbf{s})||_{\infty}$ of CMCs operating on reducible tournaments that are generated from a network-growth-based methodology with γ set to (**a**) 2.0, (**b**) 0.1, (**c**) 0.05, and (**d**) 0.01. Boxplots in blue (bottom), black (middle), and red (top) represent results where the final generated tournaments are of order 100, 500, and 1000, respectively [13]

leakage relationship extends to that between the digraph-theoretic-derived upper bounds since $||\boldsymbol{\pi} - \boldsymbol{\psi}_{\alpha}(\mathbf{s})||_1 \leq 2(1 - \frac{1}{n})$ and $||\boldsymbol{\pi} - \boldsymbol{\psi}_{\alpha}(\mathbf{s})||_{\infty} \leq (1 - \frac{1}{n})$ [13].

We now consider reducible $T_C(V_{S(n)})$ generated from the network-growth-based methodology and measure $||\boldsymbol{\pi} - \boldsymbol{\psi}_{\alpha}(\mathbf{s})||_{\infty}$ only. Figure 13.6 shows results where the temperature $1/\gamma$ for the system is increased from A to D. At a lower temperature in A, random orientations of edges generated as a result of introducing a new node at each iteration of the method are biased towards existing nodes with higher scores. This leads to more prominent transitive structures in $T_C(V_{S(n)})$ as indicated by larger Landau's index ν values. ν ranges at [8, 49], [8, 39], [7, 47] for experiments with the final generated $T_C(V_{S(n)})$ at $n = 100$, 500, and 1000, respectively. At higher temperatures, $T_C(V_{S(n)})$ generated in B-D have two strong components only (a single dominant node and a strong dominated component of $n - 1$ nodes). A higher $||\boldsymbol{\pi} - \boldsymbol{\psi}_{\alpha}(\mathbf{s})||_{\infty}$ value is a result of more leakages from the larger-sized, strong dominated component drawing out more probability mass away from the absorbing dominant node [13].

The PageRank vector $\boldsymbol{\psi}_{\alpha}(\mathbf{s})$ gives the visitation probabilities ψ_{v_i} that ranks the importance (performance) of vertices $v_i \in V_{S(n)}$ representative of the underlying

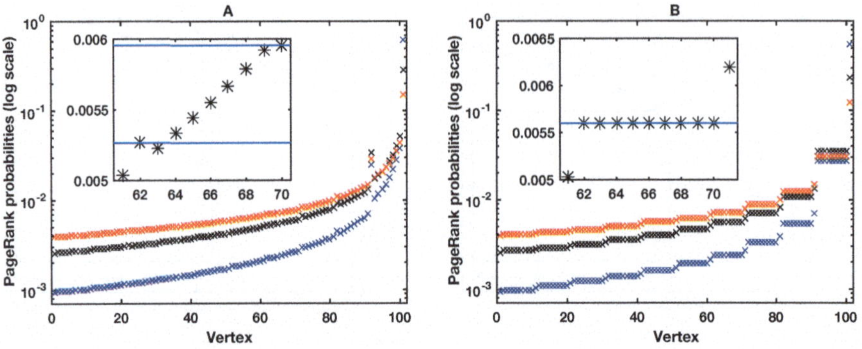

Fig. 13.7 Visitation probabilities (PageRank vector) of the CMC with restart on the vertices of reducible tournaments of order 101 (**a**) with pancyclic (maximal transitive subtournament) components of order 9 and (**b**) with regular components of order 9. Plots with blue (bottom), black (middle), and red (top) crosses are for the visitation probabilities where α is set to 0.05, 0.15, and 0.25, respectively. Insets show the visitation probabilities where $\alpha = 0.25$ for one of the strong components. Vertices are sorted according to their score sequences [13]

structure of their pairwise relations in the coevolutionary digraph $D_C(V_{S(n)})$. We PageRank index the importance of these vertices in the same manner as their score sequence for consistency. Let $\lambda : \{\psi_{v_i}\}_i^n \to \mathbb{Z}_{1\leq z\leq n}$ be the PageRanks of the vertices $v_i \in V_{S(n)}$ of $D_C(V_{S(n)})$, with visitation probabilities $\{\psi_{v_i}\}_i^n$ and $\mathbb{Z}_{1\leq z\leq n} = \{1, 2, 3, \ldots, n\}$. Vertices are ***PageRanked*** in the ascending order from the least important $\lambda(v_1) = 1$ to the most important $\lambda(v_n) = n$. Ties are handled in the usual manner through fractional (average) ranks [39]. In our experiments, we generate tournaments and strong components of odd orders so that tied ranks can be assigned with the median (integer) value. Results are summarized in Fig. 13.7.

Figure 13.7 indicates that PageRanks $(\lambda)_i^n$ of vertices is the same as their score sequences $\left(d_T^-(v_i)\right)_i^n$ even for high α setting for transitive $T_C(V_{S(n)})$. This is not surprising given that the probability mass of the absorbing node is redistributed uniformly. More important is the case involving irreducible $T_C(V_{S(n)})$ with prominent transitive structures. In the case of the pancyclic (maximal transitive subtournament) $T_C(V_{S(n)})$, visitation probabilities are sorted as $\psi_{v_2} < \psi_{v_3} < \psi_{v_4} < \cdots < \psi_{v_{n-1}} < \psi_{v_1} < \psi_{v_n}$, for reasonably low α settings. Unlike the situation with the stationary probability vector where the two highest ranks are tied since $\pi_{v_1} = \pi_{v_n}$, PageRank is able to distinguish between v_1 and v_n [13].

If such a structure is embedded within a reducible $T_C(V_{S(n)})$ (as one may encounter in real-world problems having complex structures), PageRank can be used to uncover this strong component structure for small-sized $T_C(V_{S(n)})$ where the issue of scale does not arise. For example, consider the case whereby the strong component is pancyclic (maximal transitive subtournament). Nodes v_{62} and v_{63} (inset of Fig. 13.7A) have the same out-degree globally ($d_T^-(v_{62}) = d_T^-(v_{63})$) and within the strong component, the least number of just one outgoing link locally. However, $\psi_{v_{62}} > \psi_{v_{63}}$. Both information would indicate that they are part of a strong component that dominates vertices $\{v_1, \ldots, v_{61}\}$. If the strong component is regular, all the vertices within the component have the same PageRank authorities and hence tied PageRanks (inset of Fig. 13.7b) [13].

It is known that appropriate α settings can improve the convergence rate of the Power Method to compute PageRanks (Theorem 5.1 in [41]), at the expense of increasing perturbations on PageRank authorities and subsequently the actual PageRank orderings [30]. We study the impact of α on PageRanks by computing differences between PageRanks at baseline $\alpha = 0.05$ setting and those obtained from using higher α settings. Differences are calculated through rank aggregations. We use two different distance measures. Let $T_C(V_{S(n)})$ be of odd orders n. The first measure is the ***Spearman footrule distance*** [22]

$$d_{\text{spear}}(\lambda_1, \lambda_2) = \frac{4}{n^2 - 1} \sum_{i=1}^{n} |\lambda_1(v_i) - \lambda_2(v_i)|, \tag{13.24}$$

where the normalizing constant is given by the maximum of the sum of absolute differences $\max d(\lambda_1, \lambda_2) = \frac{n^2-1}{4}$ (it is the same as with the Spearman's rank correlation [39] that calculates the distance obtained from completely opposed rankings [9]). The maximum distance corresponds to the case between transitive and regular tournaments. The second measure is the ***average rank difference***

$$d_{\text{avg}}(\lambda_1, \lambda_2) = \frac{1}{n} \sum_{i=1}^{n} |\lambda_1(v_i) - \lambda_2(v_i)|. \tag{13.25}$$

$d_{\text{avg}}(\lambda_1, \lambda_2)$ provides a useful comparison with $d_{\text{spear}}(\lambda_1, \lambda_2)$ for experiments where the n^2 factor could mask the effect of small rank differences for certain large tournaments [13].

Results summarized in Fig. 13.8 indicate that higher α settings increase the perturbations to PageRanks, which are measured by larger $d(\lambda_1, \lambda_2)$ values (e.g. the bold lines in Fig. 13.8). However, we observe that for increasing size of the generated $T_C(V_{S(n)})$, lower $d(\lambda_1, \lambda_2)$ values are obtained. These results indicate that changes to PageRanks are localized to certain vertices. For example, this is usually v_1 for pancyclic (maximal transitive subtournament) $T_C(V_{S(n)})$ in A-B. Beyond specific structures in $T_C(V_{S(n)})$, our results suggest that the presence of prominent transitive structures (A-B) and those that forces reducibility (C-D) would localize perturbations of PageRanks to a small number of vertices, leading to lower $d(\lambda_1, \lambda_2)$ values as $T_C(V_{S(n)})$ increases in size. In addition, we observe generally that $d(\lambda_1, \lambda_2)$ values are lower for reducible $T_C(V_{S(n)})$ compared to irreducible $T_C(V_{S(n)})$ at the same order n [13]. This is not unexpected given that restarts act to redistribute the probability mass from absorbing nodes uniformly to transient nodes for the PageRank random walker. In practice, one might expect that analysis has to be made on solvable coevolutionary problems represented by reducible coevolutionary digraphs. The performance of algorithms that compute PageRank

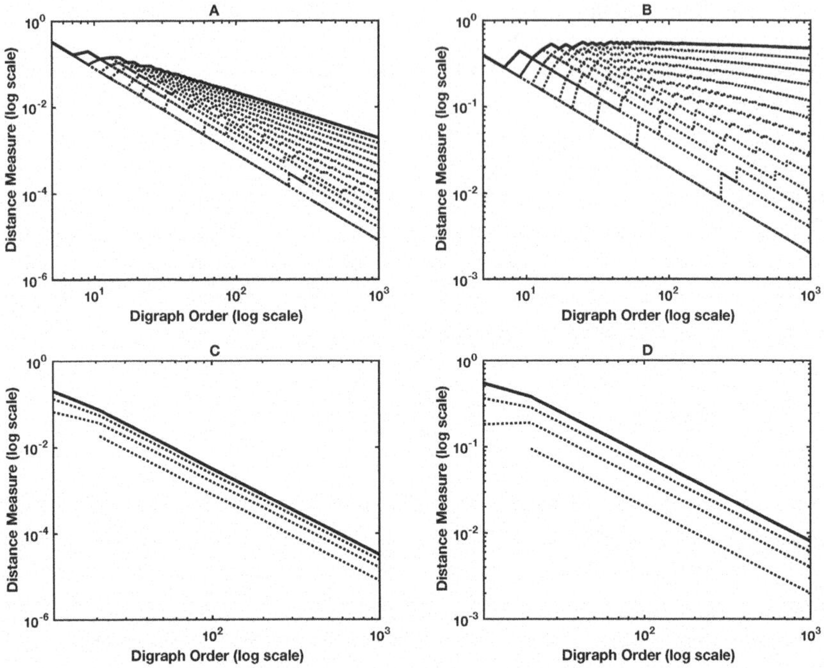

Fig. 13.8 Differences of PageRanks for coevolutionary tournaments that are irreducible having (**a**)-(**b**) pancyclic (maximal transitive subtournament) structures and reducible having (**c**)-(**d**) pancyclic (maximal transitive subtournament) components of order 9. Plots in (**a**) and (**c**) are for $d_{\text{spear}}(\lambda_1, \lambda_2)$, and (**b**) and (**d**) are for $d_{\text{avg}}(\lambda_1, \lambda_2)$. Dotted lines show difference from baseline PageRanks obtained with $\alpha = 0.05$ to PageRanks obtained from setting higher $\alpha \in [0.10, 0.90]$. Bold lines are the case where $\alpha = 0.95$ [13]

becomes crucial, and the choice then might be focused on the speed of those algorithms, e.g. convergence rates in iterative methods.

The crucial issue that remains for direct applications of PageRanking competitive coevolution is that of computational costs. Unlike the usual application areas of PageRank involving sparsely connected networks [27] for which there exist methods to speed up computation [41], coevolutionary digraphs are by definition fully connected. Regardless, the implication of our computational study is that PageRank provides a meaningful quantitative analysis that could reveal problem structures in coevolutionary digraphs as mechanism wise the PageRank random walker incorporates those qualitative information for its operations. The study is focused on laying down the foundations so that relevant quantitative analysis can be made although there are real challenges associated with large, real-world coevolutionary systems. Instead of employing PageRank for global analysis that will be computationally challenging in practice, one could instead focus on local analysis for specific subdigraphs corresponding to local neighbourhoods in the search space. This can be achieved through appropriate setting of the teleportation vector **s**. However, there might be scope to develop methods that can be employed in practice to perform relatively short runs that correspond to PageRanking strong subcomponents of the full digraph and then aggregate these runs to build a relevant and larger picture in terms of connectivities of those subcomponents.

13.4 Discussion and Conclusion

This chapter has sought to address one fundamental issue in competitive settings that deals with how problem structures can affect the performance of solutions discovered by iterative means via coevolution. Under some mild assumptions, population-one coevolutionary processes can be represented as Markov chain models operating on digraphs that capture preference relations between competing solutions to those problems. Such abstract coevolution formulation as random walks on digraphs has allowed us to study competitive coevolutionary systems in great depth. In addition to general qualitative results ensuring that the dichotomy of coevolutionary problem structures is transferred to coevolutionary search processes, i.e. (ir)reducible CMC operates on (ir)reducible coevolutionary digraphs, these qualitative characterizations of competitive coevolution are informative towards construction of tools for rigorous quantitative analysis on coevolutionary search. In particular, we have formulated the notion of expected hitting times to the absorbing class as the means to measure the speed of population-one coevolutionary search processes to find the subset of dominant solutions. We have also developed deep insights that allow us to precisely state how specific domain knowledge can impact the speed of coevolutionary search. For example, the standard random walk is a particular CMC that incorporates domain knowledge in the form of out-neighbourhoods of vertices, while a (1+1)CEA does not employ any domain knowledge and behaves as a blind search. In the worse-case comparison between the two CMCs (i.e. starting from the weakest vertex), a standard random walk has a sublinear time complexity $O(\ln n)$ but can be significantly faster if it starts the search process closer to the dominant vertex.

Another quantitative measure we have formulated for abstract coevolution is that of the stationary distribution of irreducible CMC that corresponds to visitation probabilities of vertices by coevolutionary search. Since this invariant measure requires prior qualitative knowledge (e.g. irreducibility of coevolutionary digraphs), we have established a direct and formal connection between the Markov chain model of coevolutionary processes and PageRank operating on such digraphs. We have provided existence guarantees of PageRank for any population-one coevolutionary system. We have proved that the PageRank vector is a redistribution of the stationary vector associated with the CMC on the coevolutionary digraph as a result of introducing restarts. These changes that lead to PageRank authorities from specific restart probability α setting can be quantified precisely. Crucially, the introduction of restart to coevolutionary search effectively transforms even a reducible CMC to one that is irreducible. As such, PageRank authorities constitute a refinement to the original stationary distribution measure. This lets us develop a principled approach for quantitative characterization of all coevolutionary digraphs (reducible and irreducible). A PageRank random walker characterizes a coevolutionary digraph via authorities that measure performances of individual solutions (vertices) and correspondingly their PageRanks by the (ascending) ordering of authorities. When PageRank is formulated for global analysis of the coevolutionary digraph (i.e. setting a uniform teleportation vector), both theoretical and computational results have demonstrated that PageRanks maintain the original ranks based on score sequences of vertices in reducible digraphs. Lastly, PageRank authorities with suitable normalization have a natural interpretation as visitation probabilities of vertices in the digraph by coevolutionary search processes with restart.

Although the main motivation of our study is to address the fundamental challenge arising from performing quantitative analysis on coevolutionary systems with little prior qualitative knowledge of the problem structures, our technical contributions are general and can be applied to other problem areas whereby the model one adopts is a random walk on semicomplete digraphs (an edge can be bioriented). In the context of competitive games, this would mean interactions that include draws. We have considered the setting of a restricted but a large class of population-one coevolutionary systems, which covers other learning algorithms involving self-play. However, PageRank has been applied to evolutionary systems as well whereby the structures of the underlying problems are modelled as specific reducible digraphs [33]. Beyond that, our technical approach extends the rich area of studies on random walks generalized to digraphs [17, 43] as well as corresponding models of electrical networks [2, 51]. In comparison, our study here considers non-reversible CMCs operating on both irreducible and reducible coevolutionary semicomplete digraphs, where complete oriented graphs (tournaments) are a subset. Our theoretical studies often exploit specific digraph properties (e.g. strong connectivities) to prove results for irreducible digraphs. Crucially, our approach is motivated by insights from digraph and specifically tournament theory for which we can develop a more natural formalism to study competitive coevolution in settings whereby interactions can be modelled as two-player strategic games.

We close this chapter with some remarks about the state of competitive coevolutionary studies, particularly theoretical studies. We first note that coevolutionary systems are notoriously difficult to study. One consequence is that different aspects of these systems are typically, though not always, studied in isolation using a variety of common frameworks. Even within a single framework, one could apply different methodologies given the interplay between coevolutionary systems' design, problem structures, and dynamics. For example, studies [24, 62] on coevolutionary dynamics in an evolutionary game-theoretic setting from a continuous dynamical systems framework focus on population-level analysis with respect to mixtures of pure strategies in an infinite population of agents. The focus of coevolutionary search is then on a specific mixture for which the strategies are in a state of equilibrium (e.g. Nash equilibrium). However, unlike evolutionary dynamics [63], coevolutionary dynamics can be characterized by a variety of regimes from a simple attractor (corresponding to Nash equilibrium) to periodic and even complex chaotic dynamics [24]. In the latter case, it becomes crucial to develop a more in-depth study and employ some of the sophisticated machineries in Dynamical Systems Theory such as structural stability to determine if coevolutionary dynamical systems possess the shadowing property [62]. Natural and computer-generated trajectories are pseudo-trajectories as a result of (combination) of finite population and stochastic effects in selection and through variation. Establishing Shadowing in coevolution is crucial because if pseudo-trajectories are closely followed in space and time (shadowed) by true trajectories, then the argument can be made that pseudo-trajectories constitute relevant behavioural observations of coevolutionary processes. Although we cannot prove that these coevolutionary dynamical systems possess Shadowing in general, nevertheless it is possible with additional formulation (e.g. under ***reduced dynamics*** setting where chaotic trajectories live in some invariant subset) that some form of Shadowing can be established.

Another example is the various studies within a pareto coevolution framework [21]. Here, the common aspect taken is the evaluation of individual coevolving solution using a set of test cases and then to define solution quality from the perspective of pareto dominance as in a Multi-objective Optimization Problem (MOP). However, unlike the evolutionary MOP setting, in pareto coevolution each test case is an objective that has to be optimized. The situation can be made more complex if one considers a single-population pareto coevolution whereby candidate solutions themselves constitute test cases. Regardless, there has been several fundamental studies that have contributed to the development of pareto coevolution framework. The study in [64] has introduced a problem generation methodology whereby specific attributes (e.g. intransitivity) in the test objectives can be introduced in a controlled manner. In [21], CEAs are developed that could detect and exploit gradients in the form of pareto covering order between individual layers or subsets of nondominated solutions. Theoretical underpinnings via order theory have been developed in [8] to study the emergence of geometric organization on the partially ordered solution set with respect to the family of problems introduced in [64] that can be exploited for more informative pareto coevolutionary search (comparison) of solutions.

Are there obvious connections between these frameworks, for example, that of pareto coevolution with the framework of digraph representation of coevolutionary problems and random walks for coevolutionary processes we have employed in this chapter? One close connection is that the formal methodology to uncover the structure of the pairwise comparison space in [8] and the qualitative characterization of the underlying structures in coevolutionary digraphs we first introduced in [14] are both crucial for the design and analysis of effective pareto coevolutionary search. For any competitive coevolutionary problem, qualitative knowledge of the problem is crucial towards obtaining guarantees that it can be solved because there exists a dominant subset of solutions. Its underlying reducible coevolutionary digraph would admit a decomposition over the solution set into strong components having an acyclic ordering. When one by design uses the solution set itself for evaluations in pareto coevolution, the induced pareto covering order on the solution set (search space) corresponds to this strong decomposition of the reducible digraph. This suggests a structure that can be exploited for coevolutionary search between those pareto layers (strong components) and eventually the single dominant subset of solutions.

As a further illustration of the complementary nature of these two frameworks, one can set up an inductive argument for a program that can perform pseudo-embedding of coevolved solutions:

1. Take a transitive tournament of $n \geq 4$ vertices labelled $\{v_i\}_1^n$ according to its score sequence with an ordering $s_1 < s_2 < s_3 < \ldots < s_n$.
2. Perform a single arc reversal on $v_1 v_n$ to $v_n v_1$ to obtain a pancyclic (maximal transitive subtournament) tournament.
3. The pancyclic tournament is pseudo-embedded on $\mathbb{N}^2$ with the linear realiser $L = \{()_X, ()_Y\} = \{([v_1], [v_2], [v_3], \ldots, [v_n]), ([v_n], [v_2], [v_3], \ldots, [v_1])\}$.

We use the notation $L = \{()_X, ()_Y\}$ to indicate that the pseudo-embedding can be visualized as a $\mathbb{N}^2$-space spanned along the x-axis by $()_X$ and along the y-axis by $()_Y$. The specific pancyclic structure whereby the subtournament with $n-2$ vertices $v_2, v_3, v_4, \ldots, v_{n-1}$ is always transitive allows one to set up an inductive argument and use the above program to do the pseudo-embedding for tournaments of $n \geq 4$ vertices. The inductive step can be visualized as moving the embedded line of points specified by the two ends from $\{(v_1, v_n), (v_n, v_1)\}$ to $\{(v_1, v_{n+1}), (v_{n+1}, v_1)\}$—in the latter one adds an additional point corresponding to the vertex that forms the transitive subtournament $v_2, v_3, v_4, \ldots, v_n$. It would be of interest in future studies to examine pseudo-embedding of other strong tournaments, for example, beyond $\mathbb{N}^2$-space.

There are other frameworks that have been employed for more formal studies of coevolutionary systems. In Chap. 12, we have formulated competitive coevolution as coevolutionary learning systems and made substantial use of the generalization notion and analysis approaches from Machine Learning methodologies [15, 52, 57]. Competitive coevolution has also been formulated in the context of cooptimization, whereby specific worst-case optimization analysis to compare aggregate performances of CEAs on all problem instances in a domain has been made [52, 55]. The study in [38] has performed a fundamental and deep theoretical investigation of the nature of coevolutionary interactions. Here, the study has formalized a coevolutionary setting based the notion of a ***preference function*** $f : \mathcal{A} \to 2^S$. The principle idea here is that the set $\mathcal{A}$ represents the states of information with which coevolution uses to determine which of the solutions in S are preferred (and as such form elements of the powerset 2^S of S). This requires the set $\mathcal{A}$ to be ***algebraic***, e.g. a partially ordered set $(\mathcal{A}, \subseteq)$ whereby $\mathcal{A} = 2^{S \times T}$ is a powerset with S and T being the sets of solutions and test cases, respectively. One important theoretical result in [38] is that preference functions f involving finite sets are ***convergent***. Informally, the property refers to the situation whereby for the established preference of some solutions in S at the limit of accumulation of information states in $\mathcal{A}$, there is a point (possibly) earlier in the process of the accumulation for which they are already established. The interested readers could refer to the study [38] for technical details. What we hope to highlight before closing this chapter is the observation that there are significant challenges associated with complete understanding of coevolutionary systems given the breadth and depth of available frameworks. This also dispels the notion that studies of coevolution lack rigour, and indeed, it is more the case of the opposite.

References

1. Axelrod, R.: The evolution of strategies in the iterated prisoner's dilemma. In: L.D. Davis (ed.) Genetic Algorithms and Simulated Annealing, chap. 3, pp. 32–41. Morgan Kaufmann, New York (1987)
2. Balázs, M., Folly, A.: An electric network for nonreversible markov chains. Am. Math. Mon. **123**(7), 657–682 (2016)
3. Bang-Jensen, J., Gutin, G.Z.: Digraphs Theory, Algorithms and Applications. Springer, London (2009)
4. Berman, A., Plemmons, R.J.: Nonnegative Matrices in the Mathematical Sciences. SIAM, Philadelphia (1994)
5. Bianchini, M., Gori, M., Scarselli, F.: Inside pagerank. ACM Trans. Internet Technol. **5**(1), 92–128 (2005)
6. Brin, S., Page, L.: The anatomy of a large-scale hypertextual web search engine. In: Proceedings of the 7th International World Wide Web Conference (WWW7), pp. 107–117. Brisbane, Australia (1998)
7. Brualdi, R.A.: Spectra of digraphs. Linear Algebra Appl. **432**, 2181–2213 (2010)
8. Bucci, A.: Emergent geometric organization and informative dimensions in coevolutionary algorithms. Ph.D. thesis, Brandeis University, Massachusetts (2007)
9. Burns, T., Meeker, L.D.: A mathematical model of evaluation, decision-making, and social interaction. In: J. Cochrane, M. Zeleny (eds.) Multiple Criteria Decision-Making, pp. 141–163. University of South Carolina Press, Columbia (1973)
10. Burzio, M., Pelant, J.: On complete strongly connected digraphs with the least number of 3-cycles. Discret. Math. **155**, 27–30 (1996)
11. Chellapilla, K., Fogel, D.B.: Evolution, neural networks, games, and intelligence. Proc. IEEE **87**(9), 1471–1496 (1999)
12. Chong, S.Y., Tan, M.K., White, J.D.: Observing the evolution of neural networks learning to play the game of Othello. IEEE Trans. Evol. Comput. **9**(3), 240–251 (2005)
13. Chong, S.Y., Tiňo, P., He, J.: Coevolutionary systems and pagerank. Artif. Intell. **277**, 103164 (2019)
14. Chong, S.Y., Tiňo, P., He, J., Yao, X.: A new framework for analysis of coevolutionary systems—directed graph representation and random walks. Evol. Comput. **27**(2), 195–228 (2019)
15. Chong, S.Y., Tiňo, P., Yao, X.: Measuring generalization performance in co-evolutionary learning. IEEE Trans. Evol. Comput. **12**(4), 479–505 (2008)
16. Chung, F.: Laplacians and the Cheeger inequality for directed graphs. Ann. Comb. **9**(1), 1–19 (2005)
17. Chung, F., Zhao, W.: Pagerank and random walks on graphs. In: Fete of Combinatorics and Computer Science. Bolyai Society Mathematical Studies, vol. 20, pp. 43–62. Springer, Berlin (2010)
18. Chung, K.L.: Markov Chains with Stationary Transition Probabilities. Springer, Berlin (1960)
19. Conway, J.H., Guy, R.K.: The Book of Numbers. Springer, New York (1996)
20. Cormen, T.H., Leiserson, C.E., Rivest, R.L., Stein, C.: Introduction to Algorithms, 2 edn. The MIT Press, Cambridge (2001)

21. de Jong, E.D., Pollack, J.B.: Ideal evaluation from coevolution. Evol. Comput. **12**(2), 159–192 (2004)
22. Dwork, C., Kuman, R., Naor, M., Sivakumar, D.: Rank aggregation methods for the web. In: Proceedings of the 10th International Joint Conference on World Wide Web (WWW'01), pp. 613–622. Hong Kong (2001)
23. Falgueras-Cano, J., Falgueras-Cano, J., Moya, A.: A study of the coevolution of digital organisms with an evolutionary cellular automaton. Biology **10**(11), 1147 (2021)
24. Ficici, S.G., Melnik, O., Pollack, J.B.: A game-theoretic and dynamical-systems analysis of selection methods in coevolution. IEEE Trans. Evol. Comput. **9**(6), 580–602 (2005)
25. Fogel, D.B., Hays, T.J., Hahn, S.L., Quon, J.: A self-learning evolutionary chess program. Proce. IEEE **92**(12), 1947–1954 (2004)
26. García-Pedrajas, N., Hervás-Martínez, C., Ortiz-Boyer, D.: Cooperative coevolution of artificial neural network ensembles for pattern classification. IEEE Trans. Evol. Comput. **9**(3), 271–302 (2005)
27. Gleich, D.F.: Pagerank beyond the web. SIAM Rev. **57**(3), 321–363 (2015)
28. Golub, G.H., Van Loan, C.F.: Matrix Computation, 4 edn. The Johns Hopkins University Press, Maryland (2013)
29. Grinstead, C.M., Snell, J.L.: Introduction to Probability. American Mathematical Society, Providence (1997)
30. Haveliwala, T.H.: Topic-sensitive pagerank: A context-sensitive ranking algorithm for web search. IEEE Trans. Knowl. Data Eng. **15**(4), 784–796 (2003)
31. He, J., Yao, X.: Towards an analytic framework for analysing the computation time of evolutionary algorithms. Artif. Intell. **145**, 59–97 (2003)
32. Hemasinha, R.: An algorithm to generate tournament score sequences. Math. Comput. Model. **37**, 377–382 (2003)
33. Herrmann, S., Ochoa, G., Rothlauf, F.: Pagerank centrality for performance prediction: the impact of the local optima network model. J. Heuristics **24**(3), 243–264 (2018)
34. Hillis, W.D.: Co-evolving parasites improve simulated evolution as an optimization procedure. Phys. D **42**, 228–234 (1990)
35. Horn, R.A., Johnson, C.R.: Matrix Analysis, 2 edn. Cambridge University, Cambridge (2012)
36. Iosifescu, M.: Finite Markov Processes and Their Applications. Wiley, New York (1980)
37. Jeh, G., Widom, J.: Scaling personalized web search. In: Proceedings of the 12th World Wide Web Conference (WWW'03), pp. 271–279. Budapest, Hungary (2003)
38. Jung, A., Rowe, J.E.: Convergence of preference functions. Theor. Comput. Sci. **488**, 66–77 (2013)
39. Kendall, M.G.: Rank Correlation Methods, 4 edn. Charles Griffin, London (1975)
40. Kleinberg, J.M.: Authoritative sources in a hyperlinked environment. J. ACM **46**(5), 604–632 (1999)
41. Langville, A.N., Meyer, C.D.: Inside pagerank. ACM Trans. Internet Technol. **1**(3), 335–380 (2004)
42. Levin, D.A., Peres, Y., Wilmer, E.L.: Markov Chains and Mixing Times. American Mathematical Society, Providence (2009)
43. Li, Y., Zhang, Z.: Random walks on digraphs, the generalized digraph Laplacian and the degree of asymmetry. In: Algorithms and Models for the Web-Graph (WAW 2010). Lecture Notes in Computer Science, vol. 6516, pp. 74–85. Springer, Berlin (2010)
44. Lu, X., Menzel, S., Tang, K., Yao, X.: Cooperative co-evolution-based design optimization: a concurrent engineering perspective. IEEE Trans. Evol. Comput. **22**(2), 173–188 (2018)
45. Ma, X., Li, X., Zhang, Q., Tang, K., Liang, Z., Xie, W., Zhu, Z.: A survey on cooperative co-evolutionary algorithms. IEEE Trans. Evol. Comput. **23**(3), 421–441 (2019)
46. Meyer, C.D.: Matrix Analysis and Applied Linear Algebra. SIAM, Philadelphia (2000)
47. Meyn, S., Tweedie, R.L.: Markov Chains and Stochastic Stability, 2 edn. Cambridge University, Cambridge (2009)
48. Moon, J.W.: Topics on Tournament. Holt, Rinehart and Winston (1968)

49. Ng, A.Y., Zheng, A.X., Jordan, M.I.: Link analysis, eigenvectors and stability. In: Proceedings of the 17th International Joint Conference on Artificial Intelligence (IJCAI'01), pp. 903–910. Seattle, USA (2001)
50. Norris, J.R.: Markov Chains. Cambridge University, Cambridge (1998)
51. Palacios, J.L., Gómez, E., Río, M.D.: Hitting times of walks on graphs through voltages. J. Probab. **2014** (2014). https://doi.org/10.1155/2014/852481
52. Popovici, E.: Bridging supervised learning and test-based co-optimization. J. Mach. Learn. Res. **18**(38), 1–39 (2017)
53. Popovici, E., Bucci, A., Wiegand, R.P., De Jong, E.D.: Coevolutionary principles. In: G. Rozenberg, T. Bäck, J.N. Kok (eds.) Handbook of Natural Computing, pp. 987–1033. Springer, Berlin (2012)
54. Popovici, E., De Jongl, K.: The dynamics of the best individuals in co-evolution. Nat. Comput. **5**, 229–255 (2006)
55. Popovici, E., Winston, E.: A framework for co-optimization algorithm performance and its application to worst-case optimization. Theor. Comput. Sci. **567**, 46–73 (2015)
56. Potter, M.A., Jong, K.A.D.: Cooperative coevolution: an architecture for evolving coadapted subcomponents. Evol. Comput. **8**(1), 1–29 (2000)
57. Rosin, C.D., Belew, R.K.: New methods for competitive coevolution. Evol. Comput. **5**(1), 1–29 (1997)
58. Rudolf, B., Markošová, M., Čajági, M., Tiňo, P.: Degree distribution and scaling in the connecting-nearest-neighbors model. Phys. Rev. E **85**(2), 026114 (2012)
59. Schweitzer, P.J.: Perturbation theory and finite Markov chains. J. Appl. Probab. **5**(2), 401–413 (1968)
60. Sims, K.: Evolving 3D morphology and behavior by competition. Artif. Life **1**(4), 353–372 (1994)
61. Tetali, P.: A characterization of unique tournaments. J. Comb. Theory Ser. B **72**, 157–159 (1998)
62. Tiňo, P., Chong, S.Y., Yao, X.: Complex co-evolutionary dynamics—structural stability and finite population effects. IEEE Trans. Evol. Comput. **17**(2), 155–164 (2013)
63. Vose, M.D., Liepins, G.E.: Punctuated equilibria in genetic search. Complex Systems **5**(1), 31–44 (1991)
64. Watson, R.A., Pollack, J.B.: Coevolutionary dynamics in a minimal substrate. In: Proceedings of the 2001 Genetic and Evolutionary Computation Conference (GECCO'01), pp. 702–709 (2001)
65. Werfel, J., Mitchell, M., Crutchfield, J.P.: Resource sharing and coevolution in evolving cellular automata. IEEE Trans. Evol. Comput. **4**(4), 388–393 (2000)
66. Yuan, B., Lu, X., Tang, K., Yao, X.: Cooperative coevolution-based design space exploration for multi-mode dataflow mapping. ACM Trans. Embed. Comput. Syst. **20**(3), 1–25 (2021)

Index

A

Agent-based modelling (ABM), 95
- agent-based computational economics, 97
- agent-based simulation, 95

Automatic programming, 113

B

Blackbox optimization, 36
Boolean Satisfiability Problem (SAT), 31, 155, 165

C

Coevolutionary algorithms (CEAs), 69
- competitive CEA, 71, 104
  - classical coevolutionary learning, 419
  - improved coevolutionary learning, 436
- competitive-cooperative CEA, 87, 117
- cooperative CEA, 74, 112
  - contribution-based cooperative coevolution with differential grouping, 235
  - cooperative CEA with random grouping, 206
  - cooperative coevolutionary GA, 303, 309
  - differential evolution cooperative coevolution with differential grouping, 226
  - differential evolution cooperative coevolution with random grouping, 210
  - Self-evaluation evolution, 276
- parallel cooperative CEA, 334
  - coevolutionary concurrent design method, 347, 352
  - communication frequency self-adaptation, 353
  - novelty-driven cooperative coevolution with stochastic selection, 347, 350

Coevolutionary computation, 45
Coevolutionary pathologies, 129
- cycling, 132
- disengagement, 134
- diversity maintenance techniques, 144
  - competitive fitness sharing, 145
  - implicit fitness sharing, 145
  - pareto coevolution, 146
  - reducing virulence, 145
- forgetting, 133
- mediocre stable states, 136
- overspecialization, 129

Coevolutionary problem, 129, 136
- coevolutionary digraph, 136, 447
- intransitivity, 130, 138
- irreducibility, 138
- solvability, 138

Coevolutionary systems, 45, 69
- coevolutionary learning, 69, 78, 374, 407
- coevolutionary simulation, 69, 95, 375

Competitive coevolution, 63
- cooptimization setting, 63
- supervised learning setting, 64

Complex adaptive systems, 100
Concurrent Engineering (CE), 333
- monodisciplinary product design, 336
  - single UEM design, 342

X. Yao, S. Y. Chong, *Coevolutionary Computation and Its Applications*, Natural Computing Series, https://doi.org/10.1007/978-981-96-2841-4

Concurrent Engineering (CE) (*cont.*)
multidisciplinary design optimization, 333
geometric programming, 343
overlapping UEM design, 342
quasi-separable MDO, 336
universal electric motor, 335
Cooperative coevolution
collaboration selection strategies, 85
contribution-based mechanism, 235
Divide-and-Conquer, 81, 272
perturbation-based variable grouping strategy, 84
differential grouping, 202, 224
improved differential grouping, 202, 250
problem decomposition, 81
variable grouping strategies, 82
random grouping, 201

D
Design Space Exploration (DSE), 301
Diversity measurements, 439
genotypic diversity measurement, 439
edit distance, 439
phenotypic diversity measurement, 439
entropy, 440

E
Engineering design, 107
fuzzy logic controller, 108
IIR filter design, 108
satellite-module layout design, 107
Evolutionary algorithms, 3, 7
multi-objective evolutionary algorithms, 97
nondominated immune algorithm, 117
NSGA-II, 97, 108, 302, 346
novelty based search, 348
Evolutionary computation, 3

F
Fitness evaluation, 45
fitness function, 45
objective fitness, 45
quality function, 55
subjective fitness, 45
Fitness landscape, 35

G
Generalization performance, 63
Genetic programming, 18, 99, 113

I
Interactions, 59
Internet of Things (IoTs), 116

M
Malware detection, 115
Markov chains, 447
abstract coevolution, 447, 449
(1+1)CEA, 448
standard random walk, 448
coevolutionary Markov chains, 449
coevolutionary markov chains
absorbing CMC, 453
irreducible CMC, 453
expected hitting time, 454
PageRank, 447, 466
PageRank authority, 465
random walks, 447
stationary distribution, 464
Memetic computation, 314
local search, 314
Meta-modelling, 278, 315
Multi-objective optimization, 13, 97
dynamic multi-objective optimization, 116

N
Network control system (NCS), 115
Noncooperative games, 48, 69, 97, 375
ISP routing game, 378
iterated prisoner's dilemma, 48, 78, 375
IPD with more choices, 379
IPD with reputation, 382, 394
Numbers game, 130
intransitive game, 130

O
Operations research, 110
scheduling problem, 110
transportation problem, 111

P
Parallel algorithm portfolio (PAPs), 156, 158
Parameter control, 28
automatic algorithm configuration, 154
co-evolution of parameterized search, 157, 177
generative adversarial solver trainer, 156, 160
automatic parameter tuning, 154
parameter tuning, 28
self-adaptation, 22, 28, 385

Population dynamics, 132
chaotic dynamics, 142
coevolutionary dynamical systems, 132
deterministic dynamics, 144
stochastic dynamics, 144
Populations, 11
multi-population coevolution, 58, 96
predator-prey coevolution, 58, 113
single-population coevolution, 57
two-population coevolution, 58, 110
Procedural content generation (PCG), 106

R
Real-time strategy games, 104
Representations, 14
artificial neural networks, 16
Boolean networks, 5
multilayer perceptrons, 383
binary string, 19
finite state automata, 38, 88
genotype, 14
GP tree, 18
phenotype, 14
weighted piece counters, 434

S
Search-and-replacement task, 114
Search space, 14
Selection operator, 24
$(\mu + \lambda)$, 27
(μ, λ), 27
fitness proportional, 25
linear ranking, 26
tournament, 26
Separability, 61
additively separable, 61, 75, 204
nonseparable condition, 84
Solution concept, 55
generalization performance, 410
biased sample, 414
Chebyshev's bounds, 412
comparison via statistical testing, 428
parametric testing, 422, 426
second estimation of σ_i^2, 422, 425
statistical estimator, 411
maximum expected performance, 55
Synchronous dataflow (SDF), 301
multi-mode dataflow, 303

T
Test-based cooptimization, 55
Traveling Salesman Problem (TSP), 15, 153, 164
Turing machine, 5
evolutionary turing machine, 37
universal turing machine, 38

V
Variation operator, 19
crossover, 20
mutation, 19
Cauchy mutation, 280
Gaussian mutation, 21, 280, 385
recombination, 22
Vehicle Routing Problem, 157, 180

Z
Zero-sum games, 47
Othello, 434

The manufacturer's authorised representative in the EU is Springer Nature Customer Service Centre GmbH, Europaplatz 3, 69115 Heidelberg, Germany. If you have any concerns regarding our products, please contact ProductSafety@springernature.com

Printed and bound by CPI Group (UK) Ltd, Croydon, CR0 4YY
01/10/2026
02225928-0018